HAZARDOUS MATERIALS HANDBOOK FOR EMERGENCY RESPONDERS

Contributing Authors

Randy R. Bruegman
Michael Callan
Phillip L. Currance
Glenn Levy
Gregory G. Noll
Mark Sprenger

HAZARDOUS MATERIALS

HANDBOOK FOR EMERGENCY RESPONDERS

Edited by Joe Varela

JOHN WILEY & SONS, INC.
New York • Chichester • Weinheim • Brisbane • Singapore • Toronto

Originally published as ISBN 0-442-02104-6 (hc).

Published simultaneously in Canada.

Library of Congress Cataloging-in-Publication Data

Hazardous materials handbook for emergency responders / Onguard, Inc., edited by Joe Varela.
p. cm.
Includes bibliographical references and index.
ISBN 0-471-28713-X
1. Hazardous substances—Accidents—Management—Handbooks, manuals, etc. I. Varela, Joe.
T55.3.H3H3995 1991
628.9—dc20 95-36466
CIP

Printed in the United States of America.

10 9 8 7 6 5 4 3 2

CONTENTS

Chapter 7: FLAMMABLE SOLID EMERGENCIES 261

Chapter 8: OXIDIZER EMERGENCIES 289

FOREWORD

Today, as never before in history, thousands of people from all walks of life are finding themselves involved in emergency response to hazardous materials incidents. From the volunteer firefighter in rural Iowa, to the East Coast industrial plant worker, to the emergency management volunteer in mountainous Alaska, the average citizen, worker, and responder are being challenged by hazardous materials emergencies. The challenge is to understand a very complex subject and to formulate a practical, decision making framework that will work under the pressure of a threatening catastrophe.

Very few books that deal with hazardous materials response target the common man off the street. In reality, over 90% of today's haz-mat responders fit this category. The intent of this book is not to create an exhaustive technical document. Rather, this book is written in everyday language to assist the average responder, whose overwhelming task is to sort out a very technical subject. Our goal is to equip responders with information that will allow them to make "real world" emergency response decisions that work.

The *Hazardous Materials Emergency Response Handbook* specifically addresses the nine classes of hazardous materials. Ultimately, all responses to haz-mat incidents boil down to a practical understanding of the chemistry of the materials involved. Without a knowledge of the chemicals and of how to respond safely to those chemicals, emergency responders court disaster like a person driving blindly into a snowstorm.

This book is dedicated to those individuals who step up to the dangerous task of haz-mat response each day. Their efforts help make our communities safer places to live. It is our hope that this volume will be a link in a chain of protection that will not break under the strain of a hazardous materials emergency.

ACKNOWLEDGMENTS

In OnGUARD's efforts to draft, write, and rewrite a presentable publication for emergency responders, we discovered a melting pot of authoring talent which made this an eventful and challenging process. Randy Bruegman, as the primary contributing author, provided the backbone material for this text. Two instructors of our video course, *Surviving The Hazardous Materials Incident*, also served as contributing authors and had a great influence on the content of this text. Michael Callan is an "idea factory" whose mere presence fed our efforts with his uncanny wit, humor, and unceasingly vibrant personality. Without Mike's skills it would have been impossible to present many of the technical issues in this text in an easy to understand manner. Then there is Greg Noll, a name synonymous with other words found in the dictionary like "emergency responder" and "haz-mat." In short, no haz-mat text is complete without a finely tuned technical specialist like Greg. Glenn Levy served as contributing author for Chapters 1 through 5, bringing a street-level understanding of haz-mat response. We are in debt to Phillip Currance and his meticulous approach in bringing us a medical understanding for haz-mat responders. A special thanks to Mark Sprenger, president of OnGUARD, for his unwavering vision to provide quality educational tools for front-line responders.

From our authors' original scratchings, to sifting through realms of raw information, endless phone conversations, faxes, express mailings, computer entries, and word smithing, the wheels of progress churned forward. Simply put, the synergy of working alongside some of the best has made the process rewarding.

Corresponding nationwide, our technical review committee was tasked with deadlines against their own demanding workloads and schedules. Their patient resolve always gave us renewed excitement. In alphabetical order they are:

Dr. John Bowen, Hilo, HI
George Cunning, Colorado Springs, CO
Frank Docimo, Stamford, CT
Jan Dunbar, Sacramento, CA
Rick Emery, Vernon Hills, IL
Gerald L. Grey, Redwood City, CA

Bill Hand, Houston, TX
Dave Lesak, Allentown, PA
Gus Sitas, LaPorte, CO

We are deeply grateful for the timeliness and depth of Rick Emery, Gerald Grey, and George Cunning. Special merit goes to Dr. John Bowen, who acted as the final authority and whose many comments, suggestions, and corrections made a notable difference in the content of this presentation.

In the way of photography, we would like to thank Mike Callan; Greg Noll; John Cashman, Editor for the *Hazardous Materials Newsletter*; and Michael Tessier, Deputy Fire Chief for the City of Benicia, California.

Behind the scenes and certainly not forgotten is the editorial staff who worked their magic and simply made it happen. We owe much to the consistent editorial and production direction of Joe Varela. Joe's patient persistence and sixth sense for content provided us with yet another positive forecast for excellence in weathering the deadline storms. Susan Peterson spent many hours reviewing each chapter during the final editing process. We also thank our assistant editors Sondra Chavez and Carol Brown. Jeff Northway, Brad Kobielusz, and Dan Seese assisted in the area of art and illustration.

Chapter 1

INTRODUCTION TO HAZARDOUS MATERIALS

OBJECTIVES

After studying the material in this chapter, you will be able to:

- identify the agency responsible for regulating hazardous materials
- define the various types of hazardous materials
- list hazardous material incident types
- state how to recognize and identify hazardous materials
- list information resources regarding hazardous materials

INTRODUCTION

Responding to hazardous materials incidents has been one of the most exciting and demanding areas of emergency response in the past 20 years. About 100 years ago, a tank car consisted of wooden vats on a flatbed car. The vats were filled with commodities that were in high demand, such as kerosene or animal fat oil. Today, however, tank cars and trucks are sophisticated pieces of equipment that carry complex and often dangerous materials. Thousands of new chemicals are being created yearly because of the technological advances of our time.

Emergency response agencies must effectively respond to complicated incidents. They must properly handle a variety of hazardous materials. They must have a haz-mat response plan. It is both exciting and demanding work, yet it requires training, planning, readiness, and financial backing. There are three major problems that affect hazardous materials response:

Training is the primary factor in successfully handling a haz-mat incident.

- Until the introduction of NFPA 472 and OSHA 29 CFR 1910, there were *no real recognized* national standards for hazardous materials response.
- The costs of responding to and alleviating such incidents can be overwhelming. Most departments simply don't have the necessary money to gear up for haz-mat emergencies.

Notes

- Recent changes in federal, state, and local legislation, as well as increased response dangers, demand that response methods be altered.

The goal of this reference manual is to clearly present valid information for responding to haz-mat emergencies. Whether you are a firefighter, law enforcement officer, haz-mat specialist, EMS provider, municipal worker, state employee, dispatcher, or city official, the need for excellent and consistent training is critical if haz-mat incidents are to be effectively handled. It is your job as responders to bring a hazardous situation under control. Do not become part of the problem through inadequate training. Rather, work to provide effective service to your communities.

REGULATION OF HAZARDOUS MATERIALS

The Department of Transportation (DOT) is responsible for issuing and enforcing regulations for all ground-based transportation in the nation. ***Ground-based transportation*** can be by rail, road, or highway. The DOT also regulates air transportation through the FAA. These regulations cover all hazardous and non-hazardous materials and products, and are found in the Code of Federal Regulations, Title 49 (49 CFR).

HAZARDOUS MATERIAL DEFINED

"Hazardous materials are the great equalizer."
– Mike Callan

DOT has legally defined a ***hazardous material*** as:

> A substance or material, including a hazardous substance, which has been determined by the Secretary of Transportation to be capable of posing an unreasonable risk to health, safety, and property when transported in commerce, and which has been so designated (49 CFR 171.8).

Other definitions include:

> ***Extremely hazardous materials*** – Those chemicals determined by the Environmental Protection Agency (EPA) to be extremely hazardous during a spill or release as a result of their toxicities and physical/chemical properties. Initially the EPA listed 402 chemicals in this category.

> ***Dangerous goods*** – Any product, substance, or organism included by its nature or by regulation in any of the nine hazard classes listed in the United Nations schedule of hazardous materials.

Materials that fall under DOT regulations and are designated as hazardous are required to be *placarded*, with the exception of ORMs (Other Regulated Materials). There are, however, many materials that are not defined by the DOT as hazardous, but they can adversely affect health, safety, and property when released into the environment. One such example is milk. Milk is not considered a hazardous material, but it can severely harm many forms of life if accidentally induced into waterways.

We would like to suggest a more specific definition for ***hazardous material*** that can be applied on all responses:

A material or materials accidentally released from the original container and used in a manner not originally intended. Hazardous materials include materials that are unintentionally contaminated or mixed with other chemicals, or involve some outside reactive source such as heat, light, liquids, shock, or pressure.

DOT Hazard Class Definitions

Explosive – Any substance, article, or device which is designed or is able to function by explosion, with extremely rapid release of gas and heat, unless the substance, article, or device is otherwise classed in 49 CFR 171-177. Explosives are divided into six divisions (49 CFR 173.50(b)):

- ***Division 1.1*** – Explosives that have a mass explosion hazard. A mass explosion is one which affects almost the entire load instantaneously.
- ***Division 1.2*** – Explosives that have a projection hazard but not a mass explosion hazard.
- ***Division 1.3*** – Explosives that have a fire hazard and either a minor blast hazard or a minor projection hazard or both, but not a mass explosion hazard.
- ***Division 1.4*** – Explosives that present a minor explosion hazard.
- ***Division 1.5*** – Explosives which have a mass explosion hazard but are so insensitive that there is very little probability of initiation or of transition from burning to detonation under normal conditions of transport.
- ***Division 1.6*** – Extremely insensitive articles which do not have a mass explosion hazard.

Flammable Gas – Any material which is a gas at 68°F (20°C) or less and 14.7 psi (101.3 kPa) of pressure and which (1) is ignitable when in a mixture of 13% or less by volume with air at atmospheric pressure, or (2) has a flammable range with air at atmospheric pressure of at least 12% regardless of the lower flammable limit (49 CFR 173.115(a)).

Non-Flammable, Non-Poisonous, Compressed Gas – Any material or mixture which exerts in the packaging an absolute pressure of at least 41 psia (280 kPa) at 68°F (20°C) and does not meet the definition of a flammable gas or a poisonous gas. This category includes compressed gas, liquefied gas, pressurized cryogenic gas, and compressed gas in solution (49 CFR 173.115(b)).

Gas Poisonous by Inhalation – A material which is a gas at 68°F (20°C) or less and a pressure of 14.7 psi (101.3 kPa) which (1) is known to be so toxic to humans as to pose a hazard to health during transportation, or (2) is presumed to be toxic to humans based on tests conducted on laboratory animals (49 CFR 173.115(c)).

Flammable Liquid – A liquid having a flash point of not more than 141°F (60.5°C), or any liquid with a flash point at or above 100°F (37.8°C) that is transported at or above its flash point in a bulk packaging. Exceptions are listed in 49 CFR 173.120(a).

Combustible Liquid – Any liquid that does not meet the definition of any other hazard class and that has a flash point above 100°F (37.8°C) and below 200°F (93°C) (49 CFR 173.120(b)).

Flammable Solid – Any of the following three types of materials: (1) wetted explosives, (2) self-reactive materials that are liable to undergo heat-producing decomposition, or (3) readily combustible solids that may cause a fire through friction or that have a rapid burning rate as determined by specific tests (49 CFR 173.124(a)).

Spontaneously Combustible Material – (1) A pyrophoric material. A pyrophoric material is a liquid or solid that, even in small quantities and without an external ignition source, can ignite after coming in contact with air. (2) A self-heating material. A self-heating material is a material that is liable to self-heat when in contact with air and without an energy supply (49 CFR 173.124(b)).

Dangerous When Wet Material – A material that is liable to become spontaneously flammable or to give off flammable or toxic gas as a result of contact with water (49 CFR 173.124(c)).

Oxidizer – A material that may cause or enhance the combustion of other materials, generally by yielding oxygen (49 CFR 173.127(a)).

Organic Peroxide – Any organic compound containing oxygen in the bivalent –O–O– structure and which may be considered a derivative of hydrogen peroxide, where one or more of the hydrogen atoms have been replaced by organic radicals. Exceptions are listed in 49 CFR 173.128(a).

Poisonous Material – A material, other than a gas, which (1) is known to be so toxic to humans as to afford a hazard to health during transportation, (2) is presumed to be toxic to humans based on tests conducted on laboratory animals, or (3) is an irritating material with properties similar to tear gas, which causes extreme irritation, especially in confined spaces (49 CFR 173.132(a)).

Infectious Substance – A viable microorganism or its toxin or any other agent that causes or may cause disease in humans or animals (49 CFR 173.134(a)(1)). Refer to the Department of Health and Human Services regulations, 42 CFR 72.3, for details. The terms *infectious substance* and *etiologic agent* are synonymous.

Radioactive Material – Any material that spontaneously emits ionizing radiation and that has a specific activity greater than 0.002 microcuries per gram (49 CFR 173.403(y)).

Corrosive Material – A liquid or solid that causes visible destruction or irreversible alterations in human skin tissue at the site of contact, or a liquid that has a severe corrosion rate on steel or aluminum (49 CFR 173.136(a)).

Miscellaneous Hazardous Material – A material which presents a hazard during transportation but which does not meet the definition of any other hazard class. This class includes (1) any material which has an anesthetic, noxious, or other similar property which could cause extreme annoyance or discomfort to a flight crew member so as to prevent the correct performance of assigned duties, or (2) any material which meets the definition for an elevated temperature material, a hazardous substance, or a hazardous waste (49 CFR 173.140).

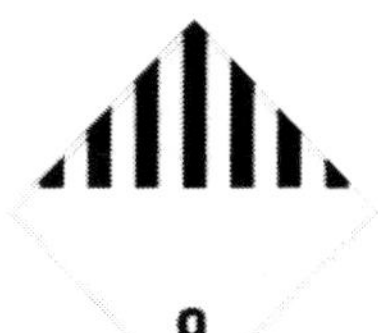

ORM-D (Other Regulated Material) – A material such as a consumer commodity which, although otherwise subject to the regulations, presents a limited hazard during transportation due to its form, quantity, and packaging. It must be a material for which exceptions are provided in the table in 49 CFR 172.101. Each ORM-D material and category of ORM-D material is listed in the table in 49 CFR 172.101 (49 CFR 173.144).

CONSUMER COMMODITY ORM-D-AIR

CONSUMER COMMODITY ORM-D

The following are additional terms used in preparation of hazardous materials for shipment (49 CFR 171.8):

Consumer Commodity – A material that is packaged and distributed in a form intended or suitable for sale through retail sales agencies for consumption by individuals for purposes of personal care or household use. This term also includes drugs and medicines.

Elevated Temperature Material – A material which (1) is a liquid and is at a temperature at or above 212°F (100°C) during transportation, (2) is a liquid with a flash point at or above 100°F (37.8°C) that is transported at or above its flash point, or (3) is a solid and is at a temperature at or above 464°F (240°C) during transportation.

Flash Point – The minimum temperature at which a substance gives off flammable vapors which will ignite when in contact with sparks or flame. Criteria are listed in 49 CFR 173.121.

Hazard Zone – One of four levels of hazard (Hazard Zones A through D) assigned to gases, as specified in 49 CFR 173.116(a), and one of two levels of hazard (Hazard Zones A and B) assigned to liquids that are poisonous by inhalation, as specified in 49 CFR 173.133(a).

Notes

Hazardous Substance – A material, including its mixtures and solutions, that (1) is listed in the appendix to 49 CFR 172.101; (2) is in a quantity, in one package, which equals or exceeds the reportable quantity listed in the appendix to 49 CFR 172.101; and (3) when in a mixture or solution, is in a concentration by weight which equals or exceeds the concentration corresponding to the reportable quantity of the material.

Hazardous Waste – Any material that is subject to the Hazardous Waste Manifest Requirements of the U.S. Environmental Protection Agency specified in 40 CFR 262.

Limited Quantity – The maximum amount of a hazardous material for which there is a specific labeling or packaging exception.

Packing Group – A grouping according to the degree of danger presented by hazardous materials. Packing Group I indicates great danger; Packing Group II, medium danger; Packing Group III, minor danger.

THE HAZARDOUS MATERIALS RESPONDER

Haz-mat responders are all persons directly or indirectly involved in the response, including private industry, dispatchers, planners, police officers, EMS providers, and firefighters. These emergency service personnel must act as a team to insure the most effective response to haz-mat situations. That assurance only comes through proper training.

Haz-Mat Responders

Hazardous Materials Responders*		
TITLE	**ACTION**	**PERSONNEL**
First Responder Awareness	• Recognize problem • Identify (if possible) • Notify more qualified responders	• Police • Some firefighters • Public works • Other field units
First Responder Operations	• ***Defensive*** skills • Contain spill (diking) • Minimize harm (evacuation, water fog, protecting in place)	• Some police • Firefighters • Industrial spill team • EMS
Technician	• ***Offensive*** operations • Plugging and patching • Controlling the spill and stopping release	• HMRTs • Industrial brigades • Emergency response teams
Specialist	• Responds with and provides support to technicians	• Skill oriented basis • "Super teams" • State • Industrial (DuPont, Dow, etc.)
On-Scene Commander (OSHA); Incident Commander	• Any incident commander above the awareness level	• Fire chiefs • Battalion chiefs • Sheriffs and deputies • State troopers • ERT supervisor

* As defined by the Occupational Safety and Health Administration, 29 CFR 1910.120, March 6, 1989.

Figure 1-1

HAZARDOUS MATERIALS INCIDENT TYPES

Haz-mat incidents can occur anywhere at any time. Responders must identify areas of potential hazard in their communities and be prepared to meet those hazards. The seven major types of incidents to which emergency personnel will respond are railway, highway, ship and barge, fixed installation, pipeline, container failures, and aircraft accidents. The *type of incident* determines the exact response or action.

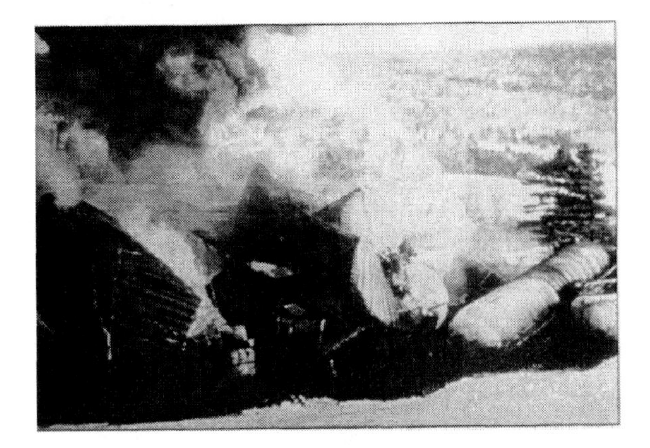

Railway. The most common reason for railway accidents is track failure or collision. The increase in highway transportation in the mid-20th century reduced the need for railroads. Track maintenance was often ignored, which resulted in an increase in the number of serious railroad accidents. In the past few years, however, railroad companies have committed time and money to improve track conditions. Other common factors causing rail emergencies include human, signal, or mechanical errors.

It is critical to know and meet rail line operators in your area *before an accident occurs.* Make sure you and the operator understand each other's accident procedures. Problems can be magnified due to the large quantities of materials involved in railway accidents and poor access to the emergency site.

Highway accidents are the most common type of transportation haz-mat incident. Highway accidents usually involve many complications and products. Highway accidents are often caused by human error and can involve injuries to several people in multiple vehicles. Weather can also cause, or intensify, highway emergencies. Highway accidents will be discussed further in subsequent chapters.

Ship and barge accidents occur on our nation's waterways. Even a minor spill in a waterway can have long term negative effects on people and the environment. Emergency responders will have a tremendous impact on these incidents.

Essential waterway information:

- Know the local waterways, rivers, streams, irrigation ditches, and lakes.
- Have a general idea of the origin and destination of the waterways.
- Always remember, water is gold in many areas.
- Waterways are easily contaminated and difficult to clean up.
- Know the owner of the waterway or the responsible party.
- Most waterways have human contact.
- Preplan agency procedures.
- Anticipate!

Fixed installation accidents occur anywhere materials are manufactured, processed, used, transported, or stored. While it is impossible to discuss all types of fixed facilities, we will cover common causes of fixed installation accidents, including:

- equipment failure
- human error
- accidental mixing of reactive products
- product or tank overflow
- physical damage to containers
- exposure to fire, water, or heat

Incidents that occur in educational or research facilities usually will not be termed "major" incidents. Such facilities, however, may have large numbers of people, and a small spill can quickly become life threatening.

Pipeline accidents occur as a result of mechanical failure and human error (digging prior to locating the pipe). Pipelines are often overlooked, yet they carry billions of gallons of product annually. Pipeline incidents are potentially dangerous to life and property.

Container failure. Generally, any material is safe in its container. There are, however, conditions in addition to mishandling or human error that may cause the container to fail if it is involved in an accident, including:

- thermal failure
  - pressure or non-pressure vent failure
  - container fails from heat or direct flame impingement
- mechanical failure
  - container is filled beyond its capacity
  - defective container
  - gasket, valve stem, or connection failure
  - physical damage to the valve or container
  - container is damaged by flying debris from another container
- chemical failure
  - poor quality control, causing contamination or impurity in the product
  - gasket, valve stem, or connection failure due to chemical action

Anticipating the dangers or reactions that may occur during container failure incidents will be discussed in following chapters.

Aircraft accidents are relatively infrequent, but they are almost always serious. Such accidents are difficult to control and can be emotional for responders. It is important to be prepared to assist personnel in dealing with the special effects that can occur.

IDENTIFICATION AND RECOGNITION

Failure to identify the material will increase the hazard.

Hazardous materials must be properly identified before any action can be taken to control an incident. Failure to properly identify the hazardous material will only increase the hazard. There are many ways to identify materials in transportation. The first was designed and implemented by the Department of Transportation (DOT). This method requires the use of placards and labels and is the most accessible to responders.

Labels

Package Labeling

Package labels are diamond shaped symbols four inches square that are attached to the package being shipped. Every DOT classified haz-mat shipment must be marked with the appropriate label or labels, unless otherwise specified. Displaying the name of the material on the package is also required. In addition, when two or more warning labels are required, they must be displayed next to each other. The hazard class or division number must appear on the bottom of the primary hazard label. The class number may **not** be displayed on a subsidiary hazard label.

Semi-Trailer Placarding

Rail Car Placarding

Placards

Placards are diamond shaped symbols 10-3/4 inches square that are applied to each side and end of a motor vehicle, rail car, freight container, or portable tank container carrying hazardous materials. (See Figure 1-2 for placarding guidelines.)

Placard

DOT Placarding Guidelines

Any transport vehicle, freight container, or rail car containing any quantity of material listed in Table 1 must be placarded. When the gross weight of all hazardous materials covered in Table 2 is less than 1,001 lbs (454 kg), no placard is required.

TABLE 1
(PLACARD ANY QUANTITY)

Hazard Class or Division	Placard Name
1.1	EXPLOSIVES 1.1
1.2	EXPLOSIVES 1.2
1.3	EXPLOSIVES 1.3
2.3	POISON GAS
4.3	DANGEROUS WHEN WET
6.1 (PGI, PIH only)	POISON
7 (Radioactive Yellow III)	RADIOACTIVE

TABLE 2
(PLACARD 1,001 LBS OR MORE)

Hazard Class or Division	Placard Name
1.4	EXPLOSIVES 1.4
1.5	EXPLOSIVES 1.5
1.6	EXPLOSIVES 1.6
2.1	FLAMMABLE GAS
2.2	NON-FLAMMABLE GAS
3	FLAMMABLE
Combustible Liquid	COMBUSTIBLE
4.1	FLAMMABLE SOLID
4.2	SPONTANEOUSLY COMBUSTIBLE
5.1	OXIDIZER
5.2	ORGANIC PEROXIDE
6.1 (PGI or II, other than PGI, PIH)	POISON
6.1 (PGIII)	KEEP AWAY FROM FOOD
6.2	NONE
8	CORROSIVE
9	CLASS 9
ORM-D	NONE

Figure 1-2

Notes

The placard color, symbols, and United Nations class numbers alert the responder in the following ways:

COLOR

- *orange* indicates explosive
- *green* indicates non-flammable
- *red* indicates flammable
- *yellow* indicates oxidizing material
- *white* indicates poisonous material
- *white with vertical red stripes* indicates flammable solid
- *blue* indicates dangerous when wet
- *white over black* indicates corrosive material
- *yellow over white* indicates radioactive material

SYMBOLS

bursting ball indicates explosive

flame indicates flammable

slash W indicates dangerous when wet

skull and crossbones indicates poisonous material

circle with flame indicates oxidizing material

cylinder indicates non-flammable gas

propeller indicates radioactive

test tubes / hand / metal indicates corrosive

word *"Residue"* indicates the product has been removed, but a harmful residue may still be present

United Nations Classification System

United Nations class or division numbers must be displayed at the bottom of placards or in the hazardous materials description on shipping papers. In certain cases, this class or division number may replace the written name of the hazard class in shipping paper descriptions. The class and division number breakdown is listed below.

Notes

The UN hazard class or division number must be displayed at the bottom of placards and labels and on shipping papers.

United Nations Classification System

Class 1	**Explosives**
Division 1.1	Explosives with a mass explosion hazard
Division 1.2	Explosives with a projection hazard
Division 1.3	Explosives with a predominant fire hazard
Division 1.4	Explosives with a minor explosion hazard
Division 1.5	Very insensitive explosives
Division 1.6	Extremely insensitive detonating substances
Class 2	**Gases**
Division 2.1	Flammable gases
Division 2.2	Non-flammable, non-poisonous compressed gases
Division 2.3	Poisonous gases
Class 3	**Flammable Liquids** **Combustible Liquids**
Class 4	**Flammable Solids** **Spontaneously Combustible Materials** **Dangerous When Wet Materials**
Division 4.1	Flammable solids
Division 4.2	Spontaneously combustible materials
Division 4.3	Materials that are dangerous when wet
Class 5	**Oxidizers** **Organic Peroxides**
Division 5.1	Oxidizers
Division 5.2	Organic peroxides
Class 6	**Poisonous Materials** **Infectious Substances**
Division 6.1	Poisonous materials
Division 6.2	Infectious substances
Class 7	**Radioactive Materials**
Class 8	**Corrosive Materials**
Class 9	**Miscellaneous Hazardous Materials**

Figure 1-3

Notes

NFPA 704 System

The National Fire Protection Association (NFPA) 704 system is designed for fixed facilities such as buildings, storage tanks, or individual rooms when additional hazardous materials identification is necessary. This system also uses a diamond symbol to identify the hazard. The diamond is divided into four differently colored sections, each identifying a type of hazard. Each hazard poses a varying degree of danger and is rated on a scale from 0 to 4. The least hazardous is rated 0, and the most hazardous is rated 4.

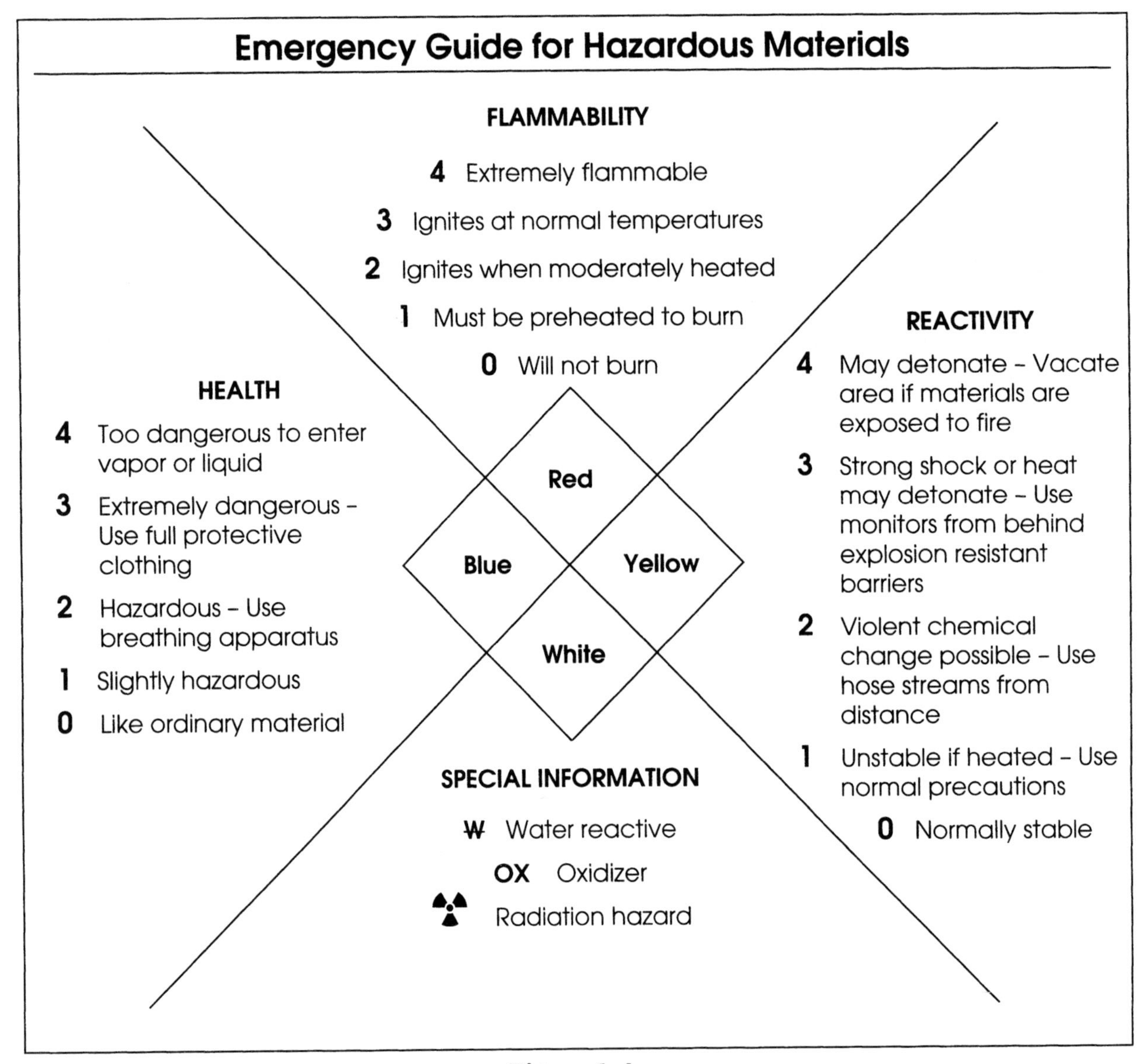

Figure 1-4

Shipping Papers

Notes

DOT regulations require shipping papers to accompany the shipment of hazardous materials and wastes. When responding to a possible haz-mat call, check for:

- shipping papers
- proper shipping name
- type of packaging
- total quantity
- product hazard information
- emergency response information
- name of shipper
- name of consignee

Shipping Paper Identification Chart

Mode of Transportation	Title of Shipping Paper	Location of Shipping Papers	Responsible Person
Highway[1]	Bill of lading	Cab of vehicle	Driver
Rail[2]	Waybill; consist	With conductor or engineer[3]	Conductor or engineer[3]
Water	Dangerous cargo manifest on barge	Wheelhouse or pipelike container	Captain or master
Air	Air bill with shipper's certification for restricted articles	Cockpit	Pilot

1. Manufacturer's data sheets are generally available from driver in addition to bills of lading.
2. STCC (Standard Transportation Commodity Code) number is used extensively on rail transportation shipping papers.
3. Many trains are dispatched without a caboose. In these cases, the conductor rides in the engine and should have the shipping papers.

Figure 1-5

Notes

Material Safety Data Sheets

Material Safety Data Sheets (MSDS) are another source of haz-mat information that can be useful in recognizing and identifying hazardous materials. Federal and state laws have mandated that local communities be notified of haz-mat use. MSDS provide a means to supply the required information to the community. The MSDS information is completed by the manufacturer and regulated by the U.S. Department of Labor, Occupational Safety and Health Administration (OSHA). Although no format for MSDS is mandated, many manufacturers now follow the American National Standards Institute (ANSI) standard entitled *Preparation of Material Safety Data Sheets*. The following information must be provided on each sheet:

- material name
- chemical formula
- common synonyms
- chemical family
- manufacturer's name
- emergency telephone number
- hazardous ingredients
- regulated exposure limits
- physical data
- fire and explosion data
- health hazard data
- reactivity data
- spill or leak procedures
- special protection information
- special precautions

Emergency response personnel should use the MSDS data available in their response area in preplanning. *Knowledge* is the key to a successful outcome in a haz-mat emergency. Sample MSDS and shipping papers are located in Appendix B.

Other Methods of Product Identification

In addition to the methods of identification mentioned above, there are numerous other ways to determine the presence of hazardous materials. Often the first responder can gain valuable information from:

- type of occupancy
- location of the incident
- type of container(s) involved
- other special markings or colors
- chemical or scientific name
- manufacturer or trade name
- obvious chemical and physical characteristics
- bystanders or responsible parties

Methods of product identification will be detailed in the chapters that follow.

INFORMATION RESOURCES

Notes

There are a number of resources, references, and agencies to assist you in preparing for a haz-mat emergency. It is advisable for emergency service personnel to have several publications available for reference. The following are valuable key resources:

Printed Resources

- *Emergency Response Guidebook*

Verbal Resources

- CHEMTREC
- CANUTEC
- National Response Center

Electronic Resources

- Computerized databases, such as CAMEO

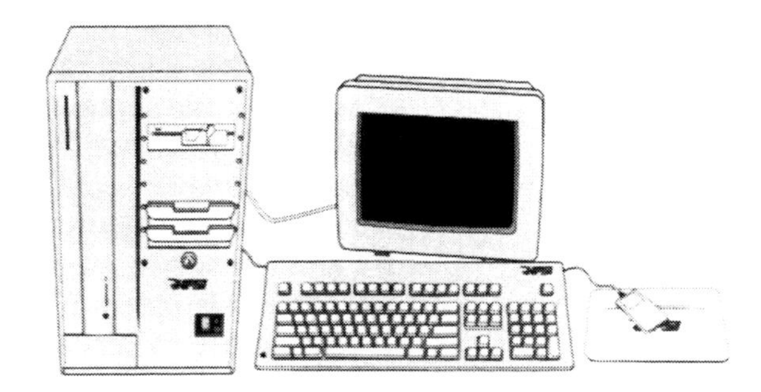

Use Your Resources

A listing of recommended resources and reference materials is located in Appendix C. The list contains telephone numbers of various agencies and technical personnel who may be able to assist as consultants.

DOT Emergency Response Guidebook

The *Emergency Response Guidebook* was developed by the U.S. Department of Transportation for firefighters, police, and other emergency service personnel for initial action involving hazardous materials. The information should guide responders in protecting themselves and the public during the initial phases of an incident.

The guidebook assists individuals in making decisions, but it does not serve as a substitute for knowledge or judgment. It gives recommendations that apply in the majority of cases. Although this document was primarily designed for use during highway or railway haz-mat incidents, it will be useful in handling incidents in other modes of transportation and at facilities such as terminals and warehouses.

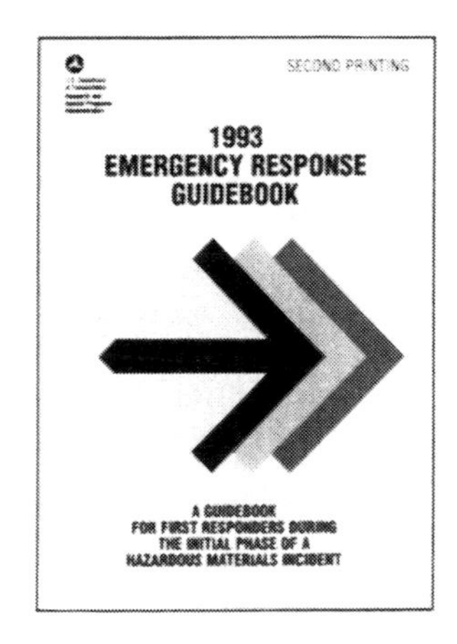

PROCEDURES FOR USING EMERGENCY RESPONSE GUIDEBOOK

First identify the material from one of the following places:

- four-digit number on the placard or the orange panel used with a placard
- four-digit number on the shipping papers or package
- name on the shipping papers, or package
- placard

After the materials are identified, locate the guide number in one of the following:

- ID numerical index
- alphabetical name index

Notes

Once the guide number is determined, double-check the ID number. If the index entry is highlighted, look for the ID number and name of the material in the Table of Initial Isolation and Protective Action Distances located in the rear of the book. Read the *numbered guide page* carefully and begin appropriate protective and response actions. If the four digit number or proper chemical name cannot be found, and the placard can be identified, go to the Placard Table and turn to the appropriate guide number. Refer to Guide 11 if only the general classification can be determined.

Materials listed in the guidebook's index are assigned to a numbered response guide section that identifies the most significant potential hazards and gives guidance for initial emergency response action. (Sample pages from the *Emergency Response Guidebook* are included in Figures 1-6, 1-7, and 1-8.) Since many materials represent similar types of hazards calling for similar action, only a few numbered response guides are required for all the materials listed. Incidents involving more than one hazardous material require more expertise. *Remember,* the material involved in an accident may, by itself, be non-hazardous. However, a combination of different materials or involvement of a single material in a fire may produce serious health, fire, or explosion hazards.

The guidebook does not individually list explosives and blasting agents by ID number. If the shipping paper or placard identifies the material as Division 1.1, 1.2, 1.3, 1.5, or 1.6 Explosives, the instructions given in Guide 46 should be followed. If the shipping paper identifies the material as Division 1.4 Explosives, the instructions in Guide 50 should be followed.

The constant changing of identification number assignments and designation of hazardous materials by regulatory authorities makes each edition of the guidebook useful for only two or three years after publication. The *Emergency Response Guidebook* is published by the DOT and is keyed to the Hazard Identification System (HIS).

Notes

Guide No.	ID No.	Name of Material
1075	22	LPG, liquefied petroleum gas
1075	22	PETROLEUM GASES, liquefied
1076	15	**PHOSGENE**
1077	22	PROPYLENE
1078	12	CHLORODIFLUOROMETHANE and CHLOROPENTAFLUORO-ETHANE MIXTURE
1078	12	CHLOROTRIFLUOROMETHANE and TRIFLUOROMETHANE MIXTURE
1078	12	DICHLORODIFLUORO-METHANE and CHLORODIFLUORO-METHANE MIXTURE
1078	12	DICHLORODIFLUORO-METHANE and DICHLORO-TETRAFLUOROETHANE MIXTURE
1078	12	DICHLORODIFLUORO-METHANE and DIFLUORO-ETHANE MIXTURE
1078	12	DICHLORODIFLUORO-METHANE and TRICHLOROFLUORO-METHANE MIXTURE
1078	12	DICHLORODIFLUORO-METHANE, TRICHLORO-FLUOROMETHANE and CHLORODIFLUORO-METHANE MIXTURE
1078	12	DICHLORODIFLUORO-METHANE and TRICHLORO-TRIFLUOROETHANE MIXTURE
1078	12	DISPERSANT GAS, n.o.s.
1078	12	REFRIGERANT GASES, n.o.s.
1078	12	TRIFLUOROMETHANE and CHLOROTRIFLUORO-METHANE MIXTURE
1079	16	**SULFUR DIOXIDE**
1079	16	**SULFUR DIOXIDE, liquefied**
1080	12	SULFUR HEXAFLUORIDE
1081	17	TETRAFLUOROETHYLENE, inhibited
1082	18	**TRIFLUOROCHLOROETHYLENE**
1082	18	**TRIFLUOROCHLOROETHYLENE, inhibited**
1083	19	TRIMETHYLAMINE, anhydrous
1085	60	VINYL BROMIDE, inhibited
1086	17	MONOCHLOROETHYLENE
1086	17	VINYL CHLORIDE
1086	17	VINYL CHLORIDE, inhibited
1086	17	VINYL CHLORIDE, stabilized
1087	17	VINYL METHYL ETHER
1087	17	VINYL METHYL ETHER, inhibited
1088	26	ACETAL
1089	26	ACETALDEHYDE
1090	26	ACETONE, and solutions
1091	26	ACETONE OILS
1092	30	**ACROLEIN, inhibited**
1093	30	ACRYLONITRILE, inhibited
1098	57	**ALLYL ALCOHOL**
1099	57	ALLYL BROMIDE
1100	57	ALLYL CHLORIDE
1104	26	AMYL ACETATES
1105	26	AMYL ALCOHOLS
1105	26	ISOAMYL ALCOHOL
1106	68	AMYLAMINES
1107	26	AMYL CHLORIDES
1108	26	n-AMYLENE
1108	26	1-PENTENE
1109	26	AMYL FORMATES
1110	26	AMYL METHYL KETONE
1110	26	METHYL AMYL KETONE
1111	27	AMYL MERCAPTANS
1112	26	AMYL NITRATE
1113	26	AMYL NITRITES

Figure 1-6. Emergency Guidebook ID Numerical Index

Notes

Name of Material	Guide No.	ID No.
ACCUMULATORS, pressurized	**12**	1956
ACETAL	**26**	1088
ACETALDEHYDE	**26**	1089
ACETALDEHYDE AMMONIA	**31**	1841
ACETALDEHYDE OXIME	**26**	2332
ACETIC ACID, GLACIAL	**29**	2789
ACETIC ACID SOLUTION, more than 10% but not more than 80% acid	**60**	2790
ACETIC ACID SOLUTION, more than 80% acid	**29**	2789
ACETIC ANHYDRIDE	**39**	1715
ACETONE, and solutions	**26**	1090
ACETONE CYANOHYDRIN, stabilized	**55**	**1541**
ACETONE OILS	**26**	1091
ACETONITRILE	**28**	1648
ACETYL ACETONE PEROXIDE	**48**	2080
ACETYL BENZOYL PEROXIDE	**48**	2081
ACETYL BROMIDE	**60**	1716
ACETYL CHLORIDE	**29**	1717
ACETYL CYCLOHEXANE SULFONYL PEROXIDE	**52**	2082
ACETYL CYCLOHEXANE SULFONYL PEROXIDE	**52**	2083
ACETYLENE	**17**	1001
ACETYLENE, dissolved	**17**	1001
ACETYLENE TETRABROMIDE	**58**	2504
ACETYL IODIDE	**60**	1898
ACETYL METHYL CARBINOL	**26**	2621
ACETYL PEROXIDE	**49**	2084
ACID, liquid, n.o.s.	**60**	1760
ACID, SLUDGE	**60**	1906
ACID BUTYL PHOSPHATE	**60**	1718
ACID MIXTURE, hydrofluoric and sulfuric acids	**59**	1786
ACID MIXTURE, nitrating	**73**	1796
ACID MIXTURE, spent, nitrating	**60**	1826
ACRIDINE	**32**	2713
ACROLEIN, inhibited	**30**	**1092**
ACROLEIN DIMER, stabilized	**26**	2607
ACRYLAMIDE	**55**	2074
ACRYLIC ACID, inhibited	**29**	2218
ACRYLONITRILE, inhibited	**30**	1093
ACTIVATED CARBON	**32**	1362
ADHESIVE	**26**	1133
ADHESIVES, containing flammable liquid	**26**	1133
ADIPONITRILE	**55**	2205
AEROSOLS	**12**	1950
AIR, compressed	**12**	1002
AIR, refrigerated liquid (cryogenic liquid)	**23**	1003
AIR BAG INFLATORS	**31**	3268
AIR BAG INFLATORS	**32**	1325
AIR BAG MODULES	**31**	3268
AIR BAG MODULES	**32**	1325
AIRCRAFT HYDRAULIC POWER UNIT FUEL TANK	**28**	3165
ALCOHOL (beverage)	**26**	1170
ALCOHOL, denatured	**26**	1987
ALCOHOL, denatured (toxic)	**28**	1986
ALCOHOL (ethyl)	**26**	1170
ALCOHOL, nontoxic, n.o.s.	**26**	1987
ALCOHOLATES SOLUTION, n.o.s. in alcohol	**26**	3274
ALCOHOLIC BEVERAGE	**26**	1170
ALCOHOLIC BEVERAGES	**26**	3065
ALCOHOLS, n.o.s.	**26**	1987
ALCOHOLS, toxic, n.o.s.	**28**	1986
ALDEHYDES, n.o.s.	**26**	1989
ALDEHYDES, toxic, n.o.s.	**28**	1988
ALDOL	**55**	2839
ALDRIN and its mixtures	**55**	2761
ALKALI METAL ALCOHOLATES, n.o.s.	**38**	3206

Figure 1-7. Emergency Guidebook Alphabetical Name Index

Notes

GUIDE 30

ERG 93

POTENTIAL HAZARDS

HEALTH HAZARDS

Poisonous; may be fatal if inhaled, swallowed or absorbed through skin.
Contact may cause burns to skin and eyes.
Runoff from fire control or dilution water may cause pollution.

FIRE OR EXPLOSION

Extremely flammable; may be ignited by heat, sparks or flames.
Vapors may travel to a source of ignition and flash back.
Container may explode violently in heat of fire.
Vapor explosion and poison hazard indoors, outdoors or in sewers.
Runoff to sewer may create fire or explosion hazard.

EMERGENCY ACTION

Keep unnecessary people away; isolate hazard area and deny entry.
Stay upwind; keep out of low areas.
Positive pressure self-contained breathing apparatus (SCBA) and chemical protective clothing which is specifically recommended by the shipper or manufacturer may be worn. It may provide little or no thermal protection.
Structural firefighters' protective clothing is **not** effective for these materials.
Isolate the leak or spill area immediately for at least 150 feet in all directions.
See the Table of Initial Isolation and Protective Action Distances. If you find the ID Number and the name of the material there, begin protective action.
Isolate for 1/2 mile in all directions if tank, rail car or tank truck is involved in fire.
CALL Emergency Response Telephone Number on Shipping Paper first. If Shipping Paper not available or no answer, CALL CHEMTREC AT 1-800-424-9300.

FIRE

Small Fires: Dry chemical, CO_2, water spray or regular foam.
Large Fires: Water spray, fog or regular foam.
Do not get water inside container.
Apply cooling water to sides of containers that are exposed to flames until well after fire is out. Stay away from ends of tanks.
For massive fire in cargo area, use unmanned hose holder or monitor nozzles; if this is impossible, withdraw from area and let fire burn.
Withdraw immediately in case of rising sound from venting safety device or any discoloration of tank due to fire.

SPILL OR LEAK

Shut off ignition sources; no flares, smoking or flames in hazard area.
Fully-encapsulating, vapor-protective clothing should be worn for spills and leaks with no fire.
Do not touch or walk through spilled material.
Small Spills: Flush area with flooding amounts of water.
Large Spills: Dike far ahead of liquid spill for later disposal.

FIRST AID

Move victim to fresh air and call emergency medical care; if not breathing, give artificial respiration; if breathing is difficult, give oxygen.
In case of contact with material, immediately flush skin or eyes with running water for at least 15 minutes.
Remove and isolate contaminated clothing and shoes at the site.
Keep victim quiet and maintain normal body temperature.
Effects may be delayed; keep victim under observation.

Figure 1-8. Example of Emergency Guidebook Number Pages

Notes

CHEMTREC

The Chemical Transportation Emergency Center (CHEMTREC) is a service that operates 24 hours a day, seven days a week, providing advice and information concerning haz-mat incidents. It is sponsored by the Chemical Manufacturers Association. Information about the sources and types of calls CHEMTREC received in 1994 is found in Figure 1-9.

CHEMTREC provides immediate response advice for those at the scene of emergencies, then promptly contacts the haz-mat shipper involved for more detailed assistance and follow-up. You may call CHEMTREC toll-free from any point in the United States (including Alaska, Hawaii, the Virgin Islands, and Puerto Rico) or from Canada at ***1-800-424-9300***. All calls to this number should be limited to *emergencies* only. The non-emergency number for CHEMTREC is 1-800-262-8200.

In the USA or Canada, call CHEMTREC at 1-800-424-9300.

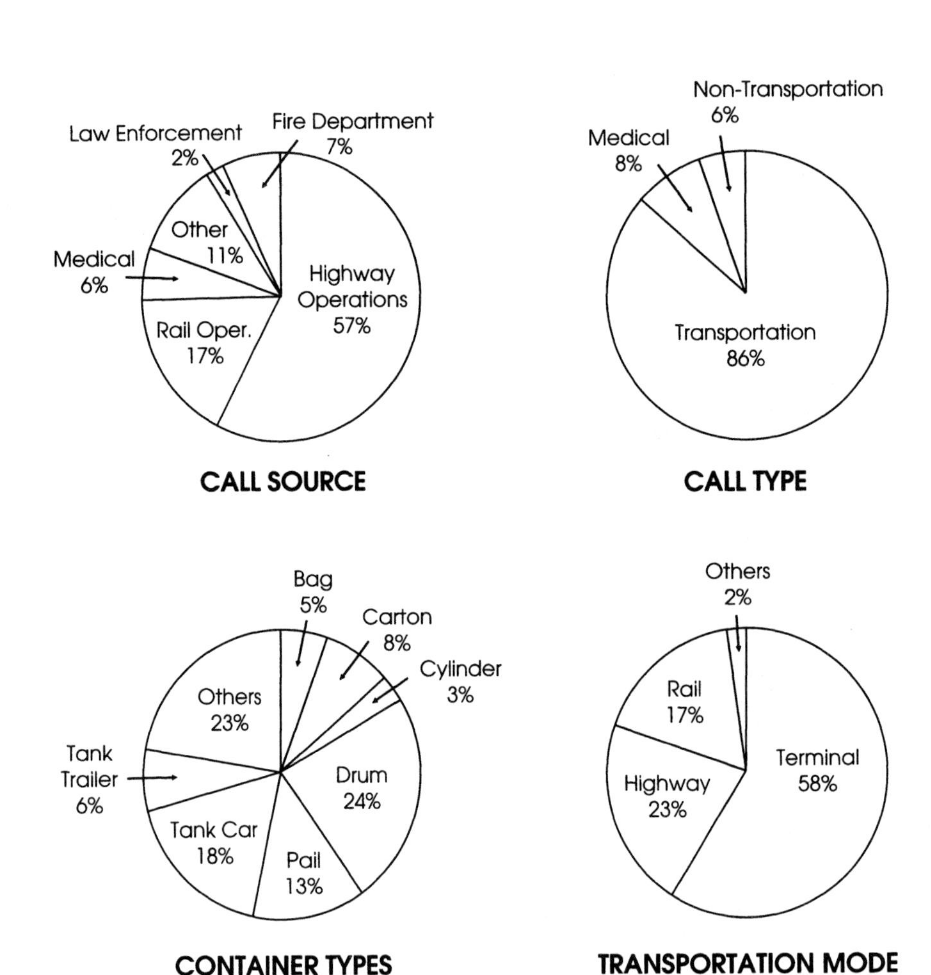

Figure 1-9. CHEMTREC Incidents, 1994

Notes

CHEMTREC maintains technical information files on over 600,000 proprietary chemicals. During emergencies, CHEMTREC can provide information regarding the effects of most chemicals on persons or the environment and can suggest methods for treatment, containment, and control of an incident. When contacting CHEMTREC, provide as much of the following information as possible:

- name, location, and telephone number of caller
- identification number or name of the product
- location of incident
- time and date of incident
- carrier's name
- rail car or truck number
- type and condition of container
- point of origin
- consignee's name and destination
- nature and extent of human injury
- nature and extent of property damage
- prevailing weather conditions
- local population information (heavily or sparsely populated)
- topography of surrounding area
- how to re-establish telephone contact with caller or other responsible party involved

Contact with CHEMTREC should be established as soon as the incident commander has arrived, surveyed the incident, and responded to the immediate needs of the situation. It is important to keep phone lines open between the shipper or manufacturer and the on-scene response leader. The CHEMTREC communication facility provides a teleconferencing bridge between their office, the experts, and you. Successful alleviation of the incident may depend upon your contact with CHEMTREC.

CANUTEC

The Canadian Transportation Emergency Center (CANUTEC) is an agency of the Canadian government located in Ottawa, Ontario, Canada. It functions similarly to its U.S. counterpart, CHEMTREC, by providing information about hazardous commodities to emergency service personnel around the clock. Statistics regarding CANUTEC's activities are found in Figures 1-10 and 1-11.

In Canada, call CANUTEC at (613) 996-6666.

Assistance with a hazardous commodity emergency in Canada can be obtained by calling CANUTEC at ***(613) 996-6666***. Provide the same type of information to CANUTEC as found in the list for CHEMTREC above. For additional information about CANUTEC, write or call:

The Canadian Transportation Emergency Center
Transport Canada
Transport of Dangerous Goods
Tower C, Place de Ville, Sixth Floor
Ottawa, Ontario, CANADA K1A 0N5
(613) 992-4624

Notes

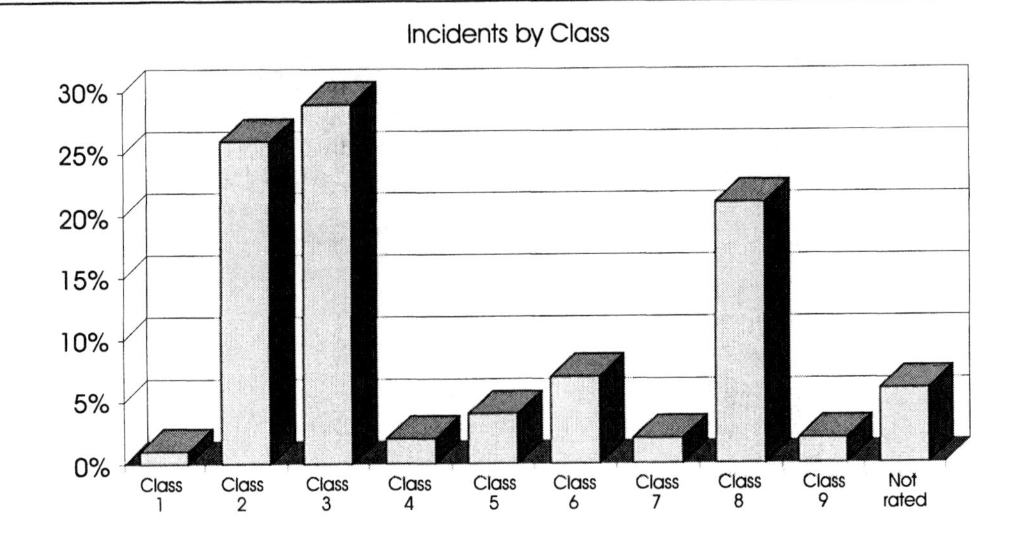

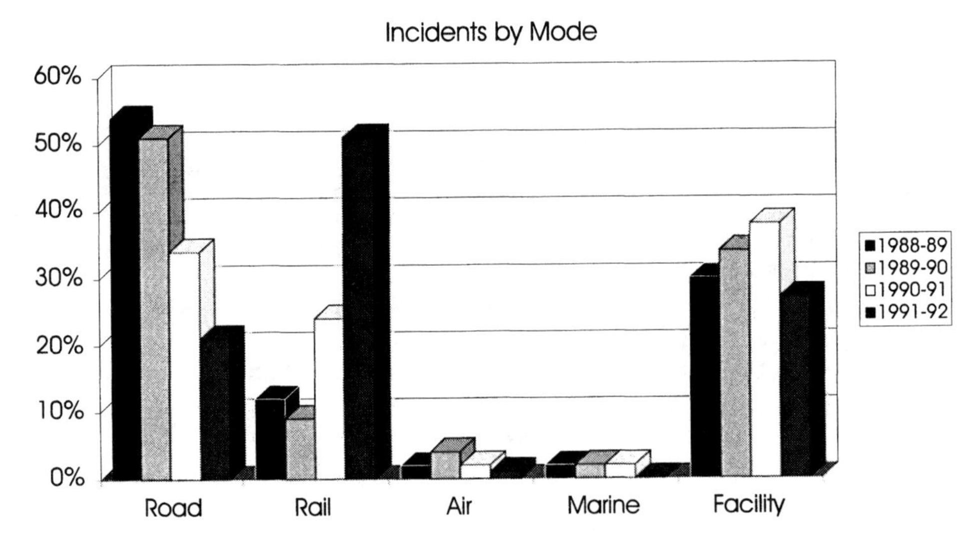

Figure 1-10. Incidents Handled by CANUTEC

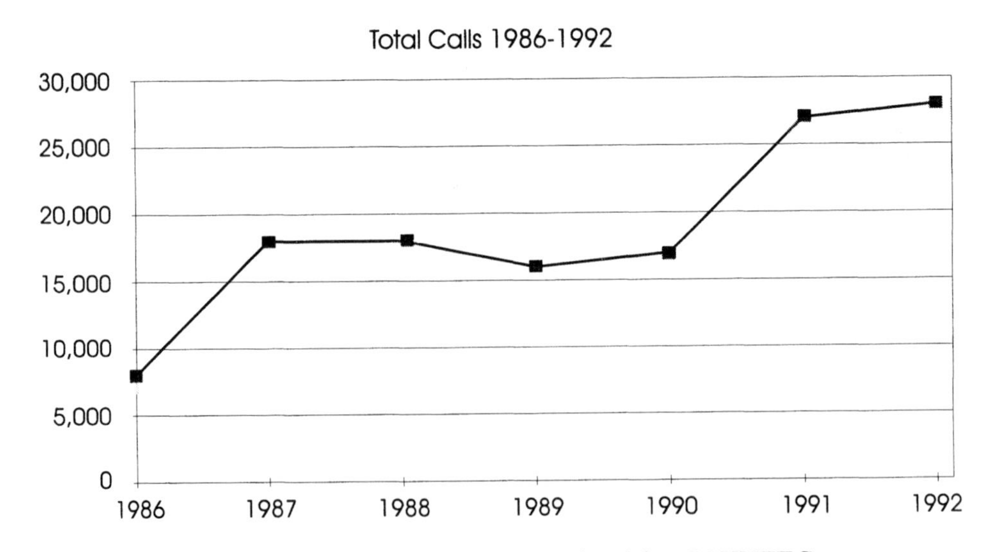

Figure 1-11. Total Calls Received by CANUTEC

National Response Center

Notes

The National Response Center is the primary federal point of contact for reporting all oil, chemical, biological, and infectious discharges into the environment.

The National Response Center (NRC) was established by the U.S. Coast Guard in 1974 to receive reports of oil spills into U.S. waters and along the adjacent shorelines. Later, the Center's functions were expanded to include receiving reports of releases of hazardous substances into the environment. Today, the NRC is the primary federal point of contact for reporting all oil, chemical, biological, and infectious discharges into the environment anywhere in the U.S. and its territories.

The NRC is staffed by Coast Guard personnel and maintains a telephone watch 24 hours a day, seven days a week. Staff members enter reports of pollution incidents into a computer database and immediately relay each report to a predesignated Federal On-Scene Coordinator. Other federal agencies may also be notified, depending upon the transportation mode involved and the severity of the incident. During major pollution incidents, the NRC is responsible for briefing senior Coast Guard, Department of Transportation, Environmental Protection Agency, and White House officials.

The National Response Center is part of a federal emergency response system known as the National Oil and Hazardous Substances Response System. A National Contingency Plan for this system is described in 40 CFR 300. The National Contingency Plan establishes three organizational levels:

- National Response Team
- Regional Response Teams
- Federal On-Scene Coordinators

There are also four special force components:

- National Strike Force
- Environmental Response Team
- Public Information Assist TeamPublic Information Assist Team
- Scientific Support Coordinators

The ***National Response Team*** consists of 15 federal agencies with interest and expertise in various aspects of emergency response to pollution incidents. It operates under the direction of the Environmental Protection Agency (EPA) and U.S. Coast Guard. The National Response Team is primarily a national planning, policy, and coordinating body and does not respond directly to incidents. It provides policy guidance prior to an incident and assistance in the form of technical advice and access to resources and equipment during an incident. The National Response Center is the contact point for activation of the National Response Team and provides facilities for the National Response Team to use in coordinating a national response action when required.

There are 13 ***Regional Response Teams***. Like the National Response Team, the Regional Response Teams are planning, policy, and coordinating bodies, and they do not respond directly to the scene. They provide assistance as requested by the Federal On-Scene Coordinator during an incident. They may also provide assistance to state and local governments in preparedness, planning, and training for emergency response.

The ***Federal On-Scene Coordinator*** coordinates all federal containment, removal, and disposal efforts and resources during an incident. This person is also the point of contact for the coordination of federal efforts with

Notes

local response. The National Response Center provides emergency response support to the Federal On-Scene Coordinator.

The Federal On-Scene Coordinator also has access to the specialized forces mentioned above:

- The Coast Guard's ***National Strike Force***. This force is composed of three strategically located strike teams. Each team is extensively trained and equipped to respond to major oil spills and chemical releases.
- The EPA's ***Environmental Response Team***. This is a group of highly trained scientists and engineers based in Edison, New Jersey, and Cincinnati, Ohio. Capabilities include multimedia sampling and analysis, hazard assessment, clean-up techniques, and technical support.
- The Coast Guard's ***Public Information Assist Team***. This is a highly skilled unit of public affairs specialists prepared to complement the existing public information capabilities of the Federal On-Scene Coordinator.
- ***Scientific Support Coordinators***. These are scientific and technical advisors provided by the National Oceanic and Atmospheric Administration. Their capabilities include contingency planning, surface/subsurface trajectory forecasting/hindcasting, and resource risk analysis.

The number of incident reports received by the National Response Center has grown from 200 in 1974 to approximately 26,300 in 1991. The growth rate is averaging 10-15% per year. Factors responsible for the growth rate include:

- increase in the number of chemicals being transported
- increase in data collection requirements due to regulatory mandates
- increase in corporate understanding of the reporting requirements
- increase in public awareness of the hazards of chemicals in the environment
- increase in public and industry awareness of the National Response Center

To handle calls, the National Response Center operates a database system known as the Incident Reporting Information System (IRIS). This system provides the NRC with the capability to collect, analyze, manage, and disseminate incident information. The NRC also operates the CAMEO database and has online access to the CHRIS and MSIS databases.

Call the National Response Center at 1-800-424-8802.

The National Response Center can be reached at the following address:

National Response Center
2100 2nd Street, S.W.
Washington, DC 20593
(202) 267-2185

To report an incident, call ***1-800-424-8802*** day or night. In the District of Columbia, call 267-2675.

CONCLUSION

Notes

What we learn, practice, and teach response crews will determine how an emergency incident will be handled. Efforts to control the incident should not *increase* the hazards. Emergency incidents require a common sense approach to ensure safety and a successful outcome. Let us determine to provide the best emergency response possible.

Most manufactured products have multiple hazards that require complicated processes of responding. The remainder of this book will take an in-depth look at how to understand and respond *safely* to a variety of haz-mat incidents.

Chapter 2

BASICS OF TOXICOLOGY AND PERSONAL PROTECTIVE EQUIPMENT

OBJECTIVES

After studying the material in this chapter, you will be able to:

- state the four routes by which chemicals can enter the body
- explain the relationship between dose and response
- explain the relationship between toxicity and hazard level
- define threshold limit value
- state diagnostic clues to chemical exposures
- list selected toxic products of combustion
- list the target organs affected by chemicals
- describe the types of respiratory, chemical, and thermal personal protective equipment available to responders
- list features to consider when selecting chemical protective clothing

TOXICOLOGICAL CONCEPTS

Health hazards to emergency response personnel during a haz-mat incident can be deadly. Dangerous exposure to harmful chemicals and long-term health effects to the responder are possible and should be considered at every incident. Responders must also consider the delayed effects (10-12 hours) personnel can exhibit as a result of exposure. Understanding toxicological concepts is necessary when handling unforeseen exposures, decontamination, equipment selection, and EMS intervention. It is critical to understand toxicity levels and diagnostic clues to chemical exposure in order to minimize the risks to personnel.

Notes

KNOW THE KILLERS!

"All things are poisonous, for there is nothing without poisonous qualities. It is only the dose which makes a thing a poison."
– Paracelsus (1493-1541)

Common Units of Measurement

ppm – Parts per million of a substance dispersed in a gas or liquid.

ppb – Parts per billion.

mmpcf – Millions of particles of a particulate per cubic foot of air.

mg/m³ – Milligrams of a substance per cubic meter of air.

Notes

Definitions of Physical Forms

The following terms describe the states of matter in which chemical contaminants may occur.

Aerosol – A suspension of liquid or solid particles in a gas. The particles are often of colloidal size (1 millimicron to 1 micron in diameter).

Dust – An air suspension of solid particles that range in size from 0.1 micron to 50 microns in diameter. A person with normal eyesight can detect dust particles as small as 50 microns in diameter. Smaller airborne particles cannot be detected by the naked eye unless strong light is reflected from the particles. Dust of respirable size (below 10 microns) cannot be seen without a microscope.

Fog – A suspension of liquid droplets in air (an aerosol).

Fume – Smoke-like particles emanating from the surface of heated metals. Also, the vapor given off from concentrated acids, from evaporating solvents, or as a result of decomposition reactions. The particles that make up a fume are extremely fine, usually less than 1.0 micron in diameter.

Gas – A state of matter having very low density and viscosity relative to liquids and solids, comparatively great expansion and contraction with changes in pressure and temperature, the ability to diffuse readily into other gases, and the ability to occupy any container completely and uniformly.

Smoke – A microscopic disperson of a solid in a gas (an aerosol). Smoke particles are usually less than 0.1 micron in diameter.

Vapor – An air dispersion of molecules of a substance that is liquid or solid in its normal state at standard temperature and pressure.

Source: *Hawley's Condensed Chemical Dictionary*

ROUTES OF CHEMICAL ENTRY

In emergency response situations, *inhalation* is the most critical entry route for chemicals into the body. The next most critical entry route is *contact with the skin*. In either case, there may be irritation of contacted tissue or absorption into the blood with possible systemic intoxication. Although the gastrointestinal tract is a potential site for absorption, the ingestion of significant amounts of chemicals is rare in the emergency situation.

Notes

Inhalation

The water solubility of a gas or vapor determines how much inhaled material reaches the lungs. Highly water-soluble gases such as ammonia, hydrogen chloride, and hydrogen fluoride quickly dissolve in the mucous membranes of the nose and upper respiratory tract, causing irritation.

At low airborne concentrations, little of the substance will reach the lungs. At high airborne concentrations, some of the gas or vapor will not be absorbed in the upper respiratory tract. However, amounts sufficient to cause severe irritation and pulmonary edema can reach the alveoli.

Insoluble gases, such as nitrogen dioxide and phosgene, are not removed by the moisture in the nose and upper respiratory tract. Insoluble gases can easily reach the outer areas of the lungs.

Gases with limited solubility, such as ozone and chlorine, cause irritation of both the upper respiratory tract and lungs. Carbon monoxide does not irritate the respiratory tract, but it is rapidly absorbed into the blood, resulting in systemic intoxication.

The particle size of aerosols determines their accessibility to small airways. If the diameter of a particle is large, impaction occurs on the pharynx or nasal cavity, preventing the particles from reaching the alveoli. Slightly smaller particles often settle within the bronchioles. Very small particles diffuse within the lung's alveoli.

Skin Contact

Four reactions may occur when a hazardous substance contacts the skin:

- The skin itself, and its lipid film and sweat, may prevent the substance from penetrating the skin.
- The substance may react with the skin surface, causing primary irritation. Acids, alkalies, and many organic solvents are most likely to cause primary irritation.
- The substance may penetrate the skin and cause sensitization. Formaldehyde, nickel, and phthalic anhydride are most likely to cause sensitization.
- Agents such as aniline, parathion, and tetraethyl lead may penetrate the skin, enter the blood, and react systemically.

Only a small number of substances are absorbed through the skin in hazardous amounts. A substance can pass through the skin by one or more of the following four routes:

- epidermal cells
- sweat glands
- sebaceous glands
- hair follicles

The main pathway into the blood is through epidermal cells, since this tissue constitutes the majority of the body's surface area. Absorption is faster through skin that is inflamed or has suffered abrasions. For this reason, chemicals that are normally not hazardous may be dangerous to individuals who are suffering from active inflammatory dermatoses.

In addition to the normal routes of skin absorption, the eyes can absorb chemicals very rapidly and present one of the fastest means of exposure.

Exposure may occur through a direct splash of a chemical into the eyes, through toxic smoke particles that are carried into the eyes, or through the absorption of toxic gases or vapors by the eyes.

Ingestion

Hazardous substances can be ingested into the body through smoking or eating with contaminated hands, or by smoking or eating in contaminated work areas. The amounts ingested are usually not acutely toxic. Exceptions are highly toxic substances, such as lead, arsenic, and mercury. Great care should be taken to avoid any ingestion of toxic materials.

Injection

Toxic substances can be injected into the body through a puncture wound or break in the skin. If protective clothing is not worn or safety precautions not followed, toxic contamination by injection can occur at a haz-mat incident. *Do not begin medical procedures involving IV therapy until the patient has been decontaminated.*

Notes

Do not begin medical procedures involving IV therapy until the patient has been decontaminated.

DOSE AND RESPONSE

Toxicology is the study of the harmful effects of a hazardous material. The most fundamental concept in toxicology is that a relationship exists between the dose (quantity) of an agent and the response in the exposed individual. Three assumptions must be considered in this relationship:

- The magnitude of the response is a function of the agent's concentration level at the biologic site.
- The concentration level at the biologic site is a function of the dose administered.
- The response and the dose are causally related. In other words, as the dose is increased, the harm is increased.

Toxicity data for a non-characterized agent are usually obtained by administering the agent into mice or rats. Results of varying doses range from death to survival. Two calculations result from statistical analysis of the data:

- ***Median lethal dose (LD_{50})*** – The dose expected to be lethal in 50% of an animal group.
- ***Median lethal concentration (LC_{50})*** – The atmospheric concentration expected to be lethal to 50% of an animal group exposed for the specified time period.

The LD_{50} or LC_{50} provides an initial index of comparative toxicity. Observation of animals provides valuable information concerning effects the agent may have on humans. Autopsies reveal which of the animal's organs were affected. Animal research often provides the only data available for many substances. In such cases, until proven otherwise, you should expect the same effects to occur in humans who are exposed to a high atmospheric concentration for a sufficient time.

The slope of the dose-response curve suggests the magnitude of the range between a no-effect level and a lethal dose (see Figure 2-1). That range is called the ***margin of safety.*** If the dose-response curve is steep, the margin of safety is small. Figure 2-1 indicates substances A and B both

Notes

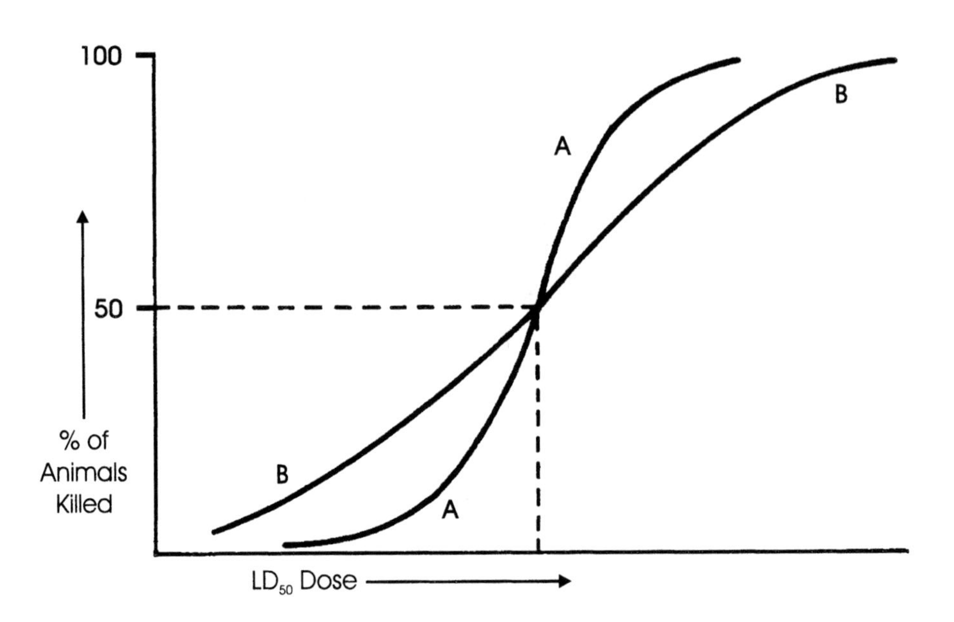

Figure 2-1. Dose vs. Response Curve

have the same LD_{50}, but substance B has a wider margin of safety than substance A.

TOXICITY AND HAZARD

Toxicity is the first factor that determines the hazard level of a substance. Toxicity is the level of a substance's ability to cause injury to biologic tissue in a given environment or situation. Toxicity depends upon:

- how the substance is absorbed, metabolized, and excreted
- how rapidly the substance acts
- warning properties of the substance

A highly dangerous substance would be an agent that is in the ***extremely toxic*** class. An extremely toxic substance would be:

- rapidly absorbed by inhalation, skin contact, or ingestion
- slowly metabolized and/or slowly excreted

Such actions would allow the agent to accumulate in the body. Accumulation produces rapid, irreversible effects or death with no warning properties. Obviously, such a substance would be hazardous to handle.

The second factor which determines a material's hazard level is its ***physical characteristics,*** or the way it will be encountered in the workplace. For example, a liquid with a high vapor pressure will reach a higher airborne concentration and will be more hazardous than an equally toxic liquid with a low vapor pressure. Factors that determine the potential for fire or explosion, such as flash point and lower flammable limit, should be considered. Any work practices that cause frequent or prolonged exposure to high levels of a toxic substance will contribute to the hazard potential.

Notes

Toxicity Classes			
Toxicity Rating	Descriptive Term	LD_{50} (wt/kg) Single Oral Dose Rats	LC_{50} (ppm) 4 Hour Inhalation Rats
1	Extremely toxic	1 mg or less	<10
2	Highly toxic	1–50 mg	10–100
3	Moderately toxic	50–500 mg	100–1,000
4	Slightly toxic	0.5–5 g	1,000–10,000
5	Practically non-toxic	5–15 g	10,000–100,000
6	Relatively harmless	15 g or more	>100,000

Figure 2-2

THRESHOLD LIMIT VALUE

The American Conference of Governmental Industrial Hygienists (ACGIH) has prepared a list of the Threshold Limit Values (TLVs) for almost 600 substances. There are three TLV categories.

- ***Threshold Limit Value/Time-Weighted Average (TLV/TWA)*** – The maximum airborne concentration of a material to which average healthy workers may be repeatedly exposed for eight hours a day, 40 hours a week, without adverse effect.
- ***Threshold Limit Value/Short-Term Exposure Limit (TLV/STEL)*** – The maximum concentration to which workers can be exposed for up to 15 continuous minutes without suffering intolerable irritation, chronic or irreversible tissue change, or narcosis of sufficient degree to increase accident proneness, impair self-rescue, or reduce work efficiency. This is provided that no more than four exposures per day are permitted within this limit with at least one hour between exposure periods, and provided that the daily TLV/TWA also is not exceeded. The STEL should be considered a maximum allowable concentration, or absolute ceiling, which is not to be exceeded at any time during the 15 minute exposure period.
- ***Threshold Limit Value/Ceiling (TLV/C)*** – The maximum concentration that should not be exceeded, even instantaneously.

Time-weighted averages permit limited exposures above the TLV, provided they are offset by equal exposures below the limit during the workday. Direct contact with any substance bearing the notation *skin* in its TLV should be avoided. The *skin* designation indicates that a substance may be readily absorbed through the skin, mucous membranes, or eyes. It is not a threshold for safe exposure.

Notes

TLVs are guidelines and are not intended as absolute boundaries between safe and unsafe concentrations.

TLVs range from 0.0002 ppm for osmium tetroxide to 5,000 ppm for carbon dioxide. TLVs for respirable dusts, such as crystalline silica and non-asbestiform talc, are expressed in terms of mmpcf and/or mg/m^3.

TLVs are revised as new information becomes available. Each year new substances are added to the TLV list. Compounds that are proven or suspected carcinogens in humans, such as benzidine or beta-naphthylamine, have no TLV value, and human exposure to these agents should be avoided.

It is important to note that TLV values are not the best guide for determining the toxicity of a substance. These values are often determined by incongruent methods and standards. TLVs are guidelines and are not intended as absolute boundaries between safe and dangerous concentrations. Exceptions are those substances which have a ceiling limit or "C."

TYPES OF EXPOSURE

Exposure may be acute or chronic.

- ***Acute*** means a single dose or exposure.
- ***Chronic*** means repeated exposure.
- ***Subacute*** means two or three exposures at a level between acute and chronic.

Exposure may be local or systemic.

- ***Local*** means action at the site of contact. A caustic material such as lye attacks at the site of contact.
- ***Systemic*** means the toxic material must pass through the skin, mucous membranes, or lungs and be transported to the action site via the bloodstream. Chlorinated solvents are a familiar example of a systemic poison. These solvents are primarily liver poisons that must be absorbed and transported to the liver through the bloodstream.

METHODS OF MEASUREMENT FOR THE RESPONDER

Toxic materials are often rated in the following terms:

- Slight, moderate, severe.
- Acute local, acute systemic, chronic local, chronic systemic.
- ***LD_{50}*** is the dosage, administered by any route except inhalation, that is necessary to kill 50% of exposed animals in laboratory tests within a specified time. It is ordinarily expressed in mg/kg body weight.
- ***LC_{50}*** is the concentration of an inhaled substance, expressed in mg/cm^3 of air, that is necessary to kill 50% of test animals exposed to it within a specified time. It is sometimes referred to as "LC_{50} INHALE."
- ***MLD*** is the minimum lethal dosage (mg/kg body weight) at which any fatalities result.

Notes

- ***IDLH*** is the atmospheric concentration that is immediately dangerous to life or health. It poses an immediate threat to life, causes irreversible or delayed adverse health effects, or interferes with a person's ability to escape from a dangerous atmosphere.
- ***Emergency response planning guideline (ERPG-2)*** is the maximum airborne concentration to which an average healthy individual can be exposed for up to one hour without experiencing irreversible or other serious health effects which could impair the person's ability to take protective action.

Emergency response personnel may be less concerned with threshold limit values (TLV) for long-term exposure than with minimum lethal dosage (MLD) or IDLH. Toxicity ratings are usually not available in only one system.

Other hazards that must be measured include:

- low oxygen levels in confined spaces and areas surrounding a leak or spill
- possibility of flash burns

Responding units must have appropriate monitoring equipment and training to assess the hazards present.

Responding units must have appropriate monitoring equipment and training to assess the hazards present.

DIAGNOSTIC CLUES TO CHEMICAL EXPOSURES

Signs and symptoms for individuals exposed to hazardous chemicals may vary greatly. An unmistakable onset can cause a patient's complaints to be quite sudden. This happens when exposed to an irritant gas such as chlorine. More often, however, the onset of signs and symptoms is vague and insidious, as occurs when exposed to a cumulative poison such as lead or mercury, or to a neurotoxic solvent such as methyl butyl ketone.

Often there is a delayed onset of acute respiratory distress, which can lead to pulmonary edema, following exposure to an irritant gas. Except for mild upper respiratory irritation at the time of exposure, a 12 hour or more symptom-free latent period may occur. The patient typically awakens short of breath, coughing, and needs prompt relief. Still another delayed reaction occurs when skin is irritated by sunshine following exposure to a photosensitizing chemical such as coal tar pitch.

The eye irritation termed ***welder's flash*** is another condition characterized by a latent reaction period. It is frequently encountered among workers near a welding site, as well as among welders themselves.

Occupational factors should be considered when encountering differences in the diagnosis of certain syndromes. Ask the patient what work he or she does. If the patient's condition is acute, ask about recent past occupations as well as present work. If the condition is chronic, as with pulmonary fibrosis, ask about the patient's work for the last several years.

Patient in Respiratory Distress

- Upper respiratory indicates mild irritants.
- Tracheobronchitis, pulmonary edema, or pneumonitis indicates severe irritants.
- Asthma indicates chemicals that are pulmonary sensitizers.

Notes

- Chronic respiratory failure indicates beryllium (granulomatosis) or fibrogenic dusts. These should be differentiated from agents causing benign pneumoconiosis or other pulmonary carcinogens.

Cyanotic Patient

- Cholinesterase inhibitors
- Methemoglobin formers act by affecting the oxygen carrying capacity of the blood.
- Asphyxiants

Unconscious Patient

- Central nervous system depressants
- Cholinesterase inhibitors
- Asphyxiants

Mild Exposure

A less critically ill person may be suspected of work related chemical exposure when certain clinical signs are present. The most frequent complaints of ***mild exposure*** are:

- unusual fatigue
- exhaustion
- irritability
- headache
- nausea
- light-headedness
- lack of coordination

When ***respiratory irritants*** are involved, the person may experience:

- excessive eye watering
- nasal irritation
- sneezing
- coughing
- tightness of the chest
- skin irritation

The diagnosis of haz-mat exposure is based upon:

- precise history of exposure to a specific toxic substance
- determination of the nature and severity of the exposure, preferably confirmed by a hygienist's evaluation of the site
- clinical signs and symptoms characteristic of the suspected toxic state
- supporting laboratory evidence of the absorption of the toxic substances in hazardous amounts

Notes

Haz-mat responders can be exposed to high concentrations of dangerous substances, and there is high risk potential. Therefore, it is important to understand the methods of measurement and the diagnostic clues that can alert you to a possible exposure problem. Chapter 3 presents a systematic approach to see that personnel are adequately screened for chemical exposure when terminating a haz-mat incident.

Laboratory Confirmation of Chemical Exposure

Several types of chemical exposure may be evaluated besides the usual clinical laboratory examinations of blood, urine, and sputum. The suspected toxin's concentration level may be measured in blood or urine, its metabolites in the urine, or an affected enzyme in the blood. Periodic biologic monitoring of substances is useful in individuals exposed to chemicals such as lead, arsenic, mercury, manganese, cadmium, fluoride, benzene (test for phenol in urine), trichloroethylene (trichloroacetic acid in urine), and organophosphates (cholinesterase in blood).

Suspected Human Carcinogens (OSHA)

Chemical Name	Common or Trade Name
2-Acetylaminofluorene	2-AAF
4-Aminodiphenyl	4-ADP
Benzidine (and its salts)	Benzidine
bis-Choromethyl ester	BCME
3,3-Dichlorobenzidine (and its salts)	CB
4-Dimethylaminoazobenzene	Methyl yellow
β-Naphthylamine	2-NA
4-Nitrobiphenyl	4-NBP
n-Nitrosodiumethylamine	Dimethylamine
β-Propiolactone	Betaprone®
Methylchloromethyl ether	CMME
α-Naphthylamine	1-NA
4,4′-Methylenebis(2-chloroaniline)	MOCA
Ethyleneimine	EI

Figure 2-3

Notes

Selected Respiratory Poisons

Poison	Signs and Symptoms
Upper Respiratory Poisons	
Hydrogen Bromide	Generally similar to HCl, but quantitative data are lacking.
Hydrogen Chloride (HCl)	At 35 ppm eye and throat irritation occurs. 100 ppm is tolerable for about an hour. Higher concentrations result in pulmonary edema.
Hydrogen Fluoride (HF)	Unlike HCl, HF cannot be tolerated even for short time periods. HF produces severe skin burns and ulcers of the upper respiratory tract.
Nitrogen Dioxide (NO_2)	Similar to SO_2, but quantitative data are lacking.
Sulfur Dioxide (SO_2)	Less than 1 ppm is detectable by taste. 3 ppm is detectable by smell. 5 to 10 ppm is highly irritating to nasal mucous membranes, throat, and glottis. Produces pulmonary edema. In actual practice, SO_2 reacts with moist air to form sulfurous and sulfuric acid mists that penetrate into the lower respiratory tract. In this event, the signs and symptoms are those of a lower respiratory poison.
Lower Respiratory Poisons	
Acrolein	Highly irritating to eyes, mucous membranes, and entire respiratory system. Quantitative data are lacking.
Bromine, Fluorine	Similar to chlorine. Quantitative data are lacking.
Chlorine	Odor detectable at 3 ppm. Immediate throat and nose irritation at 15 ppm. Hazardous at 50 ppm for even short exposures. High exposures result in pulmonary edema and audible chest rales.
Chloropicrin	Toxicity level between chlorine and phosgene.
Epichlorohydrin	Toxic agent is HCl; therefore, the signs and symptoms are similar to those of HCl.
Hydrogen Cyanide (HCN)	Death is usually rapid and essentially symptomless. Mild exposure produces headache, dizziness, suffocation, and nausea.
Hydrogen Sulfide (H_2S)	Highly toxic. Exposure to low concentrations produces conjunctivitis, vision problems, and digestive disturbances. Exposure to higher concentrations produces bronchitis and pulmonary edema.
Methyl Cyanide, Acrylonitrile	Toxic agent is HCN; therefore, the signs and symptoms are similar to those of HCN.

Notes

Selected Respiratory Poisons (cont.)

Poison	Signs and Symptoms
Lower Respiratory Poisons (cont.)	
Methyl Mercaptan, Phosphorus Pentasulfide	Toxic agent is H_2S; therefore, the signs and symptoms are similar to those of H_2S.
Phosgene	Irritating to eyes at 3 ppm. Fatal at 50 ppm for even short exposures. High exposure results in pulmonary edema, dyspnea, and chest rales.
Poisons That Affect the Oxygen Carrying Capacity of the Blood	
Aniline	Initial symptoms are due to conversion of hemoglobin to methemoglobin. Signs and symptoms are methemoglobinemia and anoxemia. Longer term exposure leads to jaundice, liver, and bladder malfunctions.
Benzene, Toluene	Initial symptoms are dizziness, mental confusion, tightening of the leg muscles, and then a stage of excitement. Continued exposure ultimately leads to a complex series of symptoms including coma.
Carbon Monoxide (CO)	Concentrations of 500 ppm for one hour are tolerable. Similar exposure to 1,000 ppm is hazardous, and 4,000 ppm is likely to be fatal. CO is particularly hazardous because of the lack of distinctive signs and symptoms.
Hydrazine, Phenylhydrazine	Symptoms are anemia and general weakness resulting from blood, liver, and kidney damage.
Nitro Compounds	Produce a wide variety of signs and symptoms resulting from anemia, jaundice, and degeneration of liver, kidney, and central nervous system. Since the hazardous properties of nitro compounds are largely those of explosive materials, the toxicity hazards have not been as extensively defined.
Other Respiratory Poisons	
Asbestos	Early stage – Shortness of breath, dry cough. Middle stage – Characteristic x-ray shadows, increasing shortness of breath, fibrosis of lung after several years. Late stage – Increasing fibrosis of lung, emphysema, and related breathing difficulties such as dyspnea, cyanosis, and anoxemia.
Alumina, Coal, Mica, Soot	Symptoms are not as well defined as those of asbestos and silica but are generally similar.

Notes

Smoke Colors and Their Meaning

Color	Meaning
Gray-white	Ordinary combustibles in early stages of burning.
Dark gray	Ordinary combustibles in later stages of burning.
Black	Hydrocarbons burning abnormally in early stages unless building contents include substantial amounts of hydrocarbon based materials.
Yellow-gray and brown-gray	Deep seated, slowly burning fire. Generally accompanied by heavy smoke stains and little or no smoke movement. Also indicates the presence of a backdraft condition.

Figure 2-4

Smoke Colors Produced by Various Combustible Substances

Color of Smoke	Combustible Substance
White	Hay, vegetable compounds
White	Phosphorus
White to gray	Benzine
Yellow to brownish-yellow	Sulfuric acid, nitric acid, hydrochloric acid
Yellow to brownish-yellow	Gunpowder
Greenish-yellow	Chlorine gas
Gray to brown	Wood, paper, cloth
Violet	Iodine
Brown	Cooking oil
Brown to black	Naphtha
Brownish-black	Lacquer thinner
Black	Turpentine, acetone
Black	Kerosene, gasoline, lubricating oil
Black	Rubber, tar, coal
Black	Foamed plastics

Note: Determination of a burning substance by smoke color is only a general method of identification. Other methods of identification must also be used.

Figure 2-5

SELECTED TOXIC PRODUCTS OF COMBUSTION

Notes

Acrolein (CH_2CHCHO)

Source	From polyolefins and cellulosics.
Type of Toxicity	Direct irritant.
Signs/Symptoms	See Selected Respiratory Poisons.
Treatment	Meticulous respiratory care. Usually minor irritation will end after exposure.

Ammonia (NH_3)

Source	Wool, silk, nylon, melamine.
Type of Toxicity	Direct irritant to respiratory structures and skin. It can exert a profoundly caustic action. Coma and convulsions are systemic aspects of toxicity.
Signs/Symptoms	Ammonia is easily identified by its pungent odor. Conjunctivitis and lacrimation are noted early as they produce temporary blindness. Restlessness, chest tightness, frothy sputum, and cyanosis with collapse may occur. These symptoms appear at concentrations greater than 1,000 ppm. At greater than 1,500 ppm, laryngospasm and immediate death may occur. Victims usually complain of intense pain in eyes, mouth, and throat, and feel they are suffocating. Inability to speak, as well as laryngeal edema with stridor may be present. Ammonia increases respiratory secretions. Skin contact produces local irritation or burns.
Treatment	Predicated on decontamination and dilution. Airways must be controlled to overcome laryngeal edema and glottic spasm. Respiratory care for pulmonary edema is crucial.

Carbon Monoxide (CO)

Source	Product of incomplete combustion.
Type of Toxicity	Lower respiratory poison which is absorbed through the alveoli of the lungs. CO acts by affecting the oxygen carrying capacity of the blood. CO is more strongly bonded to hemoglobin than oxygen. Hemoglobin is thus prevented from carrying oxygen, and cellular asphyxiation results. Concentrations of 500 ppm for one hour are tolerable. Similar exposure to 1,000 ppm is hazardous, and 4,000 ppm is likely to be fatal.
Signs/Symptoms	Shortness of breath, headache, nausea, vomiting, irritability, and confusion. In high exposures, coma, convulsions, respiratory failure, and death.
Treatment	Prompt evacuation of the victim and administration of 100% oxygen. In cases with severe exposure, hyperbaric chamber therapy can be useful.

Notes

One Hour Exposure to Carbon Monoxide

% in Atmosphere	Approximate ppm	% Carboxy-hemoglobin
0.01	100	17
0.02	200	20
0.1	1,000	60
1.0	10,000	90

Figure 2-6

Carbon Monoxide Poisoning

Carboxy-hemoglobin	Symptoms
10%	Shortness of breath on mild exertion, headache, throbbing temples.
20%	Severe headaches, nausea, and vomiting.
30%	Severe headache, irritability, impaired judgment, dizziness, nausea, and vomiting.
40–50%	Severe headache, confusion, collapse, nausea, and vomiting.
60–70%	Coma, convulsions, respiratory failure. Death if exposure continues.
80% or greater	Rapid death.

Figure 2-7

Hydrogen Chloride (HCl)

Source .... Polyvinylchloride, chlorinated acrylics, and retardant-treated materials.

Type of Toxicity .... Direct pulmonary irritant and inflamed conjunctiva of eyes. At concentrations of 15 ppm, localized irritation of the throat may occur. Pulmonary edema and laryngeal spasm may result from concentrations of greater than 100 ppm. Mists of HCl are less harmful than anhydrous HCl because the mist droplets have no hydrating action. Acid destruction occurs when HCl gas reacts with moisture in the lungs causing severe inflammation.

Signs/Symptoms ..... See Selected Respiratory Poisons.

Treatment .... The victim should receive prompt evacuation, decontamination of the skin and eyes, and administration of 100% oxygen. Patients should be observed for at least 24 hours for delayed toxicity.

Notes

Hydrogen Cyanide (HCN)

Source .... Wool, silk, polyacrylnitrile, nylon, polyurethane, paper.
Type of Toxicity .... Systemic (cellular poison). This is a fast-acting, fatal asphyxiant which induces cellular hypoxia.
Signs/Symptoms ..... See Selected Respiratory Poisons. Hydrogen cyanide may be inhaled or absorbed through the skin. Its characteristic odor of bitter almonds cannot be smelled by 30% to 50% of the population.
Treatment .... Cyanide antidote kit. Rapid treatment at the scene is required to save the victim's life.

Hydrogen Fluoride (HF)

Source .... Fluorinated resins or films.
Type of Toxicity .... Direct irritant to the skin and pulmonary tract. The fluoride ion is a direct cellular poison and interferes with calcium metabolism. It reacts with tissue proteins.
Signs/Symptoms ..... See Selected Respiratory Poisons.
Treatment .... Termination of exposure followed by decontamination through irrigation. Meticulous pulmonary care may be required for lung damage.

Hydrogen Sulfide (H_2S)

Source .... Hair, wool, meat hides, and decomposition of sulfur-containing organic materials.
Type of Toxicity .... When moisture is present, a caustic (sodium sulfide) is formed that is irritating to eyes, wet skin, and respiratory passages. Rotten egg odor is unmistakable, but low concentrations will rapidly fatigue the sense of smell. Capable of producing respiratory paralysis and death.
Signs/Symptoms ..... See Selected Respiratory Poisons.
Treatment .... Termination of exposure followed by decontamination with irrigation. Meticulous pulmonary care may be required for treatment of lung damage.

Isocyanate (Toluene Diisocyanate, $CH_3C_6H_3(NCO)_2$)

Source .... Urethane isocyanate polymers.
Type of Toxicity .... Severe irritation of eyes, gastrointestinal tract, and lungs. Acute pulmonary edema may also be noted in some persons experiencing an allergic reaction.
Signs/Symptoms ..... Irritation and inflammation of the skin, conjunctival irritation and inflammation of the eyes, nausea, vomiting, abdominal pain with inhalation, severe coughing, burning and irritation of the upper tract with a choking sensation. There is sputum production, and chronic bronchitis may result. Headaches, insomnia, euphoria, ataxia, anxiety, depression, and paranoia also may occur.

Notes

Treatment .... Irrigation and decontamination for early treatment. The asthmatic component may require theophylline treatment.

Nitrogen Oxide (N_2O)

Source .... Fabrics, cellulose nitrate, and celluloid.

Type of Toxicity .... Converted to nitric acid with hydration. Since the oxides are poorly soluble in water, delayed effects are common. Latent periods of 5 to 24 hours are possible as hydration takes place in the lungs.

Signs/Symptoms ..... See Selected Respiratory Poisons. The appearance of red-brown, orange, or copper colored gases at the fireground indicates immediate need for breathing apparatus.

Treatment .... Meticulous pulmonary care, decontamination, and irrigation.

Phosgene ($COCl_2$)

Source .... Decomposition of heated organic compounds with chlorine such as carbon tetrachloride. Chlorinated hydrocarbons and a number of plastics will produce the same. Phosgene will decompose to HCl and CO.

Type of Toxicity .... Poor water solubility leads to delayed appearance of clinical toxicity. Direct irritant of pulmonary tract.

Signs/Symptoms ..... See Selected Respiratory Poisons.

Treatment .... Irrigation after decontamination. Meticulous pulmonary care is required.

Sulfur Dioxide (SO_2)

Source .... All sulfur sources. Sulfur dioxide is a common product of oxidation of materials containing sulfur.

Type of Toxicity .... Sulfurous acid forms upon contact with moistened mucous membrane surfaces and the lungs, causing irritation.

Signs/Symptoms ..... See Selected Respiratory Poisons.

Treatment .... Decontamination through irrigation and dilution. Meticulous pulmonary care is mandatory.

TARGET ORGANS OF SELECTED CHEMICALS

Respiratory (Lungs) .... Halogen and halogen acids, hydrogen sulfide, sulfur dioxide, phosgene, hydrogen cyanide, hydrogen chloride/fluoride/bromide, nitro compounds, hydrazine, arsine, phosphine, methyl mercaptan, solvent and fuel vapors and mists, chloropicrin, carbon monoxide, phenylamine, asbestos, coal dust, talc, acrolein, acrylonitrile, epichlorohydrin, styrene.

Notes

Hepatic (Liver) .... Vinyl chloride, aromatic hydrocarbons and many derivatives, chlorinated hydrocarbons, nitroethane, nitropropane and many other nitro compounds, picric acid, pentaborane, paraquat.

Nephritic (Kidneys) .... Mercury, calcium, carbon tetrachloride, halogenated hydrocarbons, nitro compounds, paraquat, pentaborane, picric acid.

Neurologic (Nervous System) ..... Organophosphates, carbon monoxide, mercury, halogenated hydrocarbons, chlorinated hydrocarbon insecticides and solvents, methyl mercaptan, pentaborane, styrene, tetraethyl lead, rotenone.

Hematic (Blood) .... Aromatics and many derivatives, nitrochlorobenzene, nitro compounds, chlorinated hydrocarbons, carbon monoxide, methyl mercaptan, aniline, anisidine, lead, methyl Cellosolve® (2-methoxyethanol), dichloromethane, nitric oxide (NO), vinyl chloride, Warfarin®.

Skeletal (Bones) .... Fluorides, selenium, vinyl chloride.

Dermal (Skin) .... Arsenic, chromium, beryllium, other heavy metals, hexachloronaphthalene.

PERSONAL PROTECTIVE EQUIPMENT

Responders must understand the use and limitations of personal protective equipment (PPE) in order to protect themselves from contact with and exposure to hazardous materials. PPE is used when emergency operations must take place in close proximity to hazardous materials. There are three types of protective equipment to consider in a chemical emergency:

- respiratory protection
- chemical protection
- thermal protection

RESPIRATORY PROTECTION

There is a constant need for respiratory protection in a haz-mat emergency, because inhalation is one of the quickest ways to suffer harm from a haz-mat exposure. There are basically two types of respiratory protection:

- air supplied devices
- air purifying or filtering devices

Each type of respirator is designed with a specific protection factor in mind. The greatest level of protection is offered by positive pressure, supplied air devices. Filtering devices offer the lowest level of protection.

Notes

Positive pressure, supplied air respiratory protection is required for entry into atmospheres that are IDLH.

Self-Contained Breathing Apparatus

Self-contained breathing apparatus units (SCBAs) are one of two types of supplied air breathing devices. Positive pressure, supplied air respiratory protection is required for entry into atmospheres that are immediately dangerous to life or health (IDLH). The concept of IDLH was based on the hypothetical situation of a responder's not being able to escape if respirator failure occurred.

SCBA units maintain a slight positive pressure in the facepiece during both inhalation and exhalation. If a leak or failure of the seal occurs, the inside pressure should prevent the external contaminants from entering the facepiece. If a leak develops in a positive pressure SCBA, the regulator sends a continuous flow of clean air into the facepiece, preventing penetration by contaminated ambient air. It is important to follow the manufacturer's suggested guidelines for care and maintenance when using any respirator.

Positive pressure SCBAs provide the highest level of protection against airborne contaminants and oxygen deficiency. However, they have a limited air supply, usually 1/2 to 1 hour. There are some devices, such as an oxygen rebreather, that have a 4 hour duration. SCBAs are cumbersome and heavy. They limit mobility and add 20 to 30 pounds of extra weight. These features increase the chance that wearers will suffer exhaustion and heat stress in protective clothing. The facepiece may also reduce visibility.

Positive pressure SCBAs usually consist of:

- full facepiece
- exhalation valve
- breathing tube
- regulator
- air supply hose

Responders With Positive Pressure SCBA

Notes

- harness assembly
- cylinder valve
- cylinder that supplies compressed air or an oxygen generating chemical

Supplied Air Respirators

Supplied air respirators (also known as SARs or air line respirators) are the second type of supplied air breathing device. They supply air (never oxygen) to a facepiece via a hose line from a remote source. These devices can be positive or negative pressure facepieces. Air lines must not be used in an IDLH condition unless an escape SCBA is also used. The air source for air line breathing devices may be compressed air cylinders or a compressor that purifies and delivers atmospheric air.

SARs enable longer work periods than SCBAs and are less bulky. They have very practical purposes in long duration chemical emergencies and waste site activities. But there are limitations to these devices. SARs have limited use in short rescue or extrication scenarios. The air line can be damaged by cuts, abrasions, and chemical contamination. It is also vulnerable to kinking or binding, which would diminish or stop the air supply. When in use, air lines should be kept as short as possible (250 feet is the longest approved hose length for air lines). Like SCBAs, SARs reduce mobility for responders.

Positive pressure SARs usually consist of:

- full facepiece
- exhalation valve
- breathing tube
- regulator belt
- regulator
- air supply hose
- air supply
- escape SCBA tank attached to harness or belt

Air Purifying Respirators

There are many types of air purifying respirators. The air purifying respirator filters, absorbs, or adsorbs contaminants from the air, allowing breathable air to reach the facepiece. The outside air is the source of oxygen and is the key to this operation. There is no supply of safe air, as is found in SCBAs and SARs. In addition, there is no positive pressure within the facepiece. If negative pressure develops, contaminants may enter the facepiece.

Air purifying respirators cannot be used in emergency response operations unless the chemical has been identified and air monitoring has shown concentrations below the permissible exposure limit. These devices are not safe when there are hazardous chemical levels or unknown concentrations. **Never** use air purifying respirators if any of the following conditions exist:

- IDLH concentration
- oxygen deficiency
- unventilated or confined area

Notes

- unidentified contaminant
- unknown concentration
- concentration that exceeds manufacturer's maximum recommendation
- high relative humidity (may reduce the protection offered by the filtering medium)
- a chemical such as carbon monoxide or benzene that has insufficient warning properties

Respirator Selection

Selection of the proper respirator is based on several considerations:

- identity of the product
- chemical and physical properties
- known concentration
- chemicals with adequate warning properties, such as:
  - methyl ethyl ketone (MEK), which has an odor threshold limit below permissible exposure limits
  - hydrochloric acid, whose strong irritant properties make it easily detectable
  - ammonia and chlorine, whose combination of odor threshold limit and irritant properties make them easily detectable

The following physical requirements of the wearer must be considered when selecting and using any respirator:

- aptitude
- cardiovascular strength
- facial hair
- facial features
- hazard awareness and knowledge
- health
- physical condition
- training

CHEMICAL PROTECTION

There is no one chemical protective suit or material that will meet all emergency team needs. No "super suit" exists; therefore, responders should investigate the particular chemical threat and prepare for the potential problems that the threat will produce. Responders must be familiar with the styles, features, and selection considerations for chemical protective clothing (CPC).

CPC Styles

CHEMICAL VAPOR PROTECTIVE CLOTHING

A chemical vapor protective suit is a one piece garment that completely encloses the wearer. Also inappropriately referred to as an "acid suit," this ensemble includes gloves, facepiece, and boots, which are all integrated together in one unit.

Notes

Chemical vapor protective clothing should be worn when extremely hazardous substances are known or suspected to be present. If responders are dealing with an unknown or unidentified substance, then the highest level of protection (chemical vapor protective clothing) should be used. This type of CPC should also be used if there is a possibility that responders may be exposed to vapors, gases, or particulates that could harm, destroy, or be absorbed through the skin.

Chemical Vapor Protective Clothing

Chemical vapor protective suits are designed for the highest level of hazards an emergency responder may face—toxic vapors, gases, mists, or particulates in the air. The garment must be worn with the highest levels of respiratory protection, such as positive pressure SCBA or air line breathing equipment. The suit is limited by the protective material's resistance to the specific chemicals encountered.

Disposable or limited use garments usually do not meet the rigid requirement of airtightness that is needed for chemical vapor protection. However, recent manufacturer innovations are creating garments that meet the National Fire Protection Association (NFPA) and American Society for Testing and Materials (ASTM) standards for chemical vapor protective garments.

CHEMICAL SPLASH PROTECTIVE CLOTHING

Chemical splash protective clothing comes as a one piece coverall or a two piece pants and coat ensemble which may include a hood. These suits do not have a facepiece as an integral part of the suit. Also called "non-encapsulating suits," they are **not** designed to be vapor resistant, even though they may provide limited protection from vapors. Chemical splash protective clothing may be used when no chemical gases or vapors are present in concentrations that could damage or be absorbed through the skin.

A key safety factor for chemical splash protective clothing is the type of respiratory protection used. When used with supplied air breathing devices, the chemical splash suit offers a higher degree of protection than when used with other types of respiratory protection. A lesser level of protection is gained when used with air purifying respirators, and even less protection is available when used with only a face shield. The critical elements are the degree of hazard a chemical poses and its airborne concentration.

Notes

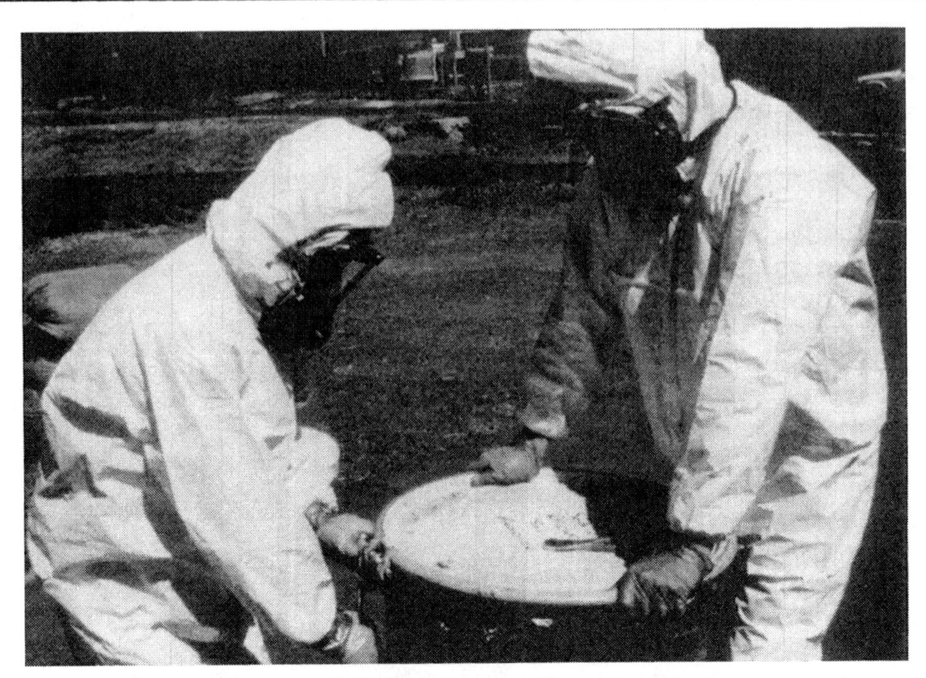

Chemical Splash Protective Clothing

To make a chemical splash garment airtight, many responders tape the wrists, ankles, and zippers. This procedure **does not** make the suit gas-tight, and modifications of this sort should **never** be assumed to be acceptable in an emergency that requires chemical vapor protective clothing. Also, such modifications can damage the garment, particularly if the CPC is reusable.

CPC Construction Features

Selecting effective chemical protective clothing for those working with hazardous materials is a critical responsibility. With so many different types of hazards, it is important that workers be provided with the proper protection. Performance requirements must be considered in choosing the appropriate protective material, according to the work activity to be performed. Agencies should carefully consider the criteria of NFPA Standards 1991 and 1992 when selecting CPC. Consider the following guidelines:

- chemical resistance
- color
- cost
- design
- durability
- ease of decontamination
- flexibility
- material type
- service life
- size
- temperature resistance

CHEMICAL RESISTANCE

Chemical resistance is the ability of a material to withstand chemical changes. A garment's chemical resistance is one of the most important features to consider when selecting CPC. The material must resist the passage of chemicals through the garment, so that contamination and injury to the wearer are minimized.

In order to make a proper selection, emergency responders must consider the chemicals that might be confronted in their community. If a transportation route is the major hazard potential, then the level of CPC protection needed may be very difficult to determine.

No one suit or material is totally impervious to all chemicals. The capability of chemical protective clothing to withstand hazardous materials is based on the clothing's resistance to:

- penetration
- degradation
- permeation

When selecting the style of CPC and the material from which the CPC is made, responders must evaluate each of these three areas long before an incident occurs.

Notes

A garment's chemical resistance is one of the most important features to consider when selecting CPC.

Penetration

Penetration is the process of chemical transport through openings in a garment. A chemical may penetrate due to design or garment imperfections. Stitched seams, buttonholes, pinholes, zippers, and woven fabrics can provide a route for chemical penetration. Methods for control can be self-sealing zippers, seams overlaid with tape, flap closures, and non-woven fabrics. Rips, tears, punctures, and abrasions will also allow penetration.

Degradation

Degradation is a chemical action involving the molecular breakdown of the material due to chemical contact. With degradation, penetration and permeation can occur. Degradation usually appears as a physical change in the material. Signs of degradation are:

- brittleness
- cracking
- discoloration
- shrinking
- softness or stickiness to the touch
- swelling

Permeation

Permeation is the rate at which a quantity of chemical passes through a protective material in a given time. Permeation takes place on a molecular level. Unlike degradation, there is no visual evidence that permeation is occurring. Permeation rate is usually expressed in micrograms of chemical permeated per square centimeter per minute of exposure ($\mu g/cm^2/min$). Factors influencing permeation of a material include:

- chemical concentration
- contact time
- humidity

Class Group	Chemical Name	Percent	Class	Physical State[1]	TYVEK® QC (TYVEK Coated with 1.25 mil Polyethylene)		TYVEK®/SARANEX® 23P		BARRICADE®	
					Breakthrough Time (min)	Permeation Rate µg/cm²/min	Breakthrough Time (min)	Permeation Rate µg/cm²/min	Breakthrough Time (min)	Permeation Rate µg/cm²/min
Cyanate	Hexamethylenediisocyanate	98	Cyanate	L	nt	nt	>480	nd	>480	nd
	*Hydrogen cyanide	99	Cyanate	G	60	111	nt	nt	108	0.53
	Methyl isocyanate	100	Cyanate	L	nt	nt	2	210	nt	nt
	Toluene-2,4-diisocyanate	80	Cyanate	L	<1	42	>480	nd	>480	nd
Element	Mercury	99+	Element	L	nt	-nt	>210	<0.01	>480	nd
	Bromine	na	Element-halogen	G	<1	High	nt	nt	9	516.7
	Chlorine gas[2]	99+	Element-halogen	G	1	18	>480	nd	>480	nd
	Chlorine (20 ppm)	20 ppm	Element-halogen	G	>480	nd	>480	nd	nt	nt
	Iodine	99.9	Element-halogen	S	>420	48	>480	nd	nt	nt
Ester	n-Amyl acetate	99	Ester	L	nt	nt	nt	nt	>480	nd
	Cellosolve acetate	98	Ester	L	nt	nt	39	1.8	nt	nt
	Ethyl acetate[2]	99	Ester	L	<1	12.7	36	6.6	>480	nd
	Methyl cellosolve acetate	98	Ester	L	nt	nt	260	1.08	nt	nt
	*Methyl parathion	57	Ester-inorganic	L	15	0.09	>120	<0.01	nt	nt
	*Methyl parathion	10	Ester-inorganic	L	>30	0.2	>240	<0.01	nt	nt
Ether	Butyl ether	99	Ether	L	nt	nt	nt	nt	>480	nd
	Diethyl ether	99	Ether	L	nt	nt	1	1.8	>480	nd

Figure 2-8. Check Permeation/Penetration Guides Before Use

- type of material
- solubility in the chemical
- temperature
- thickness of material

A key measure of permeation is ***breakthrough time***, which is expressed in minutes. Breakthrough time is the elapsed time between initial contact of a chemical with the outside surface and detection at the inside surface of the material. Permeation and breakthrough time are both important. The longer the breakthrough time, the better chemical resistance the garment has. But the amount of permeation is also critical—a larger permeation rate means greater harm when breakthrough actually occurs. The best protective material against a specific chemical is one that has a low permeation rate (if any) and a long breakthrough time. However, these actions do not always correlate, in which case a long breakthrough time is usually desired. Breakthrough times are usually published by the manufacturer of the garment or protective material, as shown in Figure 2-8.

COLOR

A brightly colored suit makes it easier to maintain visual contact between personnel who are working in vapor clouds and dimly lit sites.

COST

Cost should never be the primary consideration when selecting CPC.

The cost of CPC varies considerably. The material used, design features, style, and level of protection are all factors that influence the cost. Whenever possible, use the most protection for the least cost. However, personal safety, not cost, should be the primary consideration.

Notes

DESIGN

Consider the following design features when selecting the proper CPC:

- vapor tight or splash protection
- one or two piece construction
- sleeve features such as glove rings and "bat wing" design
- attached or unattached hoods, facepieces, gloves, and boots
- facepiece material and construction
- location of zippers, buttons, storm flaps, and seams
- pockets and Velcro® straps
- exhalation valves or ventilation ports
- compatibility with respiratory protection
- communications
- vision from inside suit

DURABILITY

CPC must be resistant to tears, punctures, and chemical action. New materials were designed with resistance and strength in mind.

EASE OF DECONTAMINATION

Ease of decontamination plays an important role when stress and fatigue affect responders. The growing concerns for complete and effective decontamination are increasing the demand for more disposable garments. Due to the limitations of decontamination, it may be necessary to discard a multiple use garment when it is contaminated by a highly toxic material. Decontamination may not stop the permeation process. Limited use or disposable garments make it easier to control any potential damage caused by reuse of a contaminated suit. It may be advisable to cover multiple use garments with disposable garments to prevent gross contamination.

FLEXIBILITY

Many materials are difficult to work in and are not pliable. Gloves and boots must be flexible to the wearer. Note that flexibility may be affected by temperature.

MATERIAL TYPE

CPC is also classified by the type of material from which it is made. These materials are grouped into two categories—elastomers and non-elastomers.

- ***Elastomers*** are plastic like materials (polymers) that can be stretched and then will return to their original shape. Most protective materials are elastomers, including Viton®, Teflon®, butyl rubber neoprene, nitrile rubber, and others.
- ***Non-elastomers*** are materials that do not have stretch ability, including Chemrel®, First Responder®, Tyvek® coated fabrics, and Barricade®.

SERVICE LIFE

Emergency responders who have a limited and infrequent number of chemical responses should choose garments with a long shelf life. Consult the manufacturer to determine a suit's recommended shelf life.

Notes

SIZE

Comfort and maneuverability are factors to consider when selecting the CPC's size. Ease of movement goes a long way in preventing accidents. Generally, a small person can wear a large size, but a large person will have problems in a small suit.

TEMPERATURE RESISTANCE

CPC must be pliable in both hot and cold weather. Be aware that the permeation rate increases at higher temperatures.

Levels of Protection

Chemical protective clothing is designed to provide specific chemical protection from a designated product. The most common way of classifying protective clothing is by the degree of protection. Many classification systems for levels of protection are in use. The EPA guidelines are the most common, with levels A, B, C, and D. These levels incorporate chemical protection as well as respiratory protection. In choosing a level of protection, consider the degree of hazard, the concentration of the chemical, and the nature of the toxicity.

LEVEL A

Level A is the highest level of protection needed, when responders are expected to encounter the greatest threat to the respiratory system, skin, and eyes. This level is usually used for high concentrations (usually above IDLH) of gases, vapors, and particulates in the air. Level A protection also provides maximum protection for the skin. This level of protection includes:

Level A Chemical Protective Clothing

Notes

- positive pressure, self-contained breathing apparatus (MSHA/NIOSH approved)
- chemical vapor protective clothing
- chemical resistant inner gloves
- chemical resistant outer gloves
- chemical resistant boots with steel toe and shank (worn over or under suit boots, depending on suit boot construction)
- two-way radio communications (intrinsically safe)
- long-john type cotton underwear (optional)
- hard hat under suit (optional)
- coveralls under suit (optional)

LEVEL B

This level is chosen when the highest level of respiratory protection is still needed, but decreased skin protection is required by the chemical hazard and concentration. This garment is chemically resistant, but it may allow passage of vapors and gases through openings at the neck, closures, or wrists. Level B protection includes:

- positive pressure, self-contained breathing apparatus (MSHA/NIOSH approved)
- chemical splash protective clothing (overalls and long sleeved jacket, coveralls, or disposable coveralls)
- chemical resistant inner gloves
- chemical resistant outer gloves
- chemical resistant outer boots with steel toe and shank
- two-way radio communications (intrinsically safe)
- coveralls under suit (optional)
- hard hat (optional)

LEVEL C

Level C represents minimum respiratory and skin protection. This level of chemical protection is used when there is a known (identified) and measured concentration, and there are no skin absorption hazards. The ensemble is worn with an air filtering respirator specifically approved for the hazardous material. Level C protection includes:

- full face, air purifying respirator (MSHA/NIOSH approved)
- chemical splash protective clothing (one piece coverall, hooded two piece suit, chemical resistant hood and apron, disposable coveralls)
- chemical resistant outer gloves
- chemical resistant inner gloves (optional)
- chemical resistant boots with steel toe and shank
- cloth coveralls inside chemical protective clothing (optional)
- two-way radio communications (intrinsically safe) (optional)
- hard hat (optional)
- escape mask (optional)

Notes

Levels B and C Chemical Protective Clothing

LEVEL D

Level D protection consists of common work clothes. This level of protection provides **no** specific respiratory or skin protection. Common work clothes should not be worn in any situation in which there are respiratory or skin hazards.

CPC Selection During Chemical Incidents

The selection of the proper protective garment is based on the nature of the chemical hazards faced. Selection of the proper CPC when facing a known chemical incident should take the following items into account:

- identity of the chemical involved
- physical, chemical, and toxicological properties of the chemical
- skin hazard of the chemical
- level of protection needed—chemical splash or chemical vapor protection
- protective garment that will provide maximum protection against permeation and degradation
- potential for heat stress

Choosing the proper style of CPC when facing incidents where the presence of hazardous substances is not certain and where chemicals are not readily identifiable is more difficult. In these situations, CPC choice should be based on visual indicators at the scene, including:

- visible emissions of gases, vapors, dust, or smoke
- potential conditions in which skin absorption is a major hazard

- indications of airborne hazards on direct-reading instruments
- presence of containers used for gases or pressurized liquids
- signs, labels, placards, or shipping papers showing highly toxic substances that could become airborne
- enclosed or poorly ventilated areas where toxic vapors, gases, and other airborne substances could accumulate

It should be noted that most CPC offers no resistance to a flash fire or flames. Cover garments, sometimes referred to as "flash suits," are available to protect CPC in the event of fire. These aluminized garments offer limited protection against flash fires.

Notes

Most chemical protective clothing offers no resistance to a flash fire or flames.

THERMAL PROTECTION

The need for thermal protection is a major limitation when facing a chemical emergency. Thermal protection may not provide sufficient chemical protection, while CPC may not provide sufficient thermal protection in toxic conditions.

Thermal protection is classified by the three major applications:

- structural firefighting clothing (firefighter full protective clothing)
- proximity suits
- fire entry suits

Structural Firefighting Protective Clothing

Notes

Structural firefighting clothing is not designed for use as chemical protective clothing.

Structural Firefighting Clothing

Structural firefighting clothing is designed to protect firefighters from common hazards encountered during fire situations. Because of the layered construction, there is moderate thermal protection. Wearers are protected from steam and moisture by a vapor barrier and are afforded some flash fire protection by virtue of the flame resistant materials, usually Nomex®, Kevlar®, or PBI®. The ensemble is usually made up of helmet, coat, pants, boots, gloves, hood of similar fire resistant material, and positive pressure, self-contained breathing apparatus (SCBA). Structural firefighting clothing is not designed for fire entry or even close proximity to heavy fire conditions, such as those found in flammable liquid fires. **Structural firefighting clothing is not designed for use as chemical protective clothing.** Additionally, when structural firefighting clothing is exposed to some chemicals, the chemicals may permanently damage the garment, causing the loss of fire retardant properties.

Proximity Suits

Proximity suits are designed for radiant heat and exposures to flame for short durations. These suits are usually aluminized garments with a hood and facepiece. They must be used with SCBA. These suits are bulky and can limit mobility. **Proximity suits are not designed for use as chemical protective clothing.**

Proximity Suits

Fire Entry Suits

Notes

As the name implies, these garments are designed to enter fire situations for a short period of time. The ensemble can withstand temperatures of up to 2,000°F (1,093°C). Because of their multi-layered construction, fire entry suits are extremely bulky and limit visibility, mobility, and dexterity. **Fire entry suits are not designed for use as chemical protective clothing.**

HEAT STRESS CONCERNS

Protective clothing helps protect responders from the hazards posed by dangerous chemicals. However, it also causes wearers to be more susceptible to heat stress, because the body loses its ability to rid itself of excess heat. The problem is intensified by the fact that responders are usually performing strenuous work while wearing protective clothing. Because of the potential for heat stress in emergency situations, it is important for responders to be able to recognize the condition and to know how to manage it properly.

The two most serious heat stress concerns are heat exhaustion and heat stroke. ***Heat exhaustion*** is a mild form of shock that can be brought on when the body's normal cooling mechanisms begin to fail. If preventative measures are not taken, heat exhaustion can lead to heat stroke. Signs and symptoms of heat exhaustion include:

- weakness
- nausea
- headache
- dilated pupils
- elevated body temperature
- rapid, shallow breathing
- weak pulse
- dizziness
- cold and clammy skin
- heavy sweating

Heat stroke is brought about when the body's cooling mechanisms become so overworked they simply shut down. Heat stroke is a serious medical emergency, and if immediate action is not taken, it can become a life threatening situation. The signs and symptoms of heat stroke include:

- cessation of sweating
- body temperature over 105°F (41°C)
- shallow breathing
- rapid pulse
- hot, dry, red skin
- nausea
- headache
- confusion
- weakness
- loss of consciousness
- convulsions

Notes

At emergency incidents, follow these guidelines for reducing the occurrence of heat stress:

- provide plenty of fluids, but avoid carbonated and caffeinated beverages
- use body cooling devices such as air cooled jackets and suits, ice vests, or water cooled vests and suits
- conduct pre- and post-incident medical monitoring
- locate rehabilitation areas in shaded or air conditioned areas
- consider providing mobile showers
- perform frequent personnel rotations

COLD STRESS

Cold exposure is a concern when dealing with incidents in cold weather conditions or when dealing with cryogenic materials or liquefied gases. Hypothermia is one of the most serious forms of cold stress. It can be brought on by any combination of cold weather, exposure to water (either by sweating or by wet clothing), wind, exhaustion, or hunger. Symptoms of hypothermia include:

- shivering
- numbness
- disorientation
- lethargy and drowsiness
- slowed breathing and pulse

CONCLUSION

Having an understanding of basic toxicology will enable emergency response personnel to approach hazardous materials incidents with an accurate appraisal of what the chemicals involved are like, what they may do, and how long responders can be exposed to such chemicals without adverse effects. Responders will be able to recognize signs of chemical exposure in others and will be able to treat such exposures appropriately. In addition, they will be able to select suitable personal protective equipment to protect themselves while working around hazardous substances. An understanding of the various types of PPE available and of the limitations of each type is also important if responders are to deal with haz-mat incidents safely.

Chapter 3

MANAGING THE HAZARDOUS MATERIALS INCIDENT

OBJECTIVES

After studying the material in this chapter, you will be able to:

- summarize the six steps for managing a haz-mat incident effectively
- state how to formulate an action plan for managing a haz-mat incident, based on the level of incident
- identify and state the function of the control zones needed in a haz-mat incident
- explain the importance of an incident command system and SOPs, and outline how to organize an IC system for the three levels of haz-mat incidents
- identify and state the function of available resources that can provide assistance at a haz-mat incident
- explain basic decontamination concepts and methods
- identify procedures to be followed during rehabilitation and medical screening
- state the purpose of post-incident analysis

INTRODUCTION

Billions of tons of hazardous materials are transported by land, sea, and air in the United States each year. Fortunately, most of these materials reach their destinations safely. Sometimes, however, an accident creates a haz-mat incident that threatens the public and environment.

In 1994 the Department of Transportation reported a total of 16,074 haz-mat incidents. These incidents resulted in 11 fatalities and 569 injuries. Highway incidents accounted for all of the fatalities and approximately 75% of the injuries. Nearly 80% of the injuries resulted from incidents involving corrosive materials, flammable and combustible liquids, and

Notes

poisonous materials. Human error was reported to be the cause of 84% of all incidents. A mere 50 hazardous materials, out of more than 2,800 identified in the Hazardous Materials Table, 49 CFR 172.101, accounted for 74% of the incidents reported. Additional details about these incidents are given in Figures 3-1 through 3-4.

Incidents at facilities that produce, manufacture, or utilize hazardous chemicals are also a concern. There are thousands of such facilities all over the United States and Canada, and spills, leaks, and fires at these facilities are not uncommon.

While we are aware that hazardous materials pose a serious threat to our health and environment, we take for granted the comforts made possible by the petroleum, nuclear, and chemical industries. It is imperative to prevent and respond properly to haz-mat incidents if we are to protect the benefits of modern life.

Haz-Mat Incident Statistics, 1994

Mode	Incidents		Deaths		Injuries	
	Number	%	Number	%	Number	%
Air	932	5.8	0	0	58	10.2
Highway	13,984	87.0	11	100.0	424	74.5
Railway	1,152	7.1	0	0	87	15.3
Water	6	0.1	0	0	0	0
TOTALS	16,074	100.0	11	100.0	569	100.0

Figure 3-1

Incident Cause by Mode, 1994

Mode	Human Error	Package Failure	Vehicle Accident/ Derailments	Other
Air	788	120	0	24
Highway	12,174	1,404	242	164
Railway	536	553	53	10
Water	4	1	0	1
TOTALS	13,502	2,078	295	199
Percent of All Incidents	84.0	13.0	1.8	1.2

Figure 3-2

Notes

Injuries and Deaths by Hazard Class, 1994			
Hazard Class	**Number of Incidents***	**Number of Injuries**	**Number of Deaths**
Flammable/Combustible Liquid	6,643	166	9
Corrosive Material	5,840	209	0
Poisonous Materials	1,430	72	0
Miscellaneous Hazardous Material	610	9	1
Combustible Liquid	417	4	0
Non-Flammable Compressed Gas	349	13	0
Oxidizer	309	19	0
Flammable Gas	171	37	1
Flammable Solid	108	8	0
Organic Peroxide	80	1	0
Poisonous Gas	52	30	0
Other Regulated Material, Class D	47	1	0
Dangerous When Wet Material	23	0	0
Spontaneously Combustible	15	0	0
Radioactive Material	10	0	0
Explosive, No Blast Hazard	4	0	0
Infectious Substance	3	0	0
Explosive, Fire Hazard	1	0	0
Very Insensitive Explosive	1	0	0
TOTALS	16,113	569	11

* Due to multiple classes being involved in a single incident, the total above may not correspond to the total in other reports.

Figure 3-3

Notes

Incidents by Top 50 Hazardous Materials, 1994

Rank	Commodity Name	Hazard Class	Incidents
1	Corrosive liquids n.o.s.	Corrosive material	1,846
2	Flammable liquids n.o.s.	Flammable/combustible liquid	1,221
3	Resin solution	Flammable/combustible liquid	506
4	Adhesives	Flammable/combustible liquid	464
5	Hydrochloric acid solution	Corrosive material	449
6	Gasoline	Flammable/combustible liquid	438
7	Sodium hydroxide solution	Corrosive material	433
8	Sulfuric acid	Corrosive material	430
9	Isopropanol	Flammable/combustible liquid	339
10	Phosphoric acid	Corrosive material	332
11	Compound cleaning liquid	Corrosive material	313
12	Paint or paint related	Flammable/combustible liquid	264
13	Potassium hydroxide solution	Corrosive material	257
14	Ink printers flammable	Flammable/combustible liquid	249
15	Ethanol	Flammable/combustible liquid	228
16	Methanol	Flammable/combustible liquid	210
17	1,1,1-Trichloroethane	Poisonous materials	209
18	Poisonous liquids n.o.s.	Poisonous materials	195
19	Caustic alkali liquid n.o.s.	Corrosive material	180
20	Petroleum distillate n.o.s.	Flammable/combustible liquid	169
21	Fuel oil #1, 2, 4, 5, 6	Flammable/combustible liquid	163
22	Compound cleaning liquid pho	Corrosive material	155
23	Dichloromethane	Poisonous materials	152
24	Compound cleaning liquid	Flammable/combustible liquid	148
25	Combustible liquid n.o.s.	Combustible liquid	145
26	Ammonia solutions 10-35%	Corrosive material	143
27	Xylenes	Flammable/combustible liquid	141
28	Hypochlorite solution 5-16%	Corrosive material	138
29	Fuel oil #1, 2, 4, 5, 6	Combustible liquid	137
30	Diphenylmethane-diisocyanurate	Poisonous materials	121
31	Formaldehyde solutions	Misc. hazardous material	119
32	Extracts flavoring liquid	Flammable/combustible liquid	118
33	Alcohols n.o.s.	Flammable/combustible liquid	117
34	Naphtha petroleum	Flammable/combustible liquid	104
35	Environmentally hazardous liquid	Misc. hazardous material	99
36	Toluene	Flammable/combustible liquid	98
37	Ammonia anhydrous	Non-flammable compressed gas	97
38	Hazardous waste solid n.o.s.	Misc. hazardous material	93
39	Hazardous waste liquid	Misc. hazardous material	89
40	Methyl methacrylate inhibited	Flammable/combustible liquid	85
41	Paint related material	Flammable/combustible liquid	84
42	Acetone	Flammable/combustible liquid	82
43	Trichloroethylene	Poisonous materials	76
44	Battery fluid acid	Corrosive material	75
45	Acetic acid solution	Corrosive material	71
45	Fire extinguishers	Non-flammable compressed gas	71
47	Coating solution	Flammable/combustible liquid	68
48	Diesel fuel	Flammable/combustible liquid	67
49	Fuel oil	Combustible liquid	65
50	Petroleum gases liquefied	Flammable gas	62
	TOTAL		11,915

Figure 3-4

MANAGING THE INCIDENT

Notes

Management of a haz-mat incident is both challenging and difficult. The haz-mat incident can often be a step into the unknown. Chemical and physical properties of hazardous materials can be stressed to maximum tolerances during a chemical emergency, causing devastating reactions. Exposure to fire, severe weather, and the mixing of chemical compounds create nightmares for the incident commander.

Preparation for haz-mat incidents is much like opening doors down an endless hallway. Each opened door can provide more insight into handling the next incident. To prepare for haz-mat situations, you should:

- train as much as possible
- preplan for incidents that are most likely to occur in your community or workplace, as well as possible incidents at local fixed storage sites
- practice—the time to learn is not during the emergency

SIX STEPS TO INCIDENT MANAGEMENT

Action taken by the first arriving haz-mat responder, police officer, or EMS provider will largely dictate the operation's overall success. Inappropriate action or failure to recognize the hazard's severity can cause the incident to escalate, thereby increasing risks to responders and the community.

Much of the existing information on haz-mat incident management addresses emergencies that are of major proportion. This has caused emergency responders to overreact to small incidents when implementing a predetermined command structure. It is also possible, however, to underestimate what procedures are needed, thereby exposing responders to greater risk and legal liability.

This book teaches a six step process for managing any haz-mat incident. These steps can be used by the first arriving emergency responder or incident commander for effective scene management. This system is tied to the acronym HAZ-MAT.

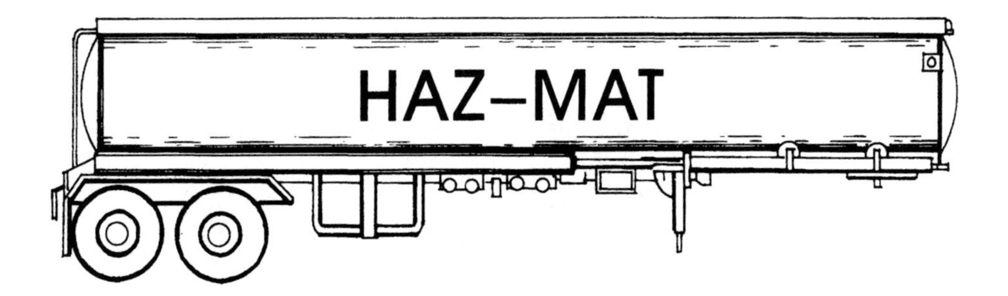

Notes

H - Hazard Identification
Recognize and identify the presence of a hazardous material.

A - Action Plan
Evaluate the situation by determining what you are going to do, what the immediate and long term needs are, and who is in charge.

Z - Zoning
Control the risk by establishing a hot zone (exclusion zone), a warm zone (contamination reduction zone), and a cold zone (support zone).

M - Managing the Incident
Establish the necessary incident command structure to handle the emergency.

A - Assistance
Determine additional resources needed, including more fire companies, haz-mat team, technical assistance, or private contractors.

T - Termination
Assess what is needed to conclude the incident, such as clean-up, decontamination, physical exams, rehabilitation, and post-incident analysis.

HAZ-MAT — HAZARD IDENTIFICATION

Before a haz-mat incident can be managed, the presence of dangerous chemicals must first be recognized. Although this sounds easy enough to do, in practice it can be difficult.

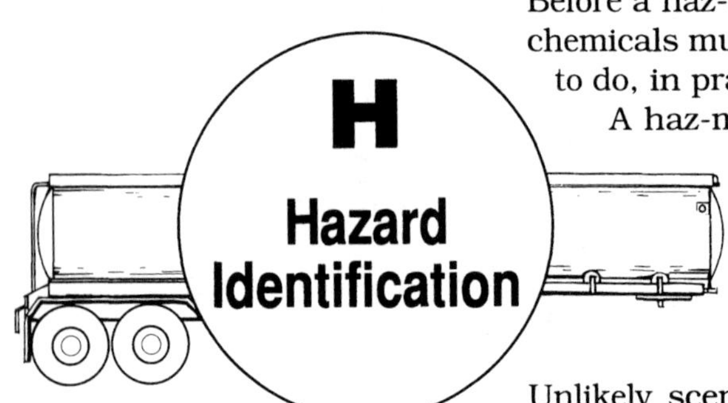

A haz-mat incident is not always a spectacular event such as a flaming tank truck or a large chlorine cloud visible from five miles away. It is often little things that tend to be overlooked, such as a small transformer in a basement with PCBs, broken pesticide cases carried on a grocery truck, or picric acid that has crystallized at the local high school.

Unlikely scenarios? Not at all! If you learn only one thing from this reference manual, it should be that every incident is a haz-mat incident until you are 100% sure that no dangerous chemicals are present.

Initial Size-Up

The most critical aspect of response to a haz-mat incident is the initial size-up of the situation. This is normally the responsibility of the first arriving emergency responder. It includes:

Every incident is a haz-mat incident until you are 100% sure that no dangerous chemicals are present.

- detection and identification of hazardous materials
- assessment of fire, explosion, and health hazards
- immediate action
- immediate follow-up actions

In size-up, as well as in all other actions, personnel safety should be the responder's primary concern. The risks of exposing personnel to danger must be weighed against the benefits. Do not risk the lives of personnel in vain attempts to recover dead bodies or to mitigate chemical emergencies that are out of control.

Detecting Hazardous Materials

Notes

Hazardous materials may be detected and identified in a number of ways. The most common are:

- preplanning to identify occupancies and locations where hazardous materials are present
- noting container shapes and sizes
- identifying standard placards or labels attached to a vehicle or container
- noting other identifying markings on buildings or containers
- reading shipping papers
- reading Material Safety Data Sheets (MSDS)
- observing the physical properties of the material

Early recognition of hazardous materials will reduce the risk to emergency personnel and allow for the initiation of proper actions to mitigate the incident.

HAZ-MAT — ACTION PLAN

An action plan begins with a good on-scene report from the first arriving responder. An effective and complete description of what is occurring at the incident will enable other incoming units to have a better understanding of the situation. One common approach involves answering the following questions:

- What do I have?
- What will it do?
- What am I doing?
- What do I need?
- Who is in command?

A

Action Plan

First Arriving Responder Communication

What do I have? The responder describes what he or she sees to the communications center and other responding units. The description may include information such as building size, construction, amount of involvement, type of occupancy, unusual conditions, and hazards. This brief report will help others picture what is occurring.

What will it do? The first arriving responder gives an assessment of what he or she feels is the potential destructive force of the chemical involved.

What am I doing and going to do? This will inform others of actions being taken. Such actions may include identifying the hazards present, initiating protective actions, advancing an attack line, or supplying water.

What do I need? Needs may include assignments given to other companies or requests for additional resources. When stating needs, remember that, "Nothing right now," is a perfectly legitimate response. It may take time to determine what the real problems are. While you are assessing problems, other responding units can be directed to areas close to the incident, but away from the front door. Use a resource pool or staging location near the incident where apparatus can wait until response assign-

Notes

ments are given. Too often the emergency scene looks like a used emergency response vehicle/apparatus lot. Keeping apparatus away until it is needed will allow easier access.

Who is in command? Everyone responding to or working at an incident needs to know that command is established. All organizations should have SOPs (standard operating procedures) that specify how command is established and transferred.

Control of the incident has begun once the first arriving responder has answered the five critical questions listed above. The first responder should immediately establish a game plan to provide for a safe, systematic approach to handling the haz-mat emergency. The original plan may not be followed throughout the incident, since haz-mat scenes are dynamic and can escalate rapidly. Therefore, the action plan should include Plan A and other alternate plans of attack if the emergency should change. If responders insist on sticking to Plan A, personnel may be killed or injured, or the incident may escalate uncontrollably.

A system to classify incidents based on severity and needed resources is effective in managing haz-mat incidents. Incident level classification will help the first arriving responder or incident commander to use appropriate control procedures.

Hazardous Materials Incident Classifications

A major incident for one agency may be easily handled by another agency. Key factors should be considered in determining the level of a haz-mat incident. These may include:

- size of agency
- preplanning
- ability of the responding agencies
- level of assistance needed
- ability to acquire additional resources

INCIDENT LEVEL CRITERIA

A three level incident classification system is outlined below. The three levels are determined by the following criteria:

- size and capability of response agency
- extent of municipal, county, and state government involvement
- extent of injuries and/or deaths
- extent of civilian evacuation needed
- availability and need of haz-mat response team
- level of technical expertise required to abate the incident

Level I Incidents

Level I incidents can be effectively managed and mitigated by first response personnel without a haz-mat response team or other special unit. Level I incidents include:

- spills that can be properly and effectively contained and/or abated by equipment and supplies immediately accessible to first responding personnel

Notes

- leaks and ruptures that can be controlled using equipment and supplies immediately accessible to first responding personnel
- fires involving toxic materials that can be extinguished and cleaned up with resources immediately available to first responding personnel
- haz-mat incidents not requiring civilian evacuation
- haz-mat incidents that can be contained and controlled using the resources of first responding personnel

Level II Incidents

Level II incidents require the special technical assistance of a haz-mat response team, industrial specialist, or government strike team. Level II incidents include:

- spills that can be properly and effectively contained and/or abated by specialized equipment and supplies immediately available to a haz-mat response team or other special unit
- leaks and ruptures that can be controlled using specialized equipment and supplies immediately available to a haz-mat response team or other special unit
- fires involving toxic and/or flammable materials that are permitted to burn for a controlled time period or are allowed to consume themselves
- haz-mat incidents that require civilian evacuation within the area of the agency that has primary jurisdiction
- haz-mat incidents where specialized technical information is required
- haz-mat incidents that can be contained and controlled using the available resources of a haz-mat response team or other specially trained unit

Level III Incidents

Level III incidents are major disasters. They include:

- spills that cannot be properly and effectively contained and/or abated by highly specialized equipment and supplies immediately accessible to a haz-mat response team or special unit
- leaks and ruptures that cannot be controlled using highly specialized equipment and supplies immediately available to a haz-mat response team
- fires involving toxic materials that are allowed to burn because water is ineffective or dangerous; or because there is a threat of large container failure; or because a large explosion, detonation, or BLEVE has occurred
- haz-mat incidents that require evacuation of civilians across jurisdictional boundaries

HAZ-MAT — ZONING

Z
Zoning

After identifying the hazard and developing an initial action plan, the next step is to isolate the hazardous material by setting up zones (see Figure 3-5). Quickly initiating site control by establishing zones reduces danger to the public and to emergency response personnel.

Setting up an elaborate site control system is a difficult task for the first responder. However, establishing the necessary hot and warm zones can be done quickly and simply.

Many responders mistakenly assume that site control is only necessary for major emergencies. This fallacy can cause an incident to escalate unnecessarily and can create management problems for even the smallest incidents.

Control zones should be used on every incident that involves hazardous materials, for two reasons:

- They are effective in gaining control of a potentially volatile situation.
- If control zones are used on every incident, whether it is a Level I, II, or III emergency, responders will be ready to establish effective site control quickly.

CONTROL ZONES

Site control is critical to effective management of a haz-mat incident. Control zones can be expanded or reduced depending on the situation. Each zone should be clearly marked. All personnel should understand agency procedures and should be thoroughly trained in the use of control zones. Remember, a site control system needs to be established early, and it should be used to some degree in every haz-mat incident. Zoning involves two protective action options, **protecting in place** and/or **evacuation**. Protecting in place is employed if it is possible to keep the public indoors safely, making a structure as airtight as possible by closing doors, windows, and ventilation systems. Evacuation is employed when the public must be moved from areas at risk to areas of safety.

Hot Zone

The hot zone is sometimes referred to as the restricted, red, or exclusion zone. This area represents danger to life or health and should be approached with extreme caution. Depending on the material involved, appropriate protective clothing and equipment are necessary to enter this zone. Establishing the size of the hot zone depends on the amount and properties of the chemical involved and the location of the incident. To determine safe distances for specific materials, refer to the DOT *Emergency Response Guidebook* or to one of the computer database systems available to emergency response personnel.

Notes

Environmental Protection Agency (EPA) Terms	Other Common Terms
Exclusion Zone	Hot, Red, or Restricted Zone
Contamination Reduction Zone (CRZ)	Warm, Yellow, or Limited Access Zone
Support Zone	Cold or Green Zone
Hot Line	Contamination Perimeter
Contamination Control Line	Safety Perimeter

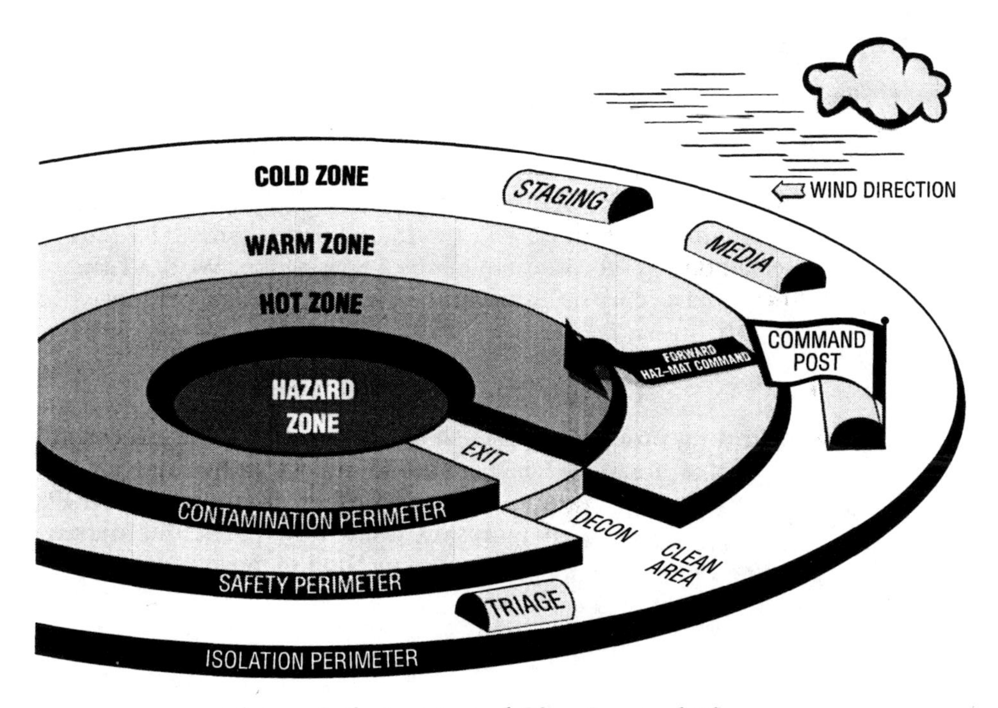

Figure 3-5. Suggested Site Control Plan

Warm Zone

The warm zone, also referred to as the limited access, yellow, or contaminated zone, is the area for site control directly outside the hot zone. It provides the forward access point to the hot zone for the necessary support personnel and equipment. It also provides escape routes and decontamination stations. Too often this site becomes the walk-through area for administrative personnel, media, and personnel stationed in the cold zone. The warm zone should be restricted to essential personnel and equipment only. All others should be kept out of the warm zone unless they are needed for hot zone support. It is easier to evacuate or reposition the warm zone when it is not congested.

Cold Zone

The cold zone, sometimes referred to as the support or green zone, is the area for site control directly outside the warm zone. This is the safe area for the incident commander, outside agencies involved in the incident, media, Red Cross, and medical personnel.

HAZ-MAT — MANAGING THE INCIDENT

M
Managing

A command system is necessary to control an emergency incident effectively. Haz-mat emergencies strain departmental command systems and resources. This chapter will not address the entire incident command (IC) system. Several IC system organization charts are provided for your review (Figures 3-7, 3-9, 3-10, and 3-11).

Before continuing, ask yourself the following questions:

- Does your organization have an IC system?
- Does your agency practice and effectively use its IC system?
- Does your organization have standard operating procedures (SOPs) for all emergencies, including haz-mat incidents?

If your answer to any of the above questions is no, you and your agency are in trouble. No matter how small or large your organization, an IC system and SOPs are essential to effective emergency operations. Without them, response to emergencies can be unorganized and will place personnel in unnecessary danger.

SOPs

Standard operating procedures (SOPs) are essential for the successful outcome of a haz-mat incident. SOPs should reflect individual agency capabilities. Each agency is unique. Too often other organizations' SOPs are blindly implemented without adapting them to a particular department's system. See Figure 3-6 for a five step method of developing SOPs.

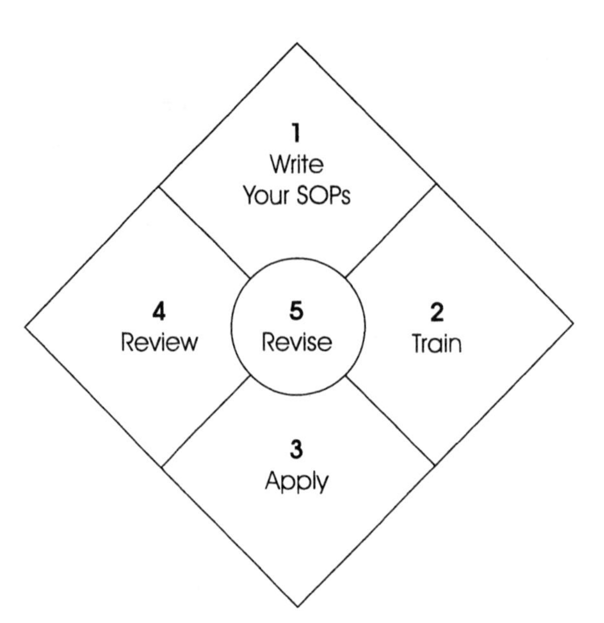

Figure 3-6. SOP Development Process

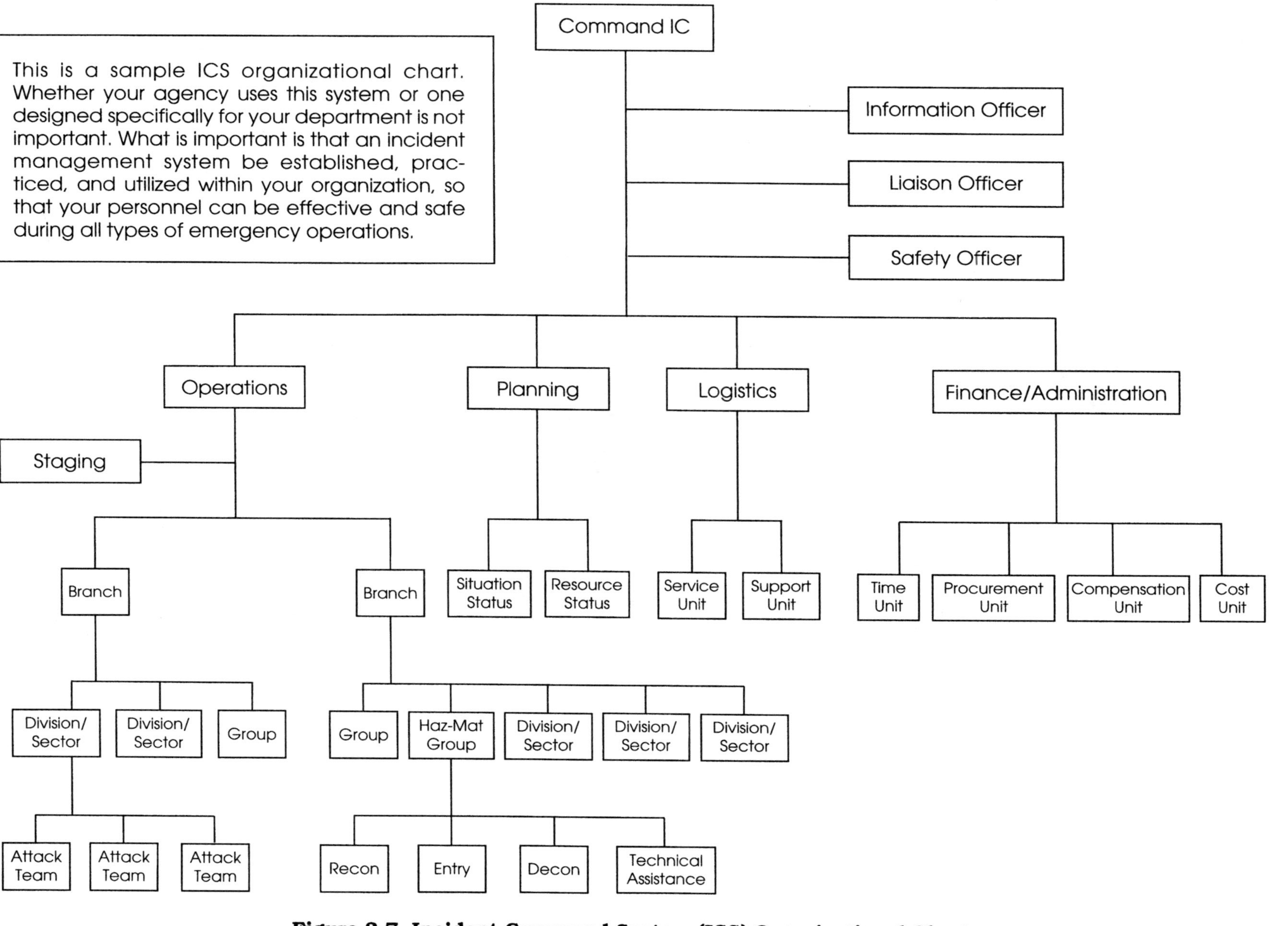

Figure 3-7. Incident Command System (ICS) Organizational Chart

Notes

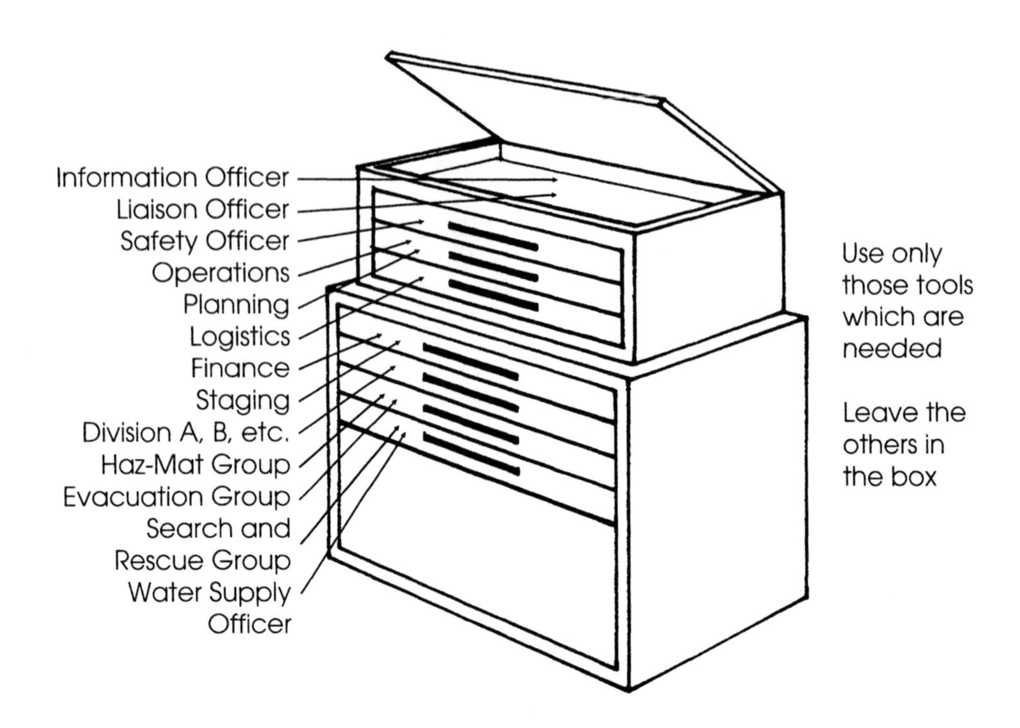

Figure 3-8. Incident Command System Toolbox

INCIDENT COMMAND

Managing a haz-mat incident begins with the first emergency responder's arrival and ends with the departure of the last piece of equipment. Command structure should be established according to the seriousness of the incident. Below, the incident command (IC) organization concept is applied to three haz-mat incident examples.

Level I Incident

This incident is a motor vehicle accident with two people injured and 30 gallons of diesel fuel spilled with no ensuing leak. The first arriving responder assumes command and may or may not relinquish control, depending upon the nature of the incident. The IC structure should be developed only as needed, as shown in Figure 3-9. In addition, it is important to *establish control zones.*

Level II Incident

This incident involves a 110 pound chlorine cylinder leaking from the valve assembly at a community swimming pool. In this situation, the first arriving responder would pass command to a higher officer and assume a supportive role, as shown in Figure 3-10. Again, it is important to *establish control zones.*

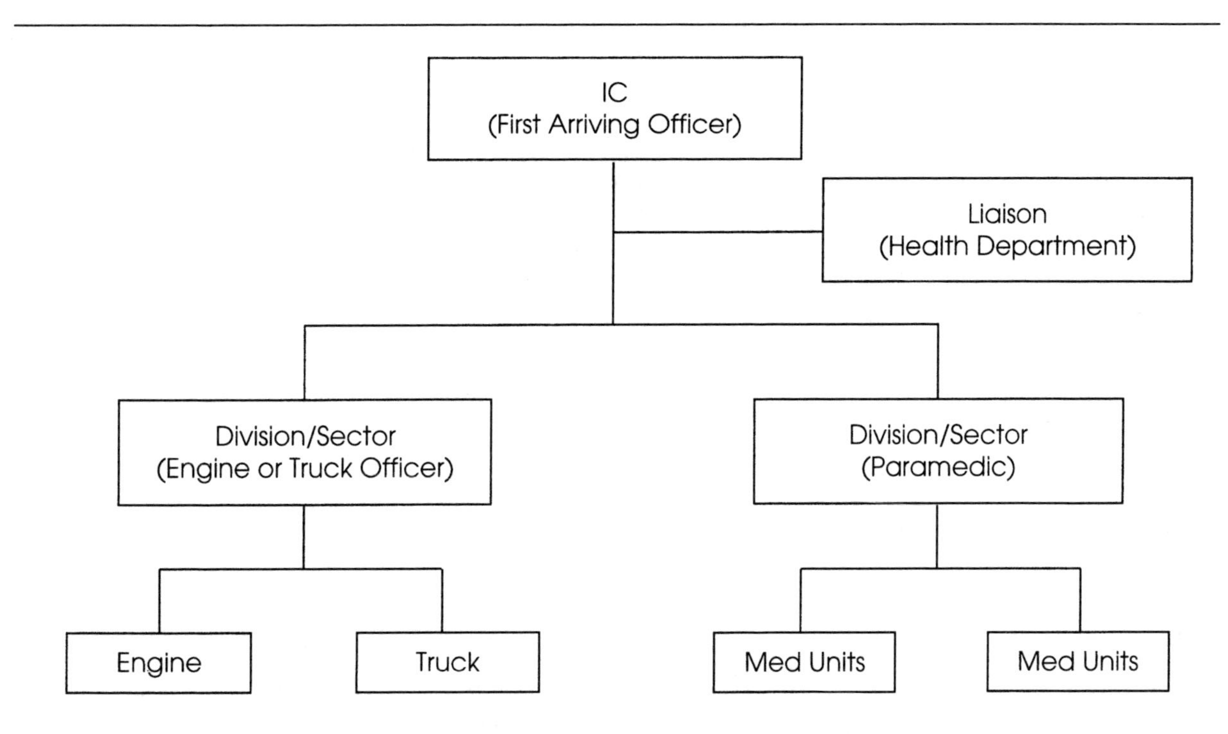

Figure 3-9. Level I Incident Command Structure

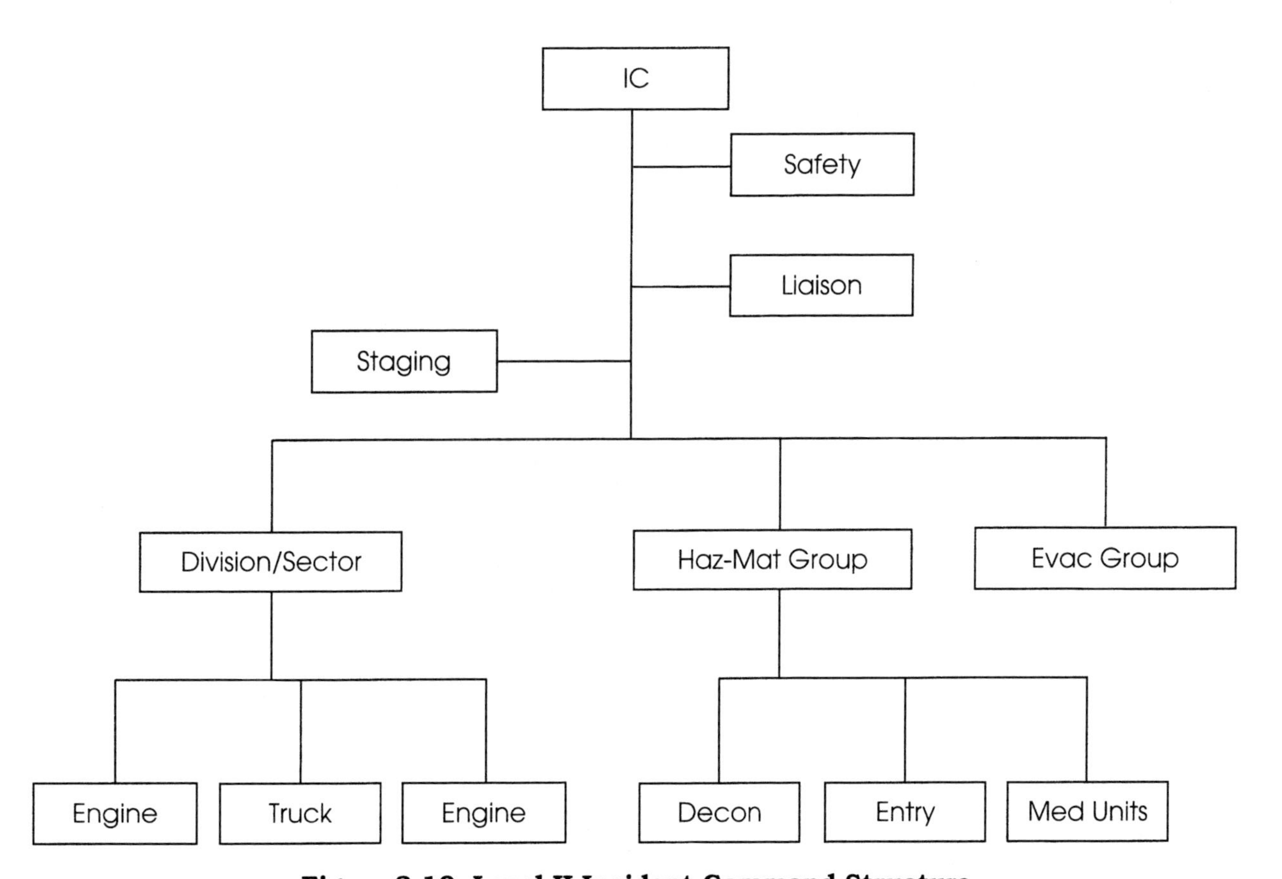

Figure 3-10. Level II Incident Command Structure

Notes

Level III Incident

This incident involves a leaking 20,000 gallon tank car of anhydrous ammonia on a rail spur in the town's industrial section. In this situation, the first arriving responder would pass command and take necessary action to protect life and property. Control zones are critical and may be established a considerable distance from the incident.

Much more response is needed during a Level II or III situation than during a Level I incident. Using a haz-mat command worksheet is helpful during such emergencies. A sample command worksheet is provided for reference (see Figure 3-12).

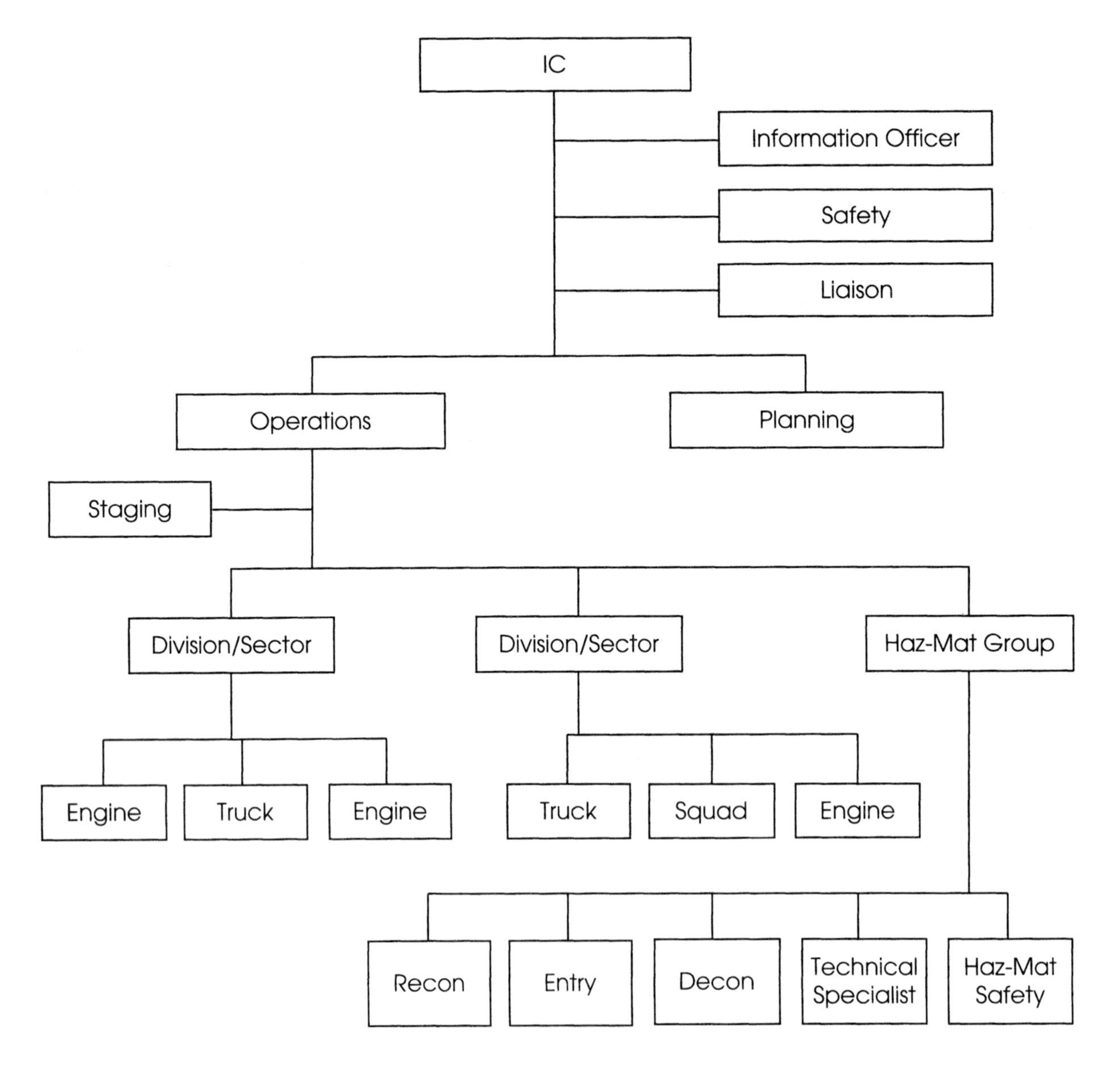

Figure 3-11. Level III Incident Command Structure

HAZARDOUS MATERIALS COMMAND WORKSHEET

LEVEL I ____ LEVEL II ____ LEVEL III ____ DOT HAZARD CLASS ____________

INCIDENT INFORMATION

Incident Name ____________

Location ____________

Time ____________

WEATHER INFORMATION			
Time:	Present Cond.	1 Hour Forecast	Long Term Forecast
Wind Direc.	N W E S	N W E S	N W E S
Wind Speed			
Rain/ Snow			

RESOURCES	
COMPANY	ASSIGNMENT

PRODUCT AND SHIPPING INFORMATION

Material Name ____________

DOT ID No. ______ Guide No. ______ Evac. Dis. ______

Flash Pt. ______ Ignition Pt. ______ Boiling Pt. ______

Lower Flam. Limit ______ Upper Flam. Limit ______

Vapor Density ______ Vapor Pressure ______

Specific Gravity ______ Water Soluble ______

Water Reactive ______ Heat/Shock Reactive ______

704 LABEL

Manufacturer/Shipper ______ Rail Car No. ______

B/L or Waybill No. ______ Origin/Destination ______

Consigner ______ Truck No. ______

SIDE 3

SIDE 2 SIDE 4

SIDE 1

In the space above, sketch the incident.

TERMINATION ACTIVITIES

- ☐ Decontamination
- ☐ Rehabilitation
- ☐ Medical Screening
- ☐ Exposure Reporting
- ☐ Incident Debriefing

CHECKLIST

- ☐ Hot Zone
- ☐ Warm Zone
- ☐ Cold Zone
- ☐ Uphill/Upwind
- ☐ Static or Moving
- ☐ Rescue
- ☐ Evacuation
- ☐ Crowd/Traffic Control

- ☐ Isolate Contaminated Area
- ☐ Diking/Diversion/Retention
- ☐ Personal Protective Clothing Required
- ☐ Decontamination
- ☐ Specialized Equipment Needed
- ☐ Medical Evaluations
- ☐ Debriefing
- ☐ Termination

AGENCIES CONTACTED:

- ☐ Law Enforcement Time ____
- ☐ Health Department Time ____
- ☐ CHEMTREC Time ____
- ☐ CANUTEC Time ____
- ☐ National Response Center Time ____
- ☐ EPA Time ____
- ☐ ____________ Time ____
- ☐ ____________ Time ____
- ☐ ____________ Time ____
- ☐ ____________ Time ____
- ☐ ____________ Time ____

Figure 3-12

HAZ-MAT — ASSISTANCE

A
Assistance

An incident commander needs adequate resources to deal with the complexities of hazardous materials. Available resources should be identified before an incident occurs. The following key resources will make incident management more successful.

Telecommunications Personnel (Dispatcher)

The dispatcher is the first link in the overall success of the operation. A well-trained dispatcher can assist emergency responders in product identification, incident assessment, and resource coordination. Critical information provided by the dispatcher will smooth the way to a safe, effective response. The dispatcher can provide information such as, "The cargo van fire that you are responding to is placarded Explosives 1.1." The dispatcher can assist in acquiring resources during a haz-mat emergency. Dispatchers should keep an updated resource list close at hand. Telecommunications personnel should have an understanding of product identification, placarding, and available resource materials prior to an incident.

Law Enforcement Officer

In most cases, a law enforcement officer is the first emergency responder on the scene. The officer should be capable of properly recognizing hazards and assessing the severity of the incident. It has been proven that the first arriving law enforcement unit can be at significant risk if the officer cannot properly recognize a hazardous material. A well-trained law enforcement officer will:

- recognize the presence of a hazardous material
- establish a hot zone to isolate the product
- initiate protective actions
- not become exposed to the incident

Because police departments usually control evacuations during haz-mat emergencies, other response agencies often are not prepared to manage evacuations. If you ask many haz-mat response teams or fire departments how they would evacuate, they will most likely say that the police department would handle evacuation. For effective scene management during an evacuation, it is vital to know:

- who is responsible for evacuation
- resources available for evacuation
- method of evacuation to be used—door to door, P.A. systems, radio broadcast, or combination
- how long it will take to evacuate one block, five blocks, etc.

Preplanning community evacuation capabilities provides the incident commander with a clear understanding of the time and resources needed to accomplish such a task. This understanding is critical for developing the incident action plan.

Local Emergency Planning Committee (LEPC)

Notes

The 1986 Emergency Planning and Community Right to Know Act requires each community to select a local emergency planning committee (LEPC). The LEPC is appointed to develop an emergency plan and gather information on chemicals in the community. LEPC representatives are a valuable resource for emergency responders as they develop information for response.

Hazardous Materials Response Team (HMRT)

Rapidly expanding technology has created a vast number of hazardous substances. Most of these substances are not considered hazardous unless they escape their containment system. Industry and the fire service have realized that a system must be in place to control haz-mat incidents. Specialized groups called emergency response teams and hazardous materials task forces were organized in the late 1970s to meet this need. The chemical industry began providing these services earlier, since it manufactured, transported, and used hazardous materials. The chemical industry is still the most knowledgeable emergency service provider.

Haz-mat incidents can be handled more competently when the special resource capability of emergency response providers keeps pace with haz-mat emergencies. Today, many emergency service managers are faced with establishing ***hazardous materials response teams (HMRTs)*** and increasing their special resource capability for haz-mat incidents.

HMRTs across the United States and Canada vary in expertise level and equipment that is available. There are plans to standardize equipment, knowledge level, and professional qualifications for HMRTs. This process began with the introduction of NFPA 472, OSHA 1910.120, and other regulations. However, for the next several years the discrepancies between HMRTs may be significant.

Hazardous Materials Response Team

Notes

The incident commander must know the assisting HMRT's capabilities and available tools/equipment to properly assign tasks during a haz-mat emergency. Frustration and a more unmanageable incident will result if a task is assigned for which the proper equipment is not available.

Agency policy should establish the operating procedures of an HMRT. If a local HMRT is not available, it is important to know the location of the closest public or private HMRT. Assess the team's capabilities and response time and establish a response agreement.

Private Technical Specialist

In many situations, a technical specialist is vital for safe management of a haz-mat emergency. The ***technical specialist*** provides specific technical expertise to the incident commander. The following scenario represents a situation in which a technical specialist would be helpful.

> You are the first to arrive at the scene of a multi-vehicle accident involving hazardous materials. A green bubbling brew and a dark amber mixture are running down the gutter from opposite directions. In addition, there are broken bags of a white powdery substance. As the chemicals mix, a large cloud of noxious fumes forms. As first responder you must:
>
> - identify the presence of a hazardous material
> - designate a hot zone
> - decide what was produced during the mixing of the chemicals
> - determine how the chemicals should be cleaned up

In this situation, incompatible chemicals that were not intended to be mixed came together, forming a new substance. A chemist with knowledge of emergency response procedures and a toxicology background would be most helpful in resolving this incident. Such technical specialists need to be identified *before* an emergency occurs. Decide what type of specialists may be needed by evaluating the haz-mat dangers present in your community or workplace. Establish a working relationship with these specialists. Train them in operational procedures to utilize their abilities effectively during large and small haz-mat incidents.

Private Contractors

One of the most difficult areas of incident management occurs after the emergency has been mitigated. This is the area of clean-up. Who is responsible for cleaning up the mess?

Due to the complexities and liabilities of hazardous waste management, clean-up is often handled by private contractors. ***Private contractors*** are privately owned companies whose personnel are trained, equipped, and insured to deal with clean-up and recovery operations. The clean-up process can be delayed hours or even days if the incident commander does not know about these companies or how to contact capable contractors.

A list of private contractors available in your area should be kept current and made available to incident commanders. This list should include the name of the key contact personnel and how they may be reached in an emergency. Before adding a private contractor to your list, review the company's capabilities, professional qualifications, and equipment.

Private Contractors

CONTRACTOR SELECTION CHECKLIST

The following is a checklist to refer to when deciding if a private contractor should be added to your list for use in haz-mat clean-up.

- Communications
  - Adequate communication facilities between field and office.
  - Sufficient personnel available to man communication facility.
  - Communications availability for workers at the spill site.
- 24-hour availability
  - Management personnel designated on a 24-hour basis.
  - Labor force and equipment available around the clock.
  - List of management available outside normal business hours.
- Trained personnel
  - Supervisory and labor force trained in haz-mat handling.
  - Trained technical force.
  - Adequate training records available for inspection.
- Industry reputation
  - List of customers utilizing clean-up services.
- Access to state approved disposal facilities
  - Use state approved, Class I facilities.
  - Licensed waste hauler.
  - If 24-hour waste facility access not available, can hazardous materials be safely and legally stored until the facility opens?
- Heavy-duty tow trucks and cranes
  - Own or have 24-hour access to reliable, heavy-duty tow and crane services that are familiar with and comply with safety standards.
- Adequate insurance coverage
  - Adequate insurance coverage for employees and other potential liabilities.

Notes

- Vacuum trucks
  - Type of trucks available—wet or dry, mild steel, stainless steel, coated, rubber-lined, etc. Diesel driven trucks and compressors are preferable to gas driven, especially when volatile chemicals are involved.
  - Number and sizes of trucks available.
  - Types of hose ends, valves, and miscellaneous fittings used. Some oil haulers, drilling field operators, and septic tank operators use aluminum fittings that are not acceptable in corrosive service.
  - Tanks ASME coded.
  - State license for liquid waste hauling.
  - Current fire marshal permits.
  - Good safety reputation.
  - Ground cables carried and used on all trucks.
  - Smoking regulations around flammables observed by drivers and operators.
  - Adequate training records available for inspection.
- Support equipment
  - Fresh air breathing apparatus
  - Construction equipment
  - Pumps
  - Hoses
  - Heavy-duty cleaning equipment
  - Containment equipment
  - Dedicated hazardous spill response equipment
  - Spark resistant tools
  - Lighting

Assistance Summary

This section has reviewed five areas of assistance that are often overlooked in haz-mat incident management. To ensure these areas are adequately covered, ask:

- Are dispatchers trained in hazardous materials incidents?
- What role will law enforcement play in haz-mat response? Will the police department assume evacuation responsibilities? If so, how will evacuation be handled?
- What are the capabilities of the available HMRT?
- Have the technical specialists most likely to be needed in your area been located?
- Have private contractors for clean-up and disposal assistance been identified?

Not all areas of assistance have been covered. Each community faces different risks with varying resources. The key factor is to anticipate your needs and establish a way to acquire resources when needed.

Notes

HAZ-MAT— TERMINATION

A systematic procedure to terminate a Level II or Level III haz-mat incident should occur after the incident has been mitigated. Unlike other emergencies, termination of a haz-mat incident is not simply a matter of picking up the equipment and returning to the base of operations. Although there are many termination procedures, this section will concentrate on the following four areas:

- decontamination
- rehabilitation
- medical screening
- post-incident analysis

T
Termination

DECONTAMINATION

There are critical decontamination procedures that should be ongoing during a haz-mat incident. Procedures to decontaminate anything leaving the hot zone and contamination perimeter must be implemented to prevent or reduce the transfer of contaminants by people and/or equipment. These procedures should include decontamination of:

- personnel
- protective equipment
- monitoring equipment
- clean-up equipment

Unless otherwise demonstrated, everything leaving the hot zone should be considered contaminated.

Responders must understand the difference between **emergency** decon and **full** decon. Emergency decontamination is the physical process of immediately reducing contamination of individuals in potentially life-threatening situations without the formal establishment of a contamination reduction corridor. Emergency decon only provides for gross decontamination, so there may still be the potential of secondary contamination and exposure to hazardous materials.

Full decontamination is the physical or chemical process of reducing and preventing the spread of contamination. Decontamination options may include rinsing equipment, personnel, etc., with large amounts of water and detergent/water solutions, or dry decontamination options such as isolation and vacuuming. If contaminants are known, then a specific detergent and/or solvent can be used to decontaminate. Figure 3-16 illustrates the minimum physical layout for personnel decontamination for a relatively small, well identified situation. Figure 3-17 illustrates the maximum physical layout for personnel decontamination during a worst case situation. Each site requires special consideration. Decontamination procedures should be modified from emergency decon to minimum layout to maximum layout based on known information.

Notes

The acronym ***I HOPE*** helps in understanding some important decontamination concepts.

I Identify

H Help or hold
O Operations
P People and equipment
E Environmental considerations

I – Identify

A hazard must be ***identified*** before action is taken to eliminate the risk of serious danger. Use reference materials to determine life hazards and characteristics of the material with which you are dealing. Be sure to use more than one reference to obtain a broader perspective on the product. Some references are easy to understand, while others are quite complex and technical. Practice using some of the references listed in Appendix C. If references contain conflicting information for a chemical, assume that the most hazardous description is correct. Remember, "Fools rush in where wise men fear to tread."

H – Help or Hold

Determine if the risks and benefits are worth a quick "in and out" ***help*** rescue or whether you should ***hold*** and use established procedures determined prior to the incident. The purpose of this reference manual is to help you consider the many factors that go into making such a decision.

O – Operations

Operations include the actual plans and procedures of decontaminating both victims and rescuers. Support personnel who were inside the warm zone must be decontaminated. The smaller the incident and the fewer people exposed, the better the outcome will be.

P – People and Equipment

Determine the appropriate number of ***people*** it will take to mitigate an incident. Be realistic—realize that the process will be lengthy if only two people are assigned to rescue 16 victims and one to run a decontamination station. If you are unsure how many people will be needed to perform each function, ask the assignment coordinator. Use people to their maximum potential. For example, once the containment crew is in chemical protective clothing (CPC), send those assisting to the initial rinse pool to aid victims being rinsed. These same people should be prepared to change out bottles for the containment crew. In addition, be aware of the ***equipment*** required for an incident. Four 30-minute air bottles are not enough for four men in CPCs for two hours!

E – Environmental Considerations

Time of day, temperature, wind speed and direction, and other ***environmental considerations*** are important factors in any incident. For example, if an incident occurs at 3:00 a.m. during a blizzard, it will be advantageous to obtain a bus, label it "DIRTY," and use it for victims. Take the victims to a warm, controlled area, such as a heated garage, where decontamination equipment and medical personnel can be stationed.

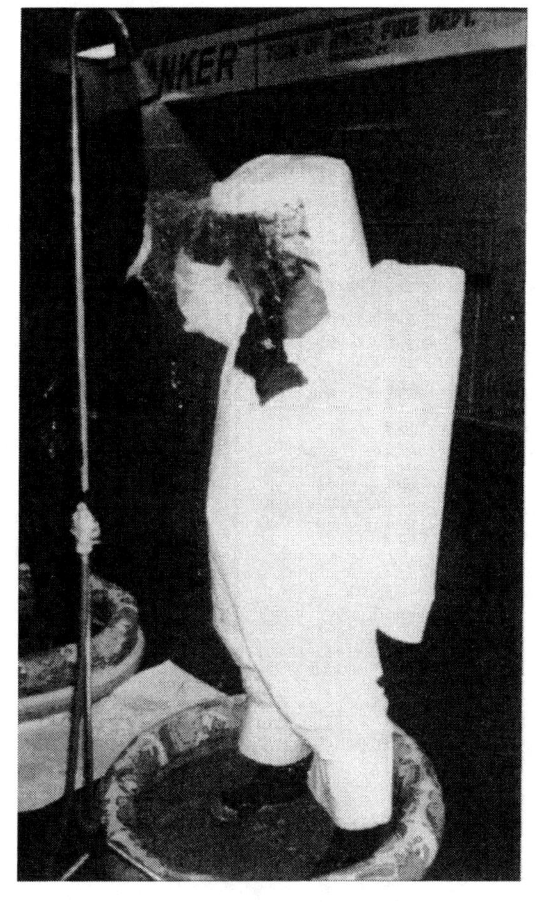

Decontamination

Initial Decontamination Steps

- Identify the material
- Isolate the area
- Deny entry

Secondary Decontamination Steps

- Establish decon officer
- Establish decon sector
- Put on correct protective clothing
- Confer with medical officer
- Make sure that persons leaving the hot zone have been cleaned

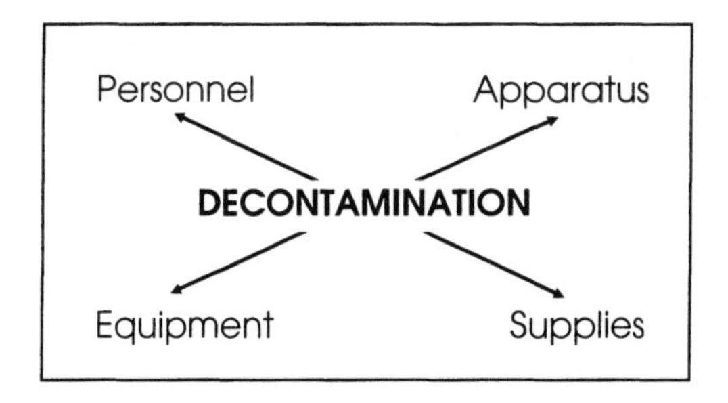

DECONTAMINATION METHODS

Dilution and Washing

Dilution ranges from rinsing off with a booster line to an eight step system of decontamination. Water usually does not change the product's chemical make-up. Adding 99 gallons of water to one gallon of a hazardous material simply makes it 100 gallons containing 1% of a hazardous material. However, using water to dilute has advantages over other means of decontamination because water is readily available. Using soap or other cleansing agents with water can remove many hazardous products. It is difficult to contain dilution runoff, but it is essential to make every effort to control the runoff. Runoff will eventually flow to some other site, transferring the problem to someone else.

Notes

Chemical Neutralization or Degradation

Chemical neutralization is advantageous because it reduces the product's hazard level. The disadvantage is the time involved to determine, obtain, and mix the correct neutralizer. Even an educated guess as to the correct neutralizing solution could cause the incident to worsen if the guess turns out to be wrong. Technical assistance from the manufacturer is usually helpful. Some neutralizing agents include household bleach, hydrated lime slurry, and liquid detergents. Guidelines for preparing decontamination solutions are given on the next page.

Absorption or Adsorption

Absorption is usually used for tools and equipment, not people. However, in a pinch, dusting a victim with dirt, sand, limestone, or other absorbing material, and later brushing it off, might reduce contamination. Absorption is an effective containment method and should be used in certain situations.

Adsorption is the chemical assimilation of a material by the surface of a solid. Activated charcoal, silica, and some clays are examples of adsorbents. Both absorbents and adsorbents pose a disposal problem after being used at a haz-mat incident.

Physical Removal

Hazardous substances can be physically removed by brushing off or wiping away visible contaminants. Vacuuming can also be used to collect materials. Special equipment may be needed that is appropriate for the material being vacuumed. Physical removal may cause dusts to become airborne, posing respiratory risks.

Solidification

Certain hazardous liquids can be solidified using commercially available products. Technical help will be needed to determine if a solidification agent is available.

Isolation and Disposal (for equipment only)

When equipment or materials cannot be safely cleaned, they should be disposed of properly. Examples of such equipment include ropes, hoses, some wood-handled tools, disposable suits, etc.

Use of Decontamination Solutions

Decontamination is not an exact science; therefore, choosing a decon solution can be controversial. Decontamination solutions are usually solutions of water and chemical compounds designed to react with and neutralize specific contaminants. The temperature of the liquid and contact time must be considered in order to assure complete neutralization.

When a ***known material*** is encountered, the chemical manufacturer or Agency for Toxic Substances and Disease Registry (ATSDR, see Appendix C) should be contacted for specific decontamination instructions. Figure 3-13 provides some basic guidelines for selecting neutralization chemicals for an identified hazard type.

Notes

The solutions described below are effective for a variety of contaminants and can usually be used on ***unknown materials***. Many departments only use detergent and water for decontamination of personnel. Decon solutions A through D should be used *for equipment only* unless otherwise advised by the ATSDR.

Guidelines for Preparing Decontamination Solutions

Warning – Decon solutions A through D should be used only for equipment unless otherwise advised by the Agency for Toxic Substances and Disease Registry (ATSDR).

Decon Solution A – A solution containing 5% sodium carbonate (Na_2CO_3) and 5% trisodium phosphate (Na_3PO_4). Mix 4 pounds of Na_2CO_3 (soda lime) and 4 pounds of commercial grade Na_3PO_4 with each 10 gallons of water. Stir until evenly mixed. These chemicals are available in most hardware stores.

Decon Solution B – A solution containing 10% calcium hypochlorite ($Ca(ClO)_2$). Mix 8 pounds of $Ca(ClO)_2$ with each 10 gallons of water. Calcium hypochlorite is available from swimming pool supply stores. Purchase it in plastic containers or transfer it from cardboard drums into clean plastic buckets marked "Oxidizer." Stir with plastic or wooden utensil until evenly mixed.

Decon Solution C – A general purpose rinse solution for decon solutions A and B. It contains 5% trisodium phosphate (Na_3PO_4). Mix 4 pounds of Na_3PO_4 with each 10 gallons of water. Stir until evenly mixed.

Decon Solution D – A dilute solution of hydrochloric acid (HCl). Mix 1 pint of concentrated HCl into 10 gallons of water. (***Caution:*** *Always* pour acid into water; *never* pour water into acid, or a violent reaction will result.) Stir with a wooden or plastic utensil until evenly mixed.

Decon Solution E – A concentrated solution of powdered detergent and water. Mix into a paste and scrub with a brush. Rinse with water.

Notes

Recommended Decontamination Solutions

	Chemical Compound	Recommended Decon Solution
1.	Inorganic acids, metal processing wastes, and polychlorinated biphenyls	A
2.	Heavy metals: mercury, lead, cadmium, etc.	A
3.	Pesticides, fungicides, chlorinated phenols, pentachlorophenol (PCP), and dioxins	B
4.	Cyanides, ammonia, and other non-acidic inorganic wastes	B
5.	Solvents and organic compounds such as trichloroethylene, chloroform, and toluene	C or A
6.	Polybrominated biphenyls (PBBs) and polychlorinated biphenyls (PCBs)	C or A
7.	Oily, greasy, unspecified wastes not suspected to be contaminated with pesticides	C
8.	Inorganic bases, alkalies, and caustic wastes	D
9.	Radioactive materials	E
10.	Infectious substances	A and B

Caution: The decontamination solutions listed above are recommended for the 10 general groups of hazardous materials. ***Always*** contact expert assistance from manufacturers, poison control centers, medical specialists, etc., to determine the best solution to use.

Figure 3-13

In-Depth Decontamination Procedure

The final stages of the decontamination process should be included in the incident's termination phase. During the final decontamination phase, all personnel and equipment entering the hot or warm zone are properly decontaminated. The incident commander must:

- account for and properly dispose of all decontamination solutions and contaminated runoff
- prepare equipment that cannot be decontaminated for proper disposal

During the final step of decontamination, local or state health officials or EPA personnel should be consulted to determine final disposal methods.

Notes

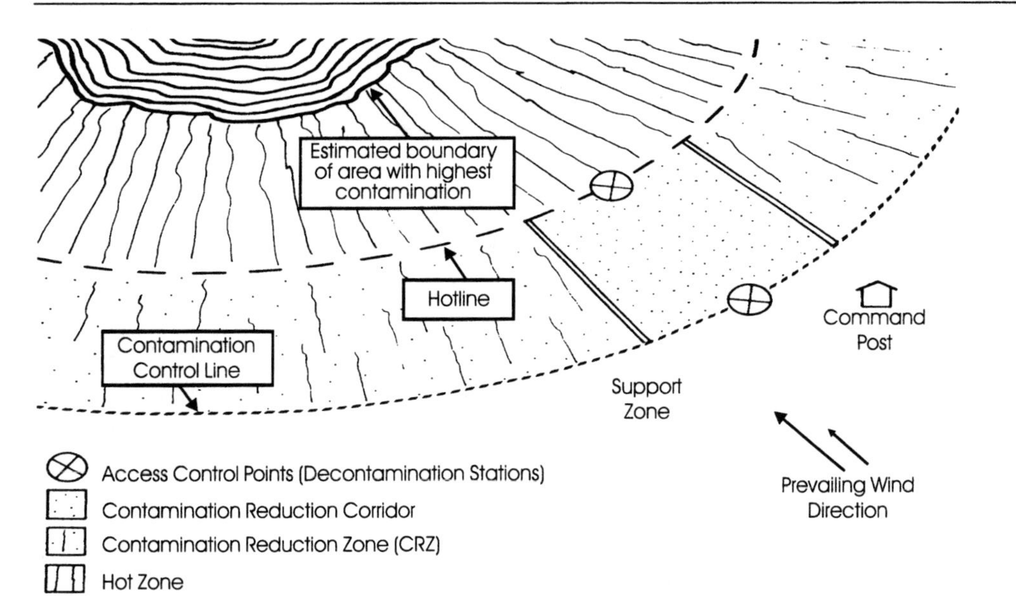

Figure 3-14. Recommended Location of Decon Stations

Special Decontamination Considerations

Instruct victims to be especially mindful of skin folds, hair, nails, hands, and feet when washing. Be aware of burns. Products that are not water soluble must be removed prior to a water wash. Solid or particle products may be brushed off prior to washing; however, take care not to inhale particles. A full rinse with water prior to removing equipment is advised.

When cleaning eyes, remove contact lenses and thoroughly irrigate. Be careful not to further damage the eyes. If topical anesthesia is used to allow for a vigorous cleaning, be aware that the victim will be unable to feel if more damage accidentally occurs.

It is important to contain runoff from decon pools. Many decon pools are inflatable, with rigid sides, and are made of either plastic or metal. Protect the bottom of plastic pools. Control the spray of hoses by using flow reducers and low spray nozzles. People will wash more thoroughly and longer if warm water is used.

Decon Considerations

- Levels of protection
- Weather
- Terrain
- Location
- Amount of decon required

Layout of the Personnel Decontamination Station (PDS)

Once the hot line section of the hot zone boundary has been established, a personnel decontamination station (PDS) should be set up. The four basic stages of decon are shown in Figure 3-15. The minimum layout of the PDS is illustrated in Figure 3-16. The minimum layout would be used for smaller incidents. A larger incident would require the maximum layout of the PDS, shown in Figure 3-17.

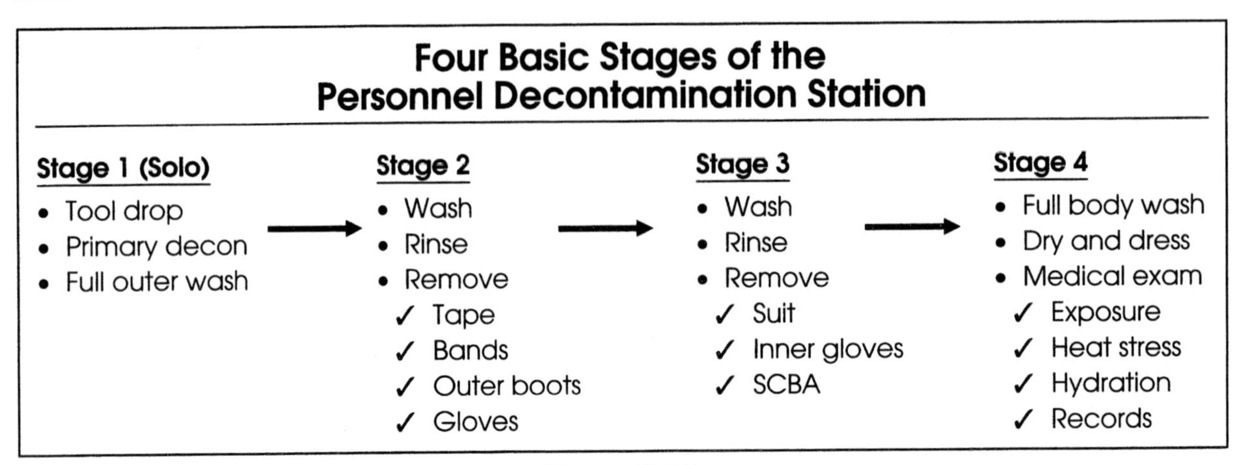

Figure 3-15

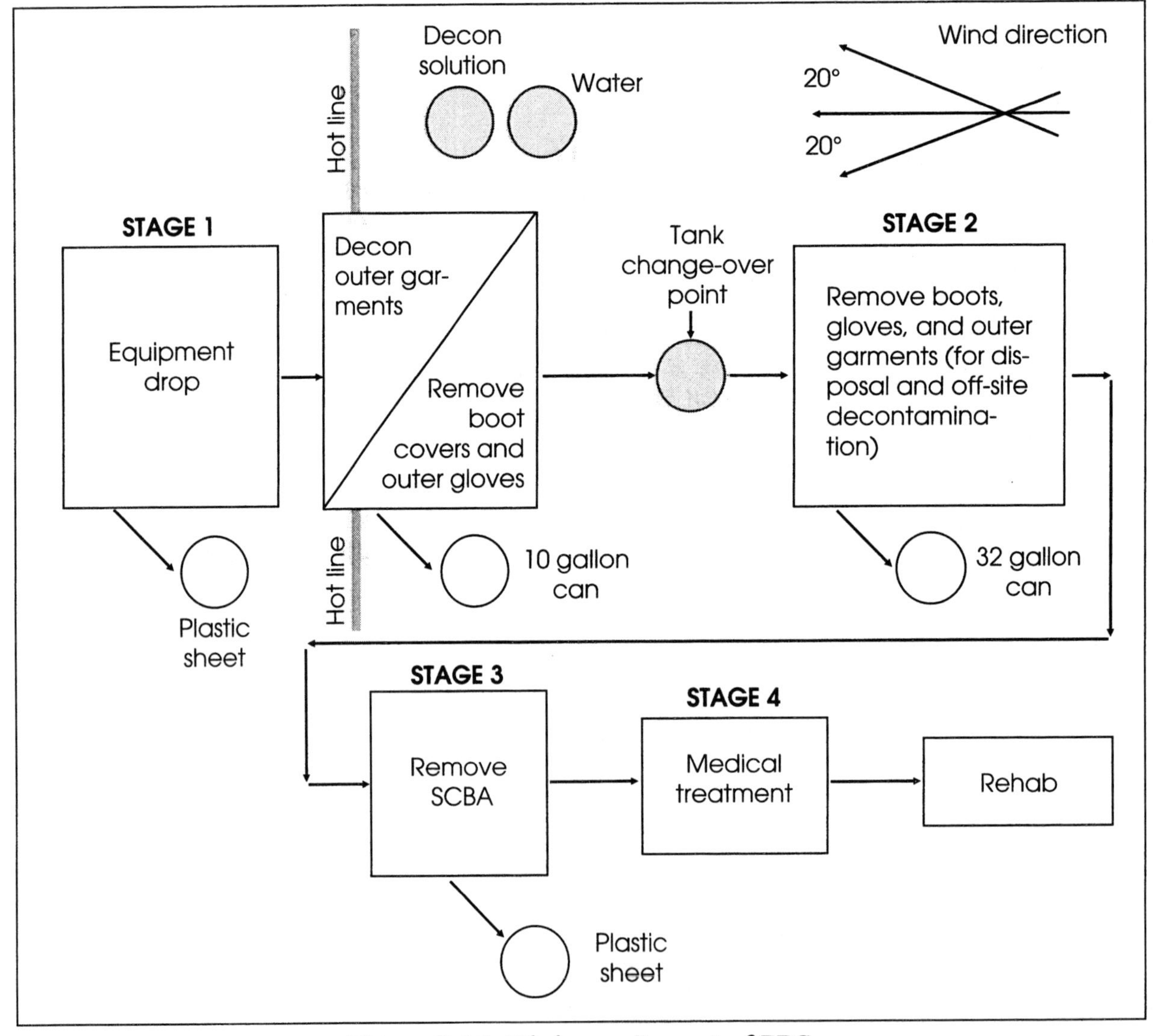

Figure 3-16. Minimum Layout of PDS

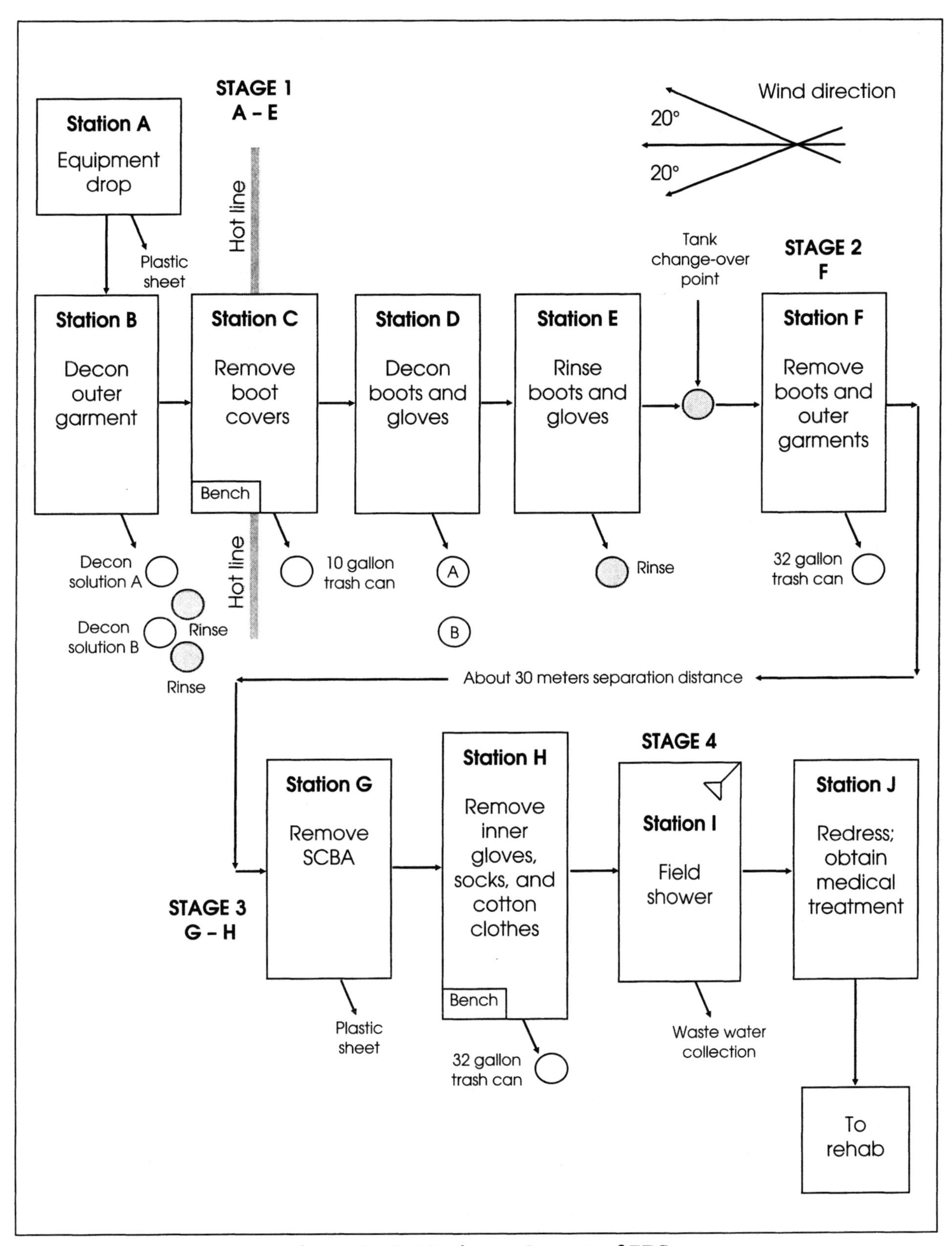

Figure 3-17. Maximum Layout of PDS

Notes

Decontamination Stations for Maximum PDS

Station A	Plastic ground sheet on which field equipment is dropped by returning members of the work party.
Station B	Washtub filled with decon solution A. Second washtub filled with rinse solution. Third washtub filled with decon solution B. Fourth washtub filled with rinse solution. Large sponge and brush for each washtub.
Station C	Bench or stool for sitting when removing boot covers. Plastic lined 10 gallon pail for disposal of boot covers.
Station D	Two 10 gallon buckets filled with decon solutions A and B.
Station E	10 gallon bucket filled with rinse solution.
Station F	Plastic lined 32 gallon trash can for rubber items.
Station G	Plastic sheet.
Station H	Bench or stool for personnel. Plastic lined 32 gallon trash can for cloth items.
Station I	Field shower set-up.
Station J	Redressing and medical treatment station. This station defines the boundary between the contamination control area and the clean area.

Figure 3-18

REHABILITATION

The welfare of personnel is often overlooked after a major emergency. After participating in a physically and emotionally demanding incident, personnel's ability to respond to another assignment should be evaluated. Often incident personnel are given a short rest period and offered food with poor nutritional value, such as donuts and cookies. After returning to the station, work will continue on cleaning the equipment to insure everything will be available for the next disaster.

It is not necessary for personnel to have a hot meal and full night of sleep after each major incident. However, during the termination phase of an incident, take time to adequately assess the ability of personnel to respond to another incident. The following procedures are advocated:

- Provide a minimum 30 minute rest period for personnel who experience extreme physical exertion or emotional pressure.

Notes

- Provide nutritional food and liquid supplements.
- Have personnel evaluated by medical personnel for possible chemical exposure and physical exhaustion. Vital signs should be taken when responders arrive at rehabilitation and as they are released for reassignment.
- Provide shower facilities and fresh clothing.
- Evaluate personnel for signs of emotional distress. This can be done by a member of the critical incident debriefing team or EMS provider.

These minimal steps will insure personnel's effectiveness and minimize long term health effects.

MEDICAL SCREENING

Medical exams and taking blood samples of personnel who were possibly exposed during the incident should be completed in the termination phase. Blood samples are often the only way to determine if personnel were exposed to the chemical involved.

It is recommended that high risk personnel, such as HMRT members, give yearly blood samples at local hospitals or a similar lab facility. The stored sample can be compared to the post-exposure sample. With such samples, the full extent of exposure can be determined while providing vital information for appropriate medical care.

Exposure Reporting

Exposure reporting, which began during the 1970s, has provided another tool for evaluating health risks encountered during emergency response. Use of exposure reporting allows an emergency responder to document and track dangerous exposure throughout his or her career. Personnel should report any incident in which they may have been exposed to a harmful substance. This includes EMS calls, fire response, and, of course, a haz-mat incident.

An excellent time to fill out the report form is during the termination phase of the incident. A sample exposure reporting form is found in Figure 3-19.

Chemical Exposure Record

INCIDENT DATA:

Name: ______________________ Company: ______________________

Date/Time: ______________________ Incident Number: ______________________

Location of Incident: ______________________

Description of Incident: ______________________

Chemical Involved: ______________________

___ Solid ___ Liquid ___ Gas ___ Vapor ___ Powder
___ Fume ___ Smoke ___ Mist ___ Fog

EXPOSURE DATA:

Type of Exposure:
___ Inhalation ___ Skin Contact
___ Ingestion ___ Through Wound

Description of Exposure: ______________________

Duration of Exposure: ______________________

Protective Equipment Worn at Time of Incident: ______________________

Quantity of Contaminant: ______________________

Extent of Contamination: ______________________

Figure 3-19

Chemical Exposure Record (cont.)

DECONTAMINATION DATA:

Method Used:

___ Dilution and Washing ___ Chemical Neutralization
___ Absorption ___ Physical Removal

Description: ______

Medical Treatment (Details of treatment, doctor, hospital, etc.): ______

Blood Sample: ______ Exposure Report: ______

RADIATION DATA:

Name of Material: ______

Estimated Amount of Material Involved: ______

Radiation Type: ______

Duration of Exposure: ______

Total Absorbed Dose: ______

______
Signature of Decontamination Officer

Figure 3-19 (cont.)

Notes

POST-INCIDENT ANALYSIS

Due to the complex and chaotic nature of emergency responsc, very few incidents go perfectly. A post-incident analysis is probably the best way to ascertain if an incident was handled properly. When used properly, the post-incident analysis can be not only a learning experience, but also a road map to the next emergency.

Post-incident analysis is a comprehensive review of the emergency. It should include input from each participating agency. This formal critique should be conducted by the agency having jurisdiction.

Unfortunately, when response does not go well, there is a tendency to place the blame on a particular agency. This blame sometimes shows up on the post-incident analysis. Conversely, the critique should not overlook problems that did exist.

The post-incident analysis should be a positive learning experience. It can be used to make future incidents more productive, safe, and manageable. A responsible party from each agency should discuss what part the agency played, why it was done, and how it may be done more effectively in the future. If deficiencies are noted during the critique, it is imperative they be addressed prior to the next incident. It will only frustrate participating agencies and personnel if problem areas are defined in the incident analysis, but not addressed and/or resolved.

CONCLUSION

Throughout the remainder of this book, response to haz-mat incidents will use the management concepts outlined in this chapter. As you progress through this manual, begin to adapt to your situation the haz-mat management concepts that are discussed, or use another system of managing a haz-mat incident that works for your organization. The important thing is to ***have a haz-mat management system in place before an emergency occurs***.

Chapter 4

EXPLOSIVE EMERGENCIES

OBJECTIVES

After studying the material in this chapter, you will be able to:

- define explosives
- describe the six classes of explosives recognized by the Department of Transportation
- explain the meaning of military markings for hazardous materials
- state basic information about common commercial and military explosives
- list the types of initiating devices for explosives
- identify three types of explosions
- explain the phases of an explosion
- identify common shipping containers for explosives
- explain how to respond to emergencies involving explosives, using the steps outlined in the acronym HAZ-MAT
- identify the types of injuries that are likely to result from accidents involving explosives

UNDERSTANDING EXPLOSIVES

INTRODUCTION

Explosives are one of the most dangerous hazard classes. Incidents involving explosives are less common than other haz-mat incidents; therefore, few responders are ever involved in such emergencies.

This section focuses on the Department of Transportation's six classes of explosives. Be aware that these six classifications do not include all materials that can explode, such as flammable gases, peroxides, etc. There are many explosive reactions that are not discussed in this section.

Notes

Special training and assistance are required to handle an explosive emergency. The following information will assist you in identifying explosive emergencies, making accurate decisions, and taking proper action.

DEFINITION OF EXPLOSIVES

The Department of Transportation (DOT) defines an ***explosive*** as:

> Any substance or article, including a device, which is designed to function by explosion (i.e., an extremely rapid release of gas and heat) or which, by chemical reaction within itself, is able to function in a similar manner even if not designed to function by explosion, unless the substance or article is otherwise classed (49 CFR 173.50(a)).

There are four basic components generally required for an explosion:

- ***Fuel*** or combustible material is needed to begin the explosive process. This material is usually a solid, but it can be a chemical mixture.
- An ***oxidizer*** provides extra oxygen not available in the air or combustible material. Extra oxygen generates the fast fire needed for an explosion.
- An ***energy source*** or initiator provides the heat, friction, or shock needed to light the combustible material and oxidizer.
- ***Confinement.*** This hot, fast fire must be contained or confined to explode.

As with the fire tetrahedron, if any of the components required for an explosion are removed, the explosion will probably not occur. Refer to 49 CFR 173.50–173.63 for specific regulations regarding transportation of explosives.

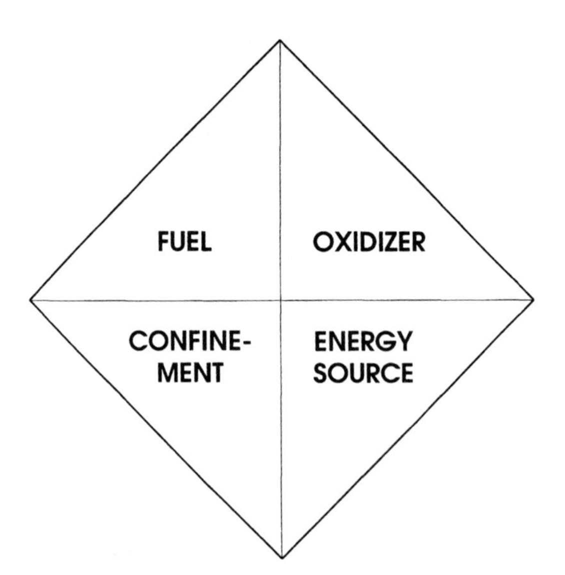

Figure 4-1. Basic Explosive Components

GENERAL CLASSIFICATIONS OF EXPLOSIVES

Notes

There are two general classifications of explosives—high explosives and low explosives.

High Explosives

High explosives function by detonation, meaning that shock and heat waves travel at speeds of greater than 3,300 feet per second and up to 29,000 feet per second. High explosives include dynamite, TNT, tetryl, and initiating explosives such as primer cord. They are usually initiated by a blasting cap, but unreasonable temperatures, shock, or friction may also cause accidental initiation. Even the initiation of a small amount of a high explosive may cause instantaneous explosion of the entire mass, since the explosive energy is transmitted as a supersonic velocity shock wave.

Low Explosives

Low explosives function by deflagration, or rapid combustion. Shock waves of low explosives travel less than 3,300 feet per second. They are ignited by a squib, powder fuse, or primer in rifle or shotgun cartridges. Low explosives include black powder, most smokeless powders, and other propellants.

Black powder may be readily ignited by a spark. The finer the powder grains, the more easily the powder may ignite as the flaming action is passed from one particle to another. This susceptibility to sparks makes transportation of black powder hazardous. The generic term, deflagrating explosives, is used for black powder or other explosives with similar composition or characteristics. Although black powder is generally considered a low explosive, it is rated as a high explosive under DOT regulations.

DOT CLASSIFICATIONS AND MARKINGS

DOT classifies two types of explosives—forbidden and acceptable. ***Forbidden*** explosives are too dangerous and unstable to be transported under any conditions. Perspective may be gained on how dangerous forbidden explosives are by considering the seriousness of accidents that have occurred with acceptable explosives.

Acceptable explosives are considered safe to transport under most conditions. Emergency responders are asked to respond to accidents involving acceptable explosives when fire or exposure conditions occur or when treatment of injuries is necessary. Explosive hazards are present at many incidents, but they often remain unrecognized.

As stated in Chapter 1, DOT classifies explosives according to six divisions (see 49 CFR 173.50(b)):

- ***Division 1.1*** consists of explosives that have a mass explosion hazard. A mass explosion is one which affects almost the entire load instantaneously. Such explosives function by detonation. Any and all quantities of Division 1.1 explosives are required to be placarded at all times in transit, whether by rail or road. Dynamite and TNT are examples of Division 1.1 explosives.

Notes

- ***Division 1.2*** consists of explosives that have a projection hazard but not a mass explosion hazard. Any quantity of Division 1.2 explosives must be placarded at all times while in transit. Rocket explosives are examples of Division 1.2 explosives.
- ***Division 1.3*** consists of explosives that have a fire hazard and either a minor blast hazard or a minor projection hazard or both, but not a mass explosion hazard. Any quantity of Division 1.3 explosives must be placarded at all times while in transit. Boosters and nitrocellulose are examples of Division 1.3 explosives.
- ***Division 1.4*** consists of explosives that present a minor explosion hazard. The explosive effects are largely confined to the package, and no projection of fragments of appreciable size or range is to be expected. An external fire must not cause virtually instantaneous explosion of almost the entire contents of the package. When transported over the road, Division 1.4 explosives require a placard only when the quantity is 1,001 pounds (454 kg) or more. When transported by rail, these explosives must be placarded at all times. Ammunition is a Division 1.4 explosive.
- ***Division 1.5*** consists of blasting agents, which are very insensitive explosives. This division is comprised of substances which have a mass explosion hazard but are so insensitive that there is very little probability of initiation or of transition from burning to detonation under normal conditions of transport. When transported over the road, Division 1.5 explosives require a placard only when the quantity is 1,001 pounds (454 kg) or more. When transported by rail, these explosives must be placarded at all times. Ammonium nitrate is an example of a Division 1.5 explosive.
- ***Division 1.6*** consists of extremely insensitive articles which do not have a mass explosive hazard. This division is comprised of articles which contain only extremely insensitive detonating substances and which demonstrate a negligible probability of accidental initiation or propagation. When transported over the road, Division 1.6 explosives require a placard only when the quantity is 1,001 pounds (454 kg) or more. When transported by rail, these explosives must be placarded at all times.
- ***Compatibility groups*** are designated by a letter beside the division number on placards and labels. DOT has assigned a compatibility group to explosive materials to prevent an increase in hazard that might result if certain types of explosives were stored or transported together. Compatibility group letters should not be confused with the old DOT classifications for explosive materials. For more information on compatibility groups, see 49 CFR 173.52.

CLASS 1 PLACARD AND LABEL REQUIREMENTS				
SUBSTANCE	**VEHICLE PLACARD**	**PACKAGE LABEL**	**QUANTITY**	**SPECIAL REQUIREMENTS**
Division 1.1	**Explosives 1.1** (black text on orange background). **Explosives A** (black text on orange background). EXPLOSIVES 1.1* 1 / EXPLOSIVES A 1 Domestic US	**Explosive 1.1** (black text on orange background). **Explosive A** (black text on orange background). 1.1* 1 / EXPLOSIVE A 1 Domestic US	Placard all quantities by rail or highway.	Must not be loaded with any other class of hazardous materials.
Division 1.2	**Explosives 1.2** (black text on orange background). **Explosives A** (black text on orange background). EXPLOSIVES 1.2* 1 / EXPLOSIVES A 1 Domestic US	**Explosive 1.2** (black text on orange background). **Explosive A** (black text on orange background). EXPLOSIVE 1.2* 1 / EXPLOSIVE A 1 Domestic US	Placard all quantities by rail or highway.	Must not be loaded with any other class of hazardous materials.
Division 1.3	**Explosives 1.3** (black text on orange background). **Explosives B** (black text on orange background). EXPLOSIVES 1.3* 1 / EXPLOSIVES B 1 Domestic US	**Explosive 1.3** (black text on orange background). **Explosive B** (black text on orange background). EXPLOSIVE 1.3* 1 / EXPLOSIVE B 1 Domestic US	Placard all quantities by rail or highway.	Must not be loaded with most other classes of hazardous materials. Division 1.3 materials may be loaded with non-flammable gases, flammable solids, and radioactive materials.

Figure 4-2

SUBSTANCE	VEHICLE PLACARD	PACKAGE LABEL	QUANTITY	SPECIAL REQUIREMENTS
Division 1.4	**Explosives 1.4** (black text on orange background). **Dangerous** (black text on white and red background). 1.4 EXPLOSIVES * 1 DANGEROUS Domestic US	**Explosive 1.4** (black text on orange background). **Explosive C** (black text on orange background). 1.4 EXPLOSIVE * 1 EXPLOSIVE C 1 Domestic US	Placard any quantity by rail; 1,001 lbs (454 kg) or more by truck.	Must be segregated from flammable gases and liquids, poisons, spontaneously combustible materials, and corrosive liquids.
Division 1.5	**Blasting Agents 1.5** (black text on orange background). **Blasting Agents** (black text on orange background). 1.5 BLASTING AGENTS * 1 BLASTING AGENTS 1 Domestic US	**Blasting Agent 1.5** (black text on orange background). **Blasting Agent** (black text on orange background). 1.5 BLASTING AGENT * 1 BLASTING AGENT 1 Domestic US	Placard any quantity by rail; 1,001 lbs (454 kg) or more by truck.	Must not be loaded with any other class of hazardous materials.
Division 1.6	**Explosives 1.6** (black text on orange background). **Dangerous** (black text on white and red background). 1.6 EXPLOSIVES * 1 DANGEROUS Domestic US	**Explosive 1.6** (black text on orange background). 1.6 EXPLOSIVE * 1	Placard any quantity by rail; 1,001 lbs (454 kg) or more by truck.	May be loaded with other hazardous materials.

Note: When required, the * on Class 1 placards and labels must be replaced with the appropriate compatibility group letter. If a subsidiary hazard is present, the subsidiary placard or label must **not** display a hazard class number. Placards identified "Domestic US" may be used in U.S. domestic highway transportation through October 1, 2001. Labels identified "Domestic US" may be seen on packages filled before October 1, 1991. All Class 1 materials must be segregated from other explosive materials according to DOT regulations (see 49 CFR 174.81(f)).

Figure 4-2 (cont.)

MILITARY MARKINGS

Notes

The ***United States military*** has established its own system for marking and identifying hazardous materials. The system uses seven symbols to mark military structures that store hazardous materials. You should become familiar with these special markings *prior* to a haz-mat incident involving military cargo. The military's three hazard symbols and four detonation and fire hazard symbols are shown in Figure 4-3. The U.S. military class 1 corresponds to DOT Division 1.1 explosives, military class 2 corresponds to DOT Division 1.2, military class 3 corresponds to DOT Division 1.3, and military class 4 corresponds to DOT Division 1.4.

Response to off-base military haz-mat incidents requires extreme caution. Military drivers may be under orders not to identify their cargo. Emergency responders may not be permitted to approach the scene when weapon systems are involved. How quickly military personnel evacuate the area may be an indication of the severity of the incident.

The ***Canadian military*** uses a similar system of identification (see Figure 4-4).

- ***Class 1 Explosives*** – Those which should be expected to quickly explode or detonate en masse after contact with fire.
- ***Class 2 Explosives*** – Those which readily ignite and burn violently, but do not necessarily explode.

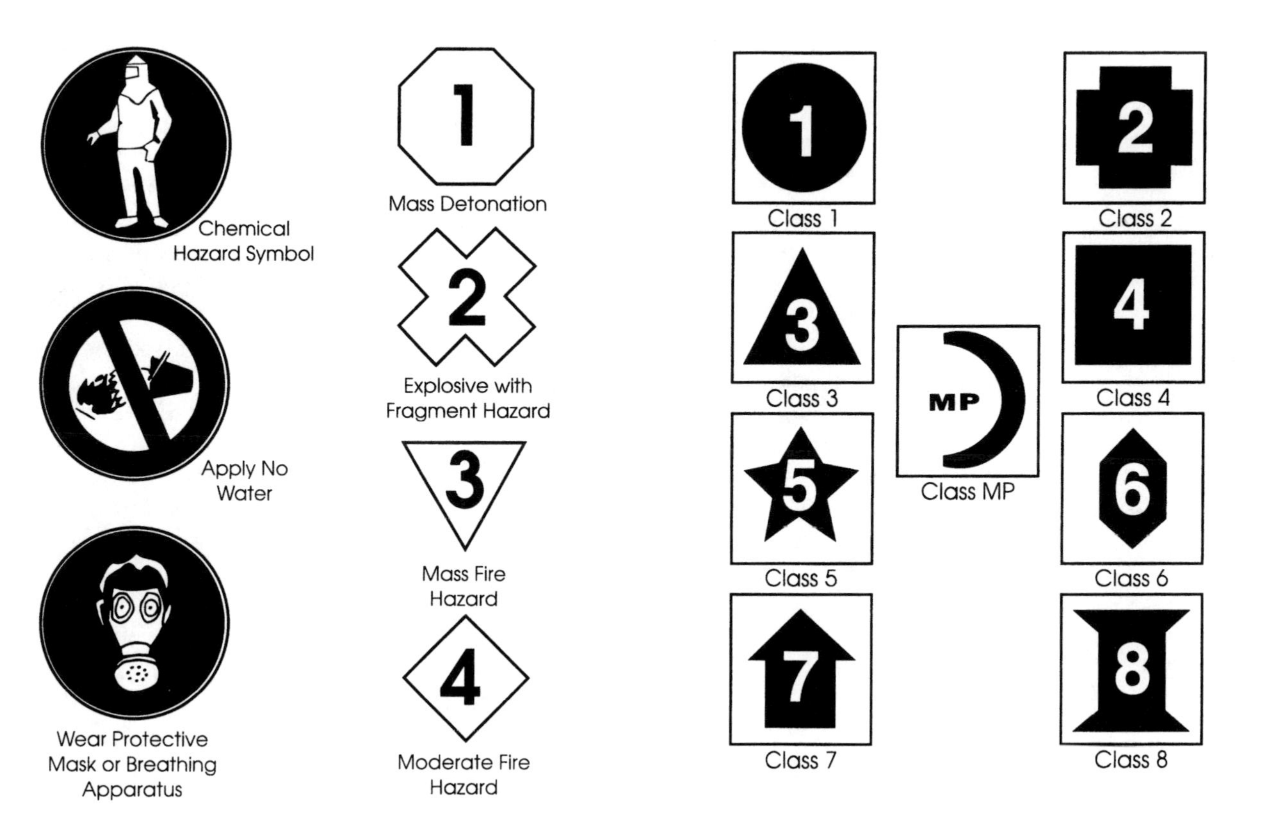

Figure 4-3. United States Military Markings

Figure 4-4. Canadian Military Explosion Class Markers

Notes

- ***Class 3 Explosives*** – Those which may explode en masse, but in contrast to Class 1 explosives, may be exposed to fire for some time before exploding. The result will be a blast and fragment hazard.
- ***Class 4 Explosives*** – Those which burn fiercely and give off dense smoke with, in some instances, toxic effects. There is no risk of mass explosion.
- ***Class 5 Explosives*** – Those containing toxic substances.
- ***Class 6 Explosives*** – Those which may be exposed to fire for some time before exploding. There is no risk of mass explosion, but small sporadic explosions will occur with increasing frequency as the fire takes hold. A fragment hazard will exist, but there is not a serious blast risk.
- ***Class 7 Explosives*** – Those involving combined flammable, toxic, and corrosive hazards. These may be exposed to fire for some time before exploding. There is no risk of mass explosion, but explosions will occur with increasing frequency as the fire takes hold. A fragment hazard will arise from pressure bursts but not from a serious blast risk. Personnel involved in firefighting shall wear protective clothing and self-contained breathing apparatus.
- ***Class 8 Explosives*** – Those in which a radiological hazard and an explosive hazard are present.
- ***Class MP*** – Substances containing metallic powders such as magnesium, aluminum, or zinc powders found either in ammunition or bulk.

COMMON USES OF EXPLOSIVES

Emergency responders encounter explosives in a variety of uses and abuses. Most incidents involving explosives occur in daily community business activities. Dynamite at a construction site, a bottle of crystallized picric acid at a school lab, a fireworks stand, black powder stored at a gun shop, or normal transport of explosives on roadways may all occur in your jurisdiction.

Unfortunately, explosives are also tools that criminals may use. Setting off an explosion with criminal intent or negligence is a crime. Motives may include vandalism, jealousy, labor disputes, sabotage, etc. Military and commercial explosives may be used in improvised bombs or other explosive devices.

Criminals may use explosives to break into a secured area or container. Terrorists may use explosives to cause senseless destruction of human life and property. Arsonists may use explosives to set or spread a fire, compound the damages of a fire, or hamper firefighting and rescue operations. Saboteurs may use explosives against material that resists other means of destruction or to create an interruption in activities after they leave the scene. Illicit drug laboratories may have chemicals that can react explosively.

COMMON COMMERCIAL EXPLOSIVES

Notes

Commercial explosives are available to the average person through commercial sources. This accessibility has caused commercial explosives to make up the majority of explosive incidents. Some common commercial explosives are listed below.

Warning: Disposal of these materials should only be attempted by trained experts.

Black Powder

Appearance .... Black powder varies from a black, fine powder to dense pellets which may be black or have a grayish-black color. Black powder is composed of 75% potassium nitrate, 15% charcoal, and 10% sulfur.

Storage .... Black powder should be stored separately from all other explosives. It has an extreme tendency to absorb moisture, which causes deterioration and affects its strength and sensitivity to ignition. Therefore, it should be stored in a dry, ventilated place.

Transportation .... Due to black powder's extreme sensitivity to heat, sparks, and friction, special care should be taken when transporting this material.

Nitroglycerin

Appearance .... Nitroglycerin is a colorless, oily liquid that is extremely sensitive to heat, shock, and friction. If the liquid becomes green in color, it should be disposed of immediately.

Storage .... Nitroglycerin is rarely kept in storage. If it is necessary to store nitroglycerin, it should be kept in earthenware containers standing in copper vessels and covered with water. The containers should be placed on wooden supports near the floor. Sawdust or other absorbent material should be kept under the containers to absorb leaks. The storage room should be kept at a low temperature, as nitroglycerin is less sensitive to shock when frozen. Nitroglycerin freezes at 55.8°F (13.2°C).

Transportation .... Nitroglycerin can be made safe to handle and transport by diluting with 30% acetone.

Other Information .... Nitroglycerin can be cleaned from floors and other surfaces with a solution of alcohol, acetone, and sodium sulfide. This solution is made by mixing 1-1/2 quarts of water, 3-1/2 quarts of denatured alcohol, 1 quart of acetone, and 1 pound of sodium sulfide (60% commercial). Be aware that heat is generated during the cleaning and neutralization process.

Notes

Dynamite

Appearance .... Generally speaking, dynamite is composed of nitroglycerin that has been combined with a porous filler desensitizer. Dynamite is manufactured in cylindrical cartridges from 7/8 inch in diameter and 8 inches in length up to 10 inches in diameter and 36 inches in length. It is encased in heavy, water resistant paper or cardboard tubes.

Storage .... Cartridges should be stored horizontally. Dynamite may emit nitroglycerin if the cartridges are stored on end. Sawdust should be kept under the dynamite to absorb leaking nitroglycerin. Sawdust should be changed frequently, and the old sawdust burned in the open air some distance from the storage site. The storage site should be dry and well ventilated. There is a tendency for the nitroglycerin to displace within the dynamite if exposed to moisture or elevated storage temperatures.

Transportation .... Dynamite is fairly safe to handle and transport. Care must be taken, however, that the product does not receive undue shock and is not mishandled.

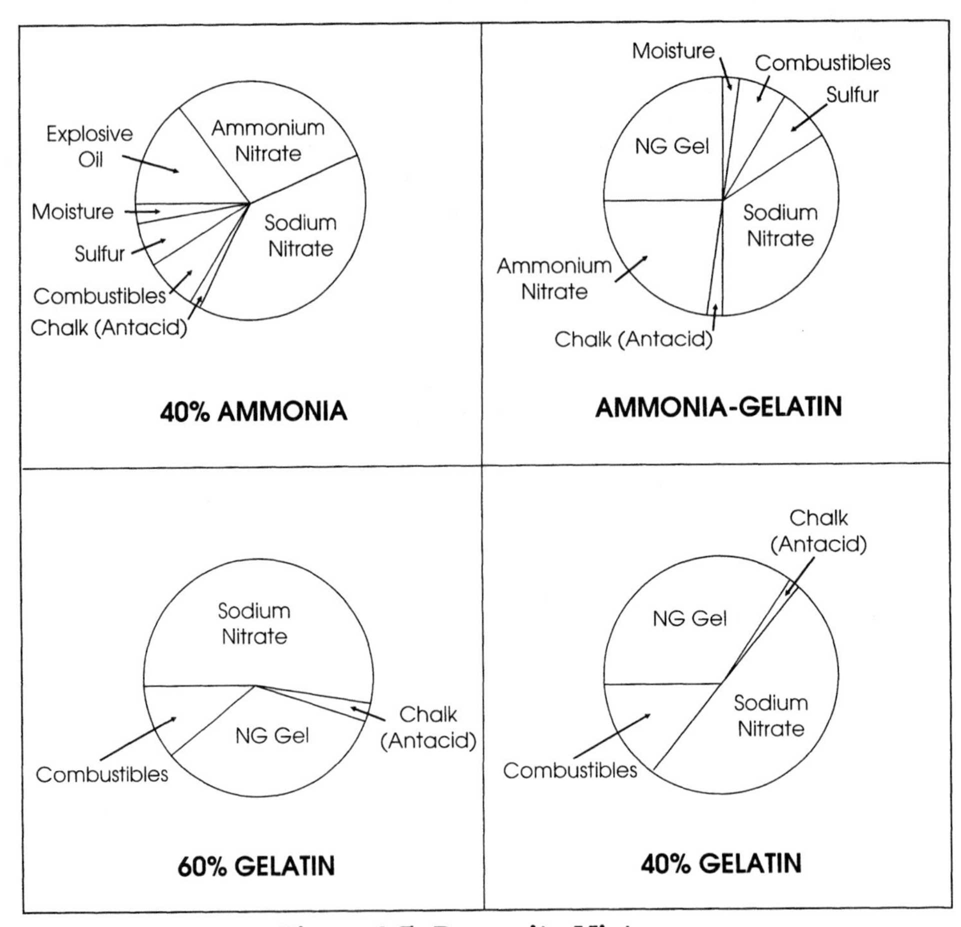

Figure 4-5. Dynamite Mixtures

Other Information .... Dynamite is considered old if an oily substance appears on the cartridge casing or if stains appear on packing cases. These changes occur when the nitroglycerin separates from the porous base, which causes the dynamite to be extremely sensitive. Old dynamite should be destroyed immediately by burning or detonation.

Notes

Ammonium Nitrate

Appearance .... Ammonium nitrate is not considered an explosive and is placarded as an oxidizer. It is mentioned here, however, because when mixed with hydrocarbon based materials, confined, and provided an energy source, it can react explosively. Ammonium nitrate is white to gray in color and crystalline in form. It is manufactured by neutralizing an aqueous solution of ammonia with nitric acid and allowing the solution to evaporate. Because ammonium nitrate is difficult to detonate by itself, it is primarily used in binary explosives, such as certain dynamites, cratering explosives, and other commercial explosives. The following list identifies the different forms and grades of ammonium nitrate:

- porous pellets or prills
- flakes
- grains
- fertilizer grade
- dynamite grade
- technical grade
- nitrous oxide grade

Storage .... Storage sites must be extremely dry and well ventilated, because ammonium nitrate readily absorbs moisture. Moisture decreases its sensitivity and effectiveness as an explosive agent. Ammonium nitrate also needs special storage precautions because it is a powerful oxidizing agent and because the chance of combustion increases when it is mixed with or sits next to a flammable material.

Transportation .... Transporting and handling ammonium nitrate is relatively safe. In the presence of moisture, however, ammonium nitrate reacts with copper to form other nitrates that are extremely sensitive to shock and impact. Therefore, brass or bronze tools should never be used when handling ammonium nitrate.

Other Information .... Laboratory analysis of residue generated from bombings indicates that ammonium nitrate has been used as an explosive agent ingredient in some cases, perhaps because it is easily obtained as common fertilizer.

Notes

COMMON MILITARY EXPLOSIVES

Military explosives are not readily available to the average person. However, they have occasionally been stolen to use in homemade bombs. These explosives should be handled with the same safety precautions used with commercial explosives. Some of the more common military explosives include:

- ***TNT*** – Light cream to rust color and usually found in 1/2 or 1 pound blocks. Fairly stable in storage.
- ***Composition B*** – A high explosive that is more effective and more sensitive than TNT. It is composed of RDX, TNT, and wax. Its shattering power and high rate of detonation cause it to be used as the main charge in certain models of torpedoes and shaped charges.
- ***Composition C-3*** – An oily, yellowish, plastic explosive which can be molded by hand. It is packaged in 2-1/4 pound blocks. It has a tendency to become hard, brittle, and difficult to mold in cold temperatures. In warm temperatures it becomes soft and exudes an oil that leaves a non-harmful deep yellow stain on the skin. C-3 catches fire easily if exposed to flame or sparks and will burn with an intense heat.

Common High Explosives

Military	Commercial
Mercury fulminate	Mercury fulminate
Lead azide	Lead azide
Diazodinitrophenol	Diazodinitrophenol
Lead styphnate	Lead styphnate
Nitromannite	Nitromannite
TNT	NG
Composition B	TETRYL
TETRYL	AN
RDX	TNT
PETN	DNT
Ammonium Picrate	Nitrostarch
Picric Acid	PETN
AN	
DNT	
EDNA	

Figure 4-6

Notes

- ***Composition C-4*** – A white plastic explosive which will not stain the skin like C-3. It is also more stable and will not stick to the hands. C-4 is packaged in 2-1/2 pound blocks.
- ***Picric Acid*** – Manufactured in light to bright yellow crystals. It is commonly found in chemistry laboratories. It was also used extensively by the Japanese during World War II as a main charge explosive filler. When contacted with metal, picric acid forms picrates, which are extremely sensitive to heat, shock, and friction. Great care should be taken when handling World War II souvenirs because such picrates may be present. Because of its extreme sensitivity, picric acid is no longer used as a main charge explosive filler.

Glossary of Explosive Abbreviations

Trade Abbreviation	Chemical Name
AN	Ammonium nitrate
AP	Ammonium picrate
BP	Black powder
DNT	Dinitrotoluene
EDNA	Ethylene dinitramine
FGAN	Fertilizer grade ammonium nitrate
LAZ	Lead azide
LS	Lead styphnate
MF	Mercury fulminate
NG	Nitroglycerin
NS	Nitrostarch
PA	Picric acid
PETN	Pentaerythritol tetranitrate
RDX	Cyclotrimethylenetrinitramine
TETRYL	Trinitrophenylmethylnitramine
TNT	Trinitrotoluene

Figure 4-7

Notes

INITIATING DEVICES

Initiating devices are designed to:

- initiate or detonate explosive charges
- supply or transmit an explosion flame
- carry a detonation wave from one point to another, or from one explosive charge to another

Electric Blasting Caps

The most common initiating device is the electric blasting cap. Both electric and non-electric blasting caps are sensitive to heat, shock, and friction. Blasting caps come in two sizes, designated 6 or 8, with number 8 being more powerful.

Electric blasting caps have lead wires that attach to an electric source. These caps use an electric current that can be initiated from a remote location. The caps consist of a metal shell loaded with several powder charges and an electrical ignition element. Electrical wires, called leg wires,

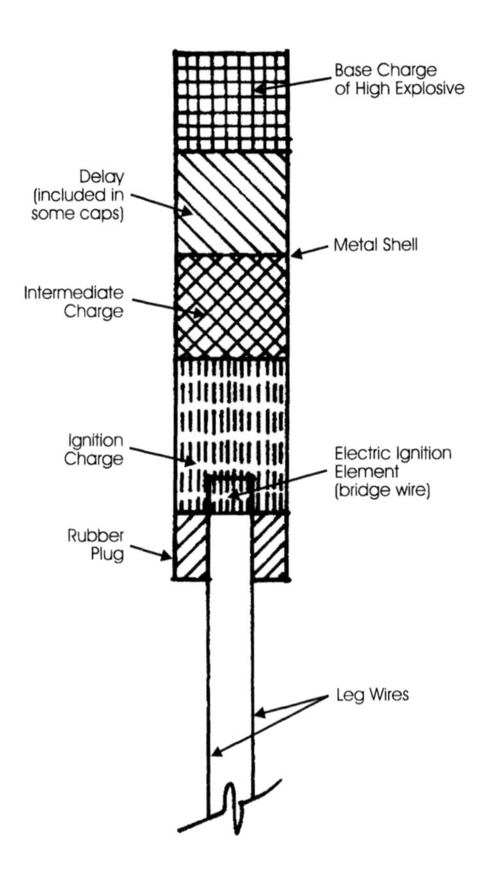

Figure 4-8. Cross Section of an Electric Blasting Cap

are attached to the elements. A rubber or sulfur plug, crimped to the metal, seals the explosives in the cap.

Caps with a delay charge slow the interval between the electric current and cap detonation. The delay can vary from a few thousandths of a second to 12 seconds. The delay enables a complete round of explosions to be fired from a single application of current. Delay caps are numbered on the basis of their delay period.

Warning: Blasting caps should never be stored with other explosives.

Non-Electric Blasting Caps

Non-electric blasting caps can be attached to one end of a safety fuse. The other end of the fuse can then be lit with an ordinary match or special igniter to initiate the explosive train. The ignition charge is designed to amplify the flame from the fuse. This flame front ignites the intermediate charge, which in turn detonates the base charge. The shell of the cap is made of aluminum or copper.

Warning: Blasting caps should never be stored with other explosives.

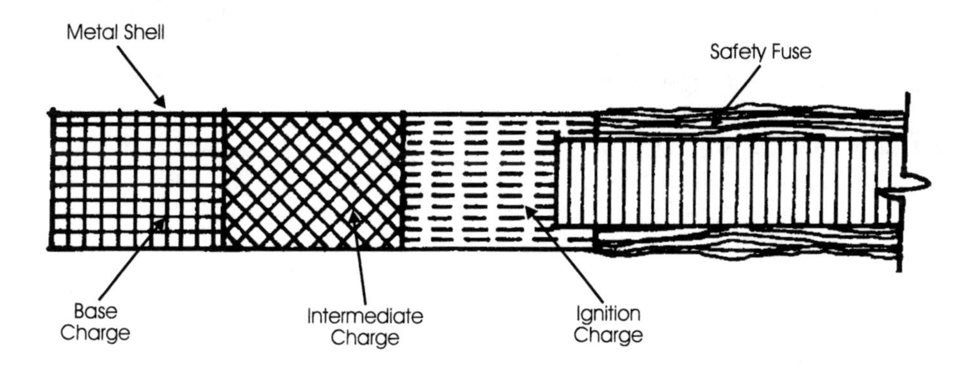

Figure 4-9. Cross Section of a Non-Electric Blasting Cap

Safety Fuses

Safety fuses are used to provide a time delay in order to detonate the main explosive charge from a safe distance. Safety fuses are made of black powder and potassium nitrate. The exterior is covered to protect against abrasion and water penetration.

Safety fuses burn at a rate of 90 to 120 seconds per yard (0.9 meter). In the United States the 120 seconds per yard (0.9 meter) material is standard.

Igniter cord is used if a series of fuses must be lit in sequence. This cord burns with a short, hot flame and ensures quick ignition. Burning rates for igniter cord vary from 8 to 20 seconds per foot (0.3 meter).

Notes

Detonating Cord

Detonating cord is a flexible cord that contains a high explosive in the core. *Be careful* not to confuse this cord with a safety fuse.

The core of the detonating cord is usually pentaerythritol tetranitrate (PETN). When PETN is used, the detonating cord will explode at the rate of 21,000 feet (6,300 meters) per second. This is a great deal faster than the 120 seconds per yard (0.9 meter) for safety fuses. In fact, if a line of detonating cord were strung from New York to California, it would take only about 15 minutes for an explosion to travel from coast to coast. A blasting cap is used to initiate the explosive train.

Shock Tube

Shock tube is another initiating device that uses PETN. However, in this application the plastic tube is coated with PETN and presents much less hazard than detonating cord. Shock tube operates by striking a plunger-like device that causes the coated plastic tube to transmit the detonating wave.

Delay Connector

When detonating cord is used, the main charge explosion is sometimes delayed so the explosion can be sequenced. The required delay is accomplished by inserting a connector in the line. Delays can range from 5 milliseconds to 25 milliseconds.

Boosters

Boosters contain a high explosive and come in varied configurations. Boosters insure primary charge detonation by increasing the intensity of the initiating charge.

TYPES OF EXPLOSIONS

There are three general types of explosions:

- mechanical
- nuclear
- chemical

Explosions vary in size and intensity depending on the amount and type of explosive used, the circumstances, and the type of explosion.

The most common type of explosion encountered is a ***mechanical explosion***. A mechanical explosion occurs when a vessel's internal pressure exceeds its ability to hold the pressure. This type of explosion commonly occurs in boilers and water heaters.

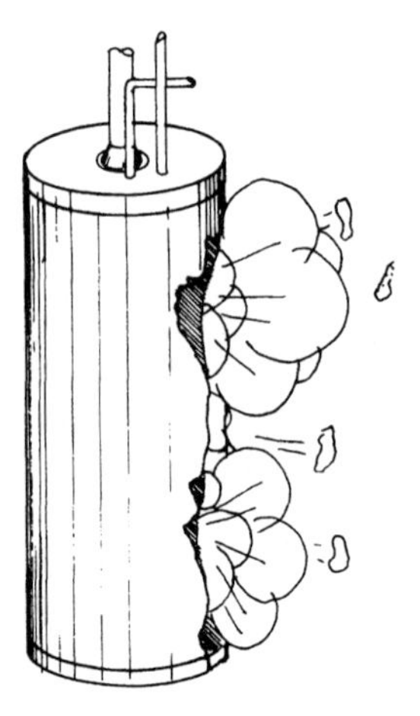

Figure 4-10
Mechanical Explosion

Notes

Chemical Explosion (Dynamite)

In a ***chemical explosion*** there is a heat and chemical reaction. A chemical explosion will occur if such a reaction is confined after initiation. An example is a Division 1.1 explosive such as dynamite.

The third type of explosion is the ***nuclear explosion***. A nuclear explosion can be generated by either fission or fusion. ***Fission*** is initiated when an atom's nucleus is split. ***Fusion*** is initiated when lighter nuclei are fused, forming an element with a heavy nucleus. Nuclear reactions produce great energy which creates the explosion.

Emergency responders usually will not know the type of explosion in an explosives incident, since most explosions travel at a minimum speed of 3,300 feet per second. This is much faster than a person can run!

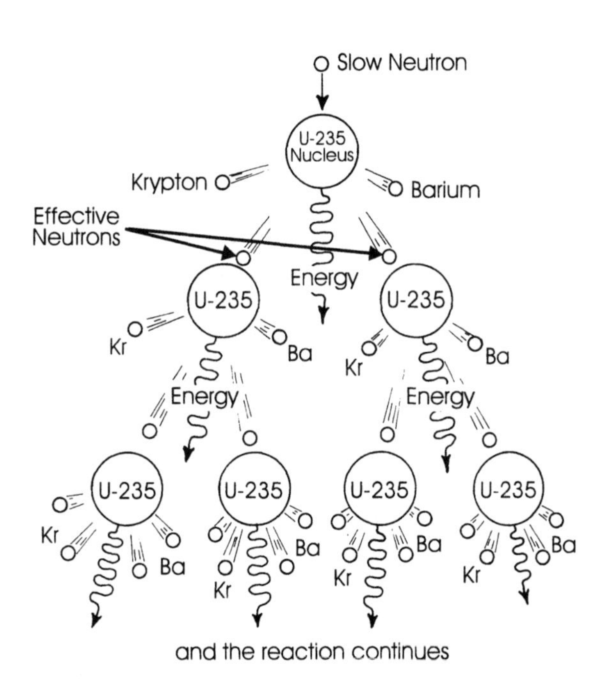

Figure 4-11. Nuclear Fission Explosion

Notes

EFFECTS OF AN EXPLOSION

An explosion has three primary or predictable phases:

- blast phase
- fragmentation phase
- thermal effect phase

The severity of the explosion depends on the type of explosive, amount of product involved, and conditions of the explosion.

Blast Phase

The first phase of an explosion is the ***blast phase***. In this phase, the rapid movement of confined combustion creates a pressure blast wave as high as 700 tons per square inch. The tremendous power of an explosion is evident when compared to a severe windstorm. A heavy wind can cause destruction to buildings, yet the wind loading criteria for many buildings is only 1 pound per square inch. The blast wave of an explosion travels outward like ripples created by a pebble dropped into water. The blast wave may travel as fast as 13,000 miles per hour. The initial blast of outward pressure is often called the ***positive phase*** of the blast. The explosive pressure can tear most objects in its path into pieces.

The second effect of the blast phase is the ***negative phase***, which is the opposite of the positive phase. The negative phase occurs when pressure from the positive phase extends outward from the explosion. This creates a vacuum in the center of the blast. This negative pressure does not last long and is quickly backfilled to the center of the explosion by outside pressure after the blast subsides. This refilling is often called ***implosion***. Implosion is the rapid equalizing of the surrounding air pressure. Loose materials can be sucked into the original blast area. Implosion may last three times longer than the actual explosion.

Fragmentation Phase

The second phase of an explosion is the ***fragmentation phase***. In this phase the container of the exploding material is destroyed. Nearby objects are torn into pieces and hurled outward. Many of these fragments will be propelled at speeds faster than bullets. As a result, injuries received from an explosion are often shrapnel type wounds.

Thermal Effect Phase

The final phase of an explosion is the ***thermal effect phase***. Thermal effect is the release of heat generated from the blast. The rate of heat release and the temperature depend on the type of explosion, the type of material involved, the distance from the blast, and the speed of the explosion.

COMMON SHIPPING CONTAINERS

Notes

Explosives are generally shipped in fiberboard boxes, many of which are treated with a wax coating to prevent contamination. Storage methods depend on use, location, and quantity. Be aware of explosive storage facilities in your area so response to an emergency will be correct! These facilities must be properly identified on run book maps. In addition, exterior storage building doors should have NFPA 704 markings.

Explosive storage facilities must be properly identified on run book maps.

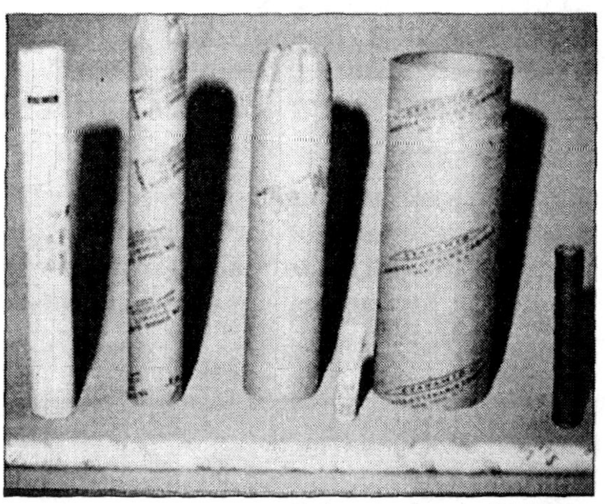

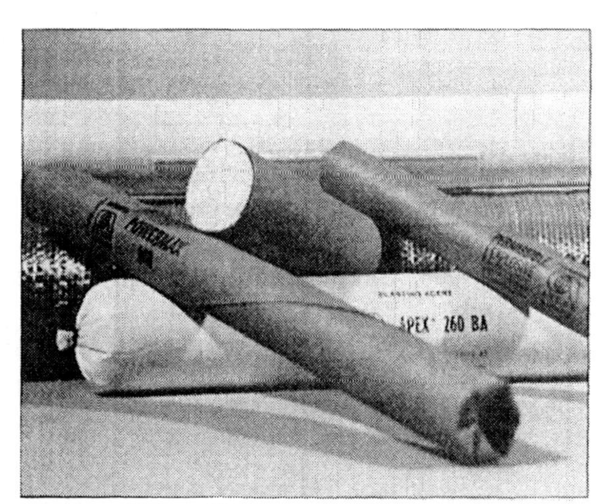

Photographs courtesy of ICI Explosives and Institute of Makers of Explosives

Dynamite is packaged in paper or fiberboard cartridges that range in diameter from 7/8 inch to 10 inches. The smaller sized cartridges are packed in fiberboard boxes for shipment. The boxes weigh approximately 50 pounds and are 1 cubic foot in volume. Large sized cartridges of dynamite are packed in heavy fiberboard tubes. These tubes meet DOT specifications and do not have to be packed in fiberboard boxes when shipped. Instead, they may be taped together in bundles weighing 50-60 pounds, or they may be shipped as a single unit.

Fuse and electric blasting caps are packed in fiberboard boxes. They are shipped in fiberboard cases. A typical blasting cap, attached cord, and shipping box are shown to the right.

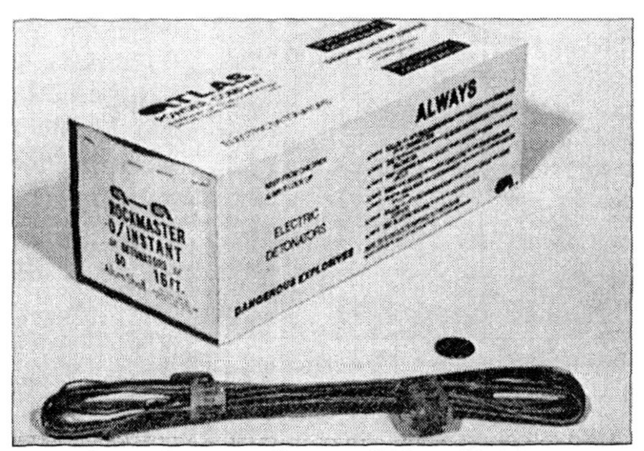

Photograph courtesy of Institute of Makers of Explosives

Notes

Photograph courtesy of Institute of Makers of Explosives

Water gels, slurries, and emulsions are packaged in plastic tubes or paper cartridges. The smaller cartridges are shipped in fiberboard boxes weighing 50-60 pounds. Large diameter cartridges are packaged in thick-walled, flexible plastic tubing which may be encased by a woven plastic sleeve or fiberboard tube.

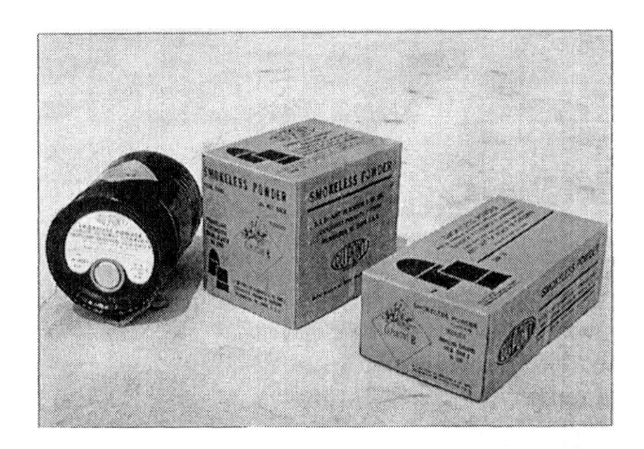

Photographs courtesy of Institute of Makers of Explosives

Smokeless powder and black powder, used for small arms ammunition, are often packaged in 1 pound metal cans. They are shipped 50 cans at a time in fiberboard cases. Black powder can also be contained in two 25 pound plastic bags and shipped in 50 pound metal kegs or 50 pound fiberboard cases.

Notes

Bulk shipments of some blasting agents, such as ***ammonium nitrate fuel oil (ANFO)***, are packed in 50 pound layered paper bags similar to cement bags. The bags have a moisture resistant plastic liner and are usually sewn at the top.

Photograph courtesy of Austin Powder Company

Detonating cord (not pictured) is normally packed in fiberboard cases on spools containing 1,000 feet of cord.

Special cargo trucks are used to ship blasting agents such as water gels, slurries, certain emulsions, and ANFO.

Photograph courtesy of Institute of Tread Corporation

Notes

If explosive materials are involved in a fire, the incident commander must immediately evacuate the area.

RESPONDING TO EXPLOSIVE EMERGENCIES

INCIDENT RESPONSE

There are a few cardinal rules when dealing with hazardous materials. One of these is the following simple rule that pertains to serious explosive incidents:

If the storage area or cargo compartment for explosive materials is involved in a fire, directly or indirectly, the incident commander must immediately evacuate the area.

These incidents are considered losers. Responders must understand the destructive potential of explosive materials.

A test conducted on May 20, 1981, by the Seattle, Washington Fire Department demonstrates the danger of materials that are often considered safe. After confiscating 5,600 pounds of illegal fireworks, the department decided to see how these materials would react to fire. A 20 foot long metal container was placed at a test site. The 5,600 pounds of fireworks were stored inside the container. One hundred twenty wooden pallets, similar to the ones used for transporting such materials, were placed five feet from the side of the container and then ignited.

It took only 2-1/2 minutes for the container to rupture after the pallets were ignited. Flaming fireworks and part of the container flew 150-200 feet into the air and 500-700 feet horizontally. In many cases, firefighters, law enforcement officers, and EMS personnel would have been within the 500 feet noted in the test.

HAZ-MAT — HAZARD IDENTIFICATION

The product and its hazards must be identified before a safe and effective action plan can be developed. Use the following steps in hazard identification:

- ***Preplanning*** for the explosive incident is critical. Know where explosives are stored and used in your community and the common transportation routes. Plan for water supply, evacuation, and life hazards.

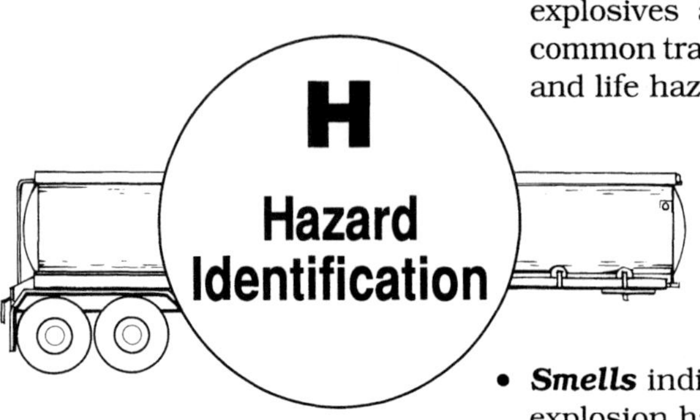

- The ***incident location*** should indicate the type of explosive, amount of explosive, and container involved.
- Note ***markings or identification*** such as placards, NFPA 704 markings, container shapes, and shipping papers.
- ***Smells*** indicate if an explosion has occurred. Remember, if one explosion has occurred, another is likely.
- ***Lightheadedness*** or ***rush*** may result from contact with nitro and nitrated explosive compounds.

Once the product has been identified, determine the possible hazards. If an explosion has not occurred, assess if conditions are right for an explosion. Consider the following:

Notes

- incident location
  - building
  - open field
  - highway
  - trailer
  - fixed site facility
- location of the explosive material
- cause of the incident
  - traffic accident
  - fire
  - train accident
- injuries
  - number
  - severity

HAZ-MAT — ACTION PLAN

If **NO FIRE** is involved, take action to:

- control ignition sources
- prevent product contamination
- control additional movement of the product
- establish a 500 foot evacuation corridor
- transfer and dispose of the load
- obtain assistance if necessary

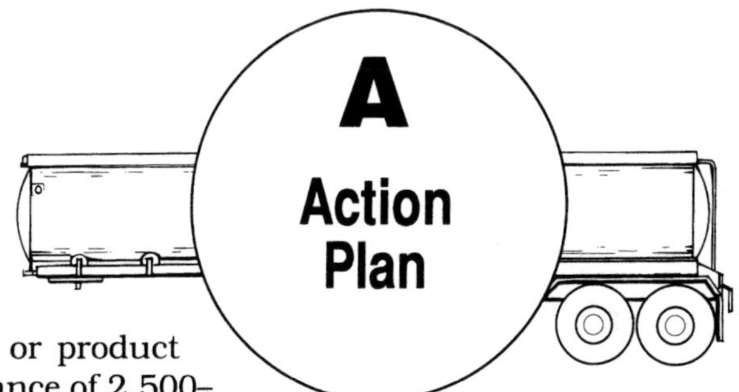

If the incident becomes unstable due to fire, leak, or product contamination, initiate a large scale evacuation to a distance of 2,500–5,000 feet in all directions, depending on the type and quantity of explosives. This area becomes the hot zone.

Never use mobile radio transmitters near electrical blasting caps, as accidental ignition can occur. An electrical energy field created in the air can trigger the cap and detonate the charge. Many agencies turn off their radios at least two blocks from a suspected explosive incident.

If there is a **FIRE that does not involve the product**, take action to:

- protect the exposed cargo (see Figure 4-12 for approximate detonation temperatures for explosives)
- extinguish the fire
- stabilize the cargo since many explosives are shock sensitive
- contain the runoff water for contaminate sampling
- consider large scale evacuation

Notes

Approximate Detonation Temperatures for Explosives

Explosive	Temperature °F	Temperature °C
Lead azide	482	250
TNT	464	240
Nitroglycerin	419	215
PETN	401	205
Mercury fulminate	349	176

Source: *Sax's*

Figure 4-12

If there is a **FIRE involved within the product/cargo area**, take action to:

- immediately withdraw emergency response personnel
- evacuate a distance of 2,500–5,000 feet in all directions to establish a hot zone
- let the product burn, remembering that an explosion is likely

Always evacuate when fire occurs in the product/cargo area. These incidents are real losers! Do not attempt to extinguish fires that involve the product at a fixed storage facility or the cargo area of a truck that carries explosives.

HAZ-MAT — ZONING

Establishing control zones and evacuation procedures should be a first consideration when dealing with explosives. The control zone distances should be based on:

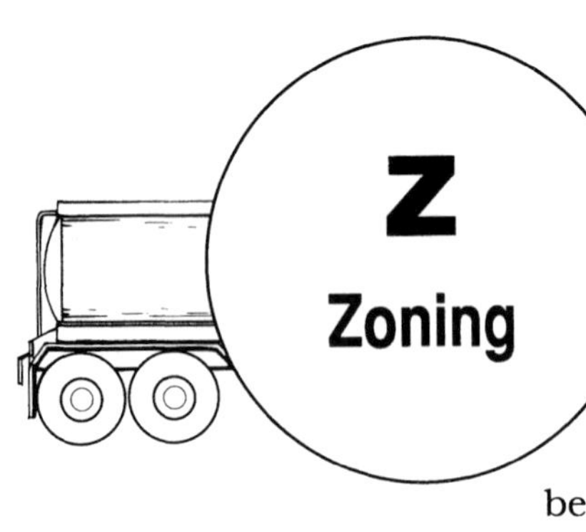

- amount of explosives
- location of the emergency
- available evacuation time

The best option is to evacuate quickly and as far away as possible. Law enforcement agencies are generally best used in securing the scene and in evacuation. *Before* an incident occurs, know which agency will conduct evacuation, how it will be done, and what resources are available.

Notes

HAZ-MAT — MANAGING THE INCIDENT

Managing explosive emergencies is difficult. If the action plan indicates extinguishing a fire that is not in the cargo area, anticipate:

- the need for additional firefighting resources, such as water supply, police department, and EMS assistance
- initiating a protective action zone of 500 feet; extend to 2,500 feet if necessary
- establishing control zones and a command post
- developing secondary operational plans if the fire is not rapidly extinguished

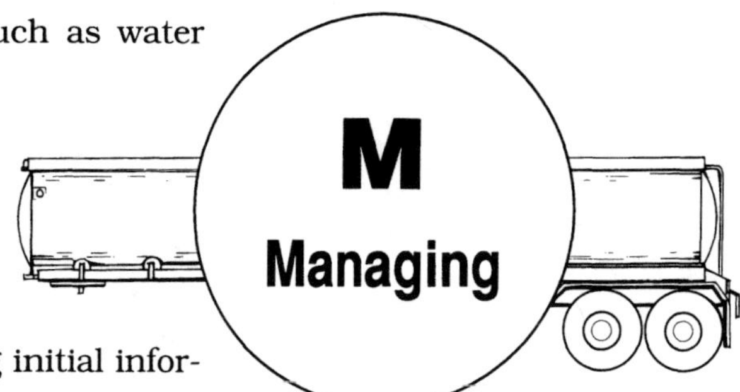

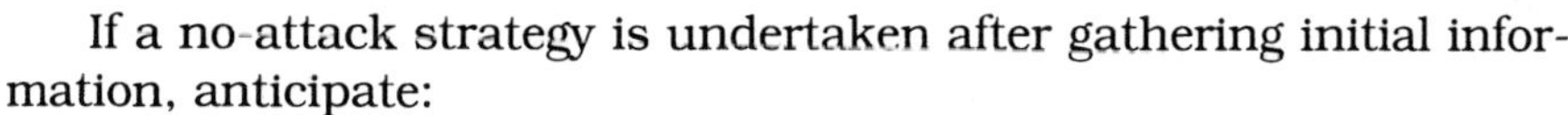

If a no-attack strategy is undertaken after gathering initial information, anticipate:

- the need for additional firefighting resources, such as water supply, police department, and EMS assistance
- initiating a hot zone of 2,500 feet
- developing secondary operational plans if no explosion occurs

Post-Explosion Scene

The post-explosion scene can be as hazardous as the pre-explosion scene, if not more so. It is likely that not all of the product was destroyed in the initial explosion. Pieces of undetonated explosives may be scattered in the blast zone. These explosives will be more sensitive because they were exposed to heat and shock from the initial explosion. If an explosion has occurred prior to your arrival, anticipate:

- a second explosion
- the need for additional firefighting resources, such as water supply, police department, and EMS assistance
- initiating a hot zone of 2,500 feet
- developing a secondary operational plan
- rescuing multiple trauma victims

Rescue Considerations

Treating injured parties after an explosion is similar to working under war conditions. Materials projected at high velocities cause serious fragmentation injuries that are subject to infection. The flame front and sharp pieces of debris such as wood and metal can travel at a rate of 1/3 mile to 6 miles per second. Thermal burns occur from the blast's explosive heat.

The primary responsibility during such incidents is to provide emergency rescue services, stabilize the incident, and have the largest number of survivors possible. However, *safety is paramount.* Immediately assess the situation to decide if personnel can be safely involved. Then establish a triage to sort patients. An assessment should always be completed, even if an incident involves only a small number of people and does not require complex EMS resources.

Notes

Assessment questions include the following:

- Where did the explosion occur? Major explosions do not usually occur in schools, homes, and shopping centers. Explosions do occur on construction sites, in mining areas, and on military bases. When responding, determine possible locations of explosives. Determine the building's occupancy, time of day, and other factors that may indicate what led up to the explosion.
- What type of explosion was it? This will determine what kinds of injuries are likely to occur.
- Is there a chance of another explosion? Have gas lines been ruptured, or are additional chemicals involved? Could another explosive device be set or activated?
- Is the building, vehicle, or area safe? Examine the entire area before committing personnel and equipment to the scene. Smoke from these incidents can be very toxic.

Once it is assessed that it is safe to respond, determine how patients will be treated, and request additional assistance. One serious patient may require as many as five rescuers. Quickly ask the following questions:

- Who is in charge? Every response agency will have a different incident command structure. *A unified command structure must be used during a multi-agency response.*
- What is the most positive approach I can take? Every responder can have a positive effect. Police and EMS providers may not be able to fight fires, but they can assist in other areas.
- Do I have enough help? Get plenty of help early. They can always be released.

Follow your state's first aid treatment guidelines and consider state certifications when providing care to patients. Remember that organized response, triage, and treatment will provide the most effective level of care for patients and response personnel alike.

General Treatment for Explosive Material Exposures

Products .... TNT, picric acid, black powder, dynamite.

Containers .... Bottles, cloth or paper bags, boxes, wooden kegs.

Life Hazard .... The primary life hazard is multi-systems trauma injuries due to the blast. Shrapnel, pressure, and thermal injuries should be anticipated. Many explosives may also cause irritation to the skin and eyes or may be highly toxic through ingestion, inhalation, or skin absorption.

Signs/Symptoms ..... The patient may exhibit respiratory distress and cyanosis. Signs of shock, an increased heartbeat, or cardiac arrhythmias may be present. The patient may have a decreased level of consciousness with headache, dizziness, drowsiness, and coma. Tinnitus may be present. In rare cases of ingestion, there may be nausea, vomiting, and/or diarrhea. Skin and eye irritation are possible. Some products may interfere with the ability of the blood to carry oxygen.

Basic Life Support .... Wear proper protective equipment and remove victim from contaminated area. Administer oxygen and remove patient's clothing. Brush or blot away any visible product and decontaminate with copious amounts of soap and water. Irrigate eyes thoroughly and continue during transport as necessary. If ingested, dilute with small quantities of water and give activated charcoal (follow advice of Poison Control and local protocols). Do not orally administer anything if the patient has a decreased level of consciousness or decreased gag reflex. Watch for shock and respiratory problems. Remember to assess for multi-system trauma and treat as appropriate.

Advanced Life Support .... Assure an adequate airway and assist ventilations as necessary. Start an IV with LR. Monitor for cardiac arrhythmias and treat as necessary. Treat signs of hypovolemic shock with IV fluids as necessary, but watch for signs of fluid overload. Administer topical anesthetic to eyes for easier irrigation and to reduce pain. Follow local protocols for all drug therapy and medical treatment.

Other Information .... A **minimum** safe distance from explosive materials incidents is 2,500 feet. Be prepared to treat multiple trauma injuries. Use caution, as secondary explosions may occur.

Notes

HAZ-MAT — ASSISTANCE

At an explosive emergency, always request immediate assistance from the local bomb specialist or regional ordnance disposal team. Even after the fire has been extinguished or an explosion has occurred, additional material may need proper disposal. Assume a hands-off approach. Never touch any suspected explosive material unless directed by an ordnance specialist.

Notes

HAZ-MAT — TERMINATION

Terminating the explosive emergency will be similar to other haz-mat responses. Use the critical termination aspects discussed in Chapter 3.

Decontamination

Decontaminating victims varies with each explosive incident. Decontaminating specific bodily injury sites is the usual procedure, as opposed to total body decontamination. Injury decontamination should be done by either EMS or hospital providers. Communication with EMS people prior to and during the incident assures effective patient decontamination.

Rehabilitation and Medical Screening

Once the explosive incident has been mitigated and the area declared safe, personnel rehabilitation can begin. This is an excellent time to determine if emergency responders can be reassigned to other duties. A complete medical screening should be administered, including injury evaluation. Vital signs should be taken once upon arrival at the rehabilitation station and again prior to being released. All information should be accurately recorded on each responder's permanent record.

Critical Incident Debriefing

An explosive emergency is one of the most stressful types of haz-mat incidents. Therefore, it is important to use the termination phase to evaluate personnel for post-incident stress. An initial incident debriefing can indicate if responders show signs of post-incident stress. If signs are present, a formal critical incident debriefing should be conducted as soon as possible. If necessary, responders should be given follow-up counseling.

Post-Incident Analysis

Explosive emergencies are not common. Therefore, it is essential to conduct a post-incident analysis with all involved parties to determine problems, lessons learned, and changes needed before the next emergency. The post-incident analysis provides an opportunity to respond more effectively during the next explosive emergency.

KEY POINTS TO REVIEW

Explosive materials come in many forms. Only those categorized by the DOT will be placarded as ***explosives***. Explosive materials are dangerous, can destroy large areas, and injure or kill everyone in the immediate area. Responding to explosive emergencies is emotional because the natural inclination is to want to do something. *Unfortunately, these emergencies are generally losers.* Remember the following important points:

- Explosives are made to blow up with no regard to location or persons in the way.

Notes

- The four primary elements for an explosion to occur are:
  - fuel
  - oxidizer
  - energy source
  - confinement
- An explosion travels very fast.
- Not all explosives will be placarded or labeled "Explosive."
- The blast effect from an explosion has both a positive and a negative pressure factor.
- There are mechanical, chemical, and nuclear explosions.

DEDICATION

Managing an incident involving explosives can be an emotional and difficult task. During this type of emergency, risk assessment is difficult at best. The basis for responders' jobs, training, equipment, and policies is to save lives. Seldom in your career will you be required to take no action to save lives. The explosives emergency, however, may be such a time.

We have talked about the hazards of explosives in this chapter. In closing, let us look at an explosives incident that required an incident commander to make difficult decisions under extreme circumstances.

On November 29, 1988, firefighters from Kansas City, Missouri, responded to a pickup truck fire at 3:43 a.m. Upon arrival, firefighters saw another fire at a construction site and requested police department assistance and another pumper. After extinguishing the pickup fire, the first engine went to assist the second at the construction site.

After hearing the dispatcher's report that explosives were on the site, Battalion Chief Marion Gerrman attempted to pull back the first responding units. Unexpectedly, a massive explosion rocked the scene. Chief Gerrman was several hundred feet away, yet his car was destroyed. Fortunately, neither he nor his driver was seriously injured.

Although Chief Gerrman called for additional resources, he quickly realized that six of his fellow firefighters were probably killed in the explosion. He made the difficult decision to take no further action until the area could be secured. This difficulty was compounded when other responding firefighters wanted to aid their fallen brothers. The wisdom of Chief Gerrman's decision was evident when a second major explosion occurred about 40 minutes later.

The decision to take no further action was a difficult, but correct choice. All emergency responders who face the hazards of explosives can learn from this tragedy. Unfortunately, this same type of incident could happen in most communities throughout the United States and Canada.

The fire service was shocked and saddened by the loss of these six firefighters. They were experienced men who cared for the people they served. It is to them and their families that this chapter is dedicated.

Chapter 5

GAS EMERGENCIES

OBJECTIVES

After studying the material in this chapter, you will be able to:

- define the terms flammable gas, compressed gas, gas poisonous by inhalation, liquefied gas, and cryogenic liquid
- state how and why gases are compressed into cylinders for commercial use
- explain the basic physical and chemical properties of gases
- summarize the placarding and labeling requirements for gases
- describe the various types and sizes of shipping containers and fixed site storage vessels for gases
- describe the basic safety features and relief valves found on tanks and cylinders of compressed, liquefied, and cryogenic gases
- describe how and why BLEVEs occur
- state the general characteristics and hazards of cryogenics
- summarize basic considerations in responding to the three categories of gas emergencies, using the acronym HAZ-MAT

UNDERSTANDING COMPRESSED AND LIQUEFIED GASES

INTRODUCTION

One of the most common groups of hazardous materials that emergency response personnel encounter is gases. Gases are used in a variety of industrial, commercial, and residential applications. The use of gases has risen dramatically during the last 15 years, increasing the risk of emergency incidents involving these materials.

Gases are normally grouped into three basic types:

- compressed
- liquefied
- cryogenic

Notes

The three types are different from each other and require different handling in the event of a leak or fire. It is important for responders to understand gases, because these substances occur in several hazard classes. Specifically, a gas can be a flammable gas (DOT hazard class 2.1); a non-flammable, non-poisonous compressed gas (hazard class 2.2); or a gas poisonous by inhalation (hazard class 2.3). However, these gases sometimes pose other hazards, such as oxidizing and corrosive potential.

DEFINITIONS

Flammable Gas (DOT Hazard Class 2.1)

The Department of Transportation (DOT) defines a ***flammable gas*** as:

> Any material which is a gas at 68°F (20°C) or less and 14.7 psi (101.3 kPa) of pressure which (1) is ignitable at 14.7 psi (101.3 kPa) when in a mixture of 13% or less by volume with air, or (2) has a flammable range at 14.7 psi (101.3 kPa) with air of at least 12% regardless of the lower limit (49 CFR 173.115(a)).

A gas must meet the DOT definition to be classified as flammable. It is possible that a gas may burn without being classified as flammable. A common example of such a gas is ammonia (see Figure 5-1). Ammonia ranks third in volume of chemicals produced in the United States. It has an ignition temperature of 1,204°F (651°C) and a flammable range from 16% to 25% in air. Because its lower flammable limit is 16% and because its flammable range is only 9 percentage points (16–25%), ammonia does not meet the DOT definition for a flammable gas.

Ammonia is usually classified as a compressed gas when shipped in the U.S. The problem with this classification system became apparent when a Shreveport, Louisiana HMRT member was killed in a flash fire involving an ammonia leak in a cold storage facility.

Compressed Gas (DOT Hazard Class 2.2)

The DOT defines a ***non-flammable, non-poisonous compressed gas*** as:

> Any material or mixture which (1) exerts in the packaging an absolute pressure of 41 psia (280 kPa) or greater at 68°F (20°C), and (2) does not meet the definition of a flammable gas or a gas poisonous by inhalation. This category includes compressed gas, liquefied gas, pressurized cryogenic gas, and compressed gas in solution (49 CFR 173.115(b)).

Many substances fitting this definition are not required to be placarded as a compressed gas. They may, however, exhibit many of the same properties as a compressed gas. The handling of compressed gases varies with the product involved, the vessel, and the form in which the gas is transported, used, or stored.

Notes

Toxic and Hazardous Properties of Ammonia		
%	ppm	Hazards
100	1,000,000	
25	250,000	Flammable Range (16–25%)
16	160,000	
3	30,000	Skin Stinging Sensation
0.5	5,000	LC_{50} – Human Respiratory Spasm
0.07	700	Blindness
0.03	300	IDLH
0.015	150	Eye Irritation
0.0035	35	STEL, PEL (OSHA)
0.0025	25	TWA
0.002	20	Threshold for Odor Detection
0.0005	5	
0	0	

Source: *Sax's*, NIOSH, CGA

Figure 5-1

Gas Poisonous by Inhalation (DOT Hazard Class 2.3)

A ***gas poisonous by inhalation*** is defined as:

> A material which is a gas at 68°F (20°C) or less and a pressure of 14.7 psi (101.3 kPa) and which (1) is known to be so toxic to humans as to pose a hazard to health during transportation, or (2) in the absence of adequate data on human toxicity, is presumed to be toxic to humans based on tests conducted on laboratory animals (49 CFR 173.115(c)).

Liquefied Compressed Gas

A ***liquefied compressed gas*** is defined as:

> A gas which, in a packaging under the charged pressure, is partially liquid at 68°F (20°C) (49 CFR 173.115(e)).

Notes

Liquefaction of many commonly used gases has become a cost effective method of transporting and storage. Any gas can be liquefied by:

- cooling the temperature enough to allow the attractive forces of the molecules or atoms to become more effective; once compressed, the gas becomes a liquid
- increasing the pressure on the gas

Cryogenic Liquid

A ***cryogenic liquid*** is defined as:

> A refrigerated liquefied gas having a boiling point colder than −130°F (−90°C) at 14.7 psi absolute (101.3 kPa) (49 CFR 173.115(g)).

HOW GAS IS COMPRESSED

A simplified explanation of how a gas is compressed follows. The product is first placed in a pressure vessel to keep it from mixing with air. Piping is used to move the product from the container to a compressor. The compressor forces the molecules of the gas closer together. The compressed molecules are piped into another container. This new container is strong enough to handle the increased pressure of the compressed molecules. The degree to which a gas can be compressed depends upon the product itself and the strength of the container. Some products can be so tightly pressurized that they become a liquid in the tank. Emergency responders must consider a product's form when developing a sound action plan.

USES OF COMPRESSED GASES

Compressed gases are used in a variety of applications. Firefighters carry compressed air to breathe during emergencies. Oxygen is commonly administered during various medical emergencies. Other professions also routinely use compressed gases. Dentists use nitrous oxide to make patients feel good before the real pain begins.

Emergency responders most commonly deal with flammable gases, because these products are extensively used in heating and manufacturing. Many natural and synthetic substances can be used in gaseous form. Many gases are compressed to provide sufficient amounts of the product for a particular use. If gas molecules were not compressed into a smaller space, cylinders would hold only enough product for a short time. An operator would have to change cylinders frequently or would need a very large cylinder to complete a given task.

PHYSICAL AND CHEMICAL PROPERTIES OF GASES

Of the three states of matter, only the gaseous state has a comparatively simple definition. A ***gas*** is a state of matter that does not have a definite shape or volume. The basic scientific laws affecting the relationship between the volume of a confined gas and its applied temperature and pressure will be discussed in this section. Emergency response personnel need a general knowledge of these laws to understand what can happen during a haz-mat emergency involving gases.

Temperature/Pressure/Volume Relationships

Pressure is created when rapidly moving gas molecules inside a container impact the container's walls. The number of impacts over a given period of time determines the pressure.

For example, assume that under normal circumstances it takes two seconds for a molecule to move from one wall of a tank to the other. If this tank is in a rollover accident, the walls of the tank may become severely bent inward. The molecules have the same velocity, but now it may take only one second for them to move from one side of the damaged tank to the other. This means the molecules experience twice as many impacts on the tank walls, which doubles the pressure inside the tank.

Notes

The number of molecular impacts over time is equal to pressure.

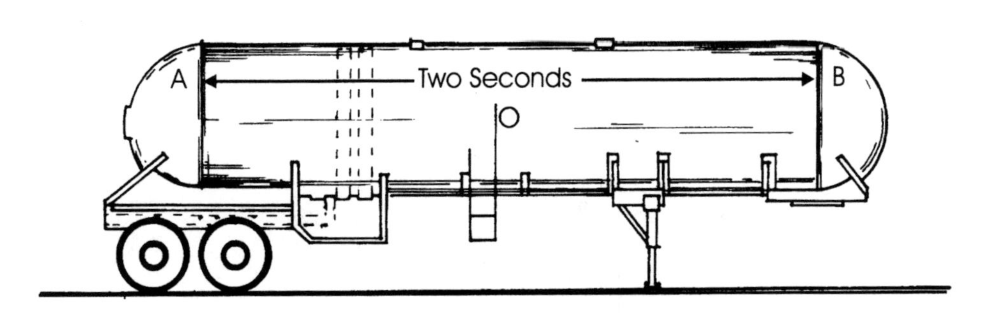

Figure 5-2. Normal Pressure Inside a Tank

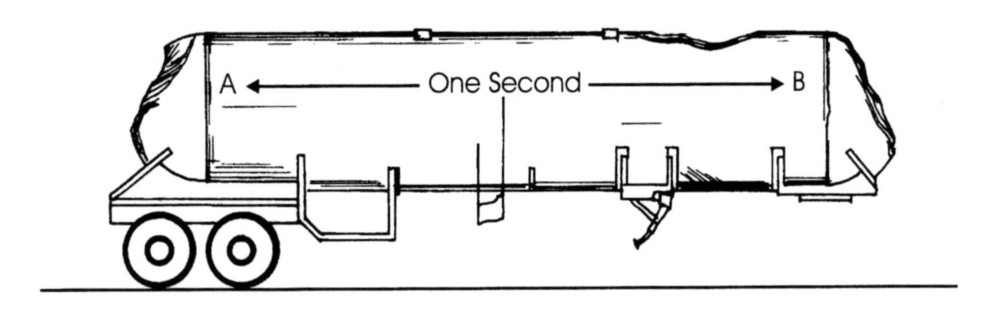

Figure 5-3. Pressure Inside a Damaged Tank

This principle was observed in 1660 by Robert Boyle, an Irish physicist. His observation is known as ***Boyle's Law***, which states:

> When the temperature and mass of a gas are kept constant, the product of the pressure and the volume is equal to a constant.

Boyle's Law can be expressed mathematically as $p \times V = constant$ at constant temperature and constant mass, where:

p = absolute pressure
V = volume

Notes

Another way to write this relationship is $p_1 \times V_1 = p_2 \times V_2$ at constant temperature, where:

p_1 = absolute pressure of a gas under one set of conditions
V_1 = volume under first set of conditions
p_2 = absolute pressure of the same gas under a second set of conditions
V_2 = volume under the second set of conditions

Pressure/Temperature Relationship

An increase in temperature increases the pressure of a confined gas.

As the temperature of a gas increases, there is a corresponding increase in molecular energy. Remember, molecular energy increases the velocity of the molecules. As the velocity increases, the number of impacts over a given time also increases. Therefore, an increase in temperature increases the pressure of a confined gas. This is an important concept for an emergency responder who finds a pressurized vessel impinged by fire.

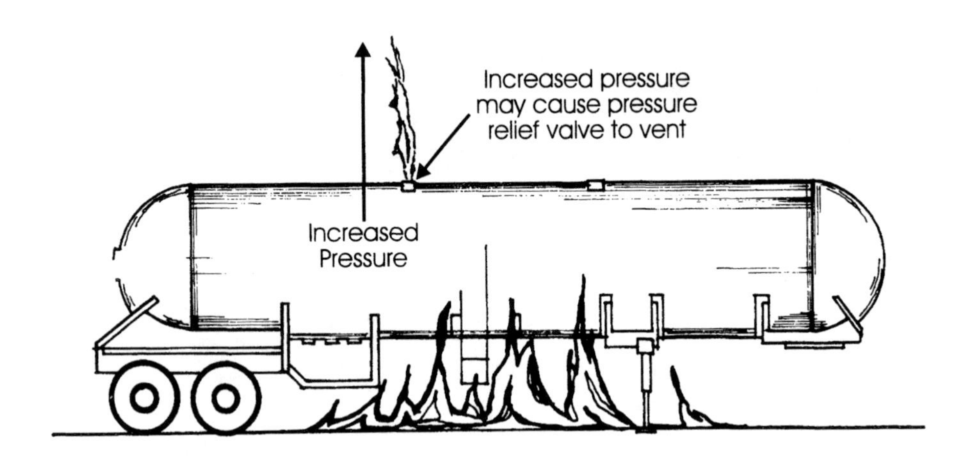

Figure 5-4. Effect of Impinging Fire on Tank Pressure

This pressure/temperature relationship is known as ***Charles' Law*** and can be stated as follows:

> If the volume of a gas is kept constant and the temperature is increased, the pressure will increase in direct proportion to the increase in absolute temperature.

Using Charles' Law, we can decrease temperature and measure the corresponding decrease in pressure. For example, if we start at 0°C and decrease the temperature to –1°C while keeping the volume constant, the pressure will decrease by 1/273 of its original value. Therefore, a compressed gas cylinder with an internal pressure of 2,200 psi at 100°C would decrease to approximately 1,610 psi at 0°C.

Vapor Pressure

When a liquid evaporates in a confined area, the vapor exerts pressure on the surroundings. When a container becomes completely saturated with vapor, ***maximum vapor pressure*** is created for that particular volume and temperature. The vapor pressures created by water at temperatures between 0°C and 100°C are shown in Figure 5-5.

Emergency response personnel need to understand that when a *liquid* is exposed to fire during a haz-mat incident, more vapor will be produced than under normal conditions. The more vapor, the greater the vapor pressure, which increases the chance of total ignition. If a *compressed or liquefied gas* is involved, a pressure relief device on the container will activate to relieve the excess pressure in the tank. If the product is flammable, it is probable that the escaping vapors will ignite and act as a blowtorch.

Over-pressurization creates another problem. A rapid increase in temperature can cause vapor pressure to increase faster than the pressure relief device is able to relieve pressure build-up. This can lead to a violent rupture of the container.

When the vapor pressure is equal to atmospheric pressure, the liquid will begin to boil. The temperature at which a liquid boils is known as the boiling point. As the atmospheric pressure changes, so does the boiling point. For example, at sea level where the atmospheric pressure is 14.7 psi (101.3 kPa), water boils at 212°F (100°C). At a higher elevation, the atmospheric pressure is lower, which causes the boiling point to be lower also. Therefore, at a high altitude accident where the container fails, the liquid will vaporize more rapidly than at sea level.

Notes

The more vapor, the greater the vapor pressure, which increases the chance of total ignition.

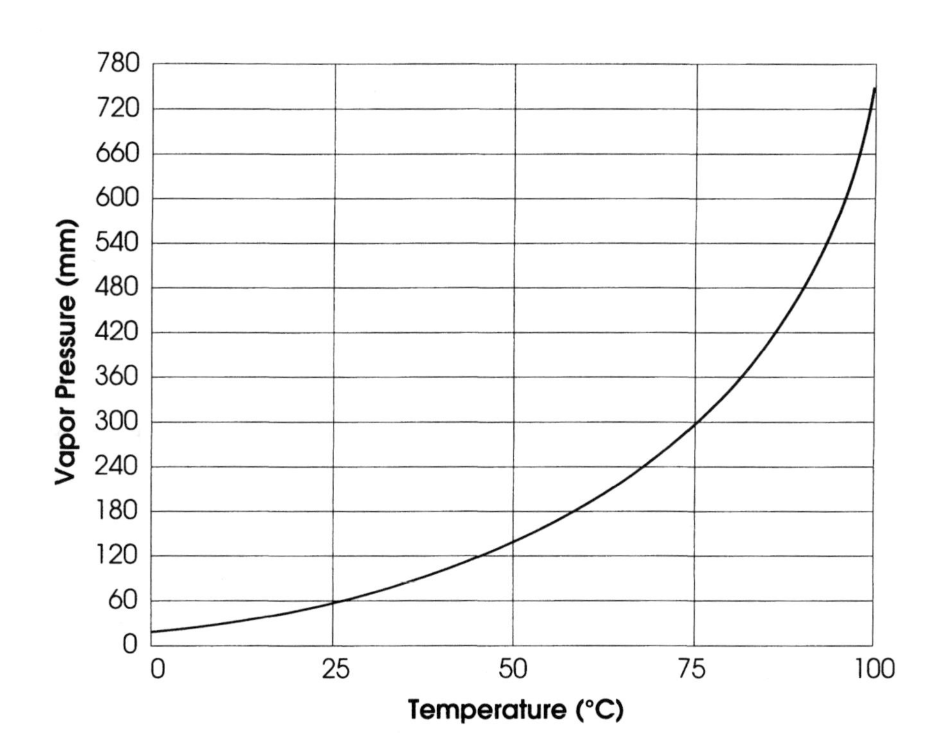

Figure 5-5. Vapor Pressure of Water

Notes

As temperature increases, so does vapor pressure.

When materials are loaded in pressurized and non-pressurized tank cars, a space is left for vapor production. This extra space is called ***outage***. A common leak in tank cars is vapor release due to overloading. This is particularly true of tank cars that are loaded in colder areas and then transported to warmer climates. Remember, as temperature increases, so does vapor pressure. Emergency responders should always consider temperature when responding to gas emergencies.

Vapor Density

Vapor Density

Vapor density refers to the weight of a gas or vapor. It compares the density of a vapor to the density of air. Air is assigned a vapor density of 1.0. A substance with a vapor density of less than 1.0 is lighter than air and will float upward into the atmosphere. A gas or vapor with a vapor density of greater than 1.0 is heavier than air and will settle into low lying areas or travel downhill. Gases that are lighter than air are listed below and can be easily remembered by the acronym HA HA MICEN. Knowing the vapor density of a gas is critical in an emergency, because vapor density affects where the gas will accumulate (i.e., in the basement or in the attic of an exposed structure) and how quickly the gas will disperse.

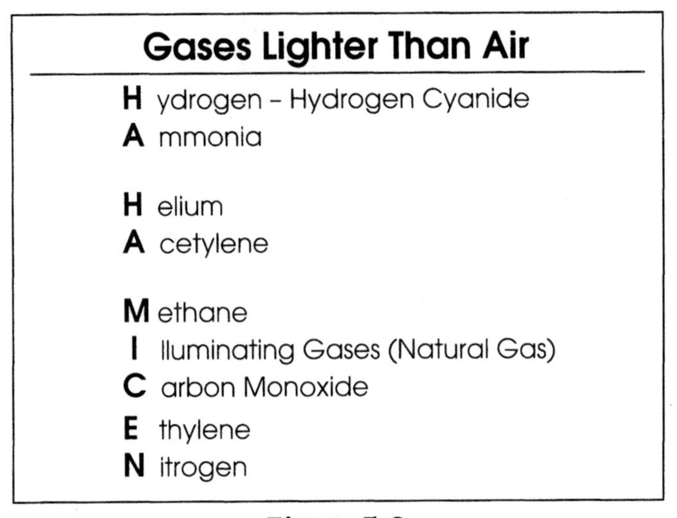

Gases Lighter Than Air

H ydrogen - Hydrogen Cyanide
A mmonia

H elium
A cetylene

M ethane
I lluminating Gases (Natural Gas)
C arbon Monoxide
E thylene
N itrogen

Figure 5-6

Boiling Point

A substance's ***boiling point*** is the temperature at which it rapidly changes to a gas, or when the rate of evaporation exceeds the rate of condensation. Responders must know that the lower the boiling point, the faster vapor is produced at a spill or fire.

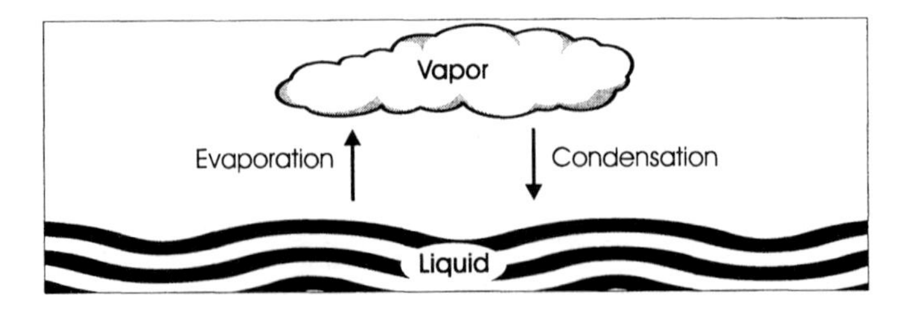

Notes

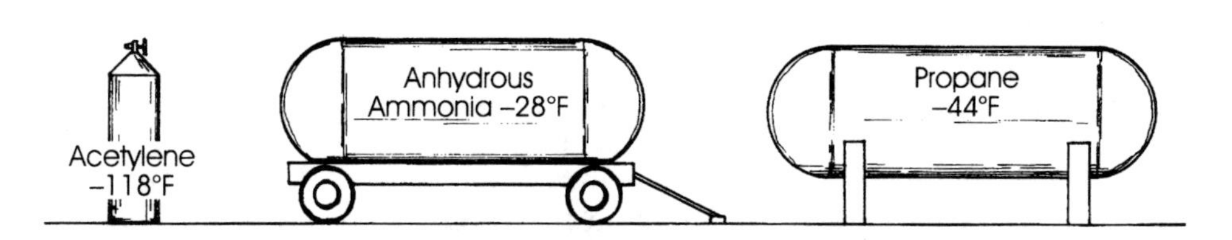

Figure 5-7. Boiling Points of Selected Gases

Water Solubility

Water solubility measures the ability of a substance to mix with water. It is important to know if a gas is water soluble. A non-soluble substance combined with water remains separate from the water. When a water soluble substance is mixed with water, a new substance is formed.

For example, when hydrogen chloride gas and water are combined, an acid is created, which in turn produces a runoff problem. Therefore, when water soluble chemicals are mixing with water, an action plan should include monitoring runoff to measure the degree of contamination.

Figure 5-8. Water Solubility of Selected Gases

Critical Point

All gases have a property known as the critical point. The ***critical point*** is the condition under which a gas can exist as a gas or a liquid. The critical point depends upon the critical temperature and the critical pressure.

Critical temperature is the maximum temperature to which a liquid or liquefied gas can be heated and still remain a liquid, regardless of the pressure exerted on it. Additional heat will cause the liquid to vaporize almost instantaneously. This is extremely important to remember if the liquid is in an enclosed container. If the liquefied gas container exceeds its critical temperature, the liquid will instantaneously convert to a gas, which may cause the container to fail violently.

Critical pressure is the maximum pressure required to liquefy a gas that has been cooled to or below its critical temperature. This means that to liquefy any gas, the gas must first be cooled to or below its critical temperature and then pressurized. Critical temperatures and critical pressures of cryogenics will be discussed later in this chapter.

Notes

Flammable Range

Flammable range means the difference between the minimum and maximum volume percentages of the material in air that forms a flammable mixture. Below the minimum volume percentage (sometimes called the lower flammable limit or LFL), the gas or vapor has too little fuel and too much air to burn. Above the maximum volume percentage (sometimes called the upper flammable limit or UFL), the gas has too much fuel and too little air. Emergency responders should become familiar with the following key points:

- Within the flammable range, a gas or vapor will burn rapidly if ignited.
- If the gas concentration is above the upper flammable limit, it has passed through the flammable range. As the gas dissipates, the concentration will once again pass through the flammable range, which poses a hazard if an ignition source is present.
- Hazards can only be determined by using appropriate monitoring equipment.

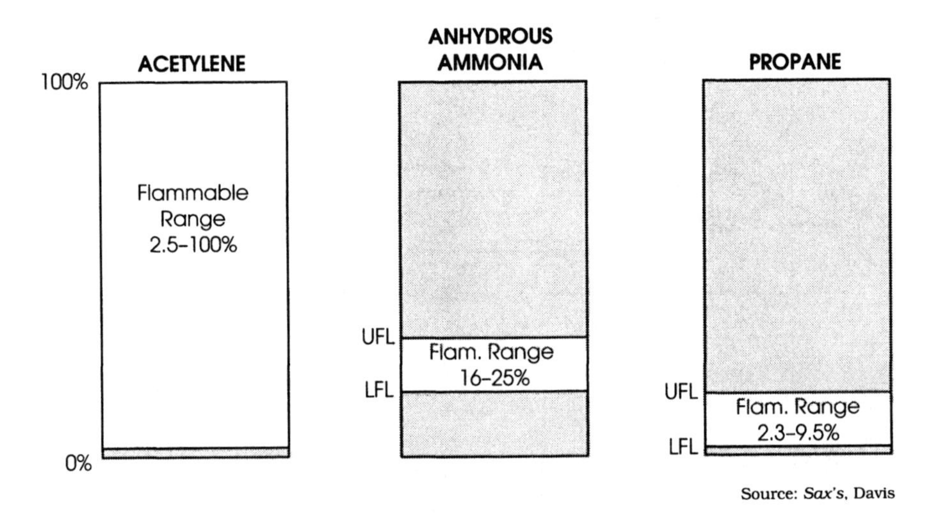

Figure 5-9. Flammable Ranges of Selected Gases

Expansion Ratio

Notes

The ***expansion ratio***, also referred to as the vapor/liquid ratio, compares the amount of gas produced when a certain volume of liquid is vaporized. Expansion ratios must be taken into consideration when developing an action plan for a gas emergency.

For example, what problems can be anticipated if a 10 gallon leak of liquid oxygen occurs at a university lab? The expansion ratio of liquid oxygen is 860 to 1. This means that the 10 gallons of liquid oxygen will produce 8,600 gallons of oxygen gas.

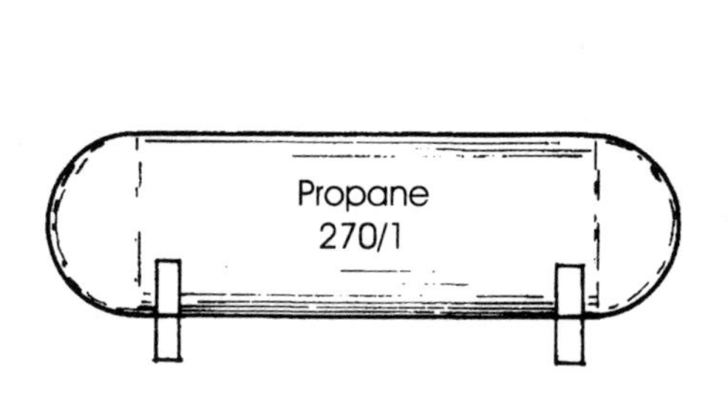

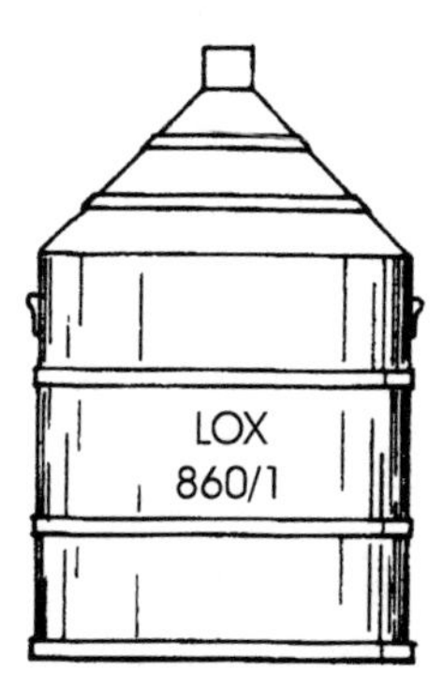

Source: CGA

Figure 5-10. Expansion Ratios of Propane and Oxygen

PLACARD AND LABEL REQUIREMENTS

Compressed and liquefied gases must be placarded according to DOT requirements. In addition, the industry follows the Compressed Gas Association (CGA) standards for identifying and labeling, as well as for storing and using compressed gases. Local gas representatives should be an excellent source of information. Any amount of a poison gas must be placarded, whether it is transported as a liquid or a gas. The required types of placards and labels for several gases are found on the following pages.

CLASS 2 PLACARD AND LABEL REQUIREMENTS

SUBSTANCE	VEHICLE PLACARD	PACKAGE LABEL	QUANTITY	SPECIAL REQUIREMENTS
Flammable Gas	**Flammable Gas** (white letters on red background). FLAMMABLE GAS 2	**Flammable Gas** (white letters on red background). If a subsidiary hazard is present, the subsidiary label must **not** display a hazard class number. FLAMMABLE GAS 2	Placard any quantity by rail; 1,001 lbs (454 kg) or more by truck.	Should not be loaded with some poisons or explosives. Should be segregated (at least 4 ft away) from radioactives and corrosives. If combined load and 5,000 lbs (2,268 kg) picked up at one facility, then a **Flammable Gas** placard is needed in addition to other placards.
Compressed Gas	**Non-Flammable Gas** (white letters on green background). NON-FLAMMABLE GAS 2	**Non-Flammable Gas** (white letters on green background). If a subsidiary hazard is present, the subsidiary label must **not** display a hazard class number. NON-FLAMMABLE GAS 2	Placard any quantity by rail; 1,001 lbs (454 kg) or more by truck.	Should not be loaded with certain explosives. If combined load and 5,000 lbs (2,268 kg) picked up at one facility, then a **Non-Flammable Gas** placard is needed in addition to other placards.
Chlorine	**Poison Gas** (black letters on white background). **Chlorine** (black letters on white background) may be used in U.S. domestic highway transportation through October 1, 2001. POISON GAS 2 CHLORINE 2 Domestic US	**Poison Gas** (black letters on white background). POISON GAS 2	Placard all quantities by rail or highway.	Should not be loaded with certain explosives. Should be segregated from certain flammables, oxidizers, and corrosives.

Figure 5-11

SUBSTANCE	VEHICLE PLACARD	PACKAGE LABEL	QUANTITY	SPECIAL REQUIREMENTS
Poison Gas	**Poison Gas** (black letters on white background). For rail transport, certain poisons require the **Poison Gas** placard to be placed against a white square background with a black border. POISON GAS 2 — INHALATION HAZARD 2 Proposed	**Poison Gas** (black letters on white background). Materials which meet the inhalation toxicity criteria must be marked "Inhalation Hazard." If a subsidiary hazard is present, the subsidiary label must **not** display a hazard class number. POISON GAS 2 — INHALATION HAZARD 2 Proposed	Placard all quantities by rail or highway.	Gases of hazard zone A should not be loaded with any explosives, flammables, oxidizers, or corrosives. Gases of other than hazard zone A should not be loaded with certain explosives and should be segregated from certain flammables, oxidizers, and corrosives.
Oxygen (gas or liquid)	**Non-Flammable Gas** (white letters on green background). **Oxygen** (black letters on yellow background) may be used in U.S. domestic highway transportation. NON-FLAMMABLE GAS 2 — OXYGEN 2 Domestic US	**Non-Flammable Gas** (white letters on green background). **Oxygen** (black letters on yellow background) may be used in U.S. domestic highway transportation. NON-FLAMMABLE GAS 2 — OXYGEN 2 Domestic US	Placard any quantity by rail; 1,001 lbs (454 kg) or more by truck.	Should not be loaded with certain explosives. If combined load and 5,000 lbs (2,268 kg) picked up at one facility, then a **Non-Flammable Gas** or **Oxygen** placard is needed in addition to other placards.

Figure 5-11 (cont.)

Notes

Storage Cylinders for Gases

SHIPPING CONTAINERS

Gases may be shipped or stored as a liquid or gas under high pressure. Storage cylinders and transporting vehicles vary in size. Cylinders can be as small as a butane cigarette lighter or as large as a 33,000 gallon tank car of liquefied petroleum gas. A tank car of this size is carrying almost 9 million cubic feet of flammable gas. Ships and barges are capable of transporting huge quantities of gases. The newer large ships can move several hundred thousand gallons of liquefied natural gas.

Industrial and Medical Cylinders

Cylinders are usually transported on open, stake-body trucks and autos. Panel trucks are also being increasingly used for the transportation of home oxygen units. Leakage usually occurs when the vehicle is involved in an accident. Gases can be identified by several means, including:

- placards
- labels
- cylinder markings
- shipping papers
- observing valve structure and cylinder type
- information obtained from the driver, gas company, or reliable representative

Compressed gas cylinders containing non-poisonous materials are labeled in accordance with either DOT requirements or Compressed Gas Association (CGA) standards (see Appendix A of CGA Pamphlet C-7). Although either label is acceptable, problems with these labeling systems do exist. For example, the non-flammable compressed gas label does not indicate all the hazards that some gases in this class pose. Therefore, labels should not be the only source of identification.

Currently, there is no uniform coloring system for identifying gas cylinders.

At one time identification of a product may have been possible by a color code. *Currently, except for gases intended for medical use, there is no uniform*

Notes

Color Coding of Medical Cylinders

Product	U.S. Cylinder Color	Canada Cylinder Color
Carbon Dioxide	Gray	Gray
Carbon Dioxide/ Oxygen	Gray and Green	Gray and Green
Cyclopropane	Orange	Orange
Ethylene	Red	—
Helium	Brown	Brown
Helium/Oxygen	Brown and Green	Brown and Green
Nitrogen	Black	Black
Nitrous Oxide	Blue	Blue
Oxygen	Green	White*
Air	Yellow*	Black and White

* Historically, white has been used in the United States and yellow has been used in Canada to identify vacuum systems. Therefore, it is recommended that white **not** be used in the United States and yellow **not** be used in Canada as a marking to identify containers for use with any medical gas.

Source: Compressed Gas Association

Figure 5-12

coloring system. The CGA and U.S. Bureau of Standards have developed a *suggested* standard for color coding cylinders to indicate the commodity contained. This system, however, is not required. See Figure 5-12 for the color coding system for medical cylinders. Department of Defense facilities also follow a uniform system.

In the medical field, compressed gas cylinders are sized from A to H. Capacity ranges from 5 cubic feet to 250 cubic feet of product, depending on pressure. Pressure ranges from just a few psi to 4,500 psi. The greater the pressure in the cylinder, the greater the chance of the cylinder's being propelled during a sizeable leak.

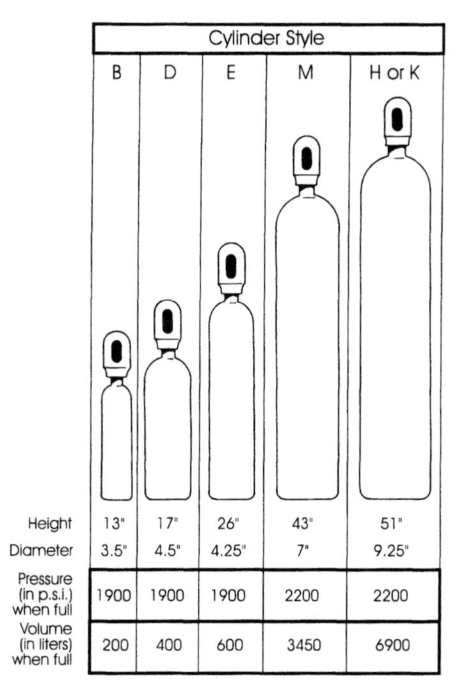

Figure 5-13. Sizes of Medical Compressed Gas Cylinders

Notes

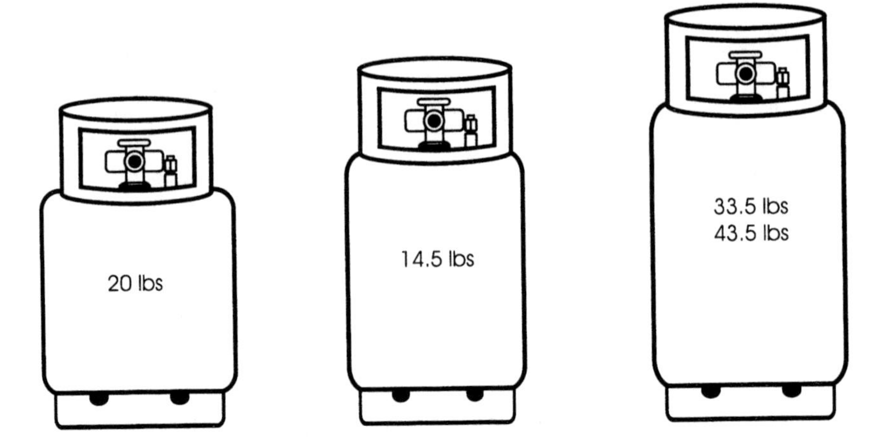

Figure 5-14. Motor Fuel Cylinders

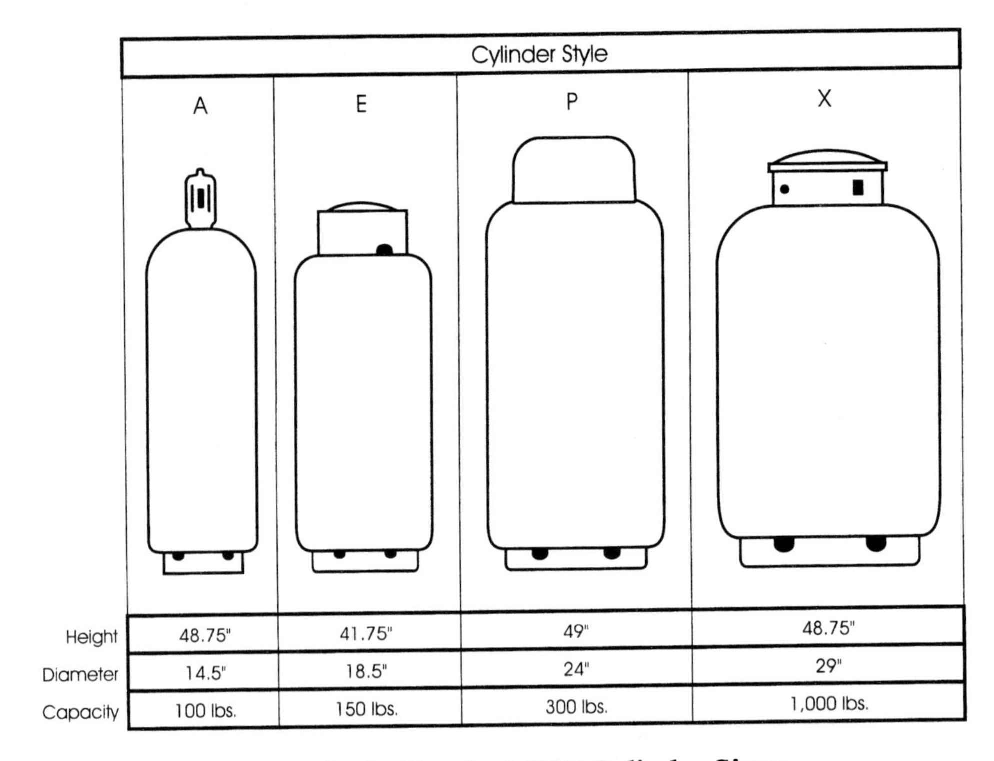

	Cylinder Style			
	A	E	P	X
Height	48.75"	41.75"	49"	48.75"
Diameter	14.5"	18.5"	24"	29"
Capacity	100 lbs.	150 lbs.	300 lbs.	1,000 lbs.

Figure 5-15. Standard DOT Cylinder Sizes

Notes

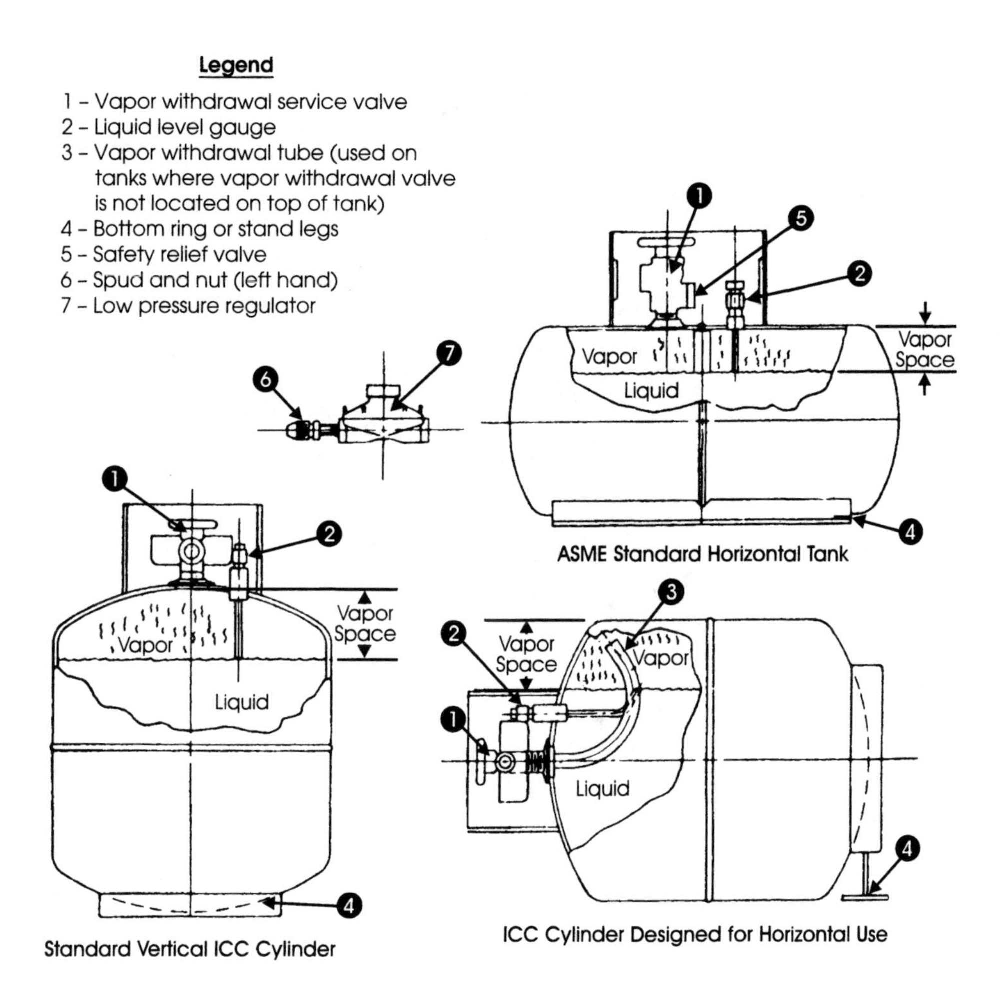

Figure 5-16. Common Terms of LP Gas Tanks

Notes

MC331 Pressure Carrier

The MC331 Pressure Carrier

The MC331 tank truck is one of the most widely used in the United States to carry liquefied gases such as propane, butane, butadiene, and anhydrous ammonia. The tank is identified by large hemispherical heads on both ends, a bolted manhole at the rear, and a guard cage around the bottom loading/unloading piping. The tanks are non-insulated, single-shell vessels, usually painted white. They may be permanently marked "Flammable" or "Compressed Gas," and they may carry an identifiable manufacturer or distributor name.

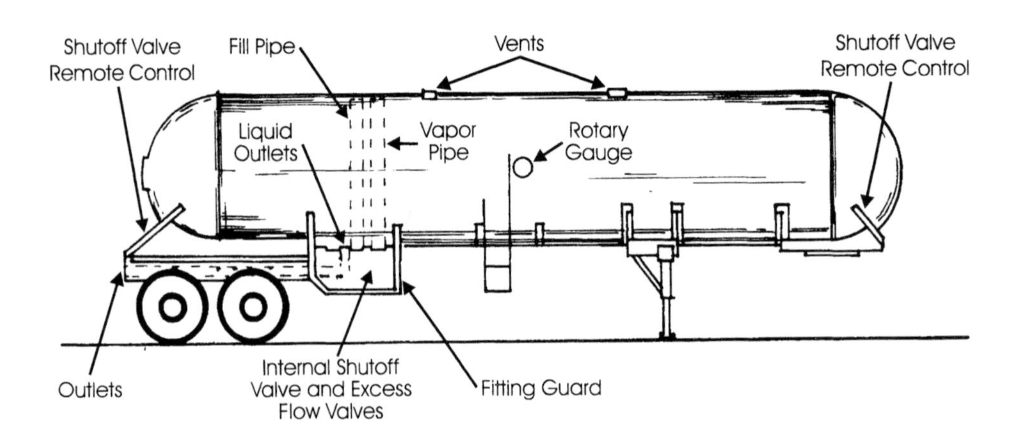

Figure 5-17. MC331 Pressure Carrier

Notes

Figure 5-18 illustrates the safety devices found on MC331 transports. The numbers below correspond to the numbers in the figure.

- Liquid discharge openings (1). Must be protected with an internal valve that can be manually, remotely, and thermally closed.
- Remote shutoff control (1A).
- Vapor openings (2). Must be protected with an internal valve that can be manually, remotely, and thermally closed.
- Remote shutoff control (2A).
- Liquid fill opening (3). May have a back flow check valve installed in the tank.

Note: Each thermal release should be within five feet of the applicable hose connection. Liquid/vapor lines may also be color coded. Drawings do not include all items required by the DOT, only those with which emergency responders should be familiar.

Hose sizes:

- Vapor: 1-1/4 inch or larger
- Liquid: 1-1/2 inch or larger

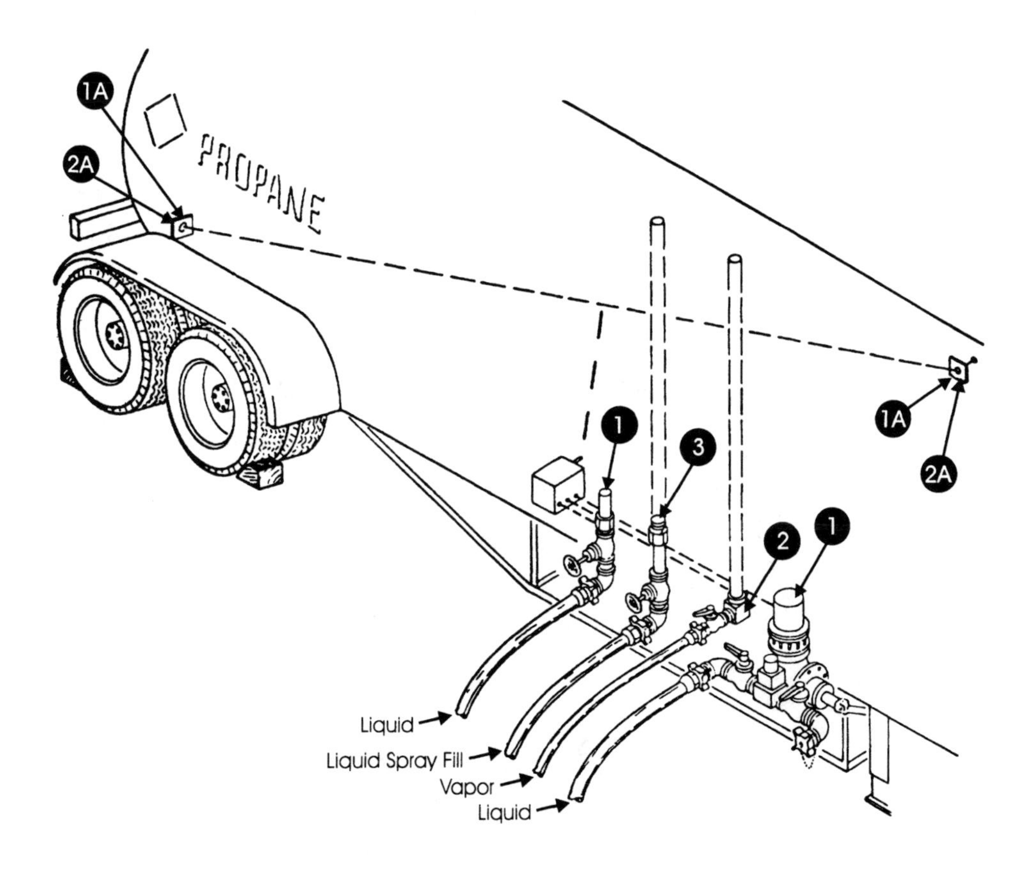

Figure 5-18. MC331 Safety Devices

Notes

MC331 transports must have the following safety devices unless emergency valves are provided at the transport end of hoses (see Figure 5-19).

- Internal valves (1). Must include manual shutoff and thermal release (temperature sensitive element) at internal valve within five feet of hose.
- Remote control to close internal valves (1A).

The plant transport unloading area must have the following safety devices:

- Concrete bulkhead or equivalent anchorage (2). Must have a predictable breakaway point to keep the valves and piping on the plant side of the connections intact.
- Vapor line (3). Must have an emergency valve with manual shutoff, thermal release, and remote control provisions.
- Remote shutoff control (3A).
- Liquid line (4). Should have either a back flow check valve that may be part of sight flow unit or manual shutoff emergency valve with thermal release and remote shutoff control.

Note: It is recommended that the hose or pipe connections be capped or plugged except during the transfer operation.

Hose sizes: Vapor: 1-1/4 inch or larger; Liquid: 1-1/2 inch or larger.

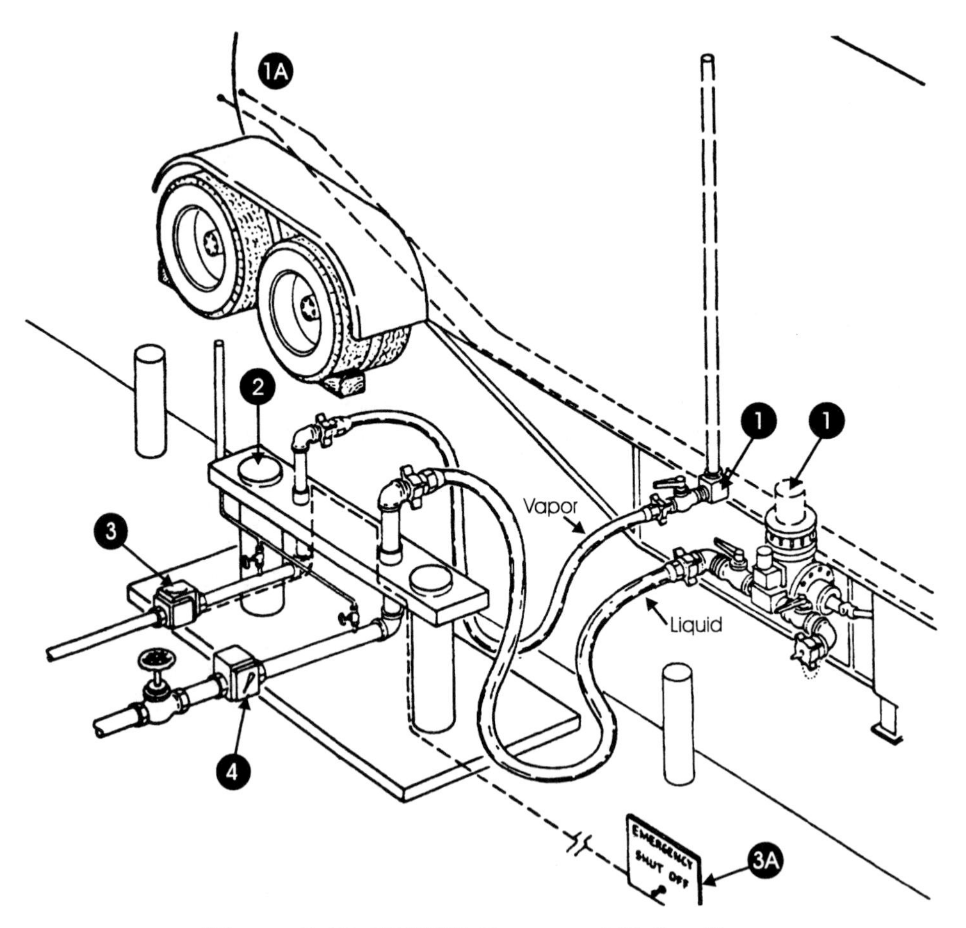

Figure 5-19. MC331 Transport Unloading

Bobtail Delivery Truck

Notes

Bobtails that are filled directly into the tank must have the safety devices listed below. The numbers correspond to the numbers in Figure 5-20.

- Liquid fill line (1). Has a back flow check valve mounted directly into the tank or an internal valve.
- Vapor line (2). Does not require emergency valve because hose diameter is less than 1-1/4 inches.

The plant loading riser must have the following safety devices:

- Concrete bulkhead or equivalent anchorage (3), or use of a predictable breakaway point that retains the valve(s) on the plant side of the connection(s) intact.
- Liquid line (4). Has a manual shutoff emergency valve and thermal release.
- Remote shutoff control (4).
- Vapor line (5). Does not need an emergency valve because the hose diameter is less than 1-1/4 inches.

Hose sizes:

- Vapor: Smaller than 1-1/4 inch
- Liquid: 1-1/2 inch or larger

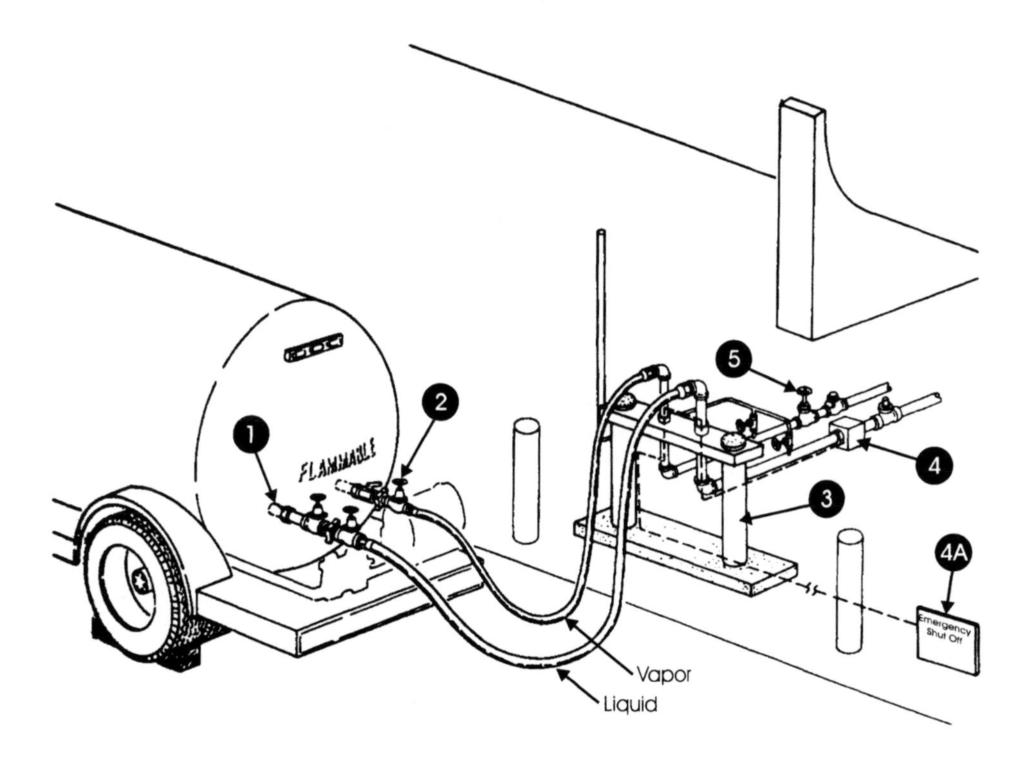

Figure 5-20. Bobtail Filling

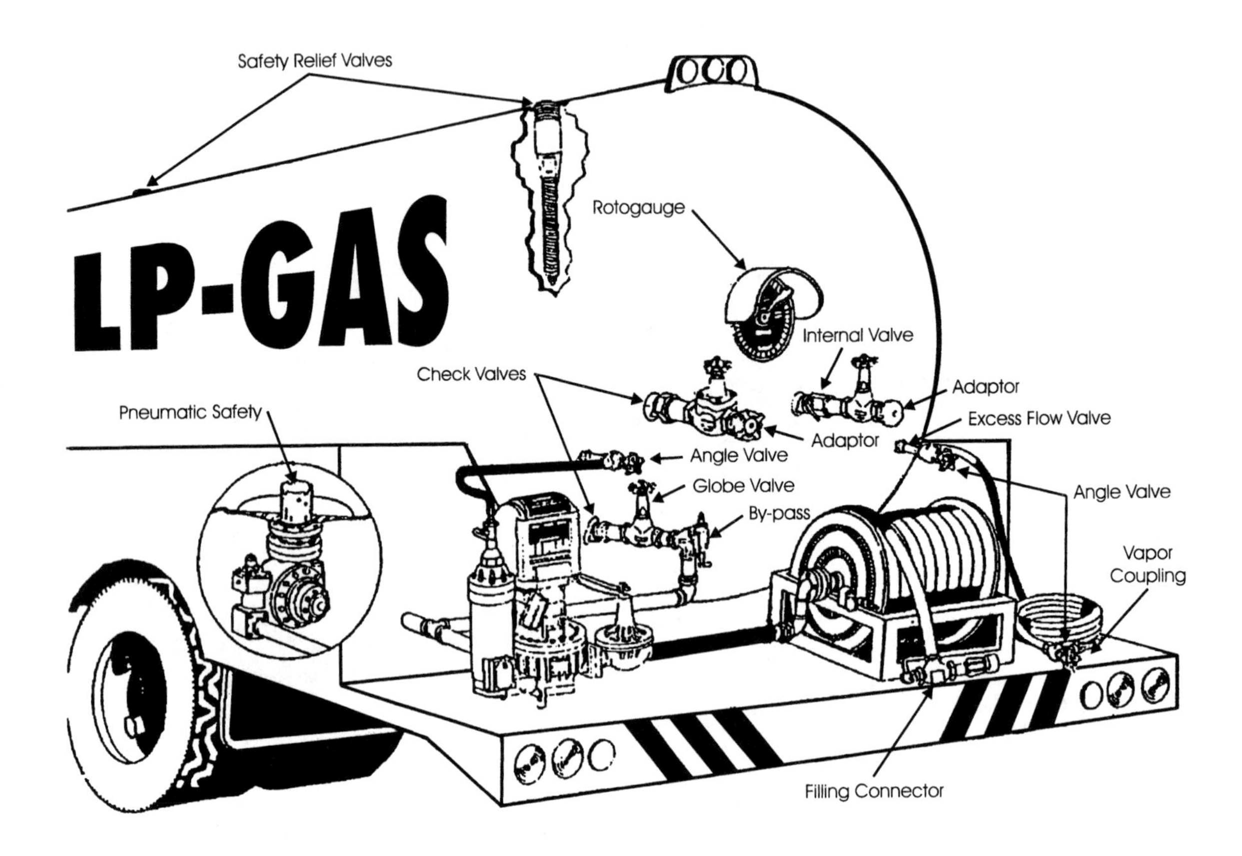

Figure 5-21. Bobtail Delivery Truck

Pressurized Rail Tank Car

Pressurized Rail Tank Cars

Pressurized rail tank cars carry flammable and non-flammable liquefied gases, poisonous gases, and other hazardous liquids. Pressurized rail tank cars nearly always contain hazardous materials. These rail cars are easily recognized by the protective housing around the manway, valves, gauging rod, and sampling well. They can have a capacity of up to 48,000 gallons; however, current regulations restrict new tank cars to 34,500 gallons. The contents of gas cars are approximately 80% liquid. Pressurized tank cars transporting flammable gases are thermally protected by either a thermal insulation and metal jacket or spray-on thermal protection. Some pressurized cars are insulated.

Do not depend upon color to identify products on rail cars. However, rail cars carrying hydrogen cyanide or hydrocyanic acid have their own designated color pattern. Tanks carrying these extremely toxic poisons are completely white in color with three red stripes, one horizontal and two vertical, three feet from each end. In an accident or derailment, this car can create a twilight zone of death around it. Always view this rail car from a distance! It is important to note that these products are also shipped in cars that are not color coded.

There are two primary visual characteristics of pressurized rail tank cars:

- There is no unloading piping under the car.
- The ends of insulated/jacketed cars are less round than the ends of thermally protected or single-shell cars with ellipsoidal ends.

Notes

CLASSES OF PRESSURE TANK CARS

DOT105 Tank Car

The DOT105 has one important feature—it is designed to meet the specifications for a single product. The specifications aid safe transport of hazardous materials through commerce.

A DOT105 tank car must be insulated (49 CFR 179.100-104). Bottom outlets are prohibited except in special cases (49 CFR 179.100-114). These cars may have additional safety features which appear in writing on the sides of the tank. The safety features are coded as follows:

- **105A** – Shelf coupler that prevents coupler override in the event two rail cars couple at high speed.
- **105S** – Shelf coupler with head shield that prevents damage to tank head in the event two tanks collide.
- **105J** – Shelf coupler, head shield, and outer jacket that prevents damage to inner tank. There is also insulation to provide thermal protection in the event of a fire and/or flame impingement.

Additional specifications for specific commodities are found in 49 CFR 179.102. Products found in this category include chlorine, liquefied flammable gas, motor fuel anti-knock compound, and hydrocyanic acid.

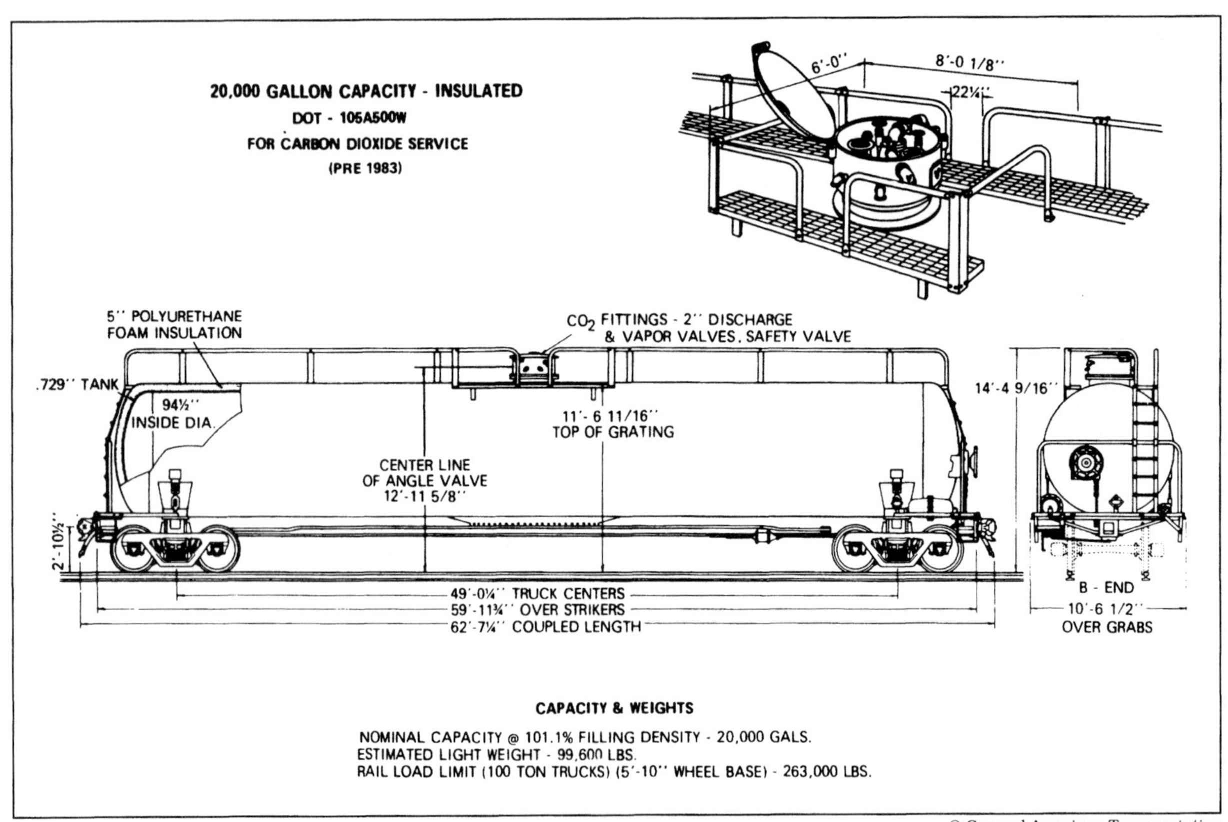

Figure 5-22. DOT105A Tank Car

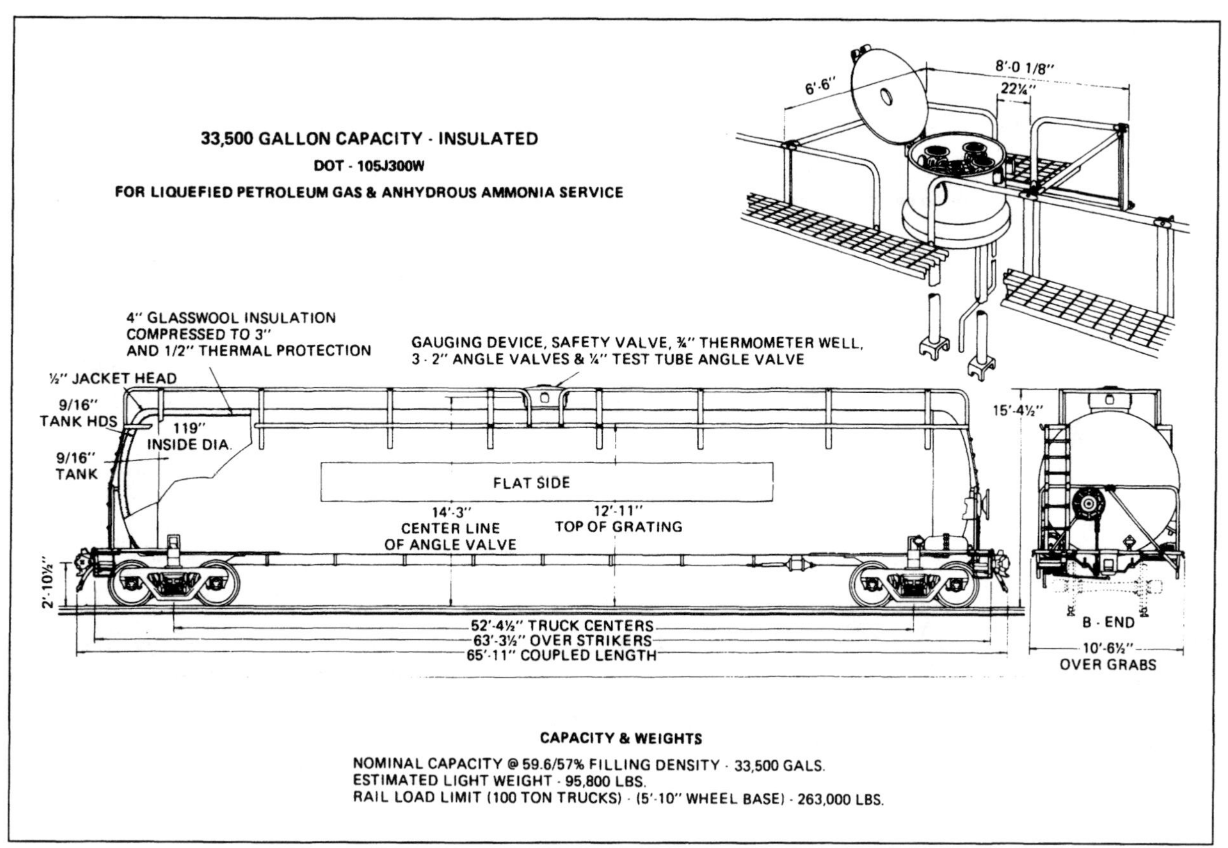

Figure 5-23. DOT105J Tank Car

DOT112 Tank Car

New and improved safety features on the DOT112 tank car are now law. As of December 31, 1980, all DOT112 tank cars must have the three safety features found on the DOT105 tank car (shelf couplers, head shield, thermal protection) if transporting flammable gases in the United States.

DOT112 tank cars may use one of two different thermal protection methods to protect the contents under fire conditions. The chosen protection must appear after the DOT112 marking.

- **DOT112T** – Indicates a spray type insulation protection. It leaves an easily identified "rough finish" on the car. If the car derails, the protection could be scratched off, leaving it vulnerable to fire.
- **DOT112J** – Indicates an outer shell or jacket surrounding the tank. The space between the two tank walls is filled with thermal insulation.

Both the DOT112T and the DOT112J aid in thermal protection. They must withstand one of the following tests:

- simulated pool fire at 1,600°F (871°C) for 100 minutes, or
- simulated torch fire at 2,200°F (1,204°C) for 30 minutes

The head shield for the DOT112T tank car is visually obvious, while the head shield for the DOT112J tank car is built into the ends of the outer jacket and is not so easily identified.

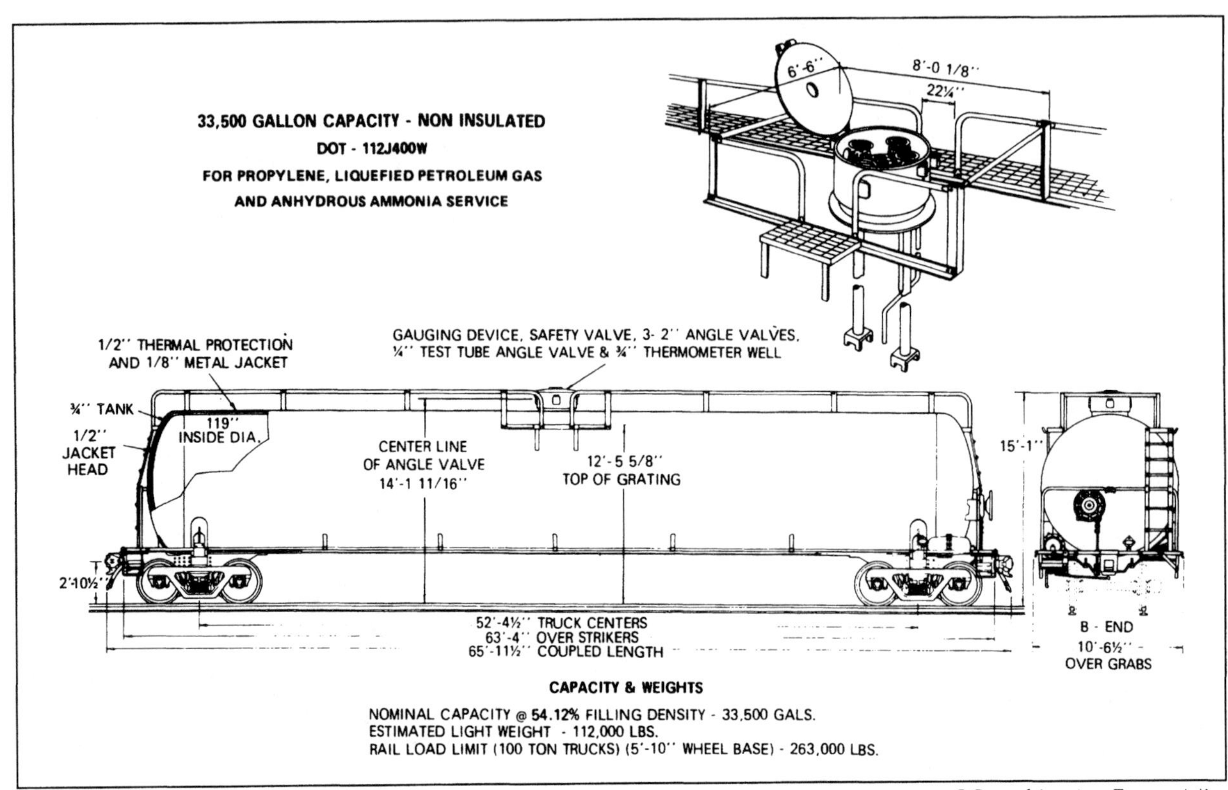

Figure 5-24. DOT112J Tank Car (Non-Insulated)

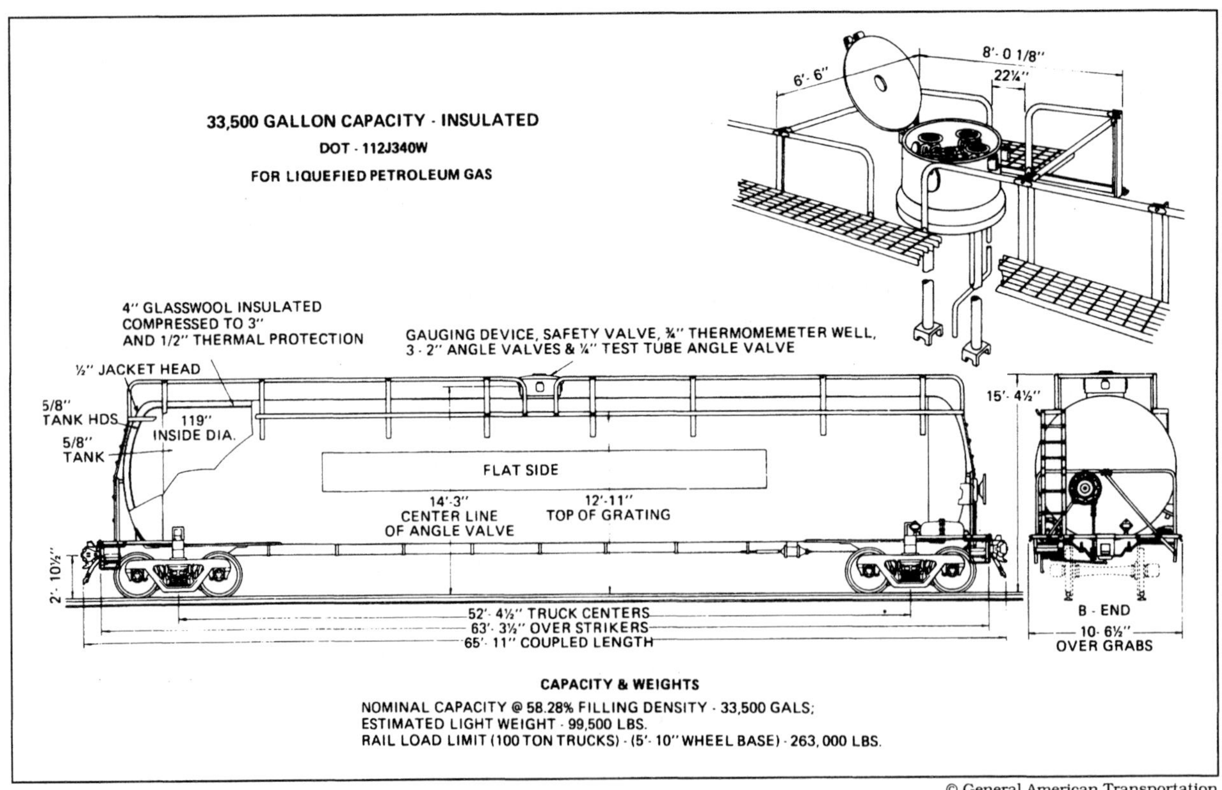

Figure 5-25. DOT112J Tank Car (Insulated)

COMMON LEAKS

Notes

There are several common leaks that can occur on pressurized tank cars. The following outlines some common leaks that emergency responders may encounter. The measures outlined here are not intended to provide complete emergency information for all high pressure tank car leaks. These measures should only be attempted by properly trained, protected, and equipped responders, such as hazardous materials technicians and specialist employees. There are other types of tank car gauging devices that are not covered here. Become familiar with the types of tank cars common in your area. In any incident, immediately contact the local railroad office, the shipper, and the nearest HMRT for assistance.

Shutoff measures are not intended to be permanent. Instead, they will help to immediately control the leak until other assistance arrives, assure time for product delivery, and allow the car to be moved to a regular maintenance facility. After all leakage repairs, obtain the car seal number and have the railroad reseal the car.

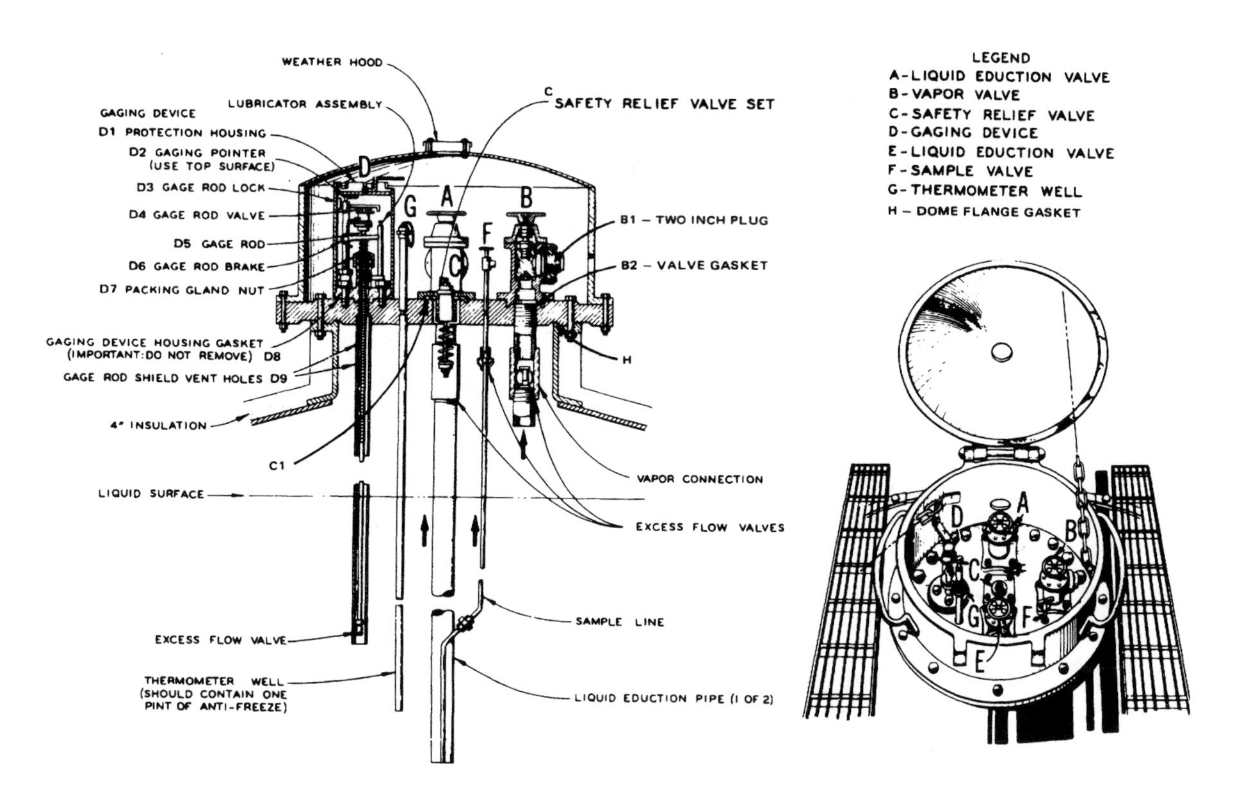

Figure 5-26. Liquid Propane Gas Tank Car Dome Fittings

Gauging Device – Point D

The protective housing (D1) must be removed before any repairs can be made. The most frequent leak involves the gauge rod valve (D4). A slight clockwise turn will easily stop the leak.

The second most frequent leak involves the packing gland nut (D7) in the gauging device. The nut may loosen, allowing product to escape. Stop the leak by tightening the packing gland nut.

Warning: The operator should never lean directly over the gauging device tube, as it can be pushed up by internal tank pressure.

Eduction Valves – Points A, E; Vapor Valve – Point B

These valves increase pressure in the tank car (the vapor valve) and remove product (valves A and E). Leakage usually occurs when the valves or 2 inch plugs are not tightened properly after loading. Use a small pipe wrench to turn the valve clockwise and/or tighten the 2 inch plug to stop leaks.

Sample Valve – Point F

If the sample valve (point F) leaks, tighten the valve by turning it clockwise. As in most valve leaks, a slight turn will stop the leakage.

Angle Valve Gaskets – Points A2, B2, and E2

These gaskets separate valves A, B, and E from the flange. Occasionally a valve will loosen and leak at the flange gasket. Tighten all four flange nuts to stop the leak. Tighten alternately and at opposite locations to get uniform pressure.

Safety Valve Gasket – Point C1

The safety valve gasket occasionally will leak like the angle valve gaskets. Tighten the four nuts as described above to stop the leak.

Safety Relief Valve – Point C

The safety relief valve prevents excess pressure build-up in the car. Faulty "O" rings may leak due to incorrect installation or normal wear. As long as the product in the tank is not water reactive, the leak can usually be reduced by filling the dome with water (approximately 10 to 15 gallons). Drainage outlets should be plugged. When water comes into contact with the dome flange, you may hear a loud popping. It presents no real danger. Be sure the safety relief valve is leaking and not relieving excess pressure.

Important: There is no excess flow valve below the safety valve. The safety valve should never be removed when there is pressure in the car.

Sample Valve Nipple – Point F

This line is 1/4 inch in diameter and usually has contact with considerable amounts of moisture. This leads to rust and corrosion, which can cause leaks.

If the sample valve nipple leaks, the nipple should be broken off completely. Use a pry bar to keep your body away from direct contact with the liquid. As the product flows from the tank, the excess flow valve will engage and stop the leak. If the excess flow valve fails, insert a tapered plug in the line to stop the leak. Then unload the car, mark it appropriately, and direct it to a maintenance shop.

Thermometer Well – Point G

Stop leaks from this source by tightening the nipple and nipple cap. Be careful not to strip the nipple threads.

Dome Flange Gasket – Point H

A leak in the dome flange gasket occurs infrequently. To stop the leak, use a large socket wrench, 1-7/8 inch to 2-1/4 inch with extension, to tighten the flange nuts in the leak area.

RAIL CAR OFF-LOADING

Notes

When a rail car is delivered to the user's location, it is normally off-loaded at an unloading riser (see Figure 5-27). Tank car unloading risers have the safety devices listed below. As an alternative, emergency valves may be installed in the tank car unloading adaptors instead of the hose end as shown. The numbers correspond to the numbers in Figure 5-27.

- Liquid hoses at the end of the tank car have an emergency valve with manual shutoff and thermal release (1).
- Remote shutoff control (1A).
- Riser ends of liquid hose connections have back flow check valves (2).
- Vapor hose has an emergency valve at each end with manual shutoff and thermal release (3).
- Remote shutoff control (3A).
- When two hoses or swivel type piping are used on tank car unloading riser, each leg of the piping should be protected by back flow check valve(s) or an emergency shutoff valve (4).

Hose sizes:

- Vapor: 1-1/4 inch or larger
- Liquid: 1-1/2 inch or larger

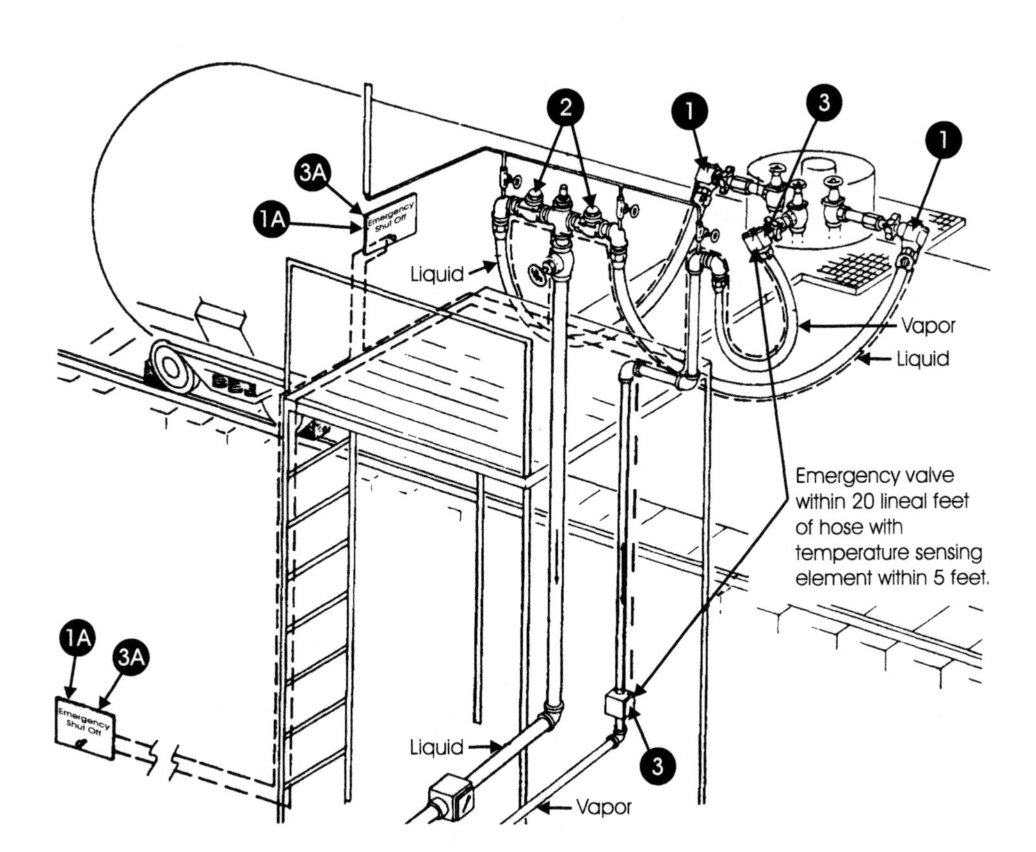

Figure 5-27. Tank Car Unloading Riser

Notes

Fixed Site Storage Tanks for Gases

FIXED SITE STORAGE VESSELS

It is important to be aware of industrial gas products and types, as well as the location of these products in your area. Obtain adequate knowledge of fixed site storage vessels to reduce confusion and increase safety during emergency operations.

High Pressure Horizontal Tanks

High pressure horizontal tanks are used to store liquid petroleum gases (LPG), high vapor pressure flammable liquids, and anhydrous ammonia. Tank size varies with use and occupancy, but it usually ranges from 500 to 30,000 gallons. Internal pressure ranges from 100 to 220 psi, depending

High Pressure Horizontal Tank

Notes

on ambient temperature. Pressure relief devices are normally set to 250 psi. They are usually single shell, non-insulated tanks that are painted white or some other highly reflective color.

Single storage containers of over 4,000 gallons capacity, or stationary multiple container systems with aggregate water capacity of more than 4,000 gallons that use a common or manifold liquid transfer line, should have the following safety devices. The numbers below correspond to the numbers in Figure 5-28.

- Concrete bulkhead or equivalent anchorage (1), or use of a predictable breakaway point.
- Emergency valve with manual shutoff at the valve (2).
- Remote shutoff control (2A).
- Thermal release within five feet of the nearest end of the hose or swivel-type piping (2B).
- Vapor line (3) does not need an emergency valve because the hose diameter is less than 1-1/4 inches.

Note:

- The hose or pipe connection should be capped or plugged except during transfer operation.
- If distance from hose to internal valve exceeds 20 lineal feet, an emergency valve must be installed at the bulkhead.

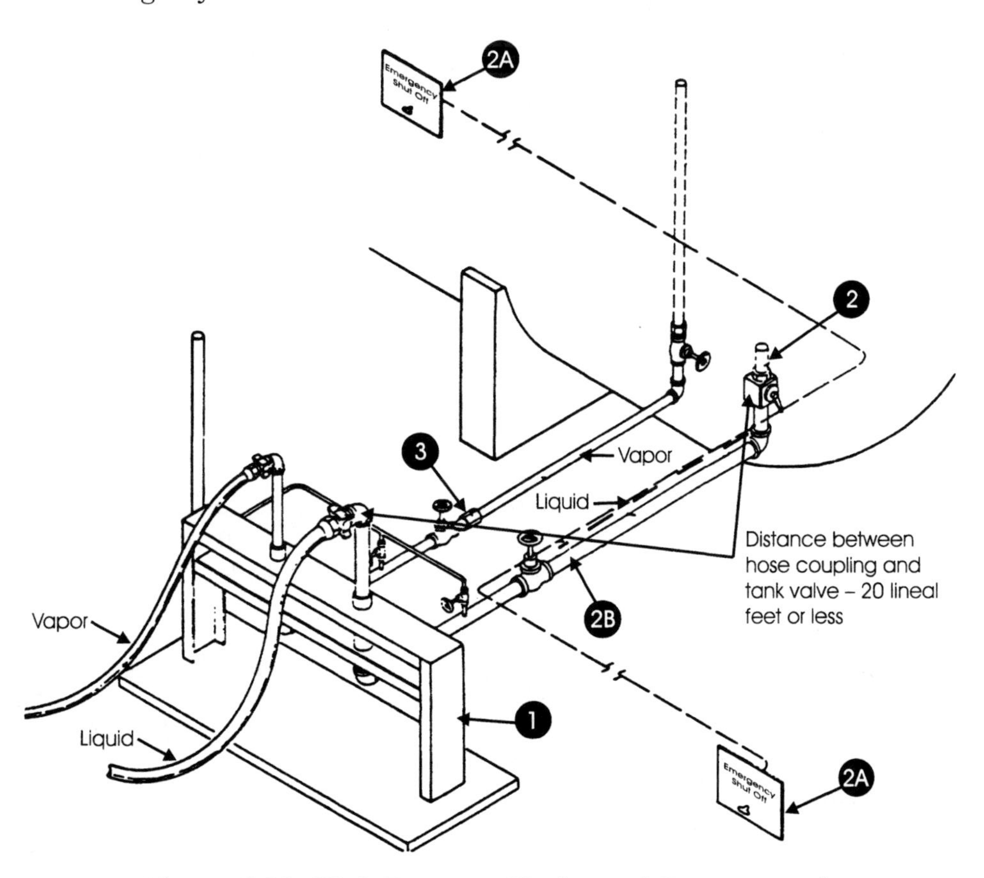

Figure 5-28. High Pressure Horizontal Storage Tank

Notes

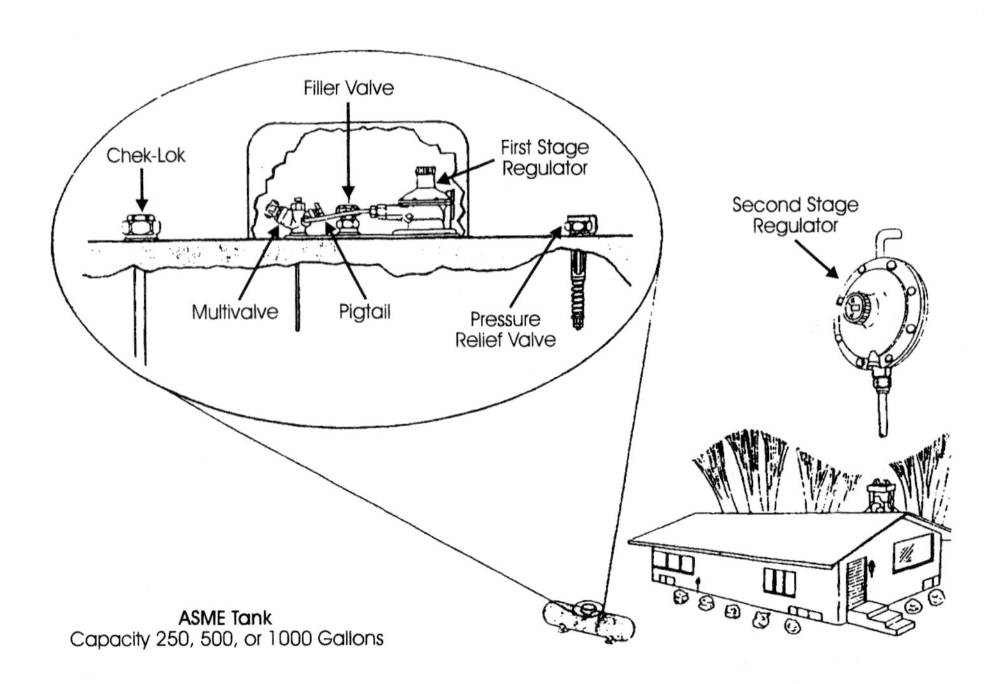

Figure 5-29. Typical Domestic Liquid Propane Gas Tank

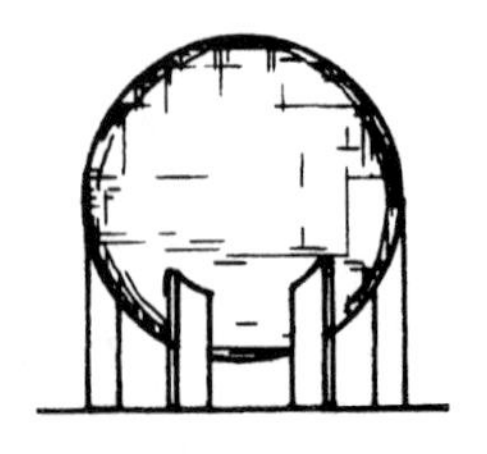

High Pressure Spherical Storage Tank

High Pressure Spherical Storage Tanks

High pressure spherical storage tanks are not common in many communities, but they are often used for storage of LP gases at production or distribution sites. These tanks are single shell vessels that are usually painted white or some other highly reflective color. They can be insulated or non-insulated. Tank capacities reach 600,000 gallons. Due to the large volume, many tanks are equipped with a water spray fire protection system.

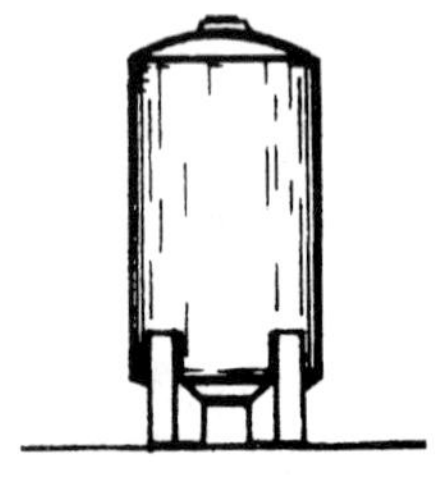

Cryogenic Liquid Storage Tank

Cryogenic Liquid Storage Tanks

Cryogenic liquid storage tanks are used to store liquid oxygen (LOX), liquid nitrogen, liquid carbon dioxide, and other cryogenic materials. The tanks are constructed similarly to the liquefied natural gas (LNG) tank (see below).

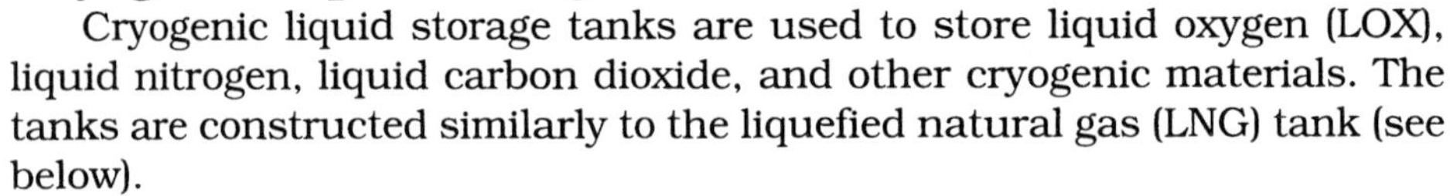

Cryogenic liquid storage tanks are usually found at industrial facilities, hospitals, gas processing facilities, labs, computer production settings, etc. Tank capacities range from 750 gallons to over 400,000 gallons. These tanks are usually painted white or some other reflective color. They are easily identified by the ice that covers adjacent piping and heat exchange equipment. Cryogenics will be discussed in greater detail later in this chapter.

Liquefied Natural Gas (LNG) Storage Tanks

Notes

Liquefied natural gas is stored at a temperature of −260°F (−162°C). This low temperature requires the use of special materials to contain and pipe the product to distribution outlets. These tanks are usually found only at production or distribution facilities.

LNG storage tanks have a double shell construction with perlite and foam insulation in between the shells. The tanks sit on foam glass blocks for support and insulation. They have electric elements for heating. LNG tanks can hold up to several hundred thousand gallons and are usually painted white or some other reflective color.

Pressure relief valves discharge to prevent excess pressurization. In addition, vacuum relief valves prevent excessive negative pressure. Tank operating pressure is usually 15.5 psia (absolute pressure; see the glossary in Appendix A), while pressure relief valves are set at 1.5 psig (gauge pressure).

All piping connections into or out of the inner tank are through the roof. Nothing penetrates the inner shell below the liquid level. LNG forms a gas when its temperature is raised and it is pumped through the pipeline. You should be familiar with the LNG storage facilities, water transport vessels, and docking areas in your community.

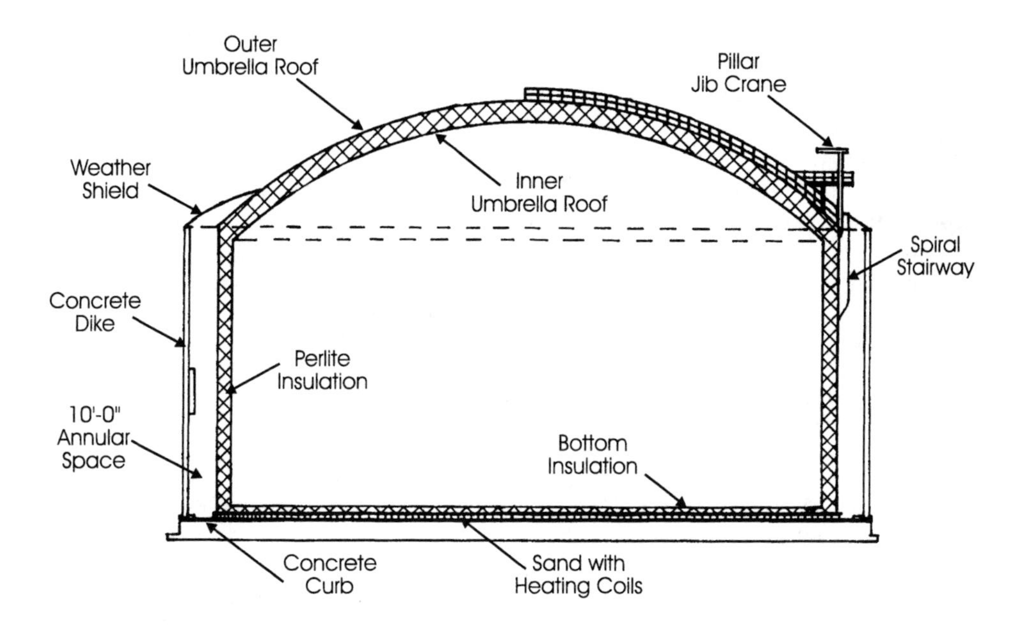

Figure 5-30. LNG Liquid Storage Tank

Notes

Gas Pipeline

PIPELINES

Pipelines are storage containers, even though they are not usually considered as such. The primary objective of this section is to offer a basic understanding of pipelines and related emergencies. It is impossible to cover all situations which could be considered a pipeline emergency. This section is general and should not be used as operating policy. Use this information as a guideline only.

Every year large volumes of products are transported via pipelines. Pipelines are considered one of the safest and most cost efficient methods of transporting large quantities of various products over great distances. One of the most common haz-mat emergencies, however, is a ruptured natural gas pipeline. This silent, unseen transporter is often overlooked by emergency personnel. A ***pipeline emergency*** is defined as:

> A situation in which leakage or uncontrolled flow of gas or liquid from a pipeline or support system has caused or may cause death or serious injury to persons or may damage property.

Pipeline incidents involve a variety of emergency professionals. Due to the large number of these incidents, interagency cooperation is necessary to reduce or eliminate hazards. Firefighters, law enforcement officials, EMS personnel, public and private utilities, industrial professionals, media, insurance companies, construction workers, health professionals, environmental specialists, and others may be called upon to work together. Often the law enforcement agency is asked to coordinate the effort.

One common product transported by pipeline is natural gas. Natural gas usually begins its pipeline travel from gas fields in the southwest United States. Natural gas is tapped from vast resources of underground deposits or as a by-product of oil extraction. The most cost efficient means of transporting this fuel is by pipeline. Other products transported by pipeline include crude oil, gasoline, diesel fuel, fuel oil, and jet fuels. More than one product is often pumped through a single pipeline. For example, 12,000 barrels of diesel may be pumped with 2,500 barrels of gasoline behind it. (*Note:* 1 barrel = 42 gallons.) Versatility is one benefit of pipeline systems.

Energy companies lay their pipelines across prairies, swamps, mountains, fields, and deserts. Local companies buy the transported product for local distribution. The product is distributed through a vast above-ground and underground system of mains, transmission lines, and gathering lines.

Low pressure systems are the most common method of transporting the product through city mains. The product may also be distributed at intermediate or high pressure, but it must be regulated down to a usable pressure.

Pipelines are generally constructed of plastic or coated mill-wrapped steel. There are some copper and iron pipes still in service. Pipes are buried 24 to 30 inches below ground in a ditch with no other pipes or cables. The most common cause of pipeline emergencies is "dig ups" by construction crews. The general location of a pipeline is usually marked, as well as a telephone number to call before digging in the area or in case of a leak.

A great deal of preplanning is required to determine location, size, pressures, products, and ownership of pipelines. Due to the large volumes of transported product, it is critical to have a sound action plan for pipeline emergencies.

Preparation Checklist for Pipeline Emergencies

- Know how to interpret pertinent information from pipeline markers. All markers are required to list the pipeline contents, pipeline owner or operator, and an emergency telephone number.
- Be aware of the pipelines in your area. Know how many there are and their locations.
- Know who owns area pipelines and how to get in touch with the responsible party.
- Have a general idea of the pipeline's origin and destination. A map showing locations and size of pipelines is helpful.
- Know your department's SOPs before an emergency.
- Preplans should include local fire, law enforcement, and other necessary agencies. Understand each agency's capabilities and responsibilities.
- Determine special resources needed to mitigate a pipeline emergency.
- Have a general idea of waterways, topography, and land use in your area. Anticipate what may happen.
- Prepare and train dispatchers to handle interagency networking during a pipeline emergency.
- Be aware that shutting off the flow of a product in a pipeline will only stop the flow at the source and may not stop the flow already in the pipeline.

Notes

Pipeline Marker

Notes

Natural Gas Pipeline System

Natural gas is pumped over hundreds of miles at various pressures. It is pumped through several pumping stations to maintain adequate volumes and pressures. Shutoff valves are located throughout the system to isolate an individual area, if necessary. Gas changes ownership from supplier to purchaser at certain points along its transport route. Purchasers include:

- cities
- industrial users
- residential users

Natural gas transmission line pressures are often in excess of 450-600 psi. Gas is moved through a series of regulators that reduce the pressure to the specific user. Normal operating pressures that the emergency responder may encounter are:

- commercial pressure, 60 psi
- residential and light uses, 1/4 – 4 psi

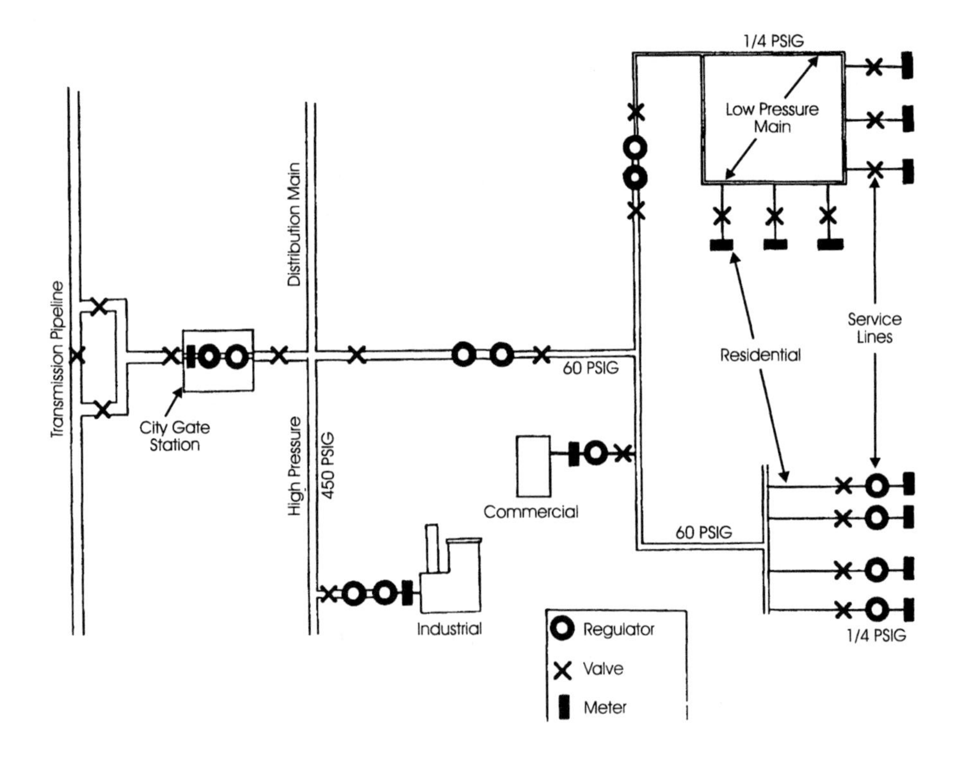

Figure 5-31. Typical Gas Distribution System

Notes

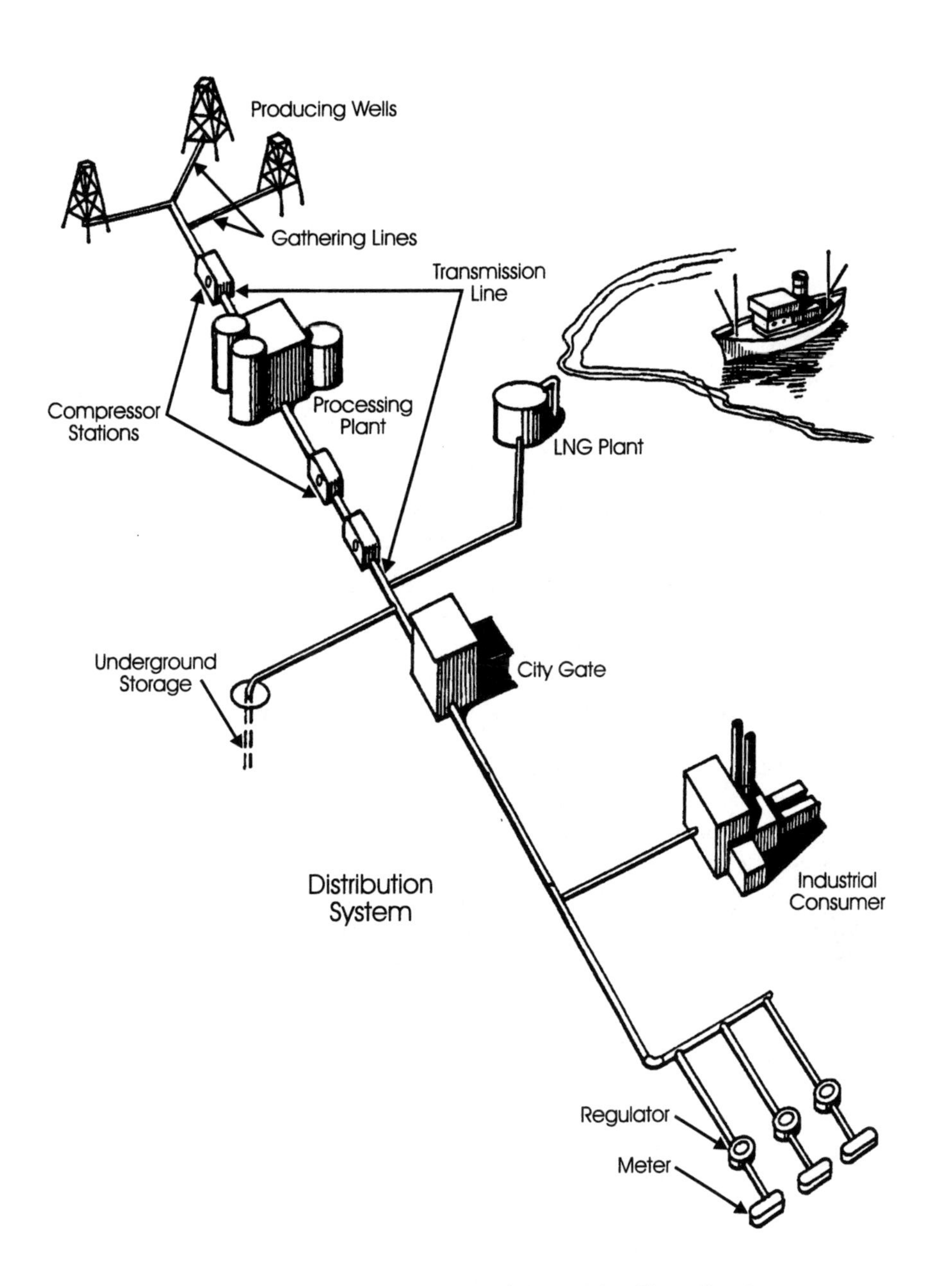

Figure 5-32. Typical Natural Gas Pipeline System

Notes

TANK AND CYLINDER SAFETY DEVICES

Responders must be familiar with the safety devices found on compressed/liquefied gas tanks and cylinders. Take time to inspect various pressure vessels to understand the inner workings and how to handle accidental leaks. There are several types of safety devices on tanks and cylinders, including:

- pressure relief devices
- valves and fittings
- gauging devices

Pressure Relief Devices

Pressure relief devices prevent rupture of normally charged cylinders that are exposed to fire. There are many types of pressure relief devices, including:

- fusible plugs
- rupture disc with fusible metal backing
- spring-loaded relief valves

This section discusses the pressure relief devices used for compressed/liquefied gas cylinder protection. The specific safety device used depends on the type of gas, DOT-rated service, test pressures, and size. The safety relief devices for pressurized gas cylinders that have been identified by the Compressed Gas Association (CGA) are listed below.

- ***Type CG-1: Rupture Disc.*** Rupture discs protect cylinders that become over-pressurized. This device is a metal disc that relieves pressure from the cylinder if the pressure reaches dangerous levels. The contents of the cylinder then exhaust through the plug opening. The burst pressure of rupture discs may not exceed the minimum DOT required test pressure of the cylinder, which is generally 5/3 of the cylinder service pressure. (See 49 CFR 173.34(d) for exceptions.)

- ***Type CG-2: Fusible Plug (165°F Nominal).*** Type CG-2 fusible plugs use a fusible alloy that melts at a temperature not more than 170°F or less than 157°F (165°F nominal). If the cylinder is exposed to fire, the fusible plug will yield or melt and release the cylinder contents. This prevents gas within the cylinder from reaching high pressure, caused by high external temperatures, and rupturing the cylinder.

- ***Type CG-3: Fusible Plug (212°F Nominal).*** Type CG-3 fusible plugs use a fusible alloy that melts at a temperature not more than 220°F or less than 208°F (212°F nominal). The fusible plug yields or melts and releases the cylinder's gas if a fire occurs.

- ***Type CG-4: Combination Rupture Disc/Fusible Plug (165°F Nominal).*** This device has a rupture disc backed by a fusible metal plug. The rupture disc burst pressure must not exceed the minimum DOT required test pressure of the cylinder: the fusible metal must yield between 157°F and 170°F (165°F nominal)). This combination safety device functions only in the presence of both excessive heat and pressure. If a fire occurs, sufficient heat is required to first melt out the fusible metal, after which the device will afford the same protection as the CG-1 rupture disc device.

Notes

- ***Type CG-5: Combination Rupture Disc/Fusible Plug (212°F Nominal).*** The construction and operation of this safety device are similar to Type CG-4, except the fusible metal backing yields or melts between 208°F and 220°F (212°F nominal).
- ***Type CG-7: Pressure Relief Valve.*** Pressure relief valves are spring-loaded and are normally closed. When the cylinder pressure exceeds the pressure setting of the relief valve, the valve opens and discharges the cylinder contents. Once the cylinder pressure decreases to the relief valve pressure setting, the valve will reseat, without leakage. Before the valve reseats, however, it vents sufficient gas to control the internal cylinder pressure. The flow rating pressure of the pressure relief valve must not be less than 75% or more than 100% of the cylinder's minimum DOT-rated test pressure. The reseating pressure must not be less than the pressure in a normally charged cylinder at 130°F. Pressure relief valves may incorporate a fusible metal to prevent premature operation of the valve. Exceptions to the above are DOT-39 cylinders. In such cylinders, the set pressure cannot exceed 80% of the minimum cylinder burst pressure and cannot be less than 105% of the cylinder test pressure.

Fusible Plugs on Chlorine Cylinders

Notes

Safety Devices for Several Gases

Safety devices used for several gases are described below. If a particular gas product is not discussed, consult a compressed gas representative for the proper information.

- ***Oxygen, Nitrogen, Helium, Argon, Air.*** These non-flammable gases are stored in cylinders under high pressure. Oxygen is an extreme oxidizer. Rupture discs are the common safety device found on industrial gas cylinders. Rupture discs with fusible metal backing (165°F nominal) are found on cylinders used in medical gas service.
- ***Carbon Dioxide, Nitrous Oxide.*** Carbon dioxide and nitrous oxide are non-flammable and are stored in cylinders as liquefied compressed gases. Cylinders used in medical gas service are protected by rupture discs backed by a fusible metal (165°F nominal). Cylinders used in industrial service are protected by rupture discs.
- ***Hydrogen.*** Hydrogen is very flammable and is stored in cylinders under high pressure. Cylinders are usually protected by rupture discs backed by a fusible metal (165°F or 212°F nominal). Rupture discs may be used on cylinders over 65 inches long.
- ***Propane, APACHI Gas.*** Propane and APACHI gas are flammable and are stored in cylinders as liquefied compressed gases. Cylinders are usually protected by spring-loaded pressure relief valves. A fusible metal plug (212°F) may be used, but only in combination with a pressure relief valve.
- ***Acetylene.*** Acetylene cylinders are protected from rupture by integrated fusible metal plugs instead of by plugs that are part of the cylinder valves. The fusible metal has a nominal yield temperature of 212°F. The fusible metal plugs cannot be repaired. Some small acetylene cylinder valves, Types B and MC, have a fusible metal plug in the valve body. These, also, cannot be repaired.
- ***Medical Gas Cylinders With Post-Type Valves.*** Post-type valves on small medical gas cylinders are usually equipped with rupture discs backed by a fusible metal (165°F nominal).

Note: Safety relief devices are not required on cylinders charged with non-liquefied gas under pressure of 300 psi or less at 70°F (21°C). Exceptions are specification 39 cylinders and cylinders used for acetylene in solution. Safety relief devices are prohibited on cylinders containing a poisonous gas/liquid or fluorine. Safety relief devices are not required on cylinders containing the following chemicals:

- methyl mercaptan
- mono-, di- or trimethylamine, anhydrous
- less than 10 pounds of nitrosyl chloride
- less than 165 pounds of anhydrous ammonia (49 CFR 173.34)

Safety Considerations

Notes

Cylinder safety devices must be maintained in proper operating condition to function correctly. Precautions are outlined below.

- Only qualified, gas supplier personnel should service pressure relief devices.
- Never tamper with safety relief devices in valves or cylinders.
- Carefully handle and store cylinders to prevent damage to the safety relief devices.
- Do not obstruct the pressure relief device. Dirt, paint, rust, or other materials prevent cylinder safety devices from functioning properly.
- Notify your supplier if fusible metal obstructs, extrudes into, or deforms a safety device. The cylinder should be removed immediately and returned to the supplier's facility for repair.
- Any problem with a safety device should be reported to the gas supplier.

Common Valves and Fittings

Illustrated below are several common valves and fittings that emergency responders may encounter in the field.

Back flow check valve always remains closed except when fuel flow from outside forces it open.

Excess flow check valve permits flow in either direction, but it closes automatically when outward flow exceeds a safe rate.

Pressure relief valve, external type, is kept closed by coil spring unless tank pressure becomes higher than working pressure.

Pressure relief valve, internal type, is not affected by accidental damage to portion outside tank.

Liquid service valve with integral excess flow check valve is found in customer bulk tanks, motor vehicle fuel tanks, and with lift truck cylinder valve.

Vapor withdrawal valve with built-in pressure relief is designed for vapor withdrawal from cylinders. It also comes with excess flow check valve and extension tube for liquid withdrawal.

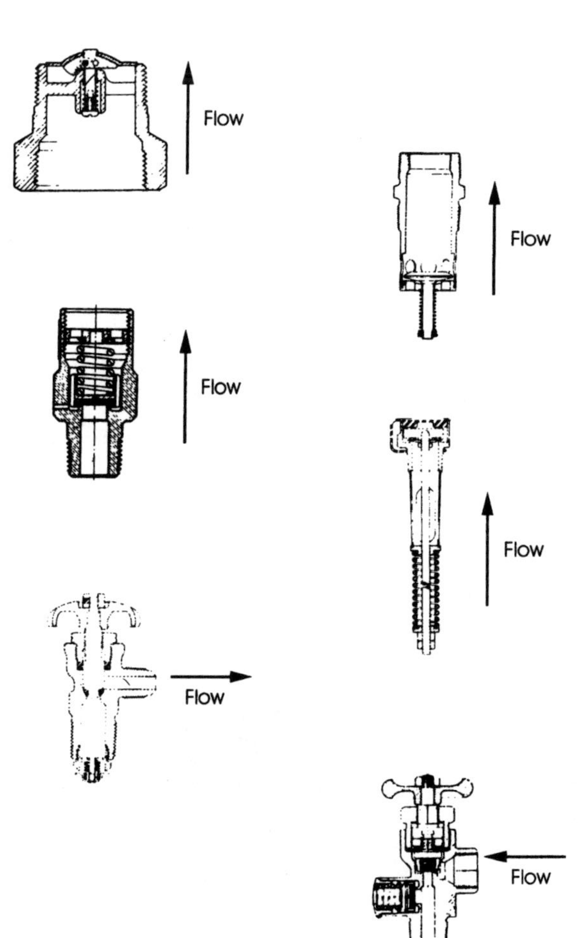

Notes

Gauging Devices

Liquid level gauge indicates fuel level by percentage in tank.

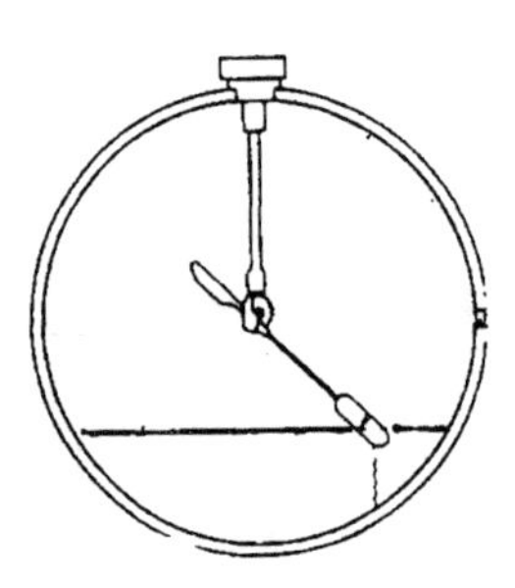

Magnetic liquid level gauging device is a rotating magnet geared to a float which affects an external needle to give a direct reading. Fixed liquid level gauge discharges liquid through bleeder when liquid level reaches end of tube.

Magnetic gauge is designed to give an instantaneous reading and operates by a remote sensor.

Rotary gauge must be operated by hand to determine liquid level.

BLEVE

BLEVE

BLEVEs (Boiling Liquid Expanding Vapor Explosions) and explosions involving liquefied gas cylinders are dramatic incidents. There is always potential for loss of life and severe property damage during such incidents.

BLEVEs occur when a contained liquid is well above its boiling point at normal atmospheric pressure, causing the container to break into two or more pieces. This phenomenon can happen to any contained liquid that meets the boiling point/pressure situation.

Many other confined products can explode when pressure builds. For example, a 55 gallon drum, half filled with water and with bungs tightly in place, can be dangerous in a fire. The fire can boil the water, increasing the pressure. Increased pressure forces the drum to expand and eventually explode. The drum can be propelled hundreds of feet. This problem magnifies if the drum contains a flammable, corrosive, or other dangerous product.

COMPRESSED GAS EXPLOSIONS

Compressed gases such as hydrogen, oxygen, carbon dioxide, argon, and nitrous oxide are placed in pressure cylinders in small quantities and in tube trailers. They are not liquefied like LP gas. Because the cylinder is filled entirely with vapor or gas, any flame impingement can easily cause the tank to fail, hurling the cylinder hundreds of feet. Most compressed gas cylinders are equipped with fusible discs or plugs to release excess pressure.

Notes

LIQUID PETROLEUM GAS (LPG) HAZARDS

Liquid petroleum gases (LPG) are flammable gases that are compressed to liquid to allow transport of greater quantities. They are usually transported in cylinder tanks painted white to reflect heat and keep pressure down. Rail tank cars painted black are also used to transport LPG; however, they are double insulated. Containers range in size from 1 gallon cylinders to 30,000 gallon rail tank cars.

LPG cars have pressure relief valves set at 250 psi. The valves open when pressure exceeds 250 psi and close when the pressure drops. A vapor space is left at the top of all LPG tanks. Flame impingement on the liquid portion of the tank is not as serious as fire near the vapor space. Flame on the vapor portion of the tank can quickly cause an explosion or BLEVE.

Similar explosions can occur with other pressurized liquid products, such as anhydrous ammonia, even though they are classified as non-flammable by the DOT. Containers are equipped with pressure relief valves.

Two types of problems can cause failure of an LPG container:

- hydrostatic destruction
- impinging fire on the vapor space

Hydrostatic Destruction

LPG fires should always be approached from the sides of the vessel, rather than from the ends.

If an overfilled vessel is impinged upon by fire, the relief valve will not be able to relieve enough pressure. The same problem can be caused by a faulty relief valve or relief valves buried in mud or hard soil due to an overturned vessel. When the vessel is 100% full of liquid and the temperature rises, the vessel can breach and fail. Projected fragments have been recorded as far as a mile away. The ends of a vessel are particularly prone to failure. This is why LPG fires should always be approached from the sides of the vessel, rather than from the ends.

Example of Hydrostatic Destruction

Impinging Fire on Vapor Space

Concentrated heat on the vapor area of a vessel will increase metal temperature enough to make it lose tensile strength. The tank will swell in the hot area. Pressure builds as the hot steel is drawn thinner. When the blister rips, the tank and contents erupt into a short-lived tower of fire.

Liquid level is important in a fire because the liquid absorbs heat. Therefore, the shell is protected from explosion if it is contacted by liquid. The part of the tank that does not have contact with liquid needs water flowing over it to keep it cool.

When vapor is released through the relief valve, the tank's pressure will decrease. The release of vapor actually cools the tank contents, which lowers the internal pressure further. If the relief valve opens and releases liquid because of overfill or an accident, the internal pressure on the tank will not change. For this reason, upset or overfilled LPG tanks are dangerous during a fire.

The tank should be cooled quickly using a sufficient amount of water. If this is not done, stay clear of the vessel and use deluge guns or flooding equipment. The rule of thumb for a large tank is, *If a minimum flow of 500 gpm on an impinged tank cannot be maintained, do not risk exposing personnel.* The actual amount of water needed will depend on the tank size. Five hundred gpm may not be sufficient to protect a 33,000 gallon rail car that is heavily involved in fire. Likewise, 500 gpm may send a small tank flying across the parking lot. The key is to maintain a film of water on the entire surface of the tank.

During the past 25 years, BLEVEs have caused loss of life and extensive property damage. Many of these accidents occurred due to lack of proper water application and a basic understanding of the physics involved. Emergency responders must quickly recognize when a tank rupture is likely. Unfortunately, there are very few signs, but there are some warning signals. Signals warning of possible tank rupture include:

- activation of the pressure relief valve

Impinging Fire on Vapor Space

Notes

Maintain a film of water on the entire surface of the tank to prevent rupture.

Notes

- increase in sounds from the pressure relief valve, indicating an increase in internal boiling and vapor production
- change in pitch from the pressure relief valve (higher and louder), indicating that the gas is escaping at a greater velocity and that internal pressure has increased
- visible steam from the tank surface, indicating a shell temperature of at least 212°F (100°C)
- increased space between the flame and pressure relief valve, indicating that internal tank pressure is probably increasing
- discoloration of the tank shell, usually in a specific area
- bulge or bubble in the tank shell
- pinging sound, indicating that metal is softened from flame impingement and stretched from internal pressure

There is no one method to mitigate the impinged compressed or liquefied gas tank emergency. Each situation will be different. Ask the following questions before establishing an action plan. Change the questions to fit the situation, location, and/or number of injuries.

- If there is a leak, where is it and how bad is it?
- What product(s) is/are involved?
- What is the shape of the container(s)?
- How much product does the container hold?
- How large is the fire?
- What is causing the fire?
- Where is the fire burning on the tank?
- How long has it been burning?
- Has the pressure relief valve activated? If so, when?
- Any changes in the fire?
- Any change in the sounds?
- How high is the flame from the pressure relief valve?

Obtain updated information while responding, and note changes. Keep track of the time between the alarm and water application when fighting any closed container fire. There are no guidelines as to when a tank has been impinged too long, yet the time factor is important in developing an action plan.

KEY POINTS TO REVIEW

By now you should have a basic understanding of:

- compressed gases
- liquefied gases
- physical and chemical properties of gases
- methods for identifying these products
- transport and storage vessels
- safety devices
- BLEVEs—how and why they occur

Before going on in this chapter, take time to review the common compressed gases found in Appendix D.

UNDERSTANDING CRYOGENICS

Notes

INTRODUCTION

The word ***cryogenic*** means very cold. The term was first used by scientists to describe any equipment, process, or product that required very low temperature refrigeration. As noted earlier, the DOT defines a cryogenic liquid as a refrigerated liquefied gas having a boiling point colder than −130°F (−90°C) at 14.7 psi (101.3 kPa) absolute. The difference between a liquefied compressed gas and a cryogenic liquid is that pressure maintains liquefied compressed gases, while refrigeration maintains cryogenics in a liquid state.

Although cryogenics have been around since the 1940s, only recently have emergency response personnel realized the hazards posed by transporting these materials. The advent of space exploration prompted the technological use of cryogenics in industry. Since that time, cryogenics have become common in every community. They are widely used in hospitals to provide oxygen for patients and as a freezing agent for certain surgeries. Cryogenics are used in the food industry for quick freezing processes. They are used in laser applications and in producing super electrical conductivity of various materials. New applications and uses are developed every year.

CHARACTERISTICS OF CRYOGENICS

There are four general characteristics that apply to all cryogenic liquids. These characteristics lead to a number of potential hazards that will be discussed later.

- The liquids are extremely cold. The boiling point for cryogenics can be −400°F (−240°C) or below.

Cryogenic Storage Facility

Notes

- Cryogenic liquids boil at ambient temperatures (surrounding air temperature). Theoretically, if outside temperatures were below the liquid's boiling point, it would remain in a liquid state.
- At normal atmospheric pressure, these materials exist as liquids. Gases must be cooled or pressurized to be converted into liquid. Cryogenics are produced by cooling.
- Cryogenics have a wide liquid/vapor ratio. For any given amount of liquid, a cryogenic produces many times more vapor.

Characteristics of Cryogenics

Temperature – All cryogenic liquids are extremely cold. Cryogenic liquids and their "boil-off" vapors can rapidly freeze human tissue. They can also cause many common materials such as carbon, steel, plastic, and rubber to become brittle or fracture under stress. The most extreme cryogenic liquids (liquid helium, liquid hydrogen, and liquid neon) can solidify air and other gases.

Pressure – Enormous pressure is produced when cryogenic liquids are vaporized in sealed containers. For example, when one unit of helium at one atmosphere of pressure vaporizes and warms to room temperature in an enclosed container, it can generate pressure greater than 14,500 psi. For this reason, pressurized cryogenic containers are protected by multiple pressure relief devices. These devices usually include a pressure relief valve for primary protection and a frangible disc for secondary protection.

Expansion – All cryogenic liquids produce large volumes of gas when they vaporize. For example, one unit of liquid nitrogen vaporizes to 695.7 units of nitrogen gas when warmed to 70°F (21°C) at 1 atmosphere of pressure. The volume expansion ratio of liquid oxygen is 860.4 to 1. Liquid neon has the highest expansion ratio of any industrial gas at 1,445 to 1.

Fogging – Most cryogenic liquids are odorless, colorless, and tasteless in the gaseous state. Liquid oxygen, however, is light blue. The cold boil-off vapors condense the moisture in the air, creating a visible fog. The fog usually extends over more area than the vaporizing gas.

Hazards – All of the cryogenic gases, except oxygen, displace breathable air. This can cause asphyxiation before a person realizes that a problem exists. Carbon monoxide adds to such hazards by its high toxicity and flammability. Liquid nitrogen and argon are chemically inactive and non-corrosive at cryogenic temperatures. Liquid oxygen is an oxidizer and enhances combustion of other materials. However, it will not burn itself. Liquid carbon monoxide, liquid natural gas, and liquid hydrogen are classified as flammable liquids and must be handled as such.

CRYOGENIC LIQUID PHYSICAL PROPERTIES

PHYSICAL PROPERTIES	Xenon (Xe)	Krypton (Kr)	Methane (CH_4)	Oxygen (O_2)	Argon (Ar)	Carbon Monoxide (CO)	Nitrogen (N_2)	Neon (Ne)	Hydrogen (H_2)	Helium (He)
Boiling point at 1 atm: °F °C	-161 -107	-244 -153	-259 -161	-297 -183	-303 -186	-312 -191	-321 -196	-411 -246	-423 -253	-452 -269
Melting point at 1 atm: °F °C	-169 -112	-251 -157	-297 -183	-361 -218	-309 -189	-341 -207	-346 -210	-416 -249	-435 -259	—[1] —
Density at boiling point and 1 atm (lb/cu ft)	191	151	26	71	87	49	50	75	4.4	7.8
Heat of vaporization at boiling point (BTU/lb)	41	46	219	92	70	93	86	37	193	10
Volume expansion ratio of liquid at boiling point and 1 atm to gas at 70°F and 1 atm	559	693	625	860	842	680	696	1,445	850	745
Flammable	No	No	Yes	No[2]	No	Yes	No	No	Yes	No

1. Helium does not solidify at 1 atm pressure.
2. Oxygen does not burn, but it supports and accelerates combustion. High oxygen atmospheres substantially increase combustion rates of other materials, and many form explosive mixtures with other combustibles. Flame temperatures in oxygen are higher than those in air.

Figure 5-33

Source: *Sax's*, CGA

Notes

THERMODYNAMICS OF CRYOGENICS

It is important to understand some of the ***thermodynamics***, or heat transfer processes, exhibited by cryogenic gases. When a gas is pressurized, its molecules increase movement, colliding with one another. This action warms the gas. The gas also compresses as it is pressurized, leaving less space between molecules. The greater the compression, the greater the action of the molecules, and the hotter the gas becomes. A good example of this is the way an SCBA air bottle warms as it is filled.

The opposite occurs if the gas is depressurized or decompressed. As the gas decompresses, the distance between the molecules increases. This slowing action decreases the number of collisions. As the number of collisions decreases, so does the amount of heat generated. The gas gets "colder," although technically it is actually losing heat. An example is the formation of dry ice when a CO_2 extinguisher discharges. The CO_2 is between 125 psi and 300 psi (depending on the type of extinguisher) inside the bottle. When discharged, pressure is decreased to atmospheric pressure. The CO_2 also decreases in temperature to at least –110°F (–79°C), cold enough to form dry ice. So, temperature and pressure determine the extent to which a gas is cooled. Remember that as the pressure of a gas is increased, the temperature also will increase. As pressure is decreased, the temperature will decrease.

All cryogenic liquids are produced through a process of alternate compression, cooling, and decompression. Typically, the gas is first pressurized to about 1,500 psi and cooled with ice water to 32°F (0°C). Then it is pressurized again to about 2,000 psi and cooled to –40°F (–40°C) with liquid ammonia. At this point the pressure is relieved and the resulting cooling causes the gas to liquefy. Some lower boiling point liquids may require pressurization above 2,000 psi.

Cryogenic Containers

CRYOGENIC LIQUID EXPLOSION AND FIRE PROPERTIES

	Oxygen (O_2)	Nitrogen (N_2)	Argon (Ar)	Helium (He)	Krypton (Kr)	Xenon (Xe)	Neon (Ne)	Methane (CH_4)	Hydrogen (H_2)	Carbon Monoxide (CO)
EXPLOSION HAZARD										
With combustible materials	Yes	No	No	No	No	No	No	No	No	No
With oxygen or air	—	No	No	No	No	No	No	Yes	Yes[1]	Yes
Pressure rupture if liquid or cold vapor is trapped	Yes	Yes	Yes	Yes	Yes	Yes	Yes	Yes	Yes	Yes
FIRE HAZARD										
Combustible	Moderate	Nil	Nil	Nil	Nil	Nil	Nil	Yes	Yes	Yes
Promotes ignition	Yes	No	No	No	No	No	No	Yes	Yes	Yes
Condenses air and expands flammable range	No	Yes	No	Yes	No	No	Yes	No	Yes	Yes
Flammable limits in air (percent by volume)	—	—	—	—	—	—	—	5–15	4–74	12–74

1. Within flammable limits.

Figure 5-34

Notes

CLASSES OF CRYOGENICS

Cryogenic gases can be divided into classes according to the type of hazard they present.

- ***Flammables.*** This group includes liquefied methane, also referred to as liquefied natural gas (LNG), and hydrogen.
- ***Oxidizers.*** This group includes liquefied oxygen, referred to as LOX, and fluorine.
- ***Toxics.*** This group includes carbon monoxide and fluorine. Carbon monoxide is also flammable, and fluorine is also an oxidizer.
- ***Non-flammables.*** Non-flammables are sometimes referred to as the inert gases. They include liquid nitrogen (LIN) and any of the noble gases (helium, neon, argon, krypton, and xenon).

HAZARDS OF CRYOGENICS

Cryogenics present four basic hazards:

- health hazards
- dangerous effects on other materials
- effects while in gaseous form
- hazards related to their large liquid/vapor ratio

Health Hazards

TISSUES

Cryogenic liquids are extremely dangerous to human tissue. The liquid or its vapors can rapidly freeze any exposed human tissue, causing a cryogenic burn. A cryogenic burn is considered to be nine times as severe as a chemical burn, and a chemical burn is nine times as severe as a thermal burn. Cryogenic burns are extremely painful and can be disfiguring, even after short exposure.

A burn's severity depends on the length of exposure, the form of exposure (vapor or liquid), and the type of exposure (splash or immersion). A small splash of the liquid can freeze localized surface tissues. Immersion in the liquid material causes the most severe damage. Immersion damages deep tissue by freezing the tissue solid. This type of injury causes damage to all surrounding tissues. For example, if a finger is immersed in a cryogenic, it freezes from the outside to the center of the bone. The tissue is so cooled that it becomes brittle in the same way certain types of plastics become brittle when exposed to extreme cold. If that same finger is then struck with a hard object, it can shatter like glass. If not struck, the tissue damage is as deep as the bone marrow. The same type of scenario can occur with an arm or leg with longer exposure.

Remember that cryogenic liquids will cool any exposed material, such as a piece of metal, to a temperature close to their own. If a person touches or picks up an exposed piece of metal with an unprotected hand, the hand tissue will freeze and may stick to the metal. Attempting to remove the hand from the metal can cause the frozen tissue to tear from the body.

CRYOGENIC LIQUID HEALTH HAZARDS

	Oxygen (O_2)	Nitrogen (N_2)	Argon (Ar)	Helium (He)	Krypton (Kr)	Xenon (Xe)	Neon (Ne)	Methane (CH_4)	Hydrogen (H_2)	Carbon Monoxide (CO)
LOCAL EFFECTS										
Cold contact injuries result from skin contact with liquid, impinging vapors, or surfaces at liquid temperature	Yes	Yes	Yes	Yes	Yes	Yes	Yes	Yes	Yes	Yes
GENERAL EFFECTS										
Inhalation of vapor or cold gas produces respiratory discomfort or difficulty	Yes	Yes	Yes	Yes	Yes	Yes	Yes	Yes	Yes	Yes
Exposure to vapor or cold gas produces frostbite or hypothermia	Yes	Yes	Yes	Yes	Yes	Yes	Yes	Yes	Yes	Yes
Inhalation of the gas can cause asphyxiation	No	Yes	Yes	Yes	Yes	Yes	Yes	Yes	Yes	Yes
Toxic	No	No	No	No	No	No	No	No	No	Yes

Figure 5-35

Notes

Cryogenic liquids produce large amounts of vapor that are capable of excluding the air from a given area.

ASPHYXIATION

Another major health hazard associated with cryogenics is asphyxiation. Asphyxiation is possible with all cryogenics except oxygen. Cryogenic liquids produce large amounts of vapor that are capable of excluding the air from a given area. This property is especially critical if a spill occurs in a confined space. As the liquid vaporizes, it dilutes the air. The vapors are cold and dense, and they hang close to the ground. The air in the immediate area will rise since it is warmer than the vapor. All of these factors combine to produce an oxygen deficient atmosphere. If the incident occurs inside a building, some areas may be totally depleted of oxygen.

If individuals enter an oxygen deficient area without wearing self-contained breathing apparatus (SCBA), they will experience subtle mental changes. Many times they are unaware of potential hazards and may not exercise good judgment. They may also fall unconscious before realizing the danger.

Effects on Other Materials

Cryogenics can cause materials to become brittle, just as they do human tissue. In fact, this is an important use of cryogenics in industry. Industrial materials that are too pliable to break or crack may be exposed to liquid nitrogen or other cryogenics. Once exposed, the material is easily ground. This process has been used on tires, foam type plastics, rubbers, and foods such as spices.

Problems arise when materials are accidentally or unintentionally exposed to cryogenic liquids. Materials that are normally pliable, durable, or even rigid become brittle and fragile. For example, if a truck's tires are exposed to a cryogenic, they become brittle and may blow out. This is particularly dangerous if the cryogenic is either flammable or an oxidizer. The blowout could initiate a fire or explosion.

EFFECTS ON METAL

Cryogenics also cause metal objects to become brittle. If the metal is stressed or receives an impact, it can lose its strength. For example, if the object happens to be a pressurized container, the stress created by its internal pressure may cause the container to fail. Or, if the object is a mechanical device with moving parts, some or all of the parts may disintegrate. This is particularly dangerous if the parts move at a high rate of speed. Pieces of the machine may be projected at high speeds as the parts disintegrate.

EFFECTS ON GASES

Cryogenic liquids may also affect other exposed gases. The cryogenic may either liquefy or solidify the other gas, depending on the boiling and freezing point of the gas. Consider the effect liquid nitrogen has on air. Liquid nitrogen has a temperature at or below −321°F (−196°C), the same as the boiling point of nitrogen. Air is composed of 78% nitrogen, 21% oxygen, and 1% other gases. Because the boiling point of oxygen is −297°F (−183°C), the liquid nitrogen is cold enough to liquefy air. After the nitrogen has boiled away, liquid oxygen is left behind.

Liquid oxygen remains for a simple reason. When two liquids with different boiling points are mixed and heated, the temperature of the mixture increases until the lower boiling point of the two is reached. After

Notes

that, any heat added to the mixture is used to change the liquid to a gas in order to keep the temperature constant (within several degrees). As the lower boiling point liquid is boiled away, the temperature will rise to the boiling point of the other liquid. This is the same principle of any distillation process, including the production of many petroleum products and liquor.

If air is exposed to liquefied hydrogen, it freezes solid. The liquefied hydrogen has a temperature of about –423°F (–253°C) (the boiling point of hydrogen). Oxygen and nitrogen freeze at –361°F (–218°C) and –346°F (–210°C), respectively. The danger is that the solidified air can either keep open or close the container's pressure relief device.

EFFECTS OF OTHER MATERIALS ON CRYOGENICS

Another consideration is the effect that other materials have on the behavior of a cryogenic liquid. Because cryogenics have such low temperatures, most other materials with which they have contact will act as a super-heated object. The cryogenic may exhibit one of several different behaviors depending on the amount involved:

- If the cryogenic slowly leaks over a smooth surface, such as concrete, it forms small balls or drops of liquid. These drops scurry over the surface the way water does on a hot skillet. The droplets are capable of rapidly moving at random and can be rather large—1/2 to 3/4 inch in diameter and possibly larger. A vapor cloud can also form rapidly. Its size depends upon the distance the drops have traveled. The movement continues until the drops either evaporate, due to the boiling liquid, or until they reach a low spot and form a pool.
- If the leak is larger, a stream of the liquid develops. The stream moves downhill and usually forms a pool in some low spot. There is generally enough liquid to considerably cool the exposed surface. Again, the liquid boils, and a vapor cloud forms. The boiling action diminishes as the surface and the bottom of the pool cool. Eventually the liquid evaporates.
- If there is a major leak, such as a major container failure or large product line rupture, a sizeable cryogenic liquid pool forms. A large vapor cloud also forms, and the liquid boils vigorously. As the exposed surface cools, the intensity of the boiling decreases, as does the rate of vapor production. If, however, some other material enters the pool, the rate of vapor production and boiling increases dramatically. The boiling may become so vigorous that the surrounding area is showered with liquid. If the material that entered the pool is a large object, its rapid cooling will cause tremendous internal stress due to uneven contraction and shrinkage. This contraction can violently crack or break the object.

Notes

It is important to realize that water, which is usually considered a cooling agent, will also react as a super-heated object with cryogenics.

EFFECT OF WATER ON CRYOGENICS

It is important to realize that water, which is usually considered a cooling agent, will also react as a super-heated object. Even if the water is close to freezing, it can be as much as 500°F (260°C) or at least 200°F (93°C) above the cryogenic's boiling point. This should be considered before using water to cool a container filled with a cryogenic liquid. If the container is damaged externally and the inner skin is exposed, water will actually heat the product. Heating the product will either initiate or increase boiling. The increasing pressure could overpower the pressure relieving device, leading to a violent rupture. (Container design and construction will be discussed later in this chapter.)

Applying water directly on the spilled liquid will also increase vapor production and dissipation of the spill. Before applying water, evaluate the consequences. If the pool is small, an "ice cap" could form over the liquid. Pressure could build and cause the ice to rupture. Also, note the vapor's direction and determine if it is moving to areas where it will prove dangerous, such as low lying inhabited areas or toward ignition sources. Control the vapors and flow while evacuating the area and eliminating the ignition sources. Also remember not to apply water to the vent area of the container. This could freeze the vent open or, more likely, closed and lead to overpressurization.

Effects of Cryogenics in Gaseous Form

The hazards of cryogenics are greatly intensified when these materials are in their gaseous form. Hydrogen and oxygen serve as good examples. Liquid oxygen is an extremely dangerous oxidizing liquid. It can cause certain organic materials to ignite spontaneously. Liquid oxygen intensifies the rate of combustion and, when mixed with petroleum products, it will produce a contact explosive. Remember that a hard hose stream may initiate the explosion. Liquid oxygen is also capable of producing an oxygen enriched atmosphere over a large area. The enriched atmosphere can cause internal combustion engines to "run away" to a high rpm rate and disintegrate.

Liquid hydrogen is flammable and is equally dangerous. Hydrogen has a flammable range from 4% to 75% and produces approximately 80,000 BTU/pound. (A BTU or ***British thermal unit*** is the amount of heat required to raise the temperature of 1 pound of water 1 degree Fahrenheit.) Liquid hydrogen burns with an almost invisible flame and is readily ignited by heat, friction, static, or electrical sparks. It also ignites spontaneously as it escapes a pressurized container. In addition, portable radio transmissions from within an area containing hydrogen gas can cause ignition. Although hydrogen is normally lighter than air, it will be heavier than the air near a spill because of the cold temperatures produced by the liquid. Liquid hydrogen may impregnate porous materials, causing them to be more flammable than normal. Damaging explosions can occur if confined liquid hydrogen is ignited.

Expansion Ratio Hazard

Notes

Another hazard associated with cryogenic liquids is the large vapor or gas to liquid ratio and the resulting vapor clouds. The gas/liquid ratio is sometimes referred to as the expansion ratio. A gas/liquid ratio compares the amount of gas produced when a given volume of liquid vaporizes. In other words, for every gallon of liquid that vaporizes, a specific amount of gas will be produced.

Cryogenic liquids have a large gas/liquid ratio. For example, a gallon of liquid nitrogen produces 696 gallons of nitrogen gas. The gas/liquid ratio is often written as 696/1. (See Figure 5-33 for specific values.)

The primary dangers associated with large gas/liquid ratios involve storage containers, transportation containers, and piping systems.

For the haz-mat responder, the primary dangers associated with large gas/liquid ratios involve storage containers, transportation containers, and piping systems. Again, if the product inside a container or pipe is heated, the liquid will produce a large amount of gas. All cryogenic containers, except dewars, are equipped with pressure relieving valves and usually frangible discs for pressure increases. If sufficient heat, however, causes the product to boil, it increases the chance that safety devices will fail.

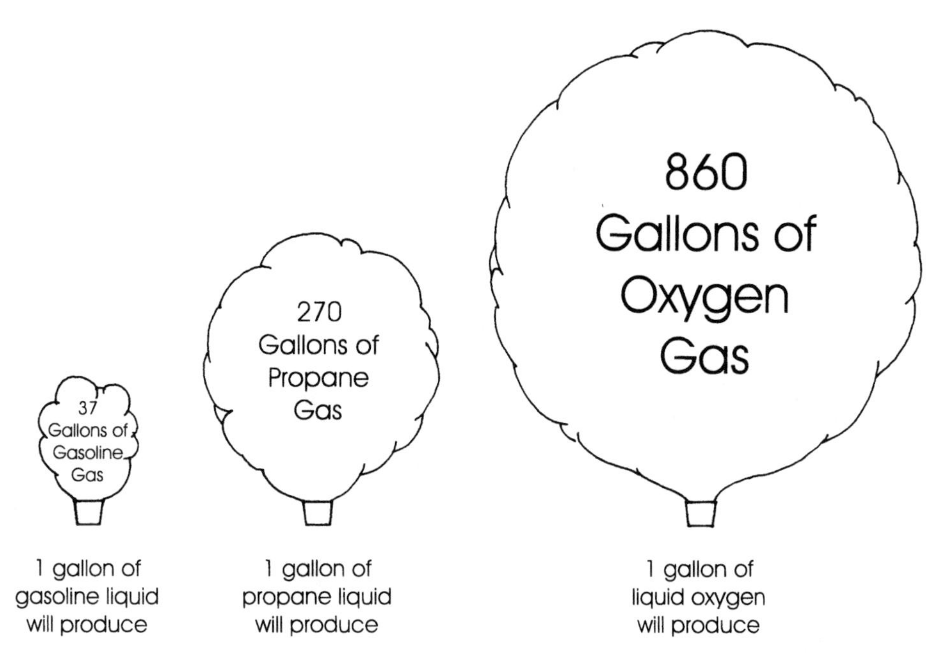

Figure 5-36. Liquid/Vapor Ratio

Notes

Vapor Cloud Hazard

A large liquid/gas ratio is not hazardous simply because a lot of gas is produced, but because the gas and its vapor cloud can be dangerous. Therefore, it is important to discuss the vapor and its effects.

Cryogenic liquids produce gases by boiling. These gases are sometimes called boil-off vapors and are considered a pure form of the liquid. The low temperature liquids produce very cold gas. In fact, the base temperature will be the same as the liquid until the gas moves from its source and is warmed by the air. Because the gas is so cold, it is very dense, remains close to the ground, and tends to flow into low lying areas. This is true even for gases that are lighter than air, such as hydrogen and helium. Some form of visible vapor cloud will always be produced. The visible cloud, however, is condensed water vapor or fog, not the gas itself. All cryogenic gases are colorless and invisible.

The visible vapors are extremely cold while in the vicinity of the liquid and are capable of rapidly freezing human tissue and causing other materials to become brittle. Although the fog will extend over an area larger than the vicinity of the liquid, it does not represent the total area the gas has covered. As the gas and air are warmed by their surroundings, the fog disappears, and the gas becomes visually undetectable. The gas and fog present two basic dangers:

- The fog is usually very thick and can totally obscure vision. It becomes difficult to locate the leak source and determine the dimensions of the spill. Low visibility also obscures objects that someone may trip over or fall into, including the liquid itself.
- The large amount of gas can create an oxygen deficient atmosphere, except in the case of liquid oxygen. Asphyxiation can occur outdoors as well as in a confined space. Therefore, evacuation of the immediate and surrounding areas is necessary, especially low lying areas.

CRYOGENIC CONTAINERS

Cryogenic liquids are stored, shipped, and handled in three types of containers: dewars, pressurized cylinders, and tanks. All three are double-walled containers with a metal outer skin or shell separated from an inner skin by either a vacuum or vacuum and insulated layer. The insulation layer prevents outside heat from reaching the product. In other words, all three containers are nothing more than efficient thermos bottles.

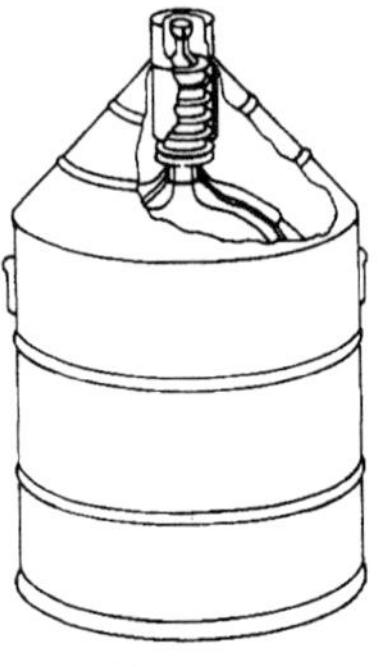

Dewar

Dewars

Dewars are small containers used to store cryogenic liquids for short periods. They contain from 5 to 200 liters of product (1 liter equals about 1 quart). The smaller dewars may be constructed of glass and closely resemble large mouth thermoses. Such containers are not pressure vessels. At most they are equipped with a dust cap that covers the neck of the container. The cap cuts down the amount of heat loss and limits ice formation that could plug the neck. Dewars that are 50 liters (about 12-1/2 gallons) or larger have a tight fitting

Notes

cap that slightly pressurizes the container. A transfer tube placed into the liquid uses pressure to push the liquid through the tube. These containers are not suitable for liquid helium or hydrogen.

Liquid Cylinder

Liquid Cylinders

Liquid cylinders are insulated vacuum jacketed containers that range in capacity from 90 to 200 liters. These cylinders are true pressure vessels that store liquefied gases under pressure in the range of 0 to 235 psi. The product may be removed either in a gaseous or liquid form. If the user dispenses gas rather than liquid, the cylinder normally operates from 0 to 180 psi. Each cylinder is equipped with a pressure relief valve and a rupture disc.

Cryogenic Tanks

Cryogenic tanks are either spherical or cylindrical. They may be permanently installed at a given location or mounted on transport trailers. The capacity of permanent tanks ranges from 500 to 420,000 gallons (liters are not used). Transportation tanks usually range from 10,000 to 14,000 gallons.

The product is withdrawn from the tank by a process that creates a slight pressure within. The maximum normal pressure within any tank is less than 25 psig. This slight pressure is created by allowing a small amount of product to pass through a ***heat exchanger***. A heat exchanger is a piping

Heat Exchanger

Notes

system with attached fins that allows a rapid transfer of heat between the product and the outside. As the product warms, it produces more gas, which increases pressure. Some transportation tanks have an ***auxiliary pump*** that generally allows the transfer of the cryogenic in its liquid form.

Auxiliary Pump

Safety Devices

All tanks are equipped with various pressure relief valves and rupture discs. They also are equipped with a variety of piping systems that involve complicated flow patterns. If the wrong valve is either opened or closed during an incident, the tank may experience too much pressure and violently rupture. *Do not open or close any valves on a cryogenic tank without specific instruction from an authorized company representative.* Remember that water will not cool a tank unless there is fire exposure.

Safety Devices

Pressure is periodically vented from tanks equipped with pressure relief devices. Between 0.4% and 3% of the contents of an average tank vaporizes each day depending on the material, container, and outside temperature. Vapor may be seen at the mouth of the vent stack once excess pressure is vented. If the outside temperature is warm, or if the tank was recently filled, this venting process may occur at short intervals. The venting by itself is of no concern.

Most cryogenic tank trucks are equipped with three safety relief devices:

- 1 inch safety relief valve that operates at 7-10 psi for normal venting as outlined above
- 2 inch safety relief valve that operates at 32 psi
- 2 inch rupture disc that operates at 61 psi

Valves

Notes

Figure 5-37. Cryogenic Rail Tank Car

HYDROSTATIC TANK DESTRUCTION AND BLEVE POTENTIAL

Two terms used on the table of physical properties and reviewed earlier were critical temperature and critical pressure. ***Critical temperature*** is the maximum temperature at which a liquid can be heated under pressure and still remain a liquid. If a liquid is heated above this temperature, it will change to a gas no matter how great the pressure. ***Critical pressure*** is the pressure that must be applied to a liquid when it reaches critical temperature to assure that the product remains a liquid.

If a liquid is heated above its critical temperature, it will change to a gas. As this change occurs, there is a drastic pressure build-up. This pressure build-up, however, will not stop the conversion to gas. The container will rupture as the pressure reaches high levels. This type of rupture is known as ***hydrostatic tank destruction***.

Can cryogenic tanks BLEVE? Yes! The same physical laws outlined earlier apply here. If fire impinges the tank's vapor space or safety devices are damaged, or both, then a BLEVE can occur.

KEY POINTS TO REVIEW

- Cryogenic liquids have a boiling point below −130°F (−90°C). They are kept liquid by cold rather than pressure.
- Cryogenic liquids will remain liquid at normal pressure.
- The hazard groups of cryogenic liquids are:
  - flammables
  - oxidizers
  - toxics
  - non-flammables
- The four hazards associated with cryogenics are:
  - health hazards
  - effects on other materials
  - effects while in gaseous form
  - large liquid/vapor ratio
- Safety relief devices include pressure relief valves and rupture discs.

Notes

RESPONDING TO GAS EMERGENCIES

Responding to a gas emergency can be complex and dangerous. These materials are found as either gas or liquid/gas as in the case of gas liquefied under pressure and cryogenic materials. In addition, gases can be flammables, non-flammables, corrosives, oxidizers, poisons, radioactives, and ORMs (Other Regulated Materials). For response purposes, the following categories will be considered:

- compressed gases
- liquefied gases
- cryogenics

RESPONDING TO COMPRESSED GAS EMERGENCIES

The majority of compressed gas emergencies involve cylinders used for small industrial processes, by consumers, and for medical applications. There are many sizes and styles of compressed gas cylinders, from the 1 pound butane cylinder used by consumers, to the 420 pound cylinder commonly used at construction sites.

The use of small compressed gas cylinders has created a potential compressed gas emergency in virtually every community in the United States and Canada. The HAZ-MAT procedure of hazard identification, action plan, zoning, managing the incident, assistance, and termination will now be applied to the compressed gas emergency.

HAZ-MAT — HAZARD IDENTIFICATION

Identify the product prior to taking action. Trucks carrying compressed gas cylinders are not placarded unless the cylinders contain a poisonous gas or the product weighs 1,001 pounds or more. Many cylinders have a label and/or the product name stenciled directly on the cylinder. In addition, many cylinders can be identified by their shape. The following is a review of the hazard identification method:

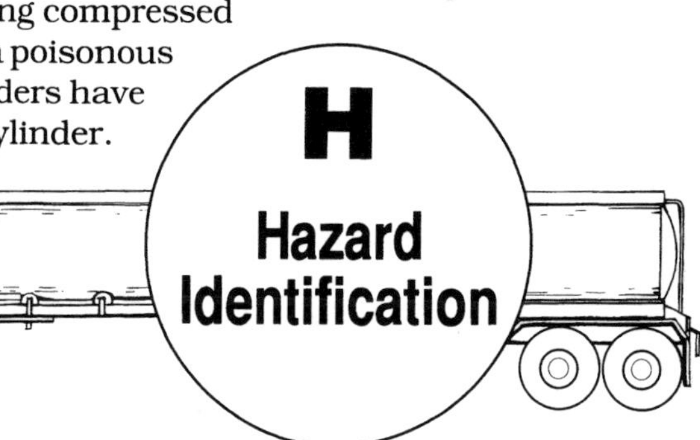

- Location of the incident
  - market
  - retail store
  - farm supply
  - roadway
  - warehouse
  - field
- Markings or identification
  - placards and/or labels
  - tank shape
  - company names
- Sounds
  - hissing
- Odor
  - burnt almonds
  - sweet smell
  - irritating to nose and eyes
- Visual indications
  - vapor clouds
  - heat waves

Notes

Once a compressed gas has been identified, evaluate the possible hazards while developing an action plan.

- Location of the leak and/or fire
  - building
  - open field
  - highway
  - trailer
  - fixed storage facility
  - confined spaces
  - below grade
- Cause of the incident
  - traffic accident
  - worker's equipment
  - chemical reaction
  - arson and/or fire
  - plane crash
  - train accident
- Injuries
  - number
  - severity
- Hazards of gases
  - oxidizer
  - flammable
  - poison, toxic
  - corrosive
- Chemical properties
  - flammable range
  - vapor density
  - reactivity
  - water solubility
- Cylinder position
  - The valve can be broken if a cylinder is accidentally knocked over. The sudden release of pressure can turn the cylinder into a fast traveling missile.
  - The safety relief/vent may not operate because of cylinder position.
- Fire hazard
  - Consider the severity of the fire, if one is present. There is no liquid in the cylinder to absorb heat if the cylinder is exposed to fire. Temperatures as low as 130°F (54°C) can activate some relief valves.
  - In case of fire, identify smoke color (light brown, black, orange).

HAZ-MAT — ACTION PLAN

When developing an action plan for a compressed gas emergency, determine:

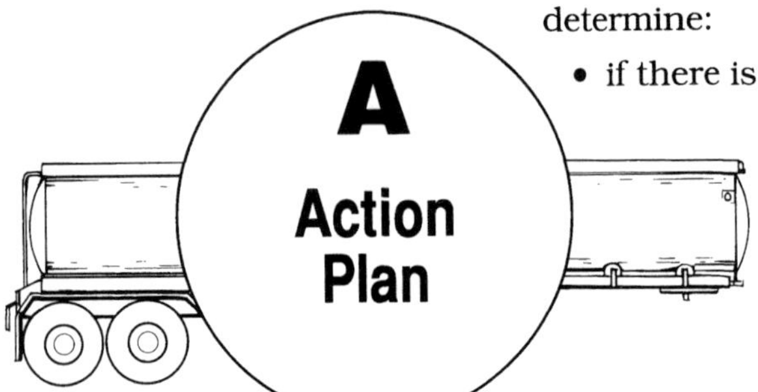

- if there is a leak, and its severity
- if the cylinder is impinged by fire

If there is a **LEAK but NO FIRE:**

- Determine the number of cylinders leaking.
- Determine what is leaking. If more than one product is leaking, will the mixture create new problems?
- Monitor the atmosphere with detection equipment to determine if the gas is flammable or if it has other detectable hazards.
- Secure the cylinders. If they cannot be secured, determine where the cylinder will travel if the valve breaks.

Notes

- Try to locate where the gas is moving. It may be necessary to shut down air handling units so that gas is not pulled into heating or air conditioning units. Secure sources of ignition if the material is a flammable or an oxidizer.
- Locate the leaking cylinders. Many times small cylinders will bleed off in a few minutes. At an outdoor location, this may be the best approach. If the cylinder is inside, try to safely move it outside.
- Determine why the cylinders are leaking.
- Determine if evacuation is necessary.
- Always use self-contained breathing apparatus (SCBA) and full protective clothing. Although the gas may be only slightly toxic, many products are asphyxiants or anesthetics. SCBAs will also protect in case of ignition.
- Remember that some gases are heavier or lighter than air. Your response and mitigation methods must adjust for these factors.

If **FIRE** is involved in the product area:

- Water may be needed to cool the cylinder or protect exposed cylinders. The heat and velocity of the fire may easily ignite nearby combustibles.
- Control the gas flow if the fire is extinguished. An extinguished cylinder may create a large pocket of gas that could reignite and create an explosion.
- Determine the position of the cylinders.
- Determine if the safety or pressure relief devices are operating. They can indicate the intensity of the fire. Do not rely entirely on these devices, since they may not be operating properly. If the devices are buried or damaged, excessive pressure may build up, causing the tank to rupture or BLEVE.

High Pressure Tube Trailer Carrying Hydrogen

Notes

- Determine if the cylinders are secure. If the cylinders are not secured, they can be easily knocked over by a hose stream, making the situation worse.
- Runoff may need to be contained. This is especially true if the materials are poisonous or corrosive.

An effective action plan is based on gathering accurate information and determining the dangers. Incidents involving small compressed gas cylinders are usually minor. Some larger compressed gas cylinders used for road transportation can create special hazards. Hydrogen is a good example. Remember, do not extinguish the fire unless the gas supply can be secured.

HAZ-MAT — ZONING

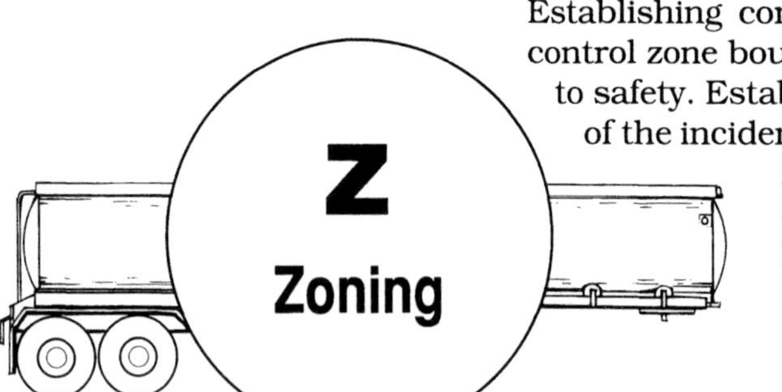

Establishing control zones is a priority. Incident size will determine control zone boundaries. People in the immediate area must be moved to safety. Establish an initial hot zone of 250 feet to help gain control of the incident. Use atmospheric metering devices to determine safe zones. Do not rely on the human nose as a gas detection device! Many gases are colorless, odorless, and tasteless. The hot zone can be expanded, and a warm and cold zone can be added if necessary.

HAZ-MAT — MANAGING THE INCIDENT

Scene Management

Every emergency requires a competent incident commander. The compressed gas emergency is no exception. The primary concern must be personnel safety. If the incident is a loser, no involvement may be the best decision. To handle the incident effectively, determine the level of incident, establish control zones, and set up the command structure.

Exploding cylinders that may be propelled into the area require that a good assessment and action plan be formulated and initiated by first arriving units. Different locations require different action. Action will be different if the emergency is in a large compressed gas facility than if it is in an auto repair shop. ***Only one person can be in charge***, and everyone should know who that person is.

Call for help early. Do not be afraid to call for more help than may be needed. Be sure to have vehicles report to the staging area and not to the scene until that specific unit is needed. Emergency vehicles may be sent back if not needed. Have an adequate water supply, and anticipate additional resources which may be needed to control the incident.

Rescue Considerations

Notes

Injuries from compressed gas emergencies will vary; however, anticipate the following types of injuries:

- Fragmentation injuries can occur if a cylinder explodes. Pieces of metal will fly in every direction at a terrific speed.
- Blunt trauma injuries will occur from spinning or projected cylinders.
- Chemical burns may occur. Inhaling corrosive fumes can cause chemical pneumonia.
- Respiratory distress is possible after inhaling toxic and non-flammable materials.
- Leaking gases can be very cold, causing thermal injury.
- Thermal burns may occur if a flammable gas ignites.
- Asphyxiation can occur from displacement of oxygen.
- Products can be toxic by absorption or inhalation.

General Treatment for Compressed and Liquefied Gas Exposures

The following are general procedures for three types of gases:

- flammable gases
- non-flammable gases
- poisonous gases

When a specific gas is identified, always research and use the recommended care and treatment procedures.

Flammable Gases

Products .... Propane, butane, hydrogen, acetylene, ethyl methyl ether.

Containers .... Compressed gas cylinders, tank trucks, tank cars.

Life Hazard .... Gases may act as simple asphyxiants, irritants, and anesthetics at high concentrations. Some attack the central nervous system, causing muscular weakness, unconsciousness, and respiratory paralysis. Liquefied forms may cause frostbite.

Signs/Symptoms ..... Patients may exhibit respiratory distress, signs of pulmonary edema, or respiratory arrest. Signs of shock and cardiac arrhythmias may be present. The patient may have a decreased level of consciousness with headache, confusion, dizziness, seizures, and coma. Nausea and vomiting may be present. Skin and eye irritation may be seen. Extensive frostbite injuries are possible with liquefied gas releases. Thermal injury may result if the gas is ignited.

Basic Life Support .... Wear proper protective equipment and remove patient from contaminated area. Administer oxygen. If patient has any signs of skin irritation, remove pa-

Notes

tient's clothing and decontaminate with copious amounts of soap and water. Irrigate eyes thoroughly and continue during transport as necessary. Watch for signs of shock, pulmonary edema, and decreased level of consciousness/seizures and treat as necessary. If frostbite occurs and there is an extended transport time, the injured area can be warmed gradually in a water bath at 104-106°F (40-41°C). Be prepared to treat thermal burns.

Advanced Life Support .... Assure an adequate airway and assist ventilations as necessary. Start IV D5W TKO if blood pressure is stable. Use replacement fluids such as LR if there are signs of hypovolemic shock or in cases of thermal burns. Watch for signs of fluid overload. Monitor for cardiac arrhythmias/pulmonary edema and treat as necessary. Seizures may require treatment with an anticonvulsant. Administer topical anesthetic to eyes for easier irrigation and to reduce pain. Follow local protocols for all drug therapy and medical treatment.

Other Information .... Use caution, as there may be an ignitable mixture of gas without a visible vapor cloud. Eliminate all sources of ignition. Liquefied gases have the potential for boiling liquid expanding vapor explosion (BLEVE).

Non-Flammable Gases

Products .... Ammonia, oxygen, nitrogen, carbon dioxide.

Containers .... Compressed gas cylinders, tank trucks, tank cars.

Life Hazard .... Gases may act as simple asphyxiants in large concentrations. Some products may be toxic and/or irritating to the skin, eyes, and mucous membranes. Lung damage may cause pulmonary edema and respiratory arrest. May burn the skin when mixed with moisture or perspiration. Liquefied forms may cause frostbite. In rare circumstances, some products have ignited and burned.

Signs/Symptoms ..... Patients may exhibit respiratory distress, signs of pulmonary edema, or respiratory arrest. Signs of shock and cardiac arrhythmias may be present. The patient may have a decreased level of consciousness with headache, confusion, dizziness, seizures, and coma. Nausea and vomiting may be present. Many products may cause severe eye, skin, and mucous membrane irritation. Burning or stinging to the eyes, nose, and throat, lacrimation, and chemical burns to moist areas of the body are possible. Extensive frostbite injuries are possible with cryogenic or liquefied gas releases.

Notes

Basic Life Support .... Wear proper protective equipment and remove patient from contaminated area. Administer oxygen. If patient has any signs of skin irritation, remove patient's clothing and decontaminate with copious amounts of soap and water. Irrigate eyes thoroughly and continue during transport as necessary. Watch for signs of shock, pulmonary edema, and decreased level of consciousness/seizures and treat as necessary. If frostbite occurs and there is an extended transport time, the injured area can be warmed gradually in a water bath at 104-106°F (40-41°C).

Advanced Life Support .... Assure an adequate airway and assist ventilations as necessary. Start IV D_5W TKO if blood pressure is stable. Use replacement fluids such as LR to treat signs of hypovolemic shock, but watch for signs of fluid overload. Monitor for cardiac arrhythmias/pulmonary edema and treat as necessary. Seizures may require treatment with an anticonvulsant. Administer topical anesthetic to eyes for easier irrigation and to reduce pain. Follow local protocols for all drug therapy and medical treatment.

Other Information .... Liquefied products may present a boiling liquid expanding vapor explosion (BLEVE) hazard. While this rapid release may not result in a typical fireball, there may be a substantial release of pressure/energy, resulting in traumatic injuries. Many of these products have an enormous expansion ratio. This will result in a large vapor cloud, increasing the chance of respiratory exposure.

Poisonous Gases

Products .... Phosgene, cyanide vapors, chlorine, phosphine, fluorine, hydrogen sulfide.

Containers .... Compressed gas cylinders, tank trucks, tank cars.

Life Hazard .... Extremely toxic. Some products can cause death immediately after inhalation. Products may be strong lung irritants. Different poisons have varying effects on the body. Some present immediate symptoms, while others have delayed effects. Poisons may be inhaled, and some may be absorbed through the skin. Generalized symptoms are discussed here, but it must be remembered that symptoms may vary markedly from product to product. Damage to the heart, lungs, kidney, and/or brain may occur. Liquefied forms may cause frostbite. Identification of the specific poison and the route of exposure is vital to determine proper treatment. Some of these products may also be flammable.

Notes

Signs/Symptoms ..... Respiratory signs and symptoms can include respiratory distress with coughing, choking, bloody sputum, pulmonary edema, and respiratory arrest. Signs of shock and cardiac arrhythmias may be present. The patient may present with a decreased level of consciousness, seizures, or coma. Severe nausea, vomiting, and abdominal pain may be present. Irritation to the eyes, mucous membranes, or skin may result. Symptoms may be immediate or delayed for hours or days. Many of these products may act as chemical asphyxiants and affect the way the body transports or utilizes oxygen. Extensive frostbite injuries are possible with liquefied gas releases.

Basic Life Support .... Wear proper protective equipment and remove patient from contaminated area. Administer oxygen. If there are any indications of skin exposure or irritation, remove patient's clothing and decontaminate with copious amounts of soap and water. Irrigate eyes thoroughly and continue during transport as necessary. Watch for signs of shock, pulmonary edema, and decreased level of consciousness/seizures and treat as necessary. If frostbite occurs and there is an extended transport time, the injured area can be warmed gradually in a water bath at 104-106°F (40-41°C).

Advanced Life Support .... Assure an adequate airway and assist ventilations as necessary. Start IV D_5W TKO if blood pressure is stable. Use replacement fluids such as LR if signs of hypovolemic shock are present. Watch for signs of fluid overload. Monitor for cardiac arrhythmias/pulmonary edema and treat as necessary. Seizures may require treatment with an anticonvulsant. Administer topical anesthetic to eyes for easier irrigation and to reduce pain. Follow local protocols for all drug therapy and medical treatment.

Other Information .... Each poison has different actions and produces different signs and symptoms. Some poisons may be flammable. Treat each according to its effect on the patient. Antidotes are available for only a small number of products. Many poisons do not have adequate warning signs. Be cautious, and do not allow responders to become victims.

HAZ-MAT — ASSISTANCE

Use multiple references to determine product hazards. Call available technical advisors for additional response information, beginning with the shipper or local distributor. Many shippers have regional response teams to assist in clean-up and product removal.

A

Assistance

HAZ-MAT — TERMINATION

Terminating a compressed gas emergency should be as systematic as the mitigation efforts. The following are critical aspects of termination that apply to a compressed gas emergency.

T

Termination

Decontamination

Decontaminating a compressed gas emergency varies, depending upon the materials involved. All clothing should be removed if personnel exposure is significant. Decontaminate individuals with plenty of water and mild soap. Special care should be taken to decontaminate protective clothing worn by first responders. The product involved may saturate protective clothing in gas emergencies. This could contaminate the wearer on a subsequent emergency or, in some cases, may react if exposed to heat or another chemical. Contain decontamination solution to aid in determining the degree of contamination. This is especially true with poisons and corrosive materials.

Medical Screening

The termination phase of an incident is an excellent time to conduct a medical screening of personnel. Inhalation of the gas is perhaps the greatest risk to responders. Early recognition of symptoms will insure the safety of personnel. Vital signs should be taken after the decontamination process and again before personnel are reassigned. Evaluation over an extended period of time may be necessary because of the delayed effect of some gases.

Incident Debriefing

Although most compressed gas emergencies are not catastrophic, the stress and emotional level of personnel should be evaluated at this time. The medical screening process is a good time to assess personnel anxiety levels. This should be done even after a small incident.

Post-Incident Analysis

Even though compressed gas incidents are fairly common, a critique of the operation should be conducted. There are lessons to learn in any incident. Evaluate the incident, define problems, and establish a plan for the next compressed gas emergency.

Notes

KEY POINTS TO REVIEW

- Compressed gases are commonly used products. They are usually found in small consumer sized cylinders, but they can be found in large cylinders for transport and also at fixed storage sites.
- Compressed gases appear in several hazard classes, with many having multiple chemical hazards.
- Most incidents involve only a single bottle or a few bottles. Preplan the compressed gas emergency with local distributors and users, and utilize their expertise.
- Cylinders can turn into projected missiles if knocked over.
- Allow the product to escape if it can be done safely.
- All compressed gas cylinders must have some type of relief device.
- In a large facility it is often difficult to determine the exact product involved.
- Always use SCBAs and full protective gear during a compressed gas emergency.

RESPONDING TO LIQUEFIED GAS EMERGENCIES

There are many liquefied compressed gases in the United States, but the three most widely used are propane, butane, and anhydrous ammonia. Most examples in this section will use one of these three gases. While response varies according to the gas involved, the key points of this chapter are applicable to most liquefied gases.

Propane, butane, anhydrous ammonia, and vinyl chloride are used by petrochemical companies, plastics firms, educational facilities, high technology production, and various industries throughout the United States and Canada. Because of the large volume of liquefied gases produced and their wide use, responders will frequently encounter incidents involving these products, which can cause personnel to lose sight of the potential dangers. Before response considerations are presented, a review will be given of the chemical properties previously discussed in this chapter for common liquefied gases.

Physical and Chemical Properties

Liquefied gases are usually in the form of a gas at normal atmospheric pressure and temperature. When pressure is used to compress the gas into a cylinder, the gas molecules are forced so close together that a liquid is formed. Once released, however, the liquefied gas greatly expands. One gallon of liquefied gas can expand 200 to 300 times its volume as it turns to gas. In a leak, it takes only a small amount of a product to generate a large amount of gas.

In most cases, the materials are shipped and stored as a liquefied gas. Response to incidents involving these products varies, as do the hazards associated with each product. Consider the specific properties of liquefied gases before responding. Study the four different liquefied gases listed in Figure 5-38:

Notes

Common Liquefied Gases					
Product	Flammable Range	Boiling Point °F	Boiling Point °C	Vapor Density	Placard
Propane	2.4 – 9.5%	-44	-42	1.5	Flammable Gas
Butane	1.8 – 8.4%	31	-1	2.0	Flammable Gas
Anhydrous Ammonia	16.0 – 25.0%	28	-2	0.6	Non-Flammable Gas (domestic); Poison Gas (international)
Vinyl Chloride	3.6 – 33.0%	7	-14	2.0	Flammable Gas

Figure 5-38

PROPANE

Propane is the most commonly used liquefied gas. It is often used for heating and cooking in rural areas and recreational vehicles, and for portable heating equipment. Its flammable range of 2.4–9.5% is considered safe for appliances. Its boiling point of –44°F (–42°C) allows it to put off enough vapors for use in most of the United States. In areas where temperatures below –44°F are experienced, a heat exchange unit is used to heat the liquid into a gas. Propane is 1.5 times as heavy as air. It sinks to the ground or, when inside structures, finds the lowest area to settle. Responders should be aware of this property, as propane vapors do not dissipate readily.

BUTANE

Butane is another common fuel. Because its boiling point is 31°F (–1°C), it will not produce sufficient vapors to support combustion when the ambient temperature is below freezing. Butane is not often found in areas that have harsh winters, but it is widely used throughout the southern states.

ANHYDROUS AMMONIA

This gas is widely used in industry as a fertilizer and a refrigerant. It is colorless, but it has an extremely pungent odor. It irritates the eyes, skin, and respiratory tract and can be fatal. It has a flammable range of 16–25% and an ignition temperature of 1,204°F (650°C). Internationally, it is shipped under a DOT "Poison Gas" label, or domestically it is shipped with a "Non-Flammable Gas" label in tank cars, tank trucks, and cylinders. Ammonia is extremely corrosive to copper, brass, zinc, and many other non-ferrous metals and alloys. It forms explosive compounds with silver and mercury.

Anhydrous ammonia is found in the pure dry gas state or as a compressed liquid. It should not be confused with ammonia hydroxide, which is anhydrous ammonia dissolved in water. Ammonia hydroxide is commonly found in homes.

Notes

Anhydrous Ammonia

Ignition Temperature	1,204.0°F (650.0°C)
Freezing Point	-107.9°F (-77.7°C)
Boiling Point	-28.1°F (-33.4°C)
Vapor Density	0.6
Vapor Pressure	5 atmospheres at 40.5°F (4.7°C) 10 atmospheres at 78.3°F (25.7°C)
Water Soluble	Extremely soluble
Specific Gravity	0.682 at -28.1°F (-33.4°C)
Flammable Range	16 - 25%
TLV	25 ppm in air

Source: *Sax's*

Figure 5-39

Anhydrous ammonia is light and rapidly diffuses in air. It is also extremely soluble in water. Water will absorb 900 times its own volume of ammonia vapor.

VINYL CHLORIDE

Vinyl chloride is a colorless, invisible gas commonly used to produce plastics, PVC, pipe, electrical insulation, and much more. Vinyl chloride has a boiling point of 7°F (–14°C) and will give off vapors under most conditions. Its special hazards include a wide flammable range of 4–22%. It is also toxic and presents an extreme health hazard to response personnel. Because it is used only in manufacturing, it presents the most danger while transported or stored. It is twice as heavy as air, which, as indicated previously, can present significant problems during mitigation efforts. Vinyl chloride is a suspected human carcinogen.

HAZ-MAT — HAZARD IDENTIFICATION

H

Hazard Identification

Because liquid compressed gases are so common, you can count on their being used in your community. These materials are found in plumber's torches and cigarette lighters. Propane is used for heat and cooking in most rural homes. Manufacturers may use anhydrous ammonia as a refrigerant and propane as a primary or secondary heat source. Farmers use anhydrous ammonia as a fertilizer. These products are found in containers ranging from the size of a small lighter to a 48,000 gallon supertanker!

As previously outlined, identification of the product and its associated hazards is necessary to a safe and effective response. Assessment should note the following:

- location of the incident to determine the type and amount of product involved

Notes

- markings or identification such as placards, labels, tank shapes, company name, etc.
- hissing or any other unusual noises
- smells and any irritation to the nose and eyes
- presence of a vapor cloud
- color of the smoke

Once a liquefied compressed gas is identified, determine the hazards.

- Identify any leaks or fire.
- Determine the cause of the incident.
- Check for injuries.
- Determine the cylinder's position.
- Determine if the container is impinged by or exposed to fire.
- Determine the physical and chemical properties.
- Think expansion ratio—this can be the greatest hazard potential.

HAZ-MAT — ACTION PLAN

The following factors should be considered when **NO LEAK or FIRE** is involved during a liquefied compressed gas accident:

- Determine the size of the container to aid in decisions concerning evacuation and resources.
- Determine the damage to the cylinder. Although the cylinders are soundly constructed, excessive movement of the tank can cause rupture.
- Determine the amount of product in the tank. The quantity will determine the necessary resources and evacuation methods. Obtain information from the driver or shipper or by checking the magnetic or liquid level gauge, if available.
- Order an immediate evacuation if the product may release.
- Move the cylinder ***if it can be done safely***. The tank is often too damaged to be moved. If so, the cylinder must be off-loaded to another tank, or bled off, before being moved.
- Determine the tank's position. It may be in a position that would hinder the operation of its relief valve.

If there is a **LEAK but NO FIRE:**

- Immediately evacuate everyone from the area, especially those downhill and downwind.
- Control all ignition sources, including response apparatus, pilot lights, cigarettes, utilities, and vehicles. Remember the flammable limits for these materials.
- Identify the product.
- Monitor the atmosphere with detection devices.
- If a rescue is necessary, determine if it can be done safely.
- Control streets and crowds.

Expect ignition to occur when a leaking flammable gas is present.

Notes

- Remember that many materials are heavier than air and will seek low lying areas.
- Be sure a water supply is available if a fog stream is used to disperse the cloud. Unmanned hose streams should be considered.
- Determine if the product is water soluble and whether runoff control is required.
- Locate the leak and determine its volume.
- Check if securing a valve will stop the leak.
- Consider wind direction.

If there is a **LEAK and FIRE:**

- Immediately evacuate everyone to at least 2,500 feet.
- Remember, *do not extinguish the fire unless the supply of gas can be stopped.*
- Determine the extent and location of the fire. Remember that flame and heat on the tank's vapor space can cause metal fatigue and BLEVE.
- Determine for how long the relief valve has operated and with how much intensity.
- Locate the source of the accident. Calculate damage in case of a BLEVE. Make good judgments. Many of these incidents will be losers and should not be attacked aggressively, if at all.
- Know how long since the accident was reported.
- Record how long it took to respond.
- Find out how long the tank has been on fire.

If you try to cool the tank, remember the following:

- Unmanned hose lines should be used to reduce danger to personnel.
- A long term water supply is necessary. Use tankers, hydrants, or draftable water source.
- A 500 gpm minimum is needed to sufficiently cool the tank. For smaller tanks, maintain a film of water over the entire tank.
- The water stream should roll over the metal to cool the tank effectively. Cool all sides of the tank if the water does not extinguish the fire.

Constantly monitor conditions. There is no sign that indicates a definite BLEVE. However, a BLEVE can be anticipated by noting changes, which include:

- Activation of the relief valve. This indicates that the tank's pressure is increasing. If the relief valve has already activated, notice whether the flame is growing. Look for signs of growing pressure, such as the flame moving farther from the top of the tank or the hissing of the relief valve getting louder. A BLEVE can still occur, regardless of a properly activated relief valve, if the internal pressure exceeds the tank's limit.

- Metal discoloring, bubbling, or bulging, which indicates a softening from the heat.
- Tank has burned a long time and the liquid level is now the vapor level.

These are dynamic and intense incidents. ***Time*** is the key word. Allow the time needed to put lines in service and set up a water supply. If time is not on your side, ***take no action***. Remember, property can be rebuilt; people cannot.

*If time is not on your side, **take no action!***

HAZ-MAT — ZONING

Whether the incident is large or small, control zones should be established immediately. Due to the expansion ratios of liquefied compressed gases, developing an ample hot zone will help control the emergency and reduce civilian injuries. A 500 foot hot zone for small leaks will provide a safety margin in case the incident changes. An incident involving larger containers, fire, or poisonous products may require a hot zone of several hundred to several thousand feet. Use monitoring devices to determine where the zones should be located. Because explosions are highly possible, keep all unnecessary personnel out of the area.

HAZ-MAT — MANAGING THE INCIDENT

Scene Management

An effective emergency operation starts with the first arriving officer. Have a system in place to control the incident, as well as a cooperative method of working with other responding agencies. The incident commander must continually monitor the size and location of control zones. Develop a command structure that not only controls the emergency response site, but also coordinates needed outside resources.

Rescue Considerations

The chemical properties of liquefied compressed gases will cause the injuries listed below. Review general care and treatment for compressed and liquefied gas injuries on pages 197 to 200.

- ***Thermal injury.*** Thermal injuries may be the biggest worry if flammables are involved. Burns may range from first degree, suffered while attempting to shut a valve, to third degree, suffered during a flash fire or BLEVE. Released liquefied gases quickly expand and cool the surrounding air, causing cold thermal injury. Pressurized gases cool drastically when released and heat when compressed.

Notes

- ***Chemical burns.*** Exposure to most liquefied gases will cause chemical burns to the skin, eyes, and mucous membranes. Gases often cause more serious burns than liquids and are more easily absorbed by some tissue. Chemical and thermal burns are similar; however, the manner of injury is different.
- ***Internal injuries/trauma.*** Shock waves from an explosion can injure anyone in the area. Personnel may be struck by flying parts, resulting in trauma injuries. Pressure waves can hurl rocks, dirt, and small objects.
- ***Asphyxiation.*** People in the immediate area may be injured or killed by leaking products that displace the oxygen.
- ***Toxicity.*** Many liquefied gases are toxic by inhalation and absorption. Rapid treatment of victims requires an understanding of toxicology.

HAZ-MAT — ASSISTANCE

Determine what assistance will be needed before an emergency. Tour local liquefied compressed gas facilities. Have knowledge of products, cylinders, and vehicles.

Know local liquefied compressed gas distributors. Obtain off-loading or technical information, even if vehicles are not locally owned. Consult Chapter 3 for additional resources. In a liquefied compressed gas emergency:

- ***Dispatcher*** can provide coordination and gather technical information and resources for the incident commander.
- ***Local law enforcement*** should be present to handle evacuation.
- Allow immediate intervention by your ***HMRT*** or other specialized response agency.
- Use a ***technical specialist*** to determine tank damage, dangers present, and methods for handling the incident.
- Use ***private contractors*** for the extensive clean-up.

HAZ-MAT — TERMINATION

Terminating the liquefied compressed gas emergency should be systematic to insure personnel safety.

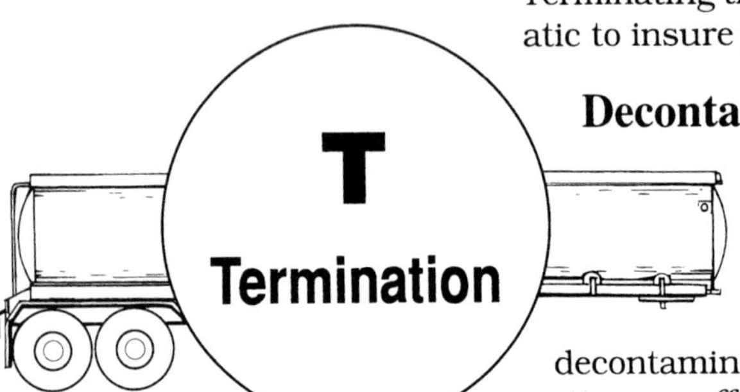

Decontamination

Decontamination of liquefied compressed gases depends on the product involved. Propane and butane usually do not require decontamination, but medical attention will be necessary. With other products, such as anhydrous ammonia, skin and clothing must be decontaminated to remove ammonia compounds. Contain decontamination runoff and contaminated soil.

Notes

Mild soap and water is a good decontamination solution for protective clothing that has been saturated during a liquefied gas incident. Always consult several reference materials and the product manufacturer for the best decontamination solution.

Other aspects of incident termination are:

- rehabilitation
- medical screening
- exposure reporting
- post-incident analysis

KEY POINTS TO REVIEW

Liquefied compressed gas emergencies are frequent haz-mat incidents. Many of these materials are flammable and/or toxic. The situation should be quickly assessed and precautions taken to keep personnel safe. Liquefied compressed gas emergencies are often intensified by emotional stress. There is no magic formula to guarantee the safe handling of these materials. However, remember the following key principles:

- Liquefied compressed gases are found everywhere, from 5 gallon barbecue cylinders to railway supertankers.
- The volume of fire on and around a tank is critical when forming an action plan.
- Note the position of the tank.
- Determine whether the pressure relief valve is operating or damaged.
- The material's weight will determine whether it will sink or rise.
- Most of the products have high expansion ratios, multiplying the difficulties of response.
- Consider the strong possibility of explosion or BLEVE before taking action.
- Immediate, large scale evacuation may be needed.
- Do not extinguish the fire unless the gas supply can be shut off.

Notes

RESPONDING TO CRYOGENIC EMERGENCIES

HAZ-MAT — HAZARD IDENTIFICATION

Cryogenics are found in research labs, medical centers, educational facilities, etc., in containers of various sizes. Remember, cryogenic materials are –130°F (–90°C) or colder. They are kept in low pressure cylinders designed similar to large thermos bottles. The uses of cryogenics range from removing warts in the local doctor's office to freezing foods in large distribution centers. Their most common use is storing oxygen for hospitals.

Noting cylinder shape and style is the easiest way to identify cryogenics. In every incident, however, note the following:

- location of the incident
- markings or identifications
- noises and smells
- tank frosting
- presence of a vapor cloud
- location of the leak and/or fire
- cause of the incident
- injuries
- position of the container
- color of the smoke

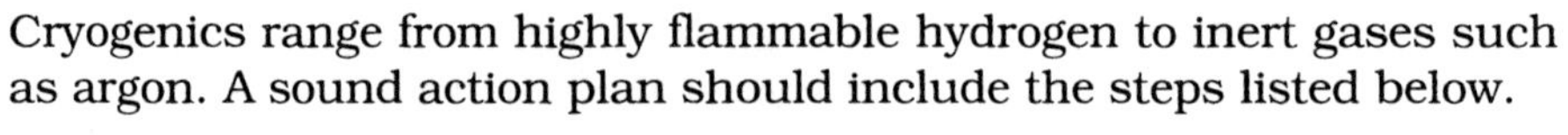

HAZ-MAT — ACTION PLAN

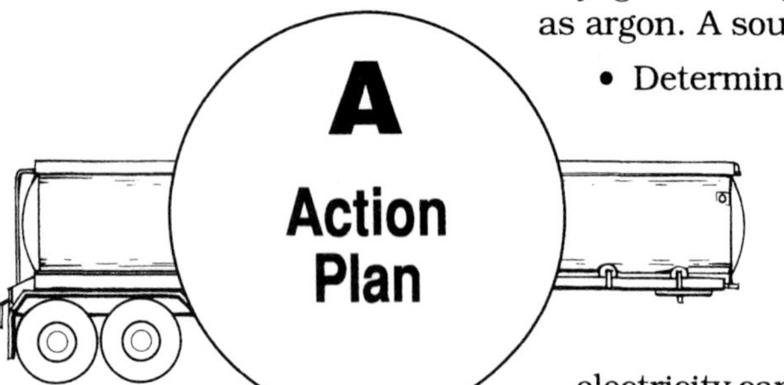

Cryogenics range from highly flammable hydrogen to inert gases such as argon. A sound action plan should include the steps listed below.

- Determine the type and size of the cylinder. Expansion ratios for these products can be higher than 900/1. Such high expansion ratios are especially dangerous inside buildings.
- Locate the cylinder.
- Determine the hazards. Some cryogenics are flammable. Hydrogen is so unstable that static electricity can cause it to ignite. Oxidizers must be controlled away from open flames or ignition sources. Always use metering devices to accurately monitor the atmosphere.
- Determine if insulation is torn from the cylinder. If so, adding water to the cylinder may raise the liquid's temperature.

If **NO FIRE** is involved in the incident, take action to:

- Control all ignition sources.
- Locate the vapor spread with metering devices.
- Determine life hazards. If necessary, evacuate the building or control access with outside assistance. Use the DOT *Emergency Response Guidebook* for protective action distances.
- Locate air intakes to the building. Restrict the product from these areas if it can be done safely.
- Contact distributor for assistance.

If **FIRE** is involved:

- Immediately protect in place or evacuate the area to the minimum distances specified by the DOT guidebook.
- Protect exposures and cool tank, if possible.
- Remember that cryogenic cylinders can BLEVE.
- Prevent or identify contamination. Some materials, such as liquid oxygen in contact with hydrocarbons, can be shock sensitive. ***Note:*** Nomex and asphalt are hydrocarbons.

HAZ-MAT — ZONING

Immediately establish control zones for civilians and responders. The distance will depend on conditions and the product involved. Initially establish a 500 foot hot zone; then use atmospheric metering devices to determine safe zones.

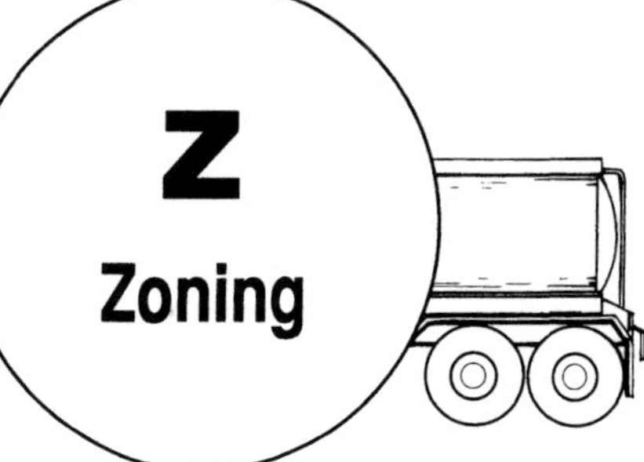

HAZ-MAT — MANAGING THE INCIDENT

Rescue Considerations

Common injuries of a cryogenic emergency are as follows:

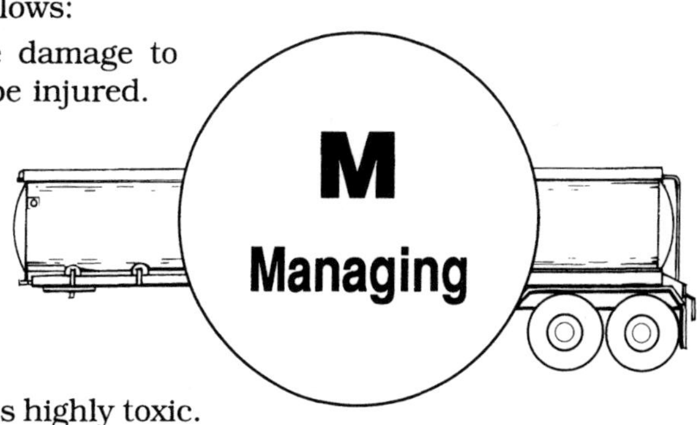

- ***Thermal injury.*** Cryogenic liquid will do severe damage to tissue. In fact, direct contact is not necessary to be injured. Exposure to the gas can severely damage eyes. Use SCBA and full protection when handling these incidents.
- ***Asphyxiation.*** A cryogenic leak can force oxygen from a confined space. Use caution when entering an area of a suspected leak. Be prepared to rescue individuals.

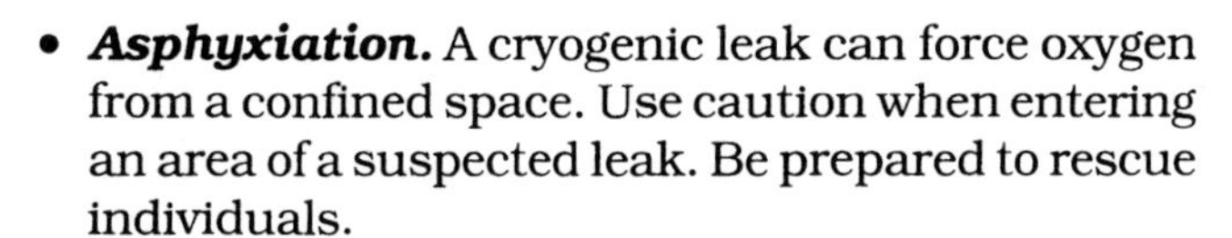

- ***Toxicity.*** Carbon monoxide, a common cryogenic, is highly toxic. Other cryogenics are also toxic.

HAZ-MAT — ASSISTANCE

The shipper, user, or distributor is an excellent resource during an emergency. Technical reference books and materials should also be consulted. Use resources to determine hazards such as:

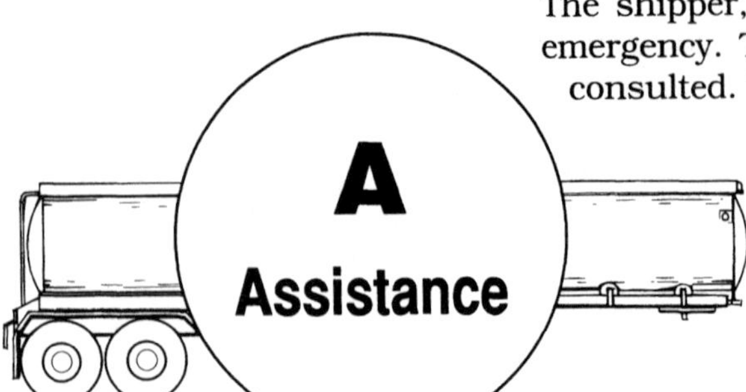

- flammability
- specific gravity
- water solubility
- toxicity

HAZ-MAT — TERMINATION

Cryogenic materials are similar to compressed and/or liquefied gases. Use these termination techniques for any haz-mat incident:

- decontamination
- rehabilitation
- medical screening
- exposure reporting
- post-incident analysis

KEY POINTS TO REVIEW

- Cryogenic materials are –130°F (–90°C) or colder.
- They are stored in low pressure, thermal insulated cylinders.
- Cylinders usually have two pressure relief systems. One vents the liquid storage space, and the other vents the vacuum space.
- Cryogenics can BLEVE if overheated and if relief valves are rendered useless.

Chapter 6

FLAMMABLE AND COMBUSTIBLE LIQUID EMERGENCIES

OBJECTIVES

After studying the material in this chapter, you will be able to:

- explain what flammable and combustible liquids are
- state the major physical and chemical properties of flammable and combustible liquids
- describe the labels and placards that are required for flammable and combustible liquids
- describe the various containers, cargo tank trucks, and rail cars used for flammable and combustible liquids
- describe the types of storage tanks used for flammable and combustible liquids
- state the types of foam and the procedures for using foam in a flammable/combustible liquid emergency
- describe the methods for dealing with environmental contamination in a flammable/combustible liquid emergency
- explain the proper response action for flammable/combustible liquid emergencies, using the acronym HAZ-MAT

UNDERSTANDING FLAMMABLE AND COMBUSTIBLE LIQUIDS

INTRODUCTION

Over 50% of this country's hazardous material responses involve flammable and combustible liquids. Motor fuels fall into this category, which brings the possibility of an incident to every community. Flammable and combustible liquids are hauled by truck, train, and barge for storage, distribution, and processing. They are often piped many miles to terminals

Notes

or industrial complexes for manufacture. These liquids are developed by chemical and petrochemical companies to meet special needs of business and industry. While some are stable products, many present new stability or reactivity problems.

DEFINITIONS

Flammable Materials

A ***flammable material*** is any solid, liquid, vapor, or gas that will ignite easily and burn rapidly. Flammable materials produce tremendous heat and may cause problems in transportation, storage, and use. Flammable solids can be found in several forms, including:

- Dust or fine powders, such as flour, cellulose, and metals.
- Materials that spontaneously ignite at low temperatures, such as white phosphorus.
- Materials that, due to their chemical structures, produce internal heat. The most common examples include animal oil, vegetable oil, and fish oil. This property is rarely seen in wet cellulose materials.
- Fibers, fabrics, and materials made of low ignition point materials.

This chapter will discuss only liquid flammable materials. Flammable solids will be covered in Chapter 7.

Flammable Liquids

It is important to have a working knowledge of the definition of flammable and combustible liquids, for several reasons:

- Weather and temperature affect how these materials react.
- Mixing these products with other materials may change the burning characteristics.
- Many materials that are considered non-volatile can become unstable under specific conditions.

The DOT's definition of a ***flammable liquid*** is:

> A liquid having a flash point of not more than 141°F (60.5°C), or any material in a liquid phase with a flash point at or above 100°F (37.8°C) that is intentionally heated and offered for transportation or transported at or above its flash point in a bulk packaging, with the following exceptions: (1) any liquid meeting the definition of a liquefied compressed gas, a compressed gas in solution, or a cryogenic liquid; and (2) any mixture having one or more components with a flash point of 141°F (60.5°C) or higher, that makes up at least 99% of the total volume of the mixture, if the mixture is not offered for transportation or transported at or above its flash point (49 CFR 173.120(a)).

Flammable liquids are classified further by the National Fire Protection Association (NFPA) as follows:

- ***Class Ia*** – Materials with a flash point **below** 73°F (22.8°C) and a boiling point **below** 100°F (37.8°C).
- ***Class Ib*** – Materials with a flash point **below** 73°F (22.8°C) and a boiling point **at or above** 100°F (37.8°C).
- ***Class Ic*** – Materials with a flash point **at or above** 73°F (22.8°C) and **below** 100°F (37.8°C).

Do not be overly concerned with these subclasses. However, it is important to understand that flammable liquids give off vapors at temperatures below 100°F and may be easily ignited. Combustible liquids also give off vapors below 100°F, but not in amounts sufficient to present an ignition hazard. A partial list of flammable and combustible liquids is given in Appendix E.

Notes

Flammable liquids give off enough vapors at temperatures below 100°F to be easily ignited.

Combustible Liquids

The DOT's definition of a ***combustible liquid*** is:

> Any liquid that does not meet the definition of any other hazard class and (1) has a flash point above 141°F (60.5°C) and below 200°F (93°C), or (2) is a flammable liquid with a flash point at or above 100°F (37.8°C). In the second case, a flammable liquid is reclassed as a combustible liquid (49 CFR 173.120(b)).

These materials are classified further by the NFPA as follows:

- ***Class II*** – Liquids with a flash point **at or above** 100°F (37.8°C) and **below** 140°F (60°C).
- ***Class III*** – Liquids with a flash point **at or above** 140°F (60°C). They may be divided as follows:
  - ***Class IIIa*** – Liquids having a flash point **at or above** 140°F (60°C) and **below** 200°F (93°C).
  - ***Class IIIb*** – Liquids having a flash point **at or above** 200°F (93°C).

Hundreds of combustible liquids are commonly transported in our communities. The primary danger associated with combustible liquids is the tremendous heat they produce in fires. The molecular size of combustible liquids makes their flash points relatively high, creating a high heat output when burning.

Notes

Common Flammable and Combustible Liquids*					
	Flash Point		Ignition Temp.		
Product	°F	°C	°F	°C	Placard
Acrolein	< 0	< -18	455	235	Flammable
Carbon Disulfide	-22	-30	257	125	Flammable
Diesel Fuel	100	38	494	257	Flammable*
Ethyl Alcohol	56	13	793	423	Flammable
Gasoline	-50	-46	536–853	280–456	Flammable
Kerosene	150	66	410	210	Flammable*
Methyl Alcohol	52	11	878	470	Flammable
Methyl Ethyl Ketone	22	-6	960	516	Flammable
Tetrahydrofuran	2	-17	610	321	Flammable
Toluene	40	4	996	536	Flammable

* A "Combustible" placard may be used for kerosene and diesel fuel in highway or rail transportation.

Source: *Sax's*

Figure 6-1

PHYSICAL AND CHEMICAL PROPERTIES OF FLAMMABLE AND COMBUSTIBLE LIQUIDS

Some physical and chemical properties, such as flammable range, apply both to gases and to vapors produced by a liquid. Others, such as critical point, apply only to gases. It is important to understand **why** flammability depends upon things such as flash point, fire point, ignition temperature, flammable range, specific gravity, vapor density, boiling point, vapor pressure, and water solubility. This information is vital for control tactics.

Flash Point

Flash point is the minimum temperature to which a liquid must be heated to produce a vapor flash if an ignition source is present. In other words, if a liquid is below its flash point, there is not enough vapor present for combustion. The vapor is considered ***too lean*** if the liquid is below its flash point. Flammable range and rich and lean vapor concentrations will be discussed later in this section.

Flash points are recorded as temperature readings in either degrees Fahrenheit or Centigrade. They may be followed by the letters "oc" (open cup) or "cc" (closed cup). Since the flash point value can vary several degrees

Notes

Flammable Liquid Emergency

depending on the calculation method used, be cautious if the temperature of the liquid in question is close to the flash point.

When determining if the temperature of a liquid is above its flash point, remember that a material is considered to be at the same temperature as its surroundings (***ambient temperature***). If the temperature of the air is 55°F, then the temperature of the material is considered to be 55°F. However, this may not be true if the material has been removed from an underground storage tank or from an insulated or heated tank.

Fire Point

Fire point is the temperature to which a material must be heated to produce enough vapor for sustained combustion. At first this definition may seem identical to that of flash point. However, when a material ignites at its flash point, there will be a vapor flash but no sustained combustion. When a material ignites at its fire point, combustion is sustained. Although exceptions do exist, the flash point and fire point are usually no more than three degrees apart. During an incident, this difference is not worth talking about.

Flammable Range

Flammable range, also known as ***flammable limits***, is the percentage of vapor needed in the air for combustion to occur (see Figure 6-2). This means that the vapors of a material must be within a certain percentage range for the material to burn. If the percent of vapor is below or above the amount necessary, the material will not burn. Anything below the lower flammable limit (LFL) is said to be ***too lean*** to burn. Anything above the upper flammable limit (UFL) is said to be ***too rich*** to burn.

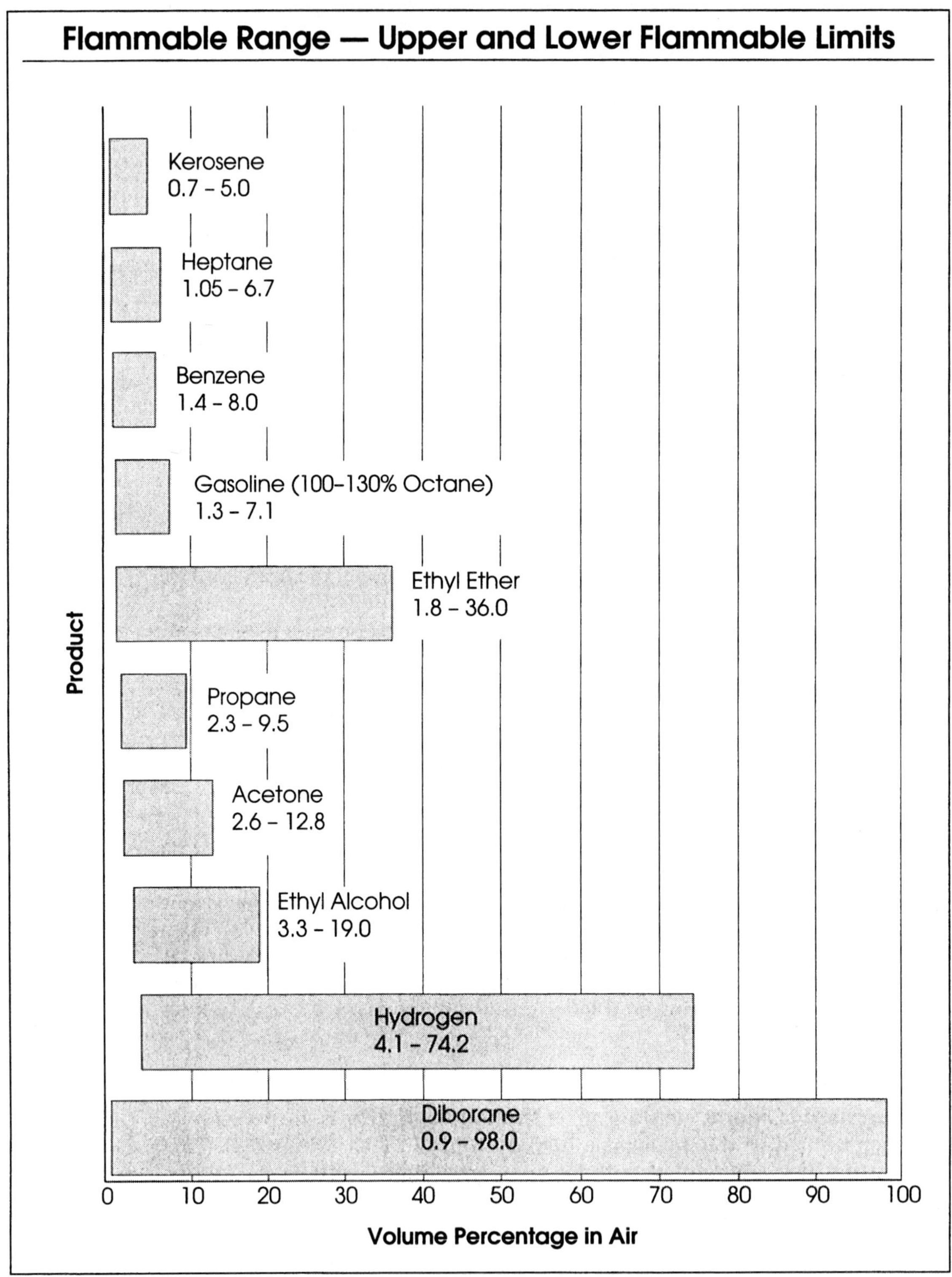

Source: *Sax's*

Figure 6-2

Boiling Point

Notes

Boiling point is the temperature at which a liquid will boil. Boiling occurs when the liquid's vapor pressure is equal to or greater than the atmospheric pressure. At temperatures below the boiling point, liquids still change to vapor but at a slower rate. This slow change of liquid to vapor is called ***evaporation***.

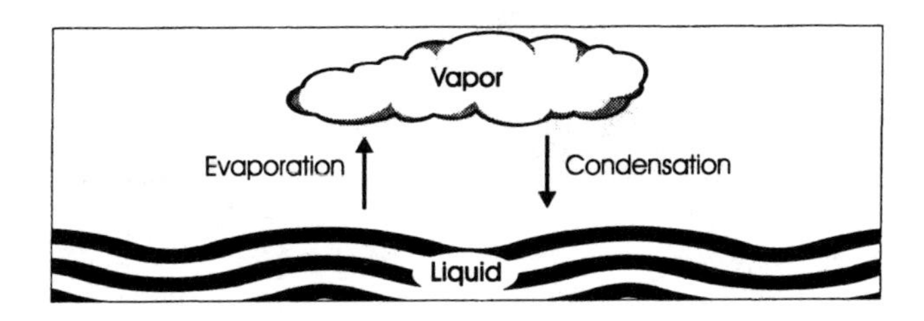

When atmospheric pressure changes, so does the boiling point. At sea level, the atmospheric pressure is approximately 14.7 psi. At 10,000 feet above sea level, the atmospheric pressure is only about 9.5 psi. Due to the decreased pressure, water boils at about 192°F (89°C) at 10,000 feet elevation, rather than at the normal 212°F (100°C).

Specific Gravity

Specific gravity compares a liquid's weight to the weight of an equal volume of water. Water has a specific gravity of 1.0. A product with a specific gravity of more than 1.0 will sink in water, while a product with a specific gravity of less than 1.0 will float in water. Most flammable liquids have a specific gravity of less than 1.0. At an incident, it is important to determine the specific gravity of a flammable/combustible liquid before designing a containment system.

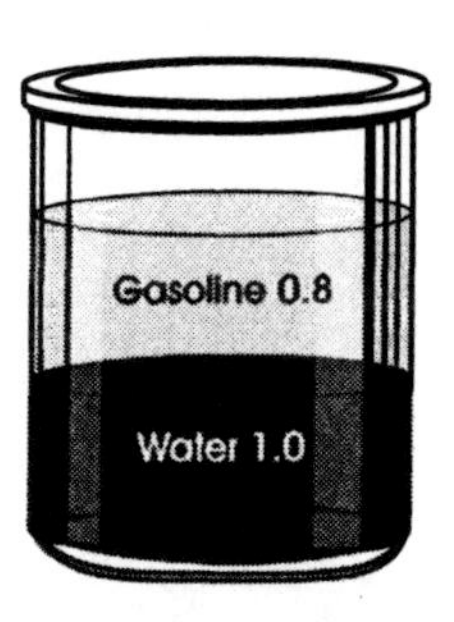

Figure 6-3
Specific Gravity of Gasoline

Boilover

A ***boilover*** is a real possibility during a storage tank fire when crude oil, which is composed of several petroleum fractions and water, is present and the flames intensify. The water settles to the bottom of the tank when stored. During a fire, a heat wave moves downward as the crude oil burns. The water at the bottom of the tank instantaneously turns to steam with an explosive force when contacted by the heat. Steam expands to 1,700 times the original volume of water, which causes burning oil to be thrown over the sides of the tank. This makes it extremely dangerous to position emergency response personnel in the diked area. Responders should seek safety when the heat wave reaches within five feet above the bottom water level. In addition, personnel working on top of the dike should be alerted to a boilover. Be aware that successive boilovers are common. Responders can determine the approximate location of the heat wave by observing the following:

- discoloration of the tank wall
- steaming when water is applied to the tank wall
- thermal sensitive tank markings

Notes

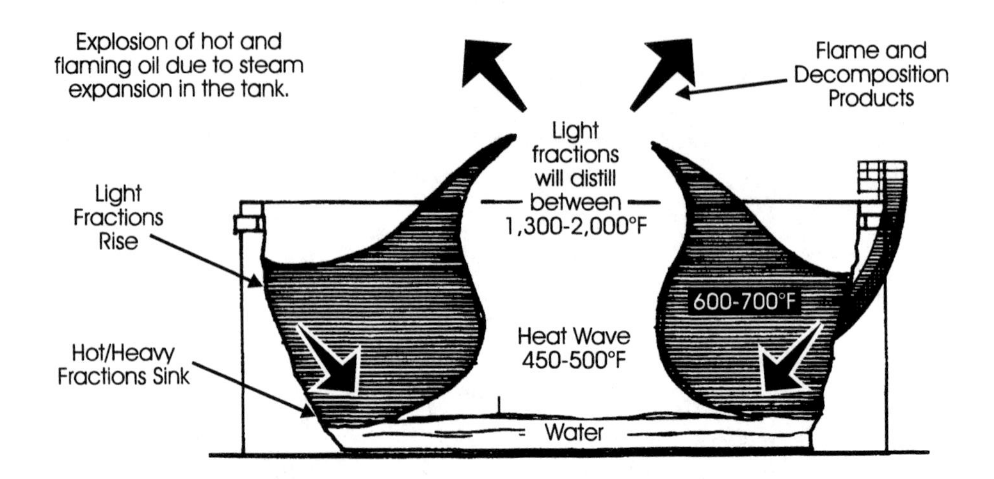

Figure 6-4. Boilover in a Storage Tank

Vapor Density

Vapor density compares the density of gas or vapor to the density of air. Air is given a vapor density of 1.0. A substance with a vapor density of less than 1.0 is lighter than air. A gas or vapor with a vapor density of more than 1.0 is heavier than air. Vapor density is usually associated with gases, but it is also important during liquid spills. It is necessary to determine a product's vapor density during a leak and/or fire in order to understand where the vapor will travel. For example, because flammable liquid vapors are heavier than air, the vapors could sink into an electrical vault, prompting an explosion.

Figure 6-5. Vapor Density of Gasoline

Radiant Heat

Notes

Radiant heat is heat that is generated by the sun, by fire, or by some other heat source. A solid surface, such as concrete or asphalt pavement, can be heated 50°F or more above the air temperature by solar radiation. For instance, if the air temperature is 85°F on a sunny day, the temperature of an asphalt surface can be over 135°F.

If a haz-mat incident occurs on a surface that has been exposed to the sun for several hours and the air temperature is 85°F, the surface heat is capable of warming the product to 135°F. If the product's flash point is 105°F, the substance will be heated above its flash point. If only air temperature is involved, the product will stay below its flash point and will present a lesser hazard.

Solar radiation can affect a product's flash point when the product is stored or transported in a dark colored container. If a product stored in a black, 55 gallon drum is exposed to the sun for some time, the product will become warmer than the surrounding temperature. That warmer temperature may be above the product's flash point.

Figure 6-6. Physical Properties of Gasoline

Adapted with permission from *Hazardous Materials for First Responders*, © 1988, Board of Regents, Oklahoma State University.

Ignition Temperature

Ignition Temperature

Ignition temperature is the minimum temperature to which a material must be heated to produce free radicals through molecular bond breakage. Combustion will take place if free radicals occur in the presence of oxygen or some other oxidizing agent. If there is no oxidizing agent but heat is still present, the free radicals remain extremely reactive. A great explosion will result if an oxidizer is introduced at a later time.

Ignition Sources

There are three ignition sources that can cause a material to reach its ignition temperature. All three ignition sources must produce temperatures equal to or above a material's ignition temperature in order for the material to ignite and experience combustion. The three ignition sources are listed below.

- ***External Ignition Sources.*** These may enter the vapor or liquid and transfer the source's heat energy directly to the material. Common examples of external sources are open flames, sparks (electrical, static, or frictional), and heated objects. Sparks, including those from static electricity, are capable of developing temperatures ranging from 2,000°F to 6,000°F.
- ***External/Internal or Autoignition Sources.*** These sources heat the vapor or liquid indirectly. The three autoignition sources are:
  - radiant heat transfer
  - convection heat transfer
  - combustion heat transfer

  All three cause heat to be transferred until the ignition temperature is reached and ignition occurs. There is no apparent external flame or spark. An example of autoignition is a skillet of oil suddenly bursting into flames when left on a stove too long.
- ***Internal or Spontaneous Ignition.*** In this case, the material or a by-product of the material produces enough heat to reach the ignition temperature. Heating may occur in two ways:
  - The biological processes of microorganisms may cause ignition. This type of ignition usually is not associated with flammable liquids, but with masses of organic materials such as straw and hay.
  - An oxidizing chemical reaction that produces heat occurs. If insulation prevents heat from dissipating to the outside, the heat will warm the material. The oxidation rate increases as more heat is added to the material, which causes an additional increase in heat within the material. This process continues until the material's ignition temperature is reached and the material begins to burn.

Pyrophoric Ignition

The DOT defines a ***pyrophoric liquid*** as:

> A liquid that, even in small quantities and without an external ignition source, can ignite within five minutes after coming in contact with air (49 CFR 173.124(b)(1)).

LABELS AND PLACARDS

Labels

All flammable liquid shipping containers are required to be labeled whether transported by road or rail.

Placards

Two placards ("Flammable" and "Combustible") are primarily used for transported flammable and combustible liquids. It should be remembered that products in other hazard classes may also be flammable. The UN/NA hazard class number is located at the bottom of the placard with the flammable liquid identification number 3.

There may be another identification number on the placard to help emergency responders quickly identify a product and its hazards. This four digit number coincides with the DOT *Emergency Response Guidebook*. In some cases the guide number will not provide a specific product identification. For example, Guide 1993 is a general identification covering cosmetics and flammable liquids that are not otherwise specified (n.o.s.).

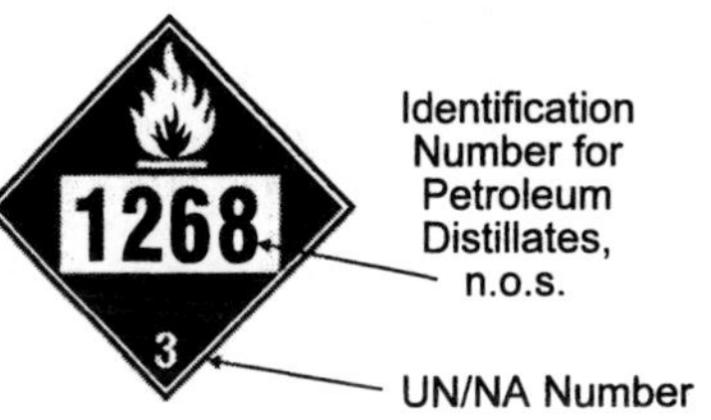

Rail Placarding

All four sides of the rail shipping container must be placarded, regardless of the quantity of material.

"Residue" Placard

Rail tank cars that contain a residue of up to 3% by volume of material must use a "Residue" placard. This placard indicates what was previously transported and indicates that some of the product may still be in the tank. Tank cars containing product residue can be just as dangerous as full cars. The residue placard is not applicable for highway transportation.

Residue Placard

Over the Road (OTR) Transportation

Placarding requirements for over the road (OTR) transportation of flammable liquids is essentially the same as for rail. The difference, however, is that the DOT **does not** require a placard for flammable liquids unless 1,001 pounds (454 kg) or more are being transported. Do not let the absence of a placard lure you into taking unsafe action. Responding to an incident involving an unplacarded load of 999 pounds of a flammable liquid could prove hazardous to your health!

Notes

STORAGE CONTAINERS

Container shapes and types vary with material and quantity. Flammable liquids are so common that they are found in containers not normally associated with hazardous materials. We will begin by discussing the smallest containers for flammable and combustible liquids.

Plastic Containers

Plastic Containers. Plastic containers store many flammable and combustible materials for use in labs, industrial settings, and the home. Two of the most common materials stored in plastic containers are isopropyl alcohol and charcoal lighter fluid. Gasoline for lawnmowers and other small engines may also be stored in plastic containers.

Glass Containers. The material most commonly stored in glass is liquor or ethyl alcohol for human consumption. These containers are found in liquor stores, restaurants, bars, and homes and can significantly endanger the safety of emergency responders if a fire breaks out. Bottles create a severe fire and injury hazard as they explode and burn materials that they fall on.

Laboratory Flasks and Beakers. The Pyrex or glass lab beaker is used in research, instructional, and manufacturing facilities. Flasks may have a rubber or cork stopper to stop the liquid and vapors from escaping. Laboratory containers are often labeled inappropriately or with a labeling system not familiar to emergency responders. Some lab technicians have no concept of the term "hazardous."

Laboratory Flasks and Beakers

Notes

Portable Tanks. Portable tanks, sometimes referred to as "totes," are a common way to transport liquid materials, such as flammable solvents, liquid fertilizers, and water treatment chemicals. These containers may have rectangular or circular cross sections and are approximately 6 feet high. Portable tanks have a 300 to 400 gallon capacity. Although referred to as non-pressure containers, these containers may have internal pressures up to 100 psi.

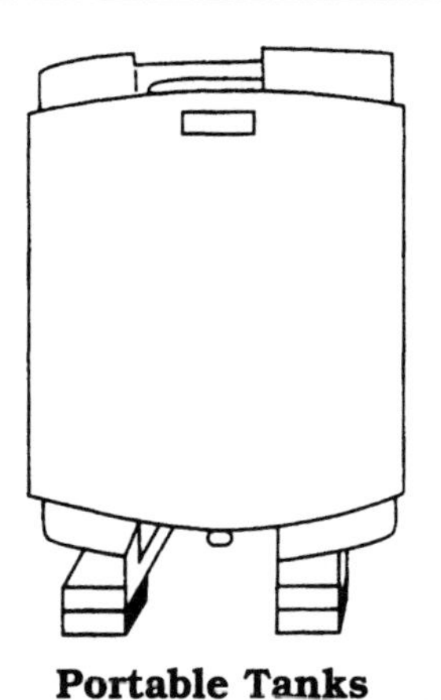

Portable Tanks

Drawing courtesy of Chemical Manufacturers Association

Metal Cans. Metal storage cans are commonly used to store gasoline and diesel in 1, 2, and 5 gallon quantities. Five gallon round metal cans are used to store paints, thinners, resins, and other products. They are found in retail outlets, buildings under construction, garages, and storage sheds. Metal cans are made of rolled and crimped steel which has been soldered or welded. They are usually not insulated or lined, and they do not have pressure relief devices. The pouring cap is generally plastic.

Metal Cans

Drums. Drums usually hold 55-85 gallons and are found at locations where large quantities of materials are used. Drums are rolled, soldered, or welded and are generally unlined. Drums are usually transported on flatbed or cargo van trucks. Many liquids are stored and transported in drums. Therefore, do not assume that all of the drums being transported or stored together contain the same material.

Drums

Notes

Liquid is taken from a drum by manually or electrically pumping the liquid from the top of the drum. The drum may be placed in a cradle to allow gravity to move the liquid. A small check valve opens a spigot, allowing the liquid to flow into a container. Many fire codes require drums to be grounded to the container being filled to reduce static electricity. Static electricity can accidentally ignite the contents. Fifty-five gallon drums have no relief valve; therefore, over-pressurization is a problem if the drums are overheated.

CARGO TANK TRUCKS

Large transport vehicles move enormous volumes of flammable and combustible liquids through our communities daily. Recognition of the vehicles most commonly used to transport flammable liquids will enable emergency responders arriving at an incident to anticipate the hazards associated with these vehicles.

MC306/DOT406 Non-Pressure Liquid Carrier

The tank truck most commonly used to transport flammable liquids is the MC306/DOT406 non-pressure liquid carrier. The MC306/DOT406 carries gasoline, alcohol, combustible liquids, fuel oils, and other similar products. All materials carried in an MC306/DOT406 must have a vapor pressure of under 3 psi. The MC306/DOT406 is easily identified by its elliptical shape. It is usually constructed of polished aluminum that has a distinctive silver appearance. Older tanks may be constructed of steel. The MC306/DOT406 has:

- longitudinal rollover protection
- valving and unloading control box under the tank
- vapor recovery piping on the right side and rear (not all MC306/DOT406s have vapor recovery)

MC306/DOT406 Non-Pressure Liquid Carrier

Notes

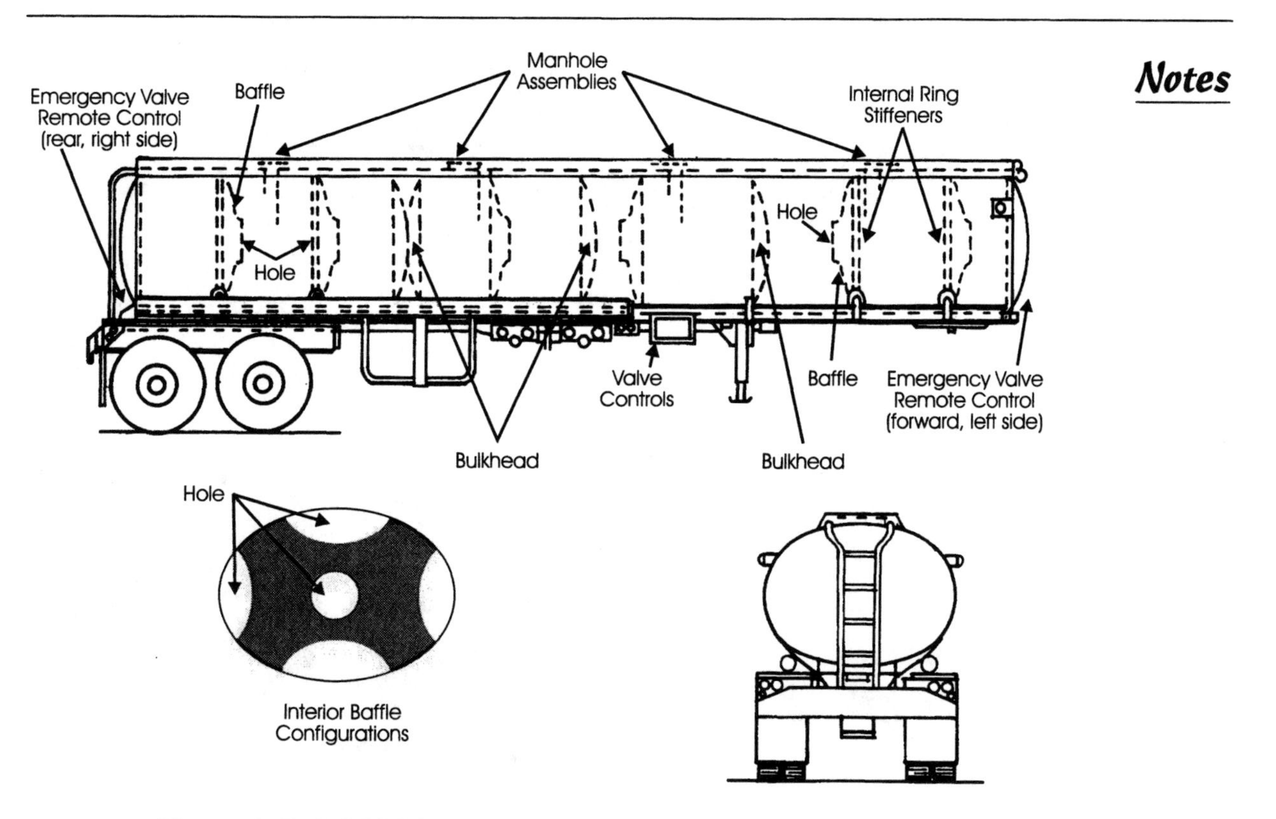

Figure 6-7. MC306/DOT406 Non-Pressure Liquid Carrier
Adapted with permission from *Hazardous Materials for First Responders*, © 1988, Board of Regents, Oklahoma State University.

- manhole assemblies on top with vapor recovery valves for each compartment
- sometimes permanent markings for compartment capacities and products

MC307/DOT407 Low Pressure Chemical Carrier

The MC307/DOT407 low pressure chemical carrier transports a variety of chemicals, including flammable corrosives and poisons. Pressure in these tanks does not exceed 40 psi. The tanks are somewhat rounded on the ends. The MC307/DOT407 tank has fusible plugs and a frangible disc or Christmas tree vent. The Christmas tree vent is a combination vacuum breather and relief device located outside the flashing box. MC307/DOT407s may be compartmentalized and may have more than two compartments. The MC307/DOT407 can be identified by several specific design features, including:

- single or double manhole assemblies protected by a flashing box
- single outlet discharge pipe at midship or rear of the tank
- possible drain hose from the manhole assembly
- some may have external ring stiffeners

Some MC307s are insulated so their rounded cross section is not apparent.

MC307/DOT407 Low Pressure Chemical Carriers

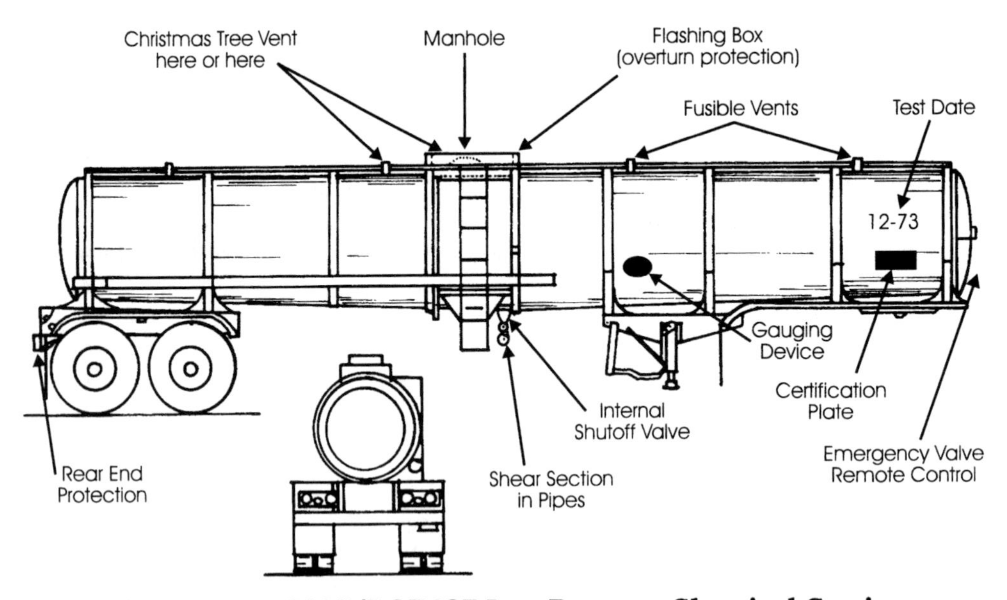

Figure 6-8. MC307/DOT407 Low Pressure Chemical Carrier

Adapted with permission from *Hazardous Materials for First Responders*, © 1988, Board of Regents, Oklahoma State University.

Safety Devices

Fire Valves. Fire valves, sometimes called internal valves, are recessed into the tank to allow fuel to flow out of the tank. These valves have a fusible link that, in a fire, will close the valve to shut off the fuel supply. The fire valve also works as a ***shear valve***. If direct mechanical damage is sustained to the valve, the valve will shear off. It will, however, leave the internal portion of the valving intact and closed, thus minimizing fuel loss.

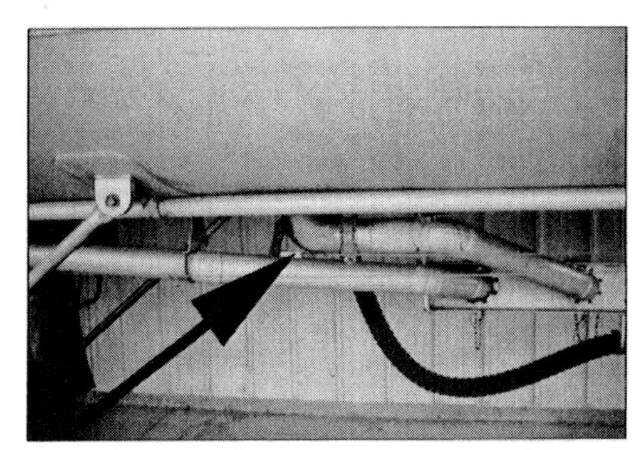

Internal Fire Valve or Shear Valve

Remote Control Shutoff. Remote control shutoffs are located on the forward left side and rear of the tank. This remote control device is activated by cable, air, or hydraulics. Fuel flow may be stopped by simply pulling a handle.

Remote Control Shutoff

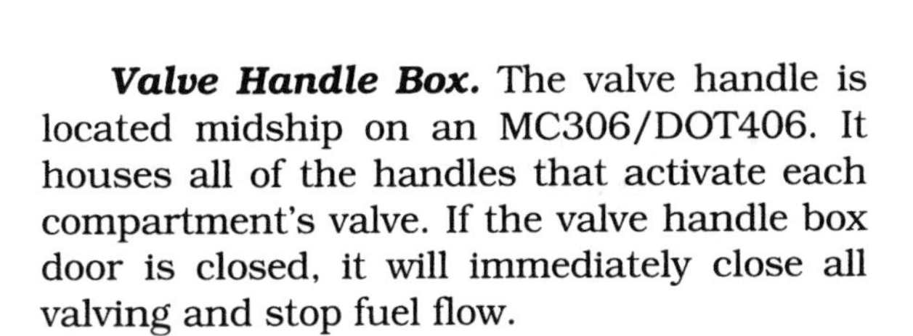

Valve Handle Box. The valve handle is located midship on an MC306/DOT406. It houses all of the handles that activate each compartment's valve. If the valve handle box door is closed, it will immediately close all valving and stop fuel flow.

Valve Handle Box

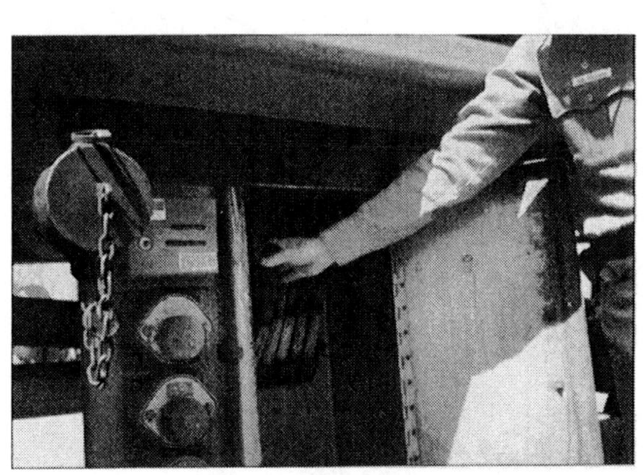

Valve Handle Box

Product Loading

Loading Rack

Product Delivery Methods

Every emergency responder should be able to quickly identify if a product is being delivered by gravity or pumping. The rule of thumb is:

- A pumping system is probably being used if the product is being delivered to an above ground tank. Small PTO pumps are commonly used.
- Gravity can be used if the product is being delivered to a below ground tank.

Pumping System

Gravity Feed Delivery

Rail Tank Car

Notes

RAIL TRANSPORT

The non-pressurized general purpose rail tank car holds from 6,500 gallons to just over 30,000 gallons of a flammable liquid. These tank cars are usually constructed of steel. However, some stainless steel, aluminum, and nickel alloy cars are in service. Rail cars are usually top loaded, and many liquid cars are equipped with bottom unloading outlets.

STORAGE TANKS, REFINING FACILITIES, AND PUMP STATIONS

Incidents at storage tanks, refineries, and pump stations can involve huge, impressive fires. It is important to preplan for incidents at these locations before they occur, as serious injury and environmental difficulties are probable. These facilities tend to be complex, and an overzealous responder can complicate problems.

Preplan for incidents at local storage tanks, refineries, and pump stations before they occur.

The most common storage tanks will be reviewed in this section. There are three storage tank classes for flammable liquids:

- atmospheric tanks of 0 to .5 psi
- low pressure tanks of .5 to 15 psi
- pressure tanks over 15 psi

Cone Roof Tank

Cone roof tanks are used to store many flammable and combustible liquids. Current hydrocarbon industry practice dictates that only combustible liquids, or those with flash points greater than 100°F, be stored in cone roof tanks. On occasion, however, emergency responders will find such liquids as crude oils and polar solvents in cone roof tanks.

Cone roof tanks have a vapor space between the liquid and the underside of the roof. An explosion will occur if the vapor in this space is ignited while in its flammable range. If the tank is designed according to API Standard 650, the roof may separate from the shell joint during an

Notes

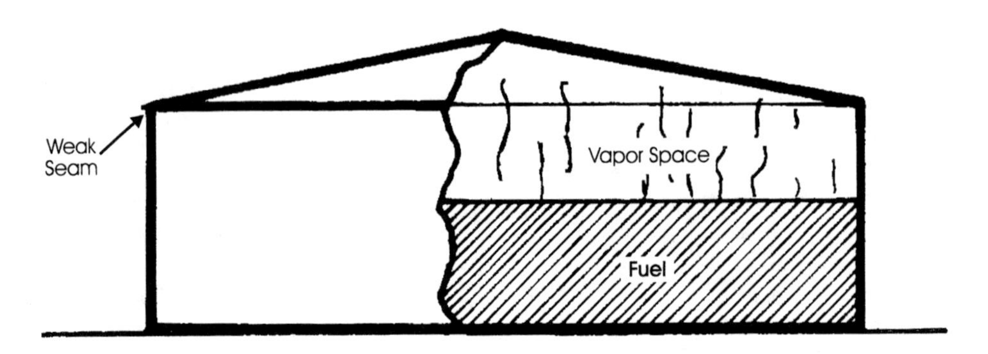

Figure 6-9. Vertical Cone Roof Tank

explosion. The roof may separate in one piece, or it may fragment, traveling considerable distances. Sometimes the roof will lift into the air and fall back into the tank. Other times only pieces of the separated roof may remain intact on the tank's top. The resulting fire will involve the entire surface area of the tank.

Vertical Flat Roof Tank

The common vertical flat roof tank varies in size and is used to store many flammable and combustible products. Flat roof tanks are similar to cone roof tanks, but they have a flat rather than cone shaped roof. Vertical flat roof tanks are usually metal and have a breather vent and an inspection cover on top. They are filled and discharged from the bottom. These tanks are often found at service stations and other similar facilities. A diked area usually surrounds these tanks to contain spills.

Open Floating Roof Tank

Open floating roof tanks are the most common tank being built today. They are usually the largest storage tank. Fire safety records for these tanks are similar to other above ground storage methods. These tanks are usually wider than tall and have a roof that actually floats on the surface of the liquid. There is no other roof covering. Looking inside is all that is needed to determine the volume of material. A ladder attached to the roof allows workers to repair the roof. Open floating roof tanks have drains to eliminate

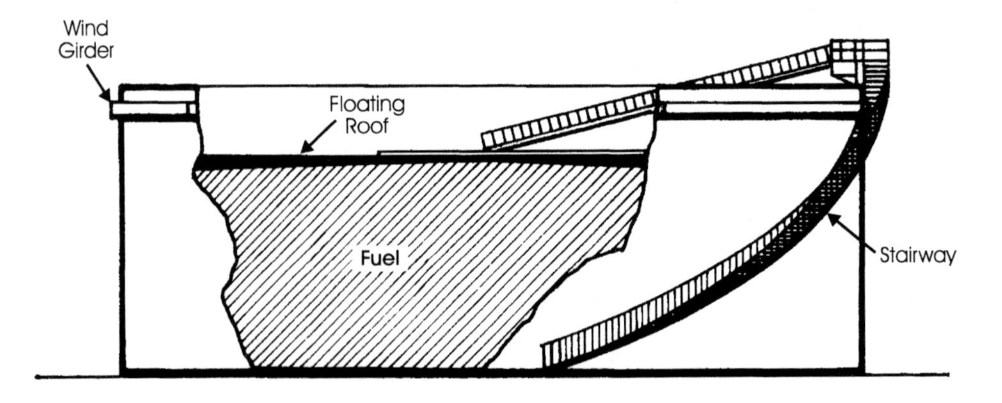

Figure 6-10. Open Floating Roof Tank

water build-up from weather and condensation. The roof has a rubber, fabric, Teflon®, or other material edge that provides a waterproof seal.

Fires in these tanks are usually confined to the annular seal area between the floating roof and the tank shell (see Figures 6-11 and 6-12). This area may consist of a pantograph or neoprene tube seal. The seal is protected by a fabric or metal weather shield. Many newer tanks have a double seal to reduce emissions. This, however, makes a fire more difficult to fight. Flames may be visible between the weather shield and the shell. The agent must be applied between the shield and the shell to extinguish the fire. Many such fires have been extinguished by using dry chemical extinguishers or hand lines from the wind girder walkway.

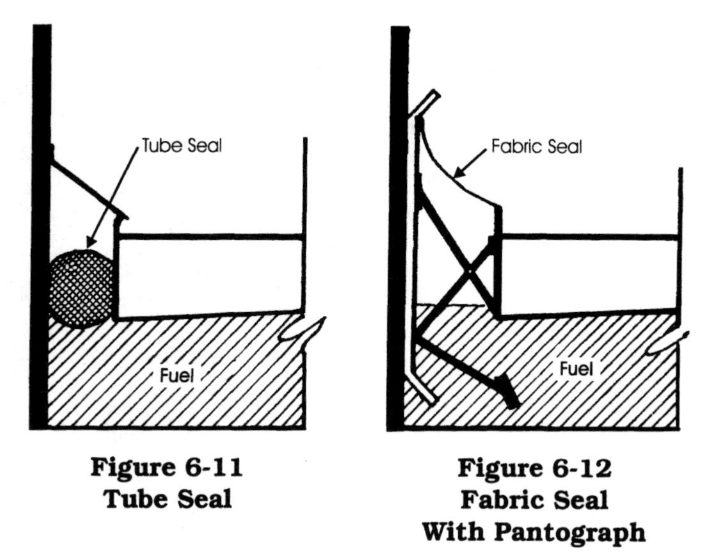

Figure 6-11
Tube Seal

Figure 6-12
Fabric Seal
With Pantograph

Covered Floating Roof Tank

The covered or internal floating roof tank consists of a cone roof tank with a weak roof-to-tank shell joint and an internal floating roof or pan. Exterior vents around the tank shell just beneath the roof joint make this type of tank easy to identify. The tank is usually free from ignitable mixtures except during the initial fill and from 18 to 25 hours afterward, depending on the product's volatility.

Covered floating roof tanks have an excellent fire safety record. However, there have been a few fires which have been extremely difficult to extinguish. Subsurface injection of foam is normally the best method to extinguish fires in these tanks. Seal or rim fires are virtually impossible to fight from the outside with portable fire extinguishing equipment. The side vents are also too small for foam streams to enter from ground level. On some occasions, the cone roof has blown off, and the floating roof or pan has sunk, leaving the entire surface area on fire. If such an explosion or fire occurs in your area, treat the fire as a cone roof tank fire and extinguish it with monitor nozzles or other topside application.

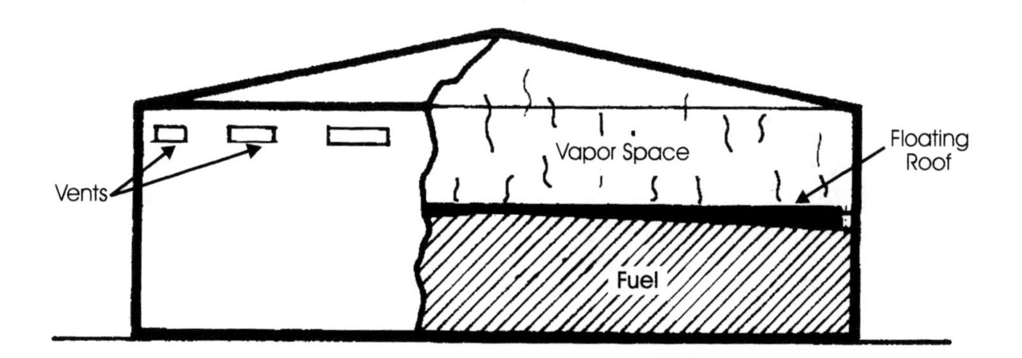

Figure 6-13. Covered Floating Roof Tank

Notes

Horizontal Tanks

Horizontal Tank

Horizontal liquid tanks are found at construction sites, farms, and bulk plant facilities. These tanks hold between 300 and 20,000 gallons. In many areas, horizontal tanks have been restricted because they pose great fire and environmental hazards. The hazards stem from the non-pressurized steel construction of the tanks. The tanks are found on the ground, on concrete saddles, and on masonry or steel supports. Tanks placed on the ground are considered the safest. The least safe are those mounted on unprotected steel supports, since such supports can collapse after 5-20 minutes of severe fire exposure.

Horizontal tanks have regular breathing vents to prevent tank distortion after filling, emptying, or temperature changes. They also need emergency vents in case of exposure to fire. In addition, each tank should have a large conventional relief valve, a hinged manhole cover that opens under low pressure, or a flat manhole cover plate aligned with the manhole flange by long bolts to permit the plate to lift under low pressure. These provisions prevent the rupture of the tank's shell or head.

In a haz-mat incident, never approach horizontal tanks from the ends.

Excessive internal pressure can develop during an exposure or ground fire if a horizontal tank is not properly vented. This pressure may damage the end of the tank. If the end is damaged, the pressure release may propel the tank forward or backward. Because of this danger, never approach horizontal tanks from the ends.

In the event of a fire, additional lines may be needed immediately to flush burning fuel away from unprotected steel supports. If the supports fail, pipes will break and the tank may split, spewing its fuel on the fire. If a tank shifts on weakened supports, it may cover the emergency vents with liquid, preventing the vents from functioning properly.

PIPELINES

Notes

Crude oil is pumped or transported from both on-shore and off-shore wells to storage facilities where it is refined. Once refined, the products are pumped through hundreds of miles of pipeline to petrochemical plants. More than one product may be transferred through a single pipeline. Products can be separated by a "plug," but often they are not. A leak of any size may yield several products, such as gasoline, diesel fuel, and crude oil.

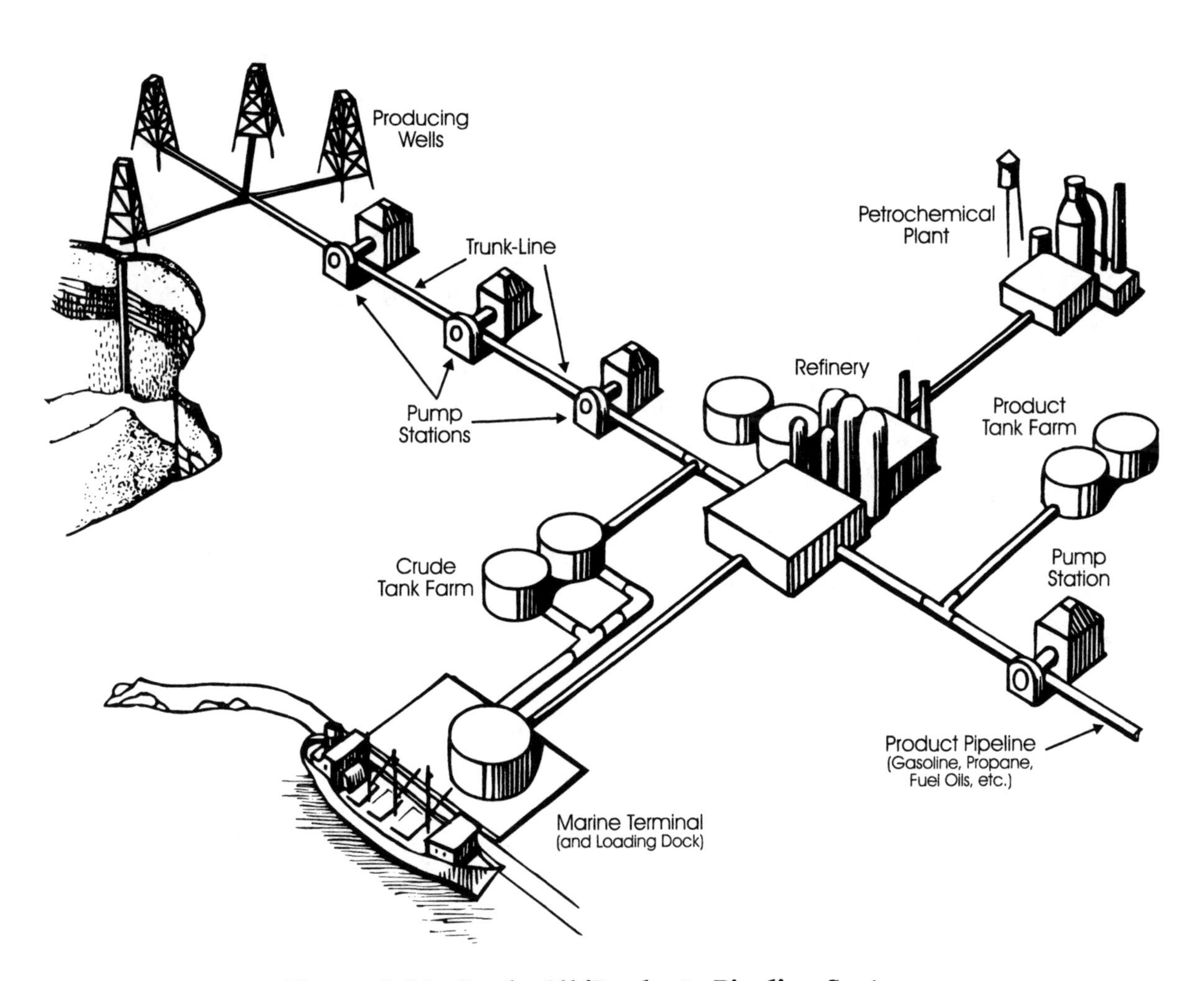

Figure 6-14. Crude Oil/Products Pipeline System

Notes

KEY POINTS TO REVIEW

Flammable and combustible liquids are quite common. Almost all production requires these materials. The number of flammable and combustible materials being transported and stored is staggering and continues to grow.

To review the material covered so far in this chapter, ask yourself these questions:

- What is a flammable/combustible liquid?
- Do you understand the physical and chemical properties of these materials?
- Are you familiar with the placarding requirements?
- Can you identify the most common storage and shipping containers?

It is important to learn all you can about responding to flammable and combustible liquid incidents. Prepare for the rest of this chapter by preplanning for a flammable/combustible incident. Answer the following questions:

- Where is a flammable/combustible liquid accident, fire, or spill likely to occur in your community or workplace?
- What time of day can such an incident be expected?
- What type(s) of products will be involved?
- What type of storage and/or transport containers will be involved?
- Who can help?
- Are the necessary resources readily available?
- How will sewers, irrigation ditches, rivers, and/or wildlife be affected?

RESPONDING TO FLAMMABLE AND COMBUSTIBLE LIQUID EMERGENCIES

Notes

INCIDENT RESPONSE

Trying to provide emergency response for flammable and combustible liquids is like trying to play the Minute Waltz on the piano with one hand. There are too many keys and not enough fingers. However, there are some preparation steps for a safe, professional response.

HAZ-MAT — HAZARD IDENTIFICATION

Identifying a flammable/combustible material is as systematic as identifying other hazard classes. Let's review the procedure for identifying the hazard:

- ***Preplanning*** for flammable liquid fires is important, because sizing up the actual incident may be difficult. Darkness or smoke may limit vision, and radiated and/or convected heat can make a close approach difficult, if not impossible. Therefore, decisions must be made based on what can be seen and what has been learned through preplanning.
- ***Location of the incident*** should indicate the type of liquid, amount of liquid, and container or vessel involved.
- Note ***markings or identification*** such as placards and tank shapes.
- ***Noises***, such as pinging or hissing from pressure relief devices, warn of potential hazards.
- ***Smells*** can also help identify a product during a leak or spill.

H
Hazard Identification

Once a flammable/combustible liquid has been identified, determine the possible hazards.

- Locate the leak or fire.
- Determine the cause of the incident.
- Find out if any injuries have been sustained.
- Determine the position of the involved container(s).
- Find out how long the container has been exposed to fire.
- Assess if there is an environmental hazard.

Notes

HAZ-MAT — ACTION PLAN

HAZARD CONTROL MEASURES

The goal of the action plan is to control the hazards of flammable/combustible liquids. The hazards associated with these incidents include:

- fire and vapor control
- environmental contamination due to runoff or spillage

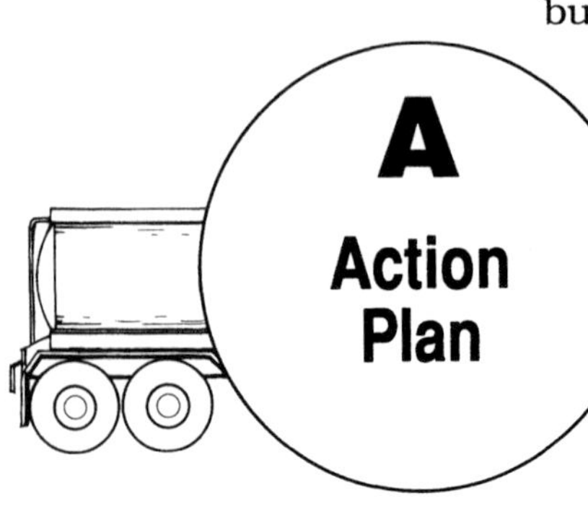

Following is a review of some of the mitigation techniques that can be used as part of the action plan.

FIRE AND VAPOR CONTROL

Fires and vapor resulting from flammable/combustible liquid incidents can be controlled by several methods, including:

- removing the oxygen supply
- breaking the chain reaction
- controlling the fuel supply
- applying water

Removing the Oxygen Supply

Removing the oxygen supply involves smothering the fire or vapors. Smothering can be used to control small or large flammable liquid fires. Hand extinguishers can be used on small fires; chemical, mechanical, or high expansion foam are appropriate for larger fires. Reignition or flashbacks from uncooled metal objects are common when dry chemicals or foam is used. The available supply of extinguishing agent may limit these efforts. Firefighting foam must be applied at the specific rate recommended by the manufacturer in order to extinguish the fire and control vapor production. Foam extinguishes flammable/combustible liquid fires in these ways:

- It excludes air from the flammable vapors.
- It prevents vapor release from the fuel surface.
- It separates the flames from the fuel surface.
- It cools by absorbing heat from the fuel surface and adjacent metal surfaces.

FOAM FACTS

Firefighting foam is one of the main substances used to remove the oxygen supply from a fire involving a flammable/combustible liquid. We will examine several basic foam facts, including:

- foam types
- basic foam terminology
- foam production
- foam solutions
- using foam successfully

FOAM TYPES

Notes

Chemical Foam

Chemical foam is formed by mixing an aqueous solution of sodium bicarbonate and stabilizer with a solution containing aluminum sulfate. Costs and large manpower requirements have rendered chemical foams obsolete. Replacement supplies are available for systems still in use.

Mechanical Foam

Mechanical or air foams are made by mixing a concentrated liquid with water and then mixing the solution with air. There are several types of foam commonly used for flammable liquid fires and spills including:

- protein foam
- fluoroprotein foam
- film forming fluoroprotein foam
- aqueous film forming foam
- alcohol resistant concentrates

Mechanical Foam

No single type of foam is perfect in every situation. Each should be selected based on the type of fuel and the nature of the incident. Proper selection depends on the type of fuel, container type (e.g., cone roof tank or spill), ambient liquid storage temperature, and other extinguishing agents used at the same time.

Protein foam has been in use since World War II. It was the first type of mechanical foam to be marketed extensively. This type of foam is produced by the hydrolysis of waste protein material (animal solids). Protein foam lacks fuel tolerance and is easily contaminated by the fuel. Therefore, most agencies now use the synthetic foams discussed below.

Fluoroprotein foam (FP) is formed by the addition of special fluorochemical surfactants to protein foam. The surfactants enhance the properties of protein foam by making it flow more easily and by providing faster flame knockdown and excellent fuel tolerance.

Like fluoroprotein foam, ***film forming fluoroprotein foam (FFFP)*** is also a combination of fluorochemical surfactants with protein foam. Unlike regular fluoroprotein foam, however, FFFP releases an aqueous film on the surface of a hydrocarbon fuel. This type of foam combines the burnback resistance of a fluoroprotein foam with an increased knockdown power resulting from the aqueous film.

Aqueous film forming foam (AFFF) is a combination of fluorochemical surfactants and synthetic foaming agents that create an aqueous film. This film is a thin layer of foam solution with unique surface energy characteristics. It spreads rapidly across the surface of a hydrocarbon fuel, causing dramatic fire knockdown.

AFFF is used for large fires involving transport or fixed site storage vessels. It is also used for small spills and leaks from such things as vehicle accidents. Some responders are reluctant to use foam on smaller incidents because of cost and lack of training in the proper application of foam. Each organization must determine whether the cost of foam and of training personnel in its use is worth the safety that foam provides.

Notes

Aqueous Film Forming Foam

AFFF is best for liquids with low flash points, such as gasoline (flash point, –50°F (–46°C)). Water fog is not effective on these low flash point liquids, but foam is an effective smothering blanket. The foam should be applied gently so that it flows slowly across the liquid's surface. AFFF has no "sticking" properties, making it completely ineffective when applied to vertical surfaces.

Alcohol resistant concentrates (ARC). Some fuels, such as gasohol products and oxygenated fuels, are not effectively extinguished by regular AFFF or fluoroprotein foams. Special "alcohol resistant" foams are available for such flammable liquids. Alcohol resistant concentrates must also be used for fighting polar based flammable liquid fires. Regular foam breaks down when used on polar solvents and water miscible liquids, such as acetone and methyl ethyl ketone. In contrast, alcohol resistant aqueous film forming foam (AR-AFFF) forms a gummy membrane (a polymeric barrier) on a polar solvent, which separates the foam from the alcohol and prevents the destruction of the foam blanket. This film seals the liquid from the air and inhibits evaporation.

Alcohol resistant foams are produced from a combination of synthetic stabilizers, foaming agents, fluorochemicals, and alcohol resistant membrane forming additives. Regular AFFF and alcohol foams may not be compatible and should not be mixed in the concentrate form. Figure 6-15 summarizes the uses of foam concentrates.

Notes

Uses of Foam Concentrates				
Foam Type	FIRES		VAPOR SUPPRESSION	
	Hydro-carbons	Polar Solvents	Hydro-carbons	Polar Solvents
Fluoroprotein	Good knockdown and excellent heat resistance	Not recom-mended	Excellent	Not recom-mended
Film Forming Fluoroprotein	Good	Not recom-mended	Good	Not recom-mended
Aqueous Film Forming	Excellent knockdown and fair heat resistance	Not recom-mended	Good	Not recom-mended
Alcohol Resistant Aqueous Film Forming	Excellent knockdown and good heat resistance	Excellent knockdown and good heat resistance	Good	Excellent

Source: National Foam

Figure 6-15

BASIC FOAM TERMINOLOGY

Foam – A stable aggregation of small bubbles, of lower density than oil or water, that shows tenacious qualities for covering horizontal surfaces.

Foam concentrate – The concentrated foaming agent obtained from a manufacturer for mixing with appropriate amounts of water and air to produce foam.

Foam solution – A homogeneous mixture of foam concentrate and water.

Polar solvent – A material that is soluble in water, such as alcohol or acetone.

Non-polar solvent – A material that is not soluble in water, such as gasoline or benzene.

Application density – The amount of foam solution (in gallons per square foot) required to cover a given fire or spill area.

Application rate – The required flow in gallons per minute (gpm) to achieve the desired application density.

Application time – The total estimated time required to apply enough foam to extinguish the fire or contain the material.

Notes

FOAM PRODUCTION

Two steps are involved in foam production:

- The ***foam concentrate*** is mixed in the proper proportions with water to produce a ***foam solution***. This is usually accomplished with a foam eductor or proportioning system.
- The ***foam solution*** is then mixed with air to form finished foam, which is then applied to the fire or spill.

FOAM SOLUTIONS

The correct percentage of ***foam concentrate*** must be added to water with some type of proportioning appliance to form the ***foam solution***. The percentage of foam concentrate can range from 1% to 10%, depending upon the type of concentrate used. Today, most foam concentrates are either 3% or 6%. 1% foam concentrate may also be found for the protection of special locations, such as off-shore oil platforms. Figure 6-16 charts the amount of foam concentrate and water needed to produce 100 gallons of foam solution.

Foam Solutions

Concentrate Type	Amount of Foam Concentrate	Amount of Water	Total Foam Solution
1% foam	1 gallon	99 gallons	100 gallons
3% foam	3 gallons	97 gallons	100 gallons
6% foam	6 gallons	94 gallons	100 gallons
10% foam	10 gallons	90 gallons	100 gallons

Figure 6-16

USING FOAM SUCCESSFULLY

After identifying the type of fuel involved and calculating the surface area of the fuel, the following factors must be quickly determined at any incident in order to use foam successfully:

- Determine the foam application density required in gpm of solution per square foot.
- Determine the application rate of foam to be applied.
- Estimate the total amount of foam solution required.
- Estimate the total gallons of foam concentrate needed to make up the required amount of foam solution.

Let's apply these four steps to a simulated 1,000 square foot spill of gasoline.

Determine Foam Application Density

Notes

Application densities vary, depending on the type of concentrate and the application device. The recommended application density for polar solvents is variable, ranging from 0.20 gpm and higher. The recommended application density for non-polar solvents ranges from 0.10 to 0.16 gpm of foam solution per square foot of surface area. Since recommended application densities vary, always use the manufacturer's recommendations. Using the example of a 1,000 square foot spill of gasoline, a non-polar solvent, the recommended density would be 0.10 gpm per square foot using AFFF.

Minimum Recommended Foam Application Density

Application type	Hydrocarbons	Polar Solvents
Fixed system application	0.10 gpm/square foot	0.20 gpm/square foot
Subsurface application in cone roof tanks	0.10 gpm/square foot	0.20 gpm/square foot
Portable application for spills	0.10 gpm/square foot (AFFF) 0.16 gpm/square foot (FFFP)	0.20 gpm/square foot
Portable application for storage tanks	0.16 gpm/square foot	0.20 gpm/square foot

Source: NFPA 11

Figure 6-17

Determine Application Rates

Knowing the recommended application density for the product involved, next determine the application rate. The following formula can be used to determine the application rate:

surface area to cover (square feet)	x	recommended density (gpm/sq ft)	=	application rate of foam solution (gpm)

Apply this formula to the example of the 1,000 square foot gasoline spill. Using AFFF with portable application devices would require the following application rate:

1,000 (square feet)	x	0.10 (gpm/sq ft)	=	100 gpm of foam solution

Estimate Total Foam Solution

To estimate the total gallons of foam solution required, the application time factor must be applied to the application rate, using the following formula:

Notes

application rate (gpm)	x	application time (minutes)	=	total foam solution (gallons)

NFPA recommends a minimum of 15 minutes of flowing foam to cover a flammable liquid spill. Using the gasoline spill as an example, the total foam solution required would be:

100 gpm foam solution	x	15 minutes	=	1,500 gallons of foam solution

Estimate Total Foam Concentrate Needed

To determine the total gallons of foam concentrate required, use the following formula:

total foam solution (gallons)	x	% of foam concentrate	=	total foam concentrate (gallons)

For illustration purposes, assume we are using 3% AFFF concentrate. Again returning to the 1,000 square foot gasoline spill, the total concentrate needed would be:

1,500 gallons total foam solution	x	3% foam concentrate	=	45 gallons of 3% foam concentrate

From the above information, the total amount of foam concentrate and water required can be calculated. Firefighting operations should not begin until an adequate supply of foam concentrate is on hand, since the fire will destroy the foam blanket if the fire is not extinguished completely with the initial application of foam.

In the event the first attempt to extinguish the fire is unsuccessful, the total calculated amount of foam concentrate will be needed for the second attempt. In other words, if the first attempt required 80 gallons of concentrate, an additional 80 gallons will be required on the second attempt. With these formulas, an action plan can be determined with the assurance that adequate resources are available to achieve extinguishment and/or control of the materials involved.

Breaking the Chain Reaction

Most small fires, such as automobile fires, small spill fires, or dip tank fires, can be extinguished by breaking the chain reaction of combustion. Dry chemical hand extinguishers or a vaporizing liquid are the best agents to use. You may be restricted by the limited supplies of extinguishing agents available. Surprisingly, even moderate sized flammable/combustible liquid fires can be controlled by portable dry chemical extinguishers. Dry chemical, although not the most popular agent, can be used until other extinguishing agents are available.

Controlling the Fuel Supply

Notes

The most common way to control the fuel supply of a burning or leaking container is to **close the appropriate valve**. Fuel that has already escaped may either be left to burn out under controlled conditions or picked up and disposed of properly. With small pipeline leaks where there is no available valve, crimp or flatten the line to stop the flow. If a pipe is sheared off, a plug can sometimes be used to stop the flow.

If the fuel supply cannot be shut off, it may be best to **flush the burning liquid to a safer location**. There will be fewer exposure problems, and the burning fuel can be better controlled. This approach may be necessary when a leak from a tank truck cannot be controlled. Flush the fuel away from the uninvolved compartments.

Understanding the specific gravity of flammable liquids will aid fire extinguishment efforts. Many of these liquids will float on water because they have a specific gravity of less than 1.0. Burning flammable or combustible liquids will float on water if hose streams are applied. This may cause a running fire where the burning liquids travel on top of the water. Exposure protection is a high priority when a potential running fire could occur.

When the incident involves liquids that are **lighter than water**, water may be your best asset. For example, water may be added to a tank with a leaking valve or hole near the bottom. The fuel will rise above the source of the leak, since the fuel floats on water. Water, rather than fuel, will then be discharged from the tank. Continuing to add water will allow the fuel level to be maintained above the leaking valve or hole until the leak is repaired or plugged. This method can also be used when the tank is on fire. ***Warning:*** Check the tank's fuel level to be sure the fuel will not overflow when water is added.

Water can be floated on top of flammable or combustible liquids that are **heavier than water**. This layer of floating water creates a vapor barrier over the surface of the fuel. This barrier aids in extinguishing a fire or suppressing flammable vapors rising from the liquid. Be careful to apply the water gently so it floats across the top of the fuel without agitating the product. Carbon disulfide fires have been successfully extinguished using this method.

The burning surface of kerosene or diesel oil type fuels should be **cooled with water fog** to a temperature below the liquid's flash point. At that temperature, there will not be enough flammable vapors for the fuel to burn. If this water spray procedure is used on heavier oils, the burning liquid will turn the water into steam, forming a froth on the surface. This froth not only cools the surface, but also cuts off the air supply by blanketing the surface.

Patching and plugging devices may be needed in leaks involving transport vehicles, such as the MC306/DOT 406, to minimize or stop the fuel flow until the product is transferred. Emergency responders can use wooden plugs, inflatable air bags, manway dome clamps (see Figure 6-18), and other devices to control the fuel supply on flammable/combustible liquid storage containers.

Notes

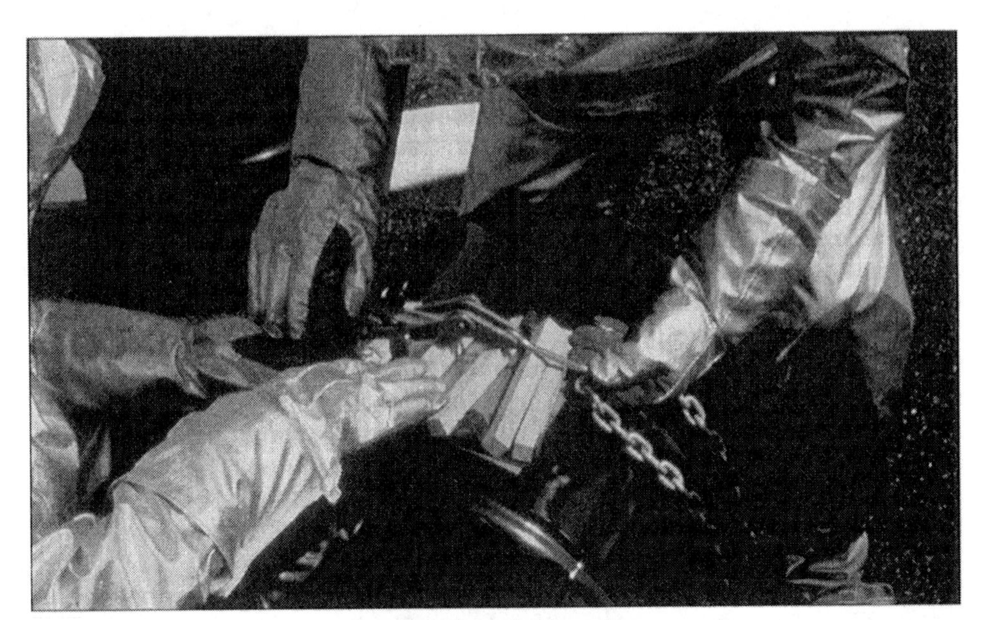

Patching and Plugging

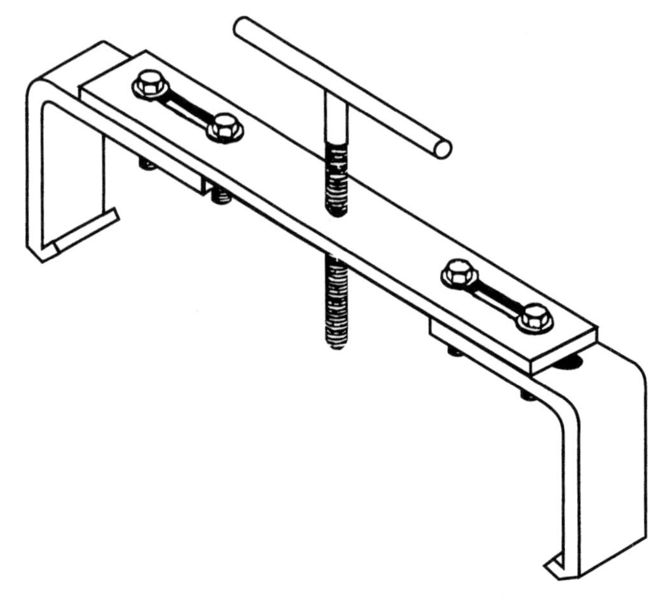

Figure 6-18. Manway Dome Clamp

Applying Water

Notes

The most effective use of water on large fires is exposure cooling, **not** extinguishment. Overuse or total reliance on water most likely will spread the burning liquid and cause heavy oils to froth. Some high flash point liquid fires, however, have been extinguished using water. The water will skim the flames off the surface, snuffing out the ignition source like blowing out a candle. This method will be successful only when there is no reignition source.

Water soluble flammable liquids can be diluted to the point that a physical change of the material occurs. When this change occurs, the flash point can be raised so that flammable vapors are no longer generated. Dilution can also be used when there is a spill but no fire. The amount of water needed to dilute varies with different products. Obtaining the large quantities of water needed to dilute the product may be difficult. Be aware that dilution will increase the volume of the product and will cause substantial runoff problems and additional environmental contamination. This method is not the best choice with flammable liquid container fires, because the water will usually cause the container to overflow, spilling the hazardous liquid.

Use water to **reduce heat output** when:

- personnel are closing valves or working close to the fire
- tank surfaces or exposures are exposed to radiant heat or direct flame impingement
- vapors from the surface of heavy oils can be reduced

Water Application on Flammable Liquids

Notes

Water can also be used to **disperse the product** when it:

- is used to move the product to a more desirable location
- will move a product so that access to control the leak is possible
- may prevent the liquid from igniting
- is used to move or disperse vapors

Water can **displace a product** when it:

- floats the liquid above a leak, interrupting the flow of the hazardous product from the leak
- is pumped into a fuel supply line to stop a line or valve leak

Water may be used to **dilute a product** when it causes a physical change in a water soluble product and:

- raises the flash point so that flammable vapors are no longer produced
- raises the flash point so that there is no fire

When using water, be careful not to move the flammable/combustible product to a more hazardous location. Also be careful that containers are not overfilled when applying water. Never direct water below a burning liquid's surface because the heat will turn the water to steam. Steam will greatly increase the volume of the water and will cause a steam explosion. The expanding water/steam will then force the liquid to slop over the sides of the tank.

ENVIRONMENTAL CONTAMINATION

Sewers and Storm Drains

There always seems to be a sewer or storm drain downhill from a large flammable/combustible liquid spill. Make every attempt to stop the flow of the liquid into the sewer. The action taken will depend on the liquid and the sewer system. If the storm sewer flows to an open waterway, controlling the flow will prevent water contamination if the liquid is toxic. It may also prevent a fire at a second location if the liquid is easily ignited.

If a closed sanitary sewer is involved, the runoff can ignite and blow manhole covers into the air. Underground services and surface roadways may be damaged. Where flammable or combustible liquids have entered enclosed sewers, large amounts of water may be effective to flush the line. Notify the water treatment plant of the involved liquid and flushing operations.

Water Contamination

If the product has entered a waterway, immediately work to reduce environmental impacts. Even if your organization does not have an HMRT, there are several ways to minimize the downstream environmental effects. Hazardous materials may be collected in a variety of ways once they enter a waterway (see Figures 6-19 through 6-21). In most cases, product containment and collection can be accomplished without using expensive equipment. Environmental effects can be reduced even if absorbent pads, booms, and other such equipment are not accessible (see Figures 6-22 through 6-27).

Notes

Set up ahead of the spill to contain all of it. Failure to account for set-up time may allow the product to pass by before containment measures are in place. Answer the following questions prior to the containment operation.

- What are the physical and chemical properties of the liquid?
- How much product has already been released into the waterway?
- What is the rate of release?
- Can the leak be stopped and how long will it take?
- How fast is the water moving?
- How long will it take to set up the containment operation?

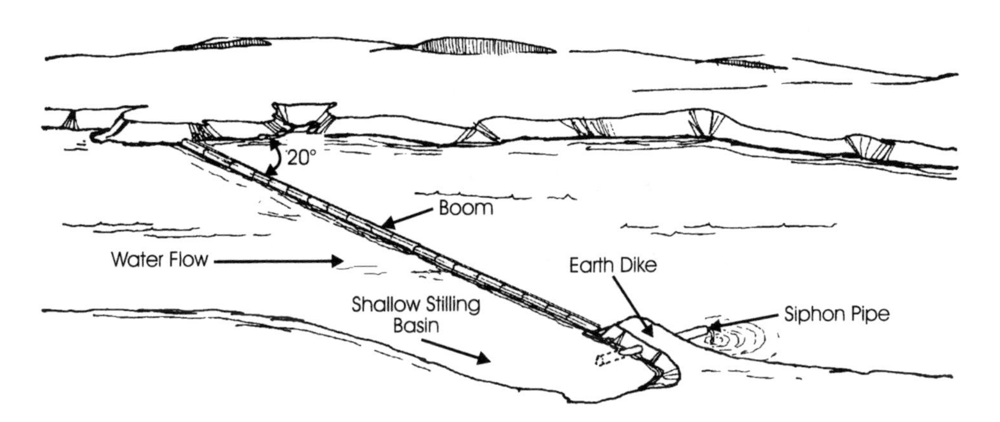

Figure 6-19. Narrow Stream Spill Containment

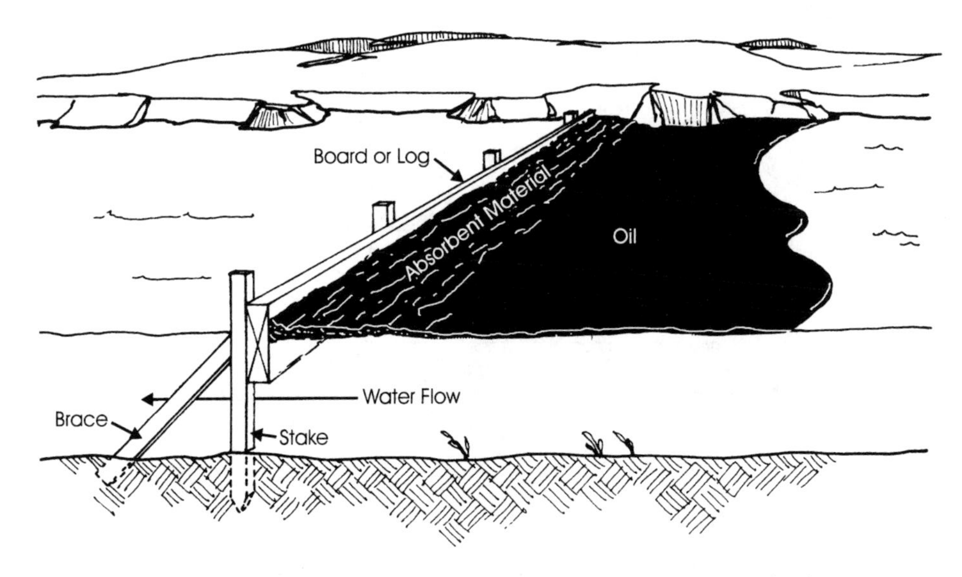

Figure 6-20. Skimming Device on Stream or Ditch

Notes

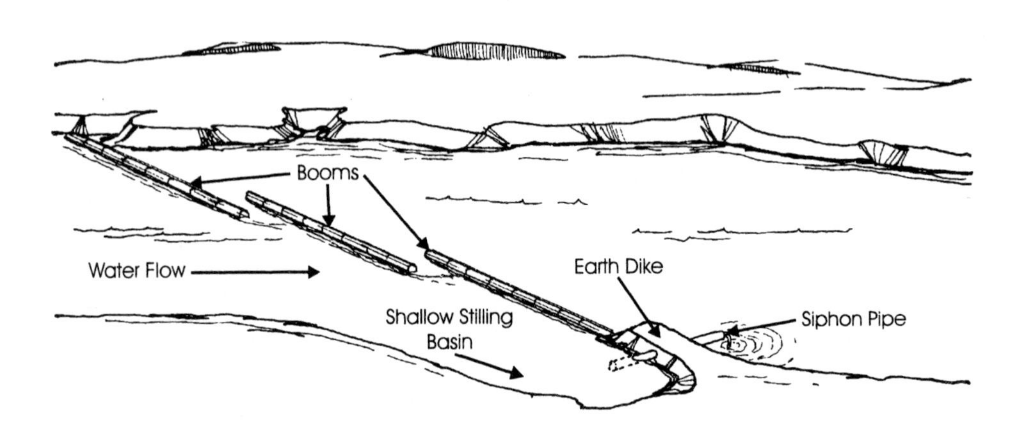

Figure 6-21. Wide Stream Spill Containment

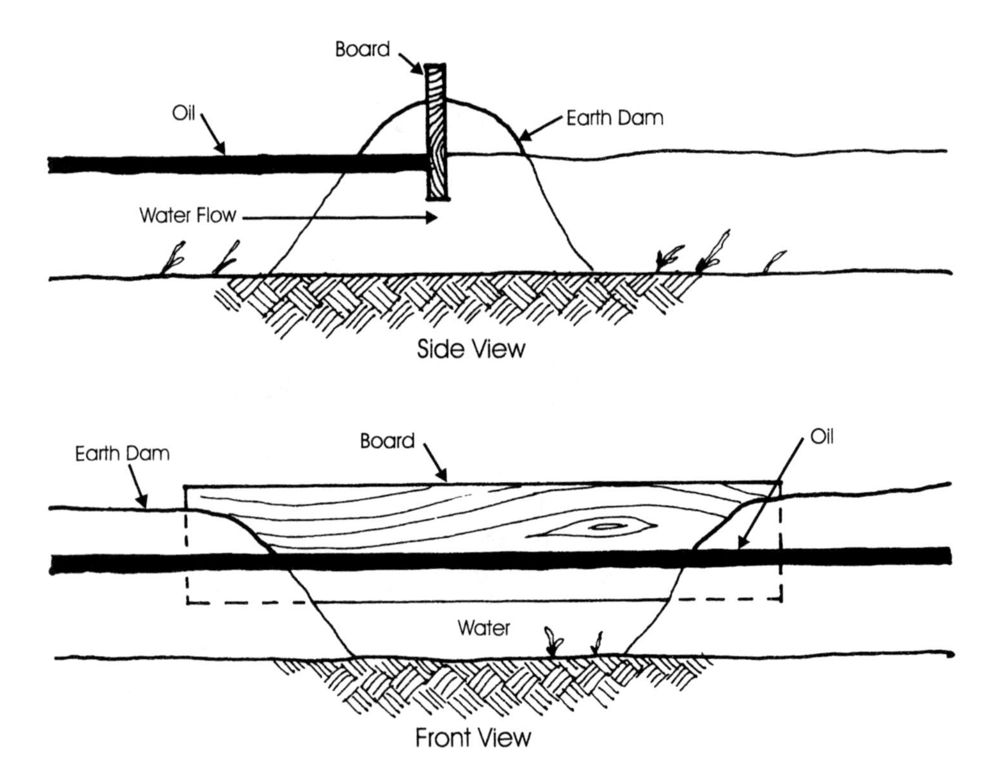

Figure 6-22. Earth Dam and Weir

Notes

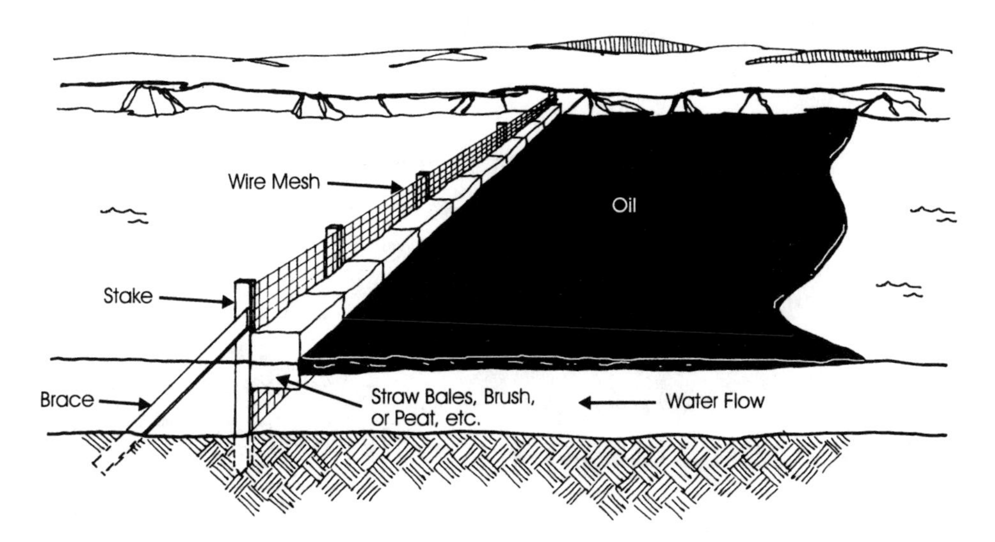

Figure 6-23. Wire Fence Filter Boom

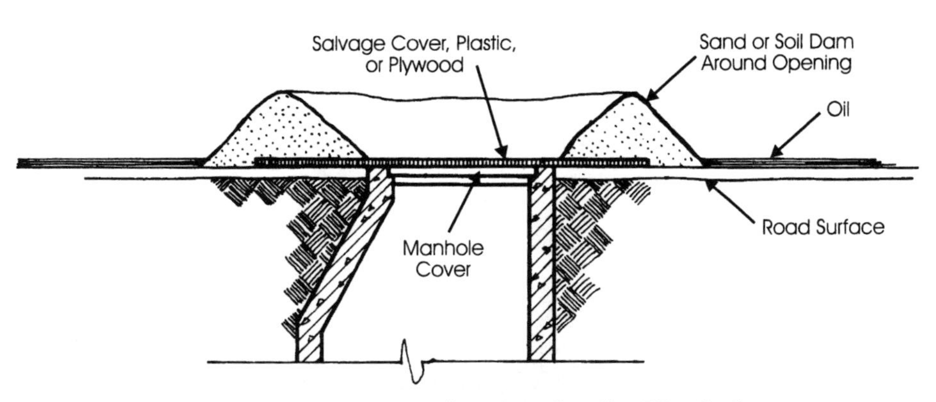

Figure 6-24. Protection Barrier for Manhole

Notes

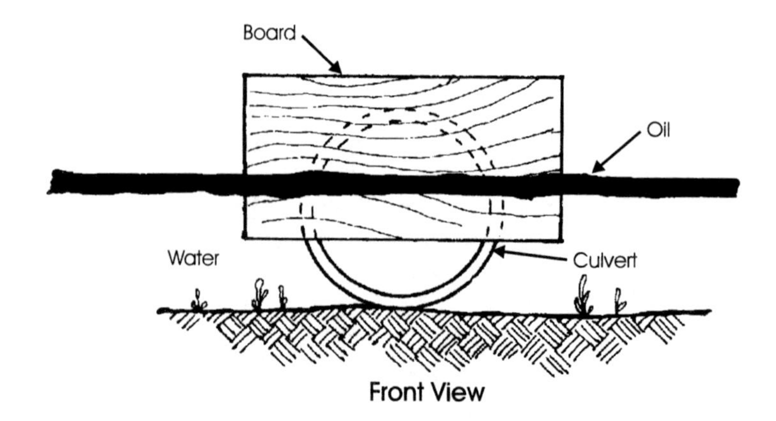

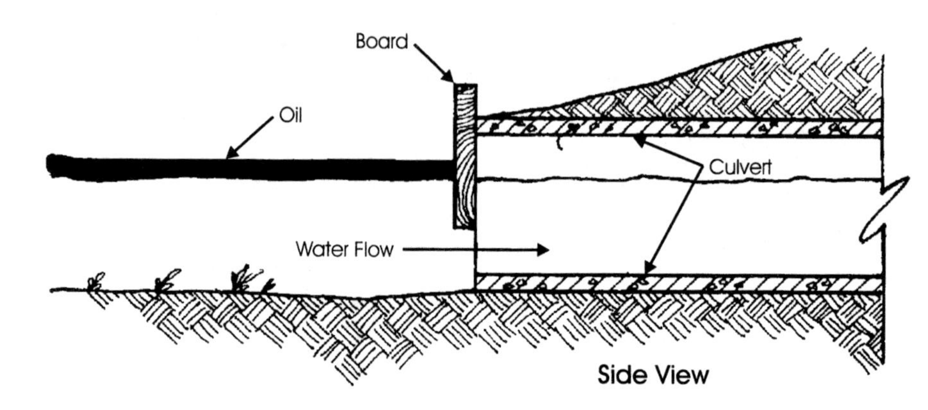

Figure 6-25. Culvert Weir

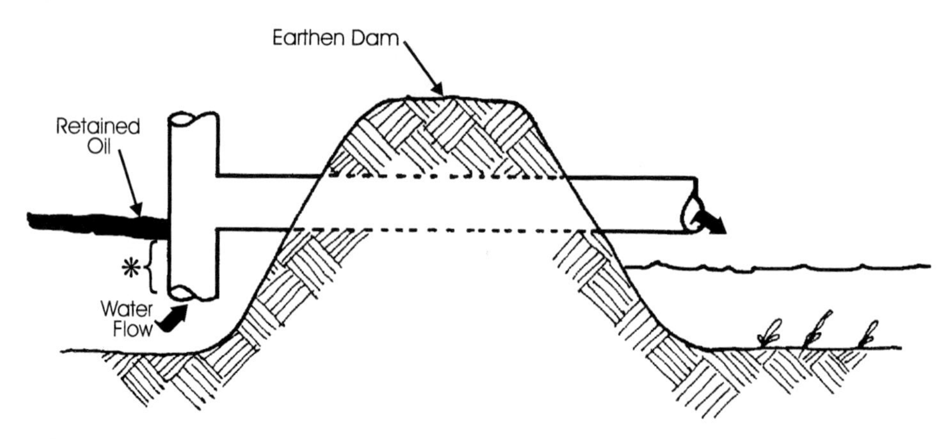

* Due to the design of the T-section, a vortex cannot develop. This is the preferred siphon arrangement when feasible.

Figure 6-26. T Siphon

Notes

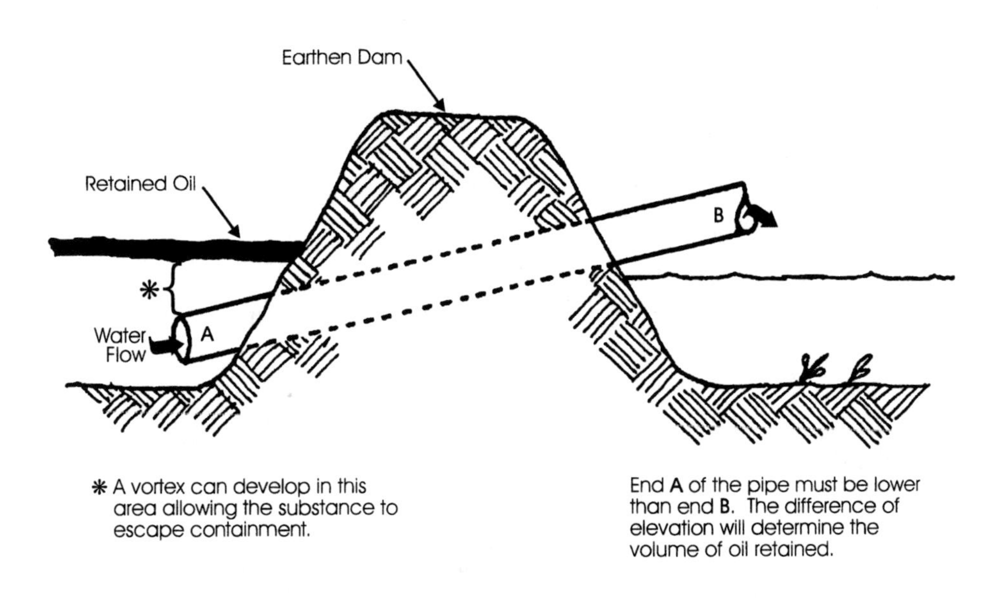

Figure 6-27. Inverted Siphon (Underflow)

ACTION PLAN QUESTIONS

If an accident involves a flammable/combustible liquid container with **NO LEAK or FIRE**, consider the following factors:

- Determine if a rescue is needed.
- Protect exposures.
- Eliminate ignition sources.
- Determine the size of the container. The size will dictate what resources are needed and the extent of evacuation.
- Determine the damage to the container. Will the container fail or develop leaks?
- Determine the liquid level in the container. Off-load product before uprighting large tankers or tank cars.
- Establish appropriate control zones if a leak is possible.
- Consider evacuation needs.

If there is a **LEAK but NO FIRE:**

- Establish control zones! Immediately evacuate everyone from the area, especially those downhill and downwind.
- Control all ignition sources around the incident. This includes responding apparatus, pilot lights, cigarettes, utilities, and vehicles.
- Determine flammable limits for these materials by using metering devices.
- Positively identify the product.
- Determine if a rescue can be performed safely. Remember, this can be a very emotional issue.

Notes

- Control streets and request crowd control.
- Determine if the product is water soluble. Control runoff if the environment is endangered.
- If the product has entered a storm sewer or waterway, minimize the environmental damage and notify water treatment facilities.
- Determine how to transfer the product to another tank.
- Apply foam or other agents as needed.

If there is a **LEAK and FIRE:**

- Order an immediate evacuation according to the type and amount of material involved.
- Control the flow of fuel, if possible.
- Keep personnel and apparatus uphill and/or upwind of the fire, if possible.
- Be sure enough water can be continuously supplied to handle the task before setting up.
- Coordinate attack, backup, and protection hose streams. Attack lines must be backed up. Personnel working at hazardous locations should be covered with a fog stream.
- Protect all exposures.
- Burning liquids must be flushed from under tanks, and unprotected steel supports must be cooled to prevent collapse and rupture.
- Apply foam or other agents as needed.
- Tanks must be cooled at points of flame impingement and above the liquid level. Direct streams along the upper side of tanks.
- When the fire has gained considerable headway in the structure of a hazardous occupancy, resort to exterior firefighting.
- Keep all apparatus headed away from the fire so it can be moved quickly.
- Use all available special firefighting equipment.
- Use unmanned streams or fixed nozzles in hazardous areas to avoid endangering personnel.
- Use a retreat signal. Have a maintained and lighted retreat path at night.
- Listen for sounds from the tank vents. A whistling sound, with growing intensity, means pressure is building because of inadequate cooling. Prepare to retreat.
- Flames which suddenly get brighter and higher signal a possible boilover. Retreat to safer positions.
- Tank vent fires that burn with a snapping, blue-red, nearly smokeless flame indicate that the fuel-air mixture may be explosive. Never cool tanks burning in this manner.
- Avoid approaching tanks from the ends.
- Be prepared for backflash until the entire area and liquids are cooled.

Notes

- Aerial apparatus can be used for a vertical attack. Be aware, however, of the danger to personnel positioned above and over the fire. Unmanned aerial streams are advisable.
- Tanks that have rolled or moved from their original position may have blocked relief valves. The relief valves may either be obstructed by the liquid or damaged by the shifting of the tank. The tank may over-pressurize or discharge liquid from the relief valve.
- There may be a vapor problem after the fire is extinguished. Maintain hose lines to dissipate these fumes, or apply additional foam for vapor control.
- After the fire is out, consider the environmental problem. Anticipate clean-up operations.

HAZ-MAT — ZONING

Establishing control zones early in the incident should be a priority. Use combustible gas meters to determine the appropriate zones. Remember, running fuel fires, heat, and smoke are a danger to personnel and bystanders. Establish a 500 foot hot zone on large leaks and fires. Develop warm and cold zones in order to stage the resources needed for large incidents. Use law enforcement personnel to maintain your control zones. Don't be afraid to expand the zones if needed.

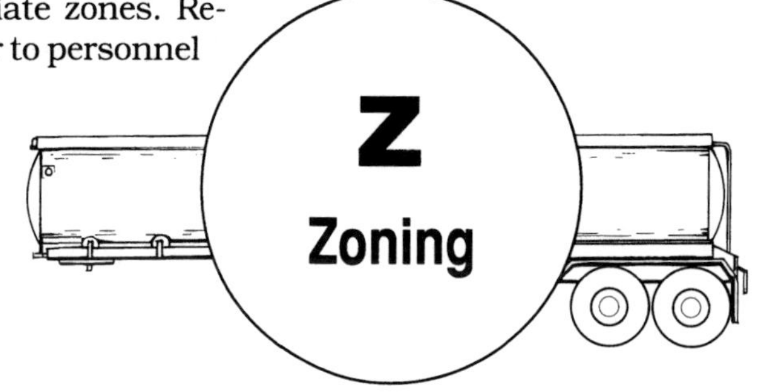

HAZ-MAT — MANAGING THE INCIDENT

Scene Management

Management difficulty depends upon the amount of product, container type, and environmental considerations. A flammable liquid incident can significantly burden resources of even the largest emergency service delivery system.

The key to successful scene management is preplanning basic strategy, tactical operations, and resource needs. Two separate incidents involving the same material can vary drastically. In preplanning, think through several possible scenarios that may occur with a material. Be prepared for whatever occurs. For example, an overturned gasoline tank truck should be attacked differently in the center of town, as opposed to on the interstate in a rural area.

The complexities of tanks, valves, and, in many cases, the on-site fire protection systems, demand that the incident commander know local facilities to ensure safe, effective fire ground operation.

Notes

Tactical Considerations

- Rescue.
- Protect exposures.
- Confine the product. Confinement should include controlling spills and cooling tanks and supports.
- Stop the fuel flow that supplies the fire.
- Extinguish or allow the fire to burn out.
- Cool any possible reignition sources.
- Maintain standby crews and lines.
- Use gas detection equipment to check for flammable/combustible vapors in the surrounding area.
- Establish environmental control measures.
- Decontaminate.

RESCUE SIZE-UP

- Rescue must be considered first.
- Transportation incidents usually involve rescue.
- Rescue those in vehicles, structures, or complexes.
- Rescue those in exposed areas.
- Determine the extent and method of evacuation.

EXPOSURE ASSESSMENT

- What is the construction, distance, occupancy, and location of adjacent exposures?
- Are other vehicles, tanks, equipment, or buildings in the complex exposed?
- Determine which vehicles or structures are exposed during transportation incidents.
- Are other vehicles, tanks, equipment, or structures outside the fire area exposed?
- Are overhead communication or power transmission lines exposed?
- Are people and property exposed to downwind vapors?
- Are people and property exposed to liquids flowing downhill?
- Are weeds and grass exposed?
- Are streams or sewers exposed?

ON-SITE EQUIPMENT AND LOCATION

- Note shutoffs for electrical power to site, including overhead lines.
- Note shutoffs for natural gas lines to buildings.
- Locate fuel supply valves from tanks, loading racks, pumps, or pipelines.
- Check for remote control shutoff valve in pump house.
- Check for water and extinguishing system control valves.
- Determine piping layout, internal exposure, and color code of piping.

Notes

FUELS

- What types of flammable/combustible liquids are involved?
- How much of each type of fuel is involved?
- Consider the physical and chemical properties of these materials when developing your overall strategy.

Rescue Considerations

Exposure to flammable/combustible liquids produces varying injuries. Vapor exposure can cause shortness of breath, headaches, nausea, vomiting, narcotic like effects, slurred speech, and other problems. Many of these products accumulate in the liver and kidneys. Use self-contained breathing apparatus and appropriate protective clothing to protect from exposure to these materials.

Assist medical personnel with victims. Assess a victim's exposure time so that aggressive and proper treatment can begin. Exposure to some liquids can cause severe chemical burns. In addition, the absorbing properties of some chemicals can create acute medical problems. Anticipate second and third degree burns and severe respiratory distress if fires occur.

Flammable and combustible liquids are often toxic and corrosive to the skin and mucous membranes. They can also act as a central nervous system depressant.

As an example, consider the effects of gasoline and kerosene. Acute exposure at high concentrations can cause dizziness, central nervous system depression, coma, collapse, and death. A concentration of 500 ppm is not considered safe for even short periods of time. Chronic inhalation of highly concentrated vapors of leaded gasoline can cause lead poisoning. Acute exposure or ingestion of the product can lead to chemical pneumonitis, which may predispose the patient to secondary bacterial infections. Pulmonary edema and hemorrhage are also common. Skin contact with gasoline can cause irritation and dermatitis. In fact, almost all solvents defat the skin, resulting in dryness, irritation, and dermatitis.

General Treatment for Flammable and Combustible Liquid Exposures

Products .... Gasoline, kerosene, methanol, isopropyl alcohol.

Containers .... Cans, drums, tank trucks, tank cars, bulk storage tanks.

Life Hazard .... Central nervous system depression that may result in respiratory arrest. Pulmonary edema may be found. Many products can cause myocardial irritability leading to cardiac arrhythmias and cardiovascular collapse. Can be toxic through ingestion, inhalation, and skin absorption. Can cause damage to skin and mucous membranes. Thermal burns may result if product is ignited.

Signs/Symptoms ..... If vapors are inhaled, a burning of the chest, respiratory distress, and pulmonary edema may be present. Cardiac arrhythmias and cardiovascular collapse may result. The patient may experience a decreased level of consciousness with tinnitus, weakness, disorientation, headache, drowsiness, lack of coordination, seizures, and possibly coma.

Notes

Products may cause irritation and burns to skin, eyes, and mucous membranes. If ingested, nausea, vomiting, diarrhea, and gastrointestinal pain may result.

Basic Life Support .... Wear proper protective equipment and remove patient from contaminated area. Administer oxygen and remove patient's clothing. Gently blot any visible liquid from the skin and decontaminate with copious amounts of soap and water. Irrigate eyes thoroughly and continue during transport as necessary. Watch for signs of shock, pulmonary edema, and decreased level of consciousness/seizures and treat as necessary. If ingested, dilute with small quantities of water and give activated charcoal (follow advice of Poison Control and local protocols). Do not orally administer anything if the patient has a decreased level of consciousness or decreased gag reflex. Be prepared to treat thermal injuries. Cover burn areas with sterile dressings after decontamination.

Advanced Life Support .... Assure an adequate airway and assist ventilations as necessary. Start IV D5W TKO if blood pressure is stable. Use replacement fluids such as LR if signs of hypovolemic shock are present or in cases of thermal burns. Watch for signs of fluid overload. Monitor for cardiac arrhythmias/pulmonary edema and treat as necessary. Seizures may require treatment with an anticonvulsant. Administer topical anesthetic to eyes for easier irrigation and to reduce pain. Follow local protocols for all drug therapy and medical treatment.

Other Information .... **Do not give emetics to induce vomiting.** Avoid epinephrine (unless in cardiac arrest) due to possible sensitized condition of the myocardium. Be prepared to treat thermal burns. Products in closed containers may present a boiling liquid expanding vapor explosion (BLEVE) hazard.

Notes

HAZ-MAT — ASSISTANCE

The flammable/combustible emergency will require assistance from many of the persons and agencies listed below.

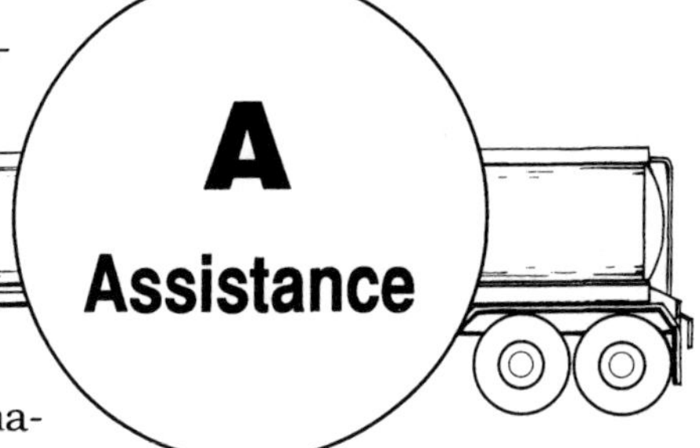

- ***Dispatcher*** can provide coordination and gather technical information and needed resources. This will assist the incident commander on the scene.
- ***Local law enforcement*** can assist in scene control and evacuation coordination.
- ***HMRT*** can provide technical expertise. They can also undertake on-site tactical duties, such as patching, plugging, diking, foam application, and decontamination.
- ***On-site technical expert*** can help determine the type and amount of material involved. This person can also determine which fire protection system would be most beneficial and can recommend a strategic location for equipment. On large tank farm fire emergencies, the on-site expert can help acquire outside resources, such as foam, diking materials, and private clean-up contractors. Do not overlook this valuable person.
- ***State and local health departments, Environmental Protection Agency, U.S. Coast Guard, and other government agencies*** should be notified on many incidents. Use the resources available through these agencies.
- ***Tow trucks, vacuum trucks, and other equipment*** can upright cylinders or containers and off-load the product. Operators should be directed by the command post. Many equipment operators are not aware of hazards and may take action that could seriously endanger themselves and others. Think safety and take command.
- ***Clean-up teams*** will be necessary after most incidents. Coordinate with the appropriate facilities representative and anticipate clean-up needs. Your community's environment will be healthier if clean-up efforts are started immediately after stabilizing the incident.

Notes

HAZ-MAT — TERMINATION

Termination of a flammable/combustible emergency should be systematic. Review proper termination actions:

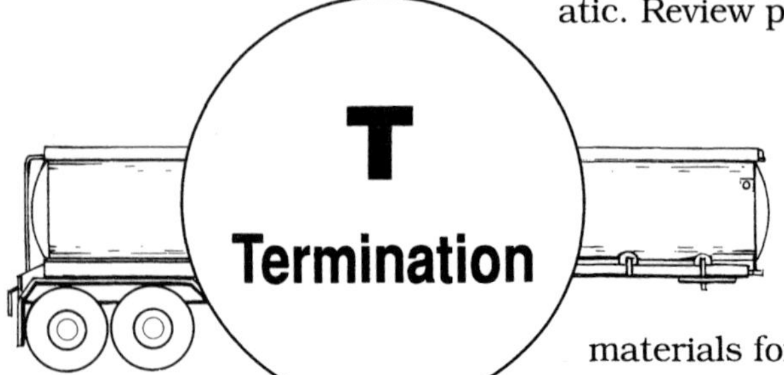

Decontamination

Decontamination is often overlooked in flammable and combustible emergencies. Take time to decontaminate personnel, protective clothing, and equipment. Mild soap and water is an adequate decontamination solution for many products. However, consult reference materials for the most effective decontamination solution.

Rehabilitation and Medical Screening

Large flammable/combustible liquid incidents can be physically demanding and emotionally draining. Evaluate personnel to decide if they are able to continue with current duties or if they should be reassigned. The evaluation should include:

- rest period
- medical evaluation
- exposure report
- initial incident debriefing

Post-Incident Analysis

Although the flammable/combustible emergency is common, a post-incident analysis should be conducted with all involved agencies. Discuss problems that were encountered and prepare a more effective response for the next such emergency.

KEY POINTS TO REVIEW

Over 50% of haz-mat responses involve flammable/combustible liquids. Consider the physical properties of these materials when developing action plans. Understand foam and its proper application prior to an incident. It may be best to allow a fire to burn. Use dikes to control spills and reduce incident size. Anticipate dangers that personnel may encounter. Be aware that control of many incidents may be beyond your ability and/or resources.

- Preplanning for the most likely incidents can enable you to deal successfully with the incident.
- Anticipate your needs prior to the incident.
- Use foams to extinguish a fire effectively and to control vapor.
- There are thousands of flammable/combustible materials. Consult reference materials before taking action.
- These liquids are often toxic and irritating to the skin and mucous membranes. They can act as a central nervous system depressant.
- Know container design and operating principles.
- Be aware of possible environmental impacts. Take action to minimize the effects.

Chapter 7

FLAMMABLE SOLID EMERGENCIES

OBJECTIVES

After studying the material in this chapter, you will be able to:

- define flammable solids, spontaneously combustible materials, and dangerous when wet materials
- identify the types of flammable solids, together with the properties and hazards of each type
- state the physical and chemical properties of the most common metallic and non-metallic flammable solids
- state fire extinguishing guidelines for each of the major flammable solids
- describe common shipping and storage containers for flammable solids
- state response considerations for flammable solid emergencies, using the acronym HAZ-MAT

UNDERSTANDING FLAMMABLE SOLIDS

INTRODUCTION

Flammable solids are found throughout the United States and Canada. These materials are used in vacuum tubes, as laboratory reagents, as fuel additives, and as sources for nuclear materials. They are also common in foundries and metallurgy shops.

The flammable solid hazard class contains many materials that have varying degrees of hazard, from minimal to severe. Some materials may be water reactive (dangerous when wet), such as calcium carbide. Others, such as phosphorus, may be pyrophoric materials that spontaneously ignite when exposed to air.

Notes

There are also flammable or combustible dusts. These are not regulated by the Department of Transportation (DOT) because they are rarely hazardous during transportation. Their hazards do appear, however, in fixed facilities where the dusts are either processed or generated.

Although flammable solid materials are common, emergencies involving them are not. Emergency responders must be prepared for a flammable solid emergency, however, because some of these materials may produce toxic by-products and/or explode when exposed to air, water, and other chemicals.

DOT CLASSIFICATIONS

Flammable solid – Any of the following three types of materials: (1) wetted explosives that when dry are explosives of Class 1 other than those of compatibility group A, which are wetted with sufficient water, alcohol, or plasticizer to suppress explosive properties; (2) self-reactive materials that are liable to undergo, at normal or elevated temperatures, a strongly exothermal decomposition caused by excessively high transport temperatures or by contamination; and (3) readily combustible solids that may cause a fire through friction, such as matches, or any solids or metal powders that have a rapid burning rate, as determined by specific tests (49 CFR 173.124(a)).

Spontaneously combustible material – (1) A pyrophoric material. A pyrophoric material is a liquid or solid that, even in small quantities and without an external ignition source, can ignite within 5 minutes after coming in contact with air. (2) A self-heating material. A self-heating material is a material that, when in contact with air and without an energy supply, is liable to self-heat (49 CFR 173.124(b)).

Dangerous when wet material – A material that, by contact with water, is liable to become spontaneously flammable or to give off flammable or toxic gas (49 CFR 173.124(c)).

Many products fall within these DOT definitions for the flammable solid hazard class. As previously mentioned, flammable or combustible dusts are not regulated by the DOT. These materials should never be overlooked during an emergency operation.

FLAMMABLE SOLID TYPES

Metallic Flammable Solids

Metals are chemical elements that have luster, can conduct heat and electricity, have tensile strength, and will form oxides. See Figure 7-1 for a comparison of the physical properties of metals and non-metals. Many metals have hazardous properties, and nearly all metals will burn under certain conditions.

Notes

Comparison of the Physical Properties of Metals and Non-Metals

Metals	Non-Metals
All are solids except mercury, cesium, and gallium	May be solids, liquids, or gases
High density	Low density
Metallic luster	No metallic luster
Malleable	Non-malleable
Ductile	Brittle (solids)
Good conductors of heat and electricity	Usually poor conductors of heat and electricity
Opaque	Transparent or translucent

Figure 7-1

Flammable metals include:

- alkali metals
  - sodium
  - potassium
  - lithium
  - rubidium
  - cesium
  - francium
- alkaline earth metals
  - magnesium
  - calcium
  - beryllium
  - strontium
  - barium
  - radium
- other metal groups
  - aluminum
  - titanium
  - zinc
- radioactive metals
  - uranium
  - plutonium
  - thorium

Some of these metals and their compounds, alloys, or mixtures can spontaneously ignite when exposed to air. Others will not ignite until they are extensively heated, but will then burn with incredible intensity.

All of these metals will react violently if water is applied to them while they are on fire! Some metals, such as the alkali metals, can be water reactive even before they are ignited. In either case, the metal reacts with the oxygen molecule in water, and there is a subsequent release of hydrogen gas. The heat of the reaction then ignites the hydrogen gas. An accumulation of hydrogen can explode and throw flaming, molten pieces of the metal into the air. It should be obvious that a metallic flammable solid emergency can be very dangerous.

All metallic flammable solids will react violently if water is applied to them while they are on fire!

Notes

Non-Metallic Flammable Solids

Non-metallic flammable solids have unusually high burning rates, require special extinguishing techniques, and/or produce extremely toxic fumes while burning. Examples include camphor (ignition temperature 871°F (466°C)), red phosphorus (ignition temperature 500°F (260°C)) and sulfur (ignition temperature 450°F (232°C)). White phosphorus is extremely hazardous in that it will spontaneously ignite in contact with air at temperatures of 86°F (30°C) or above.

Flammable or Combustible Dusts

Flammable or combustible dusts are particulate materials that burn when mixed with air. These materials are usually some form of an ordinary combustible material, such as sawdust, grain dust, plastic dust, coal dust, and flour. Common dusts and their relative explosion potential are listed in Figure 7-2.

Dusts and Their Relative Explosion Potential

Weak

- Coffee
- Cottonseed hulls
- Kelp
- Moss
- Onions
- Polyvinyl chloride
- Tin
- Triozane
- Urea formaldehyde
- Zinc

Moderate

- Alfalfa
- Coal (low volatile)
- Cocoa
- Hydrogen reduced iron
- Rice
- Skimmed milk
- Tantalum
- Uranium

Strong

- Chromium
- Coal (high volatile)
- Crude rubber
- Peanut hulls
- Phenol formaldehyde
- Polyethylene
- Silicon
- Soy protein
- Walnut shells

Severe

- Aluminum
- Cork
- Cornstarch
- Epoxy resin
- Flour
- Magnesium
- Pea flour
- Pitch
- Stearic acid
- Sugar
- Titanium
- Zirconium

Note: The degree of hazard depends on particle size and moisture. Large aluminum particles may be much less dangerous than very fine dried onion powder.

Figure 7-2

COMMON METALLIC FLAMMABLE SOLIDS

Notes

The major groups of metallic flammable solids that will be discussed in this chapter are:

- alkali metals
- alkaline earth metals
- aluminum group of metals
- titanium group of metals
- zinc group of metals
- radioactive pyrophoric metals

ALKALI METALS

Alkali metals are considered to be the most reactive of all metals. A summary of the physical and chemical properties of the alkali metals is found in Figure 7-3.

Lithium

Lithium is a silvery, light metal that floats on certain liquids. It reacts violently with moisture, acids, and oxidizers. Burning lithium produces dense, toxic fumes of lithium oxide and lithium hydroxide. Lithium reacts with water to produce highly combustible and explosive hydrogen gas. It also reacts with nitrogen.

Lithium is used in the manufacture of vacuum tubes, aircraft fuels, high performance batteries, pharmaceuticals, alloys, and metals.

LITHIUM (Li)
UN 1415
Dangerous
When Wet

704

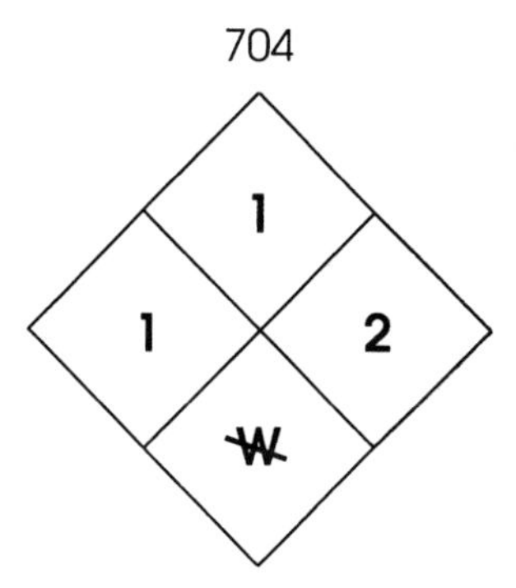

Physical and Chemical Properties of the Alkali Metals

Metal	Symbol	Melting Point °F	Melting Point °C	Ignition Temp. °F	Ignition Temp. °C	Boiling Point °F	Boiling Point °C	Reactive With O_2	Reactive With H_2O	Reactive With Halogen	Distribution
Lithium	Li	354	179	354	179	2,403	1,317	Yes	Yes	Yes	Common in industry
Sodium	Na	208	98	239	115	1,619	881	Yes	Yes	Yes	Common in industry and labs
Potassium	K	146	64	SC*	SC*	1,425	774	Yes	Yes	Yes	Common in industry
Rubidium	Rb	102	39	—	—	1,270	688	Yes	Yes	Yes	Rare
Cesium	Cs	82	28	SC*	SC*	1,301	705	Yes	Yes	Yes	Rare

* Certain forms can be spontaneously combustible.

Figure 7-3

Source: *Sax's*

Notes

Fire Extinguishing Guidelines

- Do not use dry chemicals containing sodium salts, because they can chemically react with lithium.
- Do not use water or foam.
- Do not use carbon dioxide or halogenated hydrocarbons, because these substances can react violently with alkali metals.
- Use graphite based dry chemical extinguishing agents (such as Lith-X®, manufactured by Ansul Fire Protection Company of Marinette, Wisconsin), lithium chloride, magnesium oxide, sand, or other suitable **dry** powders, such as lime, on small fires.

Note: Halogens are usually in diatomic form and include fluorine, chlorine, bromine, and iodine. Halogenated hydrocarbons are organic molecules that have chemically combined with one or more of the halogens.

Sodium

SODIUM (Na)
UN 1428
Dangerous
When Wet

704

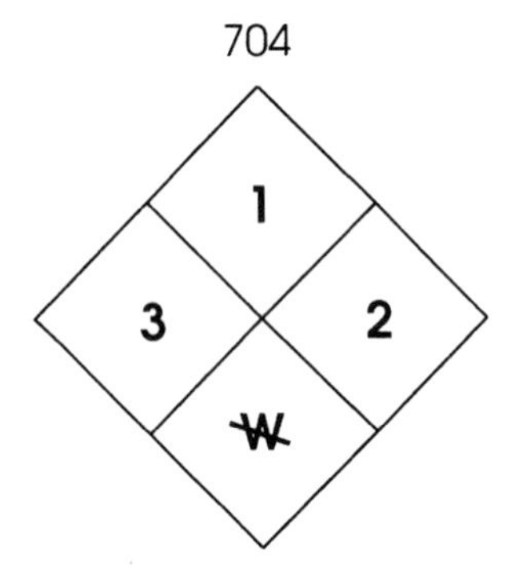

Sodium is a light, soft, silver-white metal. When sodium reacts with water, it produces hydrogen gas and sodium hydroxide. In most situations, the hydrogen gas will react explosively. Intense heat is also produced when sodium reacts with halogens, halogenated hydrocarbons, and acids. Sodium can cause **severe thermal and chemical burns**. The melting point of sodium is approximately 208°F (98°C). However, sodium/potassium alloys are more pyrophoric and have lower melting points than the pure elemental metals.

Sodium is used in producing alloys, descaling metals, and purifying molten metals. It is also used as a reduction agent, as a laboratory reagent, and as a coolant in nuclear reactors.

Fire Extinguishing Guidelines

- Do not use water, foam, soda acid, carbon dioxide, or halogenated extinguishing agents.
- Use sodium carbonate or sodium chloride based powders (such as Met-L-X®) on small pieces of burning sodium. (Met-L-X® is a fine sodium chloride powder with additives to increase its flow rate. It is manufactured by the Ansul Fire Protection Company.) Graphite, soda ash, and other appropriate powders are also effective.

Potassium

POTASSIUM (K)
UN 2257
Dangerous
When Wet

704

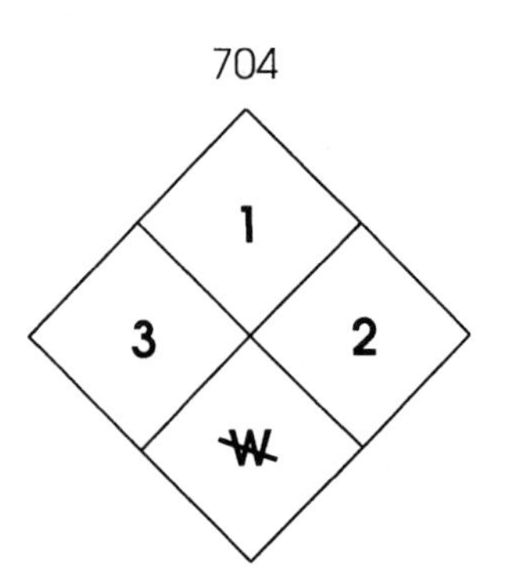

Potassium is usually in the form of cubic, silver-metallic crystals. It is spontaneously combustible in moist air or water. It reacts violently with moisture to form potassium hydroxide and hydrogen gas. Intense heat and explosions are possible during combustion. The by-products of potassium metal combustion are very toxic because of potassium's strong alkaline properties. Great care should be taken to avoid either internal or external contamination.

The most common commercial use of potassium is in the production of a sodium/potassium alloy called NAK, a heat exchange fluid.

Fire Extinguishing Guidelines

- Do not, under any circumstance, apply water or any moist substance to a potassium fire.

Notes

- Do not apply sand.
- Do not use ordinary fire extinguishers, such as carbon dioxide, foam, or halogenated extinguishing agents.
- Use sodium carbonate or sodium chloride based powders, dry graphite, soda ash, or other appropriate dry powder on small pieces of burning potassium. These powders tend to stick to the burning metal and smother the fire by sealing off oxygen.

Rubidium

RUBIDIUM (Rb)
UN 1423
Dangerous
When Wet

Rubidium is a reactive, silvery, and slightly radioactive metal. It ignites spontaneously in air and burns with a blue flame that leaves a dark brown to black residue. Rubidium reacts similarly to sodium. It decomposes water to form a strong alkali that will destroy human flesh.

Rubidium is not as common as lithium, sodium, and potassium. It is used in photocells and vacuum tubes. It is also used as a heat transfer fluid in space vehicles and in thermoelectric equipment.

Fire Extinguishing Guidelines

- Do not use water, carbon dioxide, or halogenated extinguishing agents.
- Use graphite, soda ash, or suitable dry powder.

Cesium

CESIUM (Cs)
UN 1407
Dangerous
When Wet

Cesium is pyrophoric and the most reactive of all metals. A hydrogen explosion occurs when cesium reacts with cold water. Cesium burns without a visible flame and with little smoke. **Cesium is the strongest base known and will destroy flesh on contact.**

Like rubidium, cesium is not as commonly used as other alkali metals. It is used, however, in photocells, vacuum tubes, ion propulsion systems, plasma for thermoelectric conversion, atomic clocks, and rocket propellants.

Fire Extinguishing Guidelines

- Do not use water, carbon dioxide, or halogenated extinguishing agents.
- Use graphite, soda ash, or suitable dry powder.
- Use dry chemical extinguisher on small fires.
- Move container from fire area if it can be done without risk.

Notes

BERYLLIUM (Be)
UN 1567
Poison

704

1
4 0

ALKALINE EARTH ELEMENTS

The alkaline earth elements share many of the reactivity hazards of the alkali metals, but to a lesser degree. A summary of the physical and chemical properties of the alkaline earth elements is found in Figure 7-4.

Beryllium

Beryllium is a light, steel-like metal that is less flammable than other metals. However, the powder can explode if mixed with air. Beryllium is a suspected human carcinogen and should be regarded as extremely toxic. Fatalities can occur for approximately 25% of those who have skin contact with beryllium or who inhale its dust or fumes. Symptoms, which include skin irritation, burning eyes, persistent ulcers, coughing, shortness of breath, and weight loss, can be delayed. Early fluorescent tubes contained beryllium and should be handled as a poison.

Fire Extinguishing Guidelines

- Do not use water.
- Extinguish with any inert dry powder, graphite, or dry sand.
- Personnel exposed to a beryllium fire should be decontaminated thoroughly, including hair and fingernails.
- All equipment and clothing should be washed down.

Physical and Chemical Properties of the Alkaline Earth Metals

Metal	Symbol	Melting Point °F	°C	Ignition Temp. °F	°C	Boiling Point °F	°C	Reactive With O_2	H_2O	Halogen	Distribution
Beryllium	Be	2,332	1,278	—	—	5,378	2,970	Yes	Yes	Yes	Common in industry
Magnesium	Mg	1,204	651	883	473	2,012	1,100	Yes	Yes	Yes	Common in industry
Calcium	Ca	1,548	842	1,454	790	2,703	1,484	Yes	Yes	Yes	Common in industry
Strontium	Sr	1,395	757	—	—	2,491	1,366	Yes	Yes	Yes	Pyro-technics industry
Barium	Ba	1,337	725	—	—	2,984	1,640	Yes	Yes	Yes	Electron tube industry
Radium	Ra	1,292	700	—	—	3,159	1,737	Yes	Yes	Yes	Rare

Source: *Sax's*

Figure 7-4

Magnesium

Magnesium Dust – The most serious hazard associated with magnesium dust or powder is that it can spontaneously explode. In fact, almost any damp metal dust can spontaneously ignite. A small fire can ignite any amount of suspended magnesium, causing an immediate explosion. Initiating effective firefighting under these circumstances is impossible, and evacuation and withdrawal are in order.

Magnesium Chips or Ribbons – The ribbon form of magnesium ignites quickly. Loose chips from sawing, drilling, or lathe turning readily ignite, but compact piles of chips are more difficult to ignite. Fine chips in water may ignite spontaneously.

Magnesium Solid – Magnesium found in ingots, sheets, or castings over 1/8 inch thick does not ignite easily. If on fire, however, it is difficult to find an extinguishing agent that will not cause a violent reaction. Magnesium burns intensely with a brilliant white flame. Moisture increases the pyrophoric tendencies. Hydrogen gas will be produced if water is applied during a fire, creating an explosion hazard.

Magnesium has a variety of uses. It is a common constituent of light alloys used in aircraft, automobiles, and other motorized vehicles. It is used in flares and single use flash bulbs. In addition, magnesium is used in the manufacture of precision instruments, machinery, television sets, incendiary bombs, and heating appliances.

Fire Extinguishing Guidelines

- Do not use foam, carbon dioxide, or halogenated extinguishing agents.
- Smother small fires with dry graphite, sodium chloride, soda ash, lime, or other suitable powder.
- Move container from fire area if it can be done without risk.
- For large fires, withdraw from the area and allow the magnesium metal to burn itself out, if possible.

Magnesium fire. Note the reaction as water is applied.

Notes

MAGNESIUM (Mg)
UN 1869
Flammable Solid

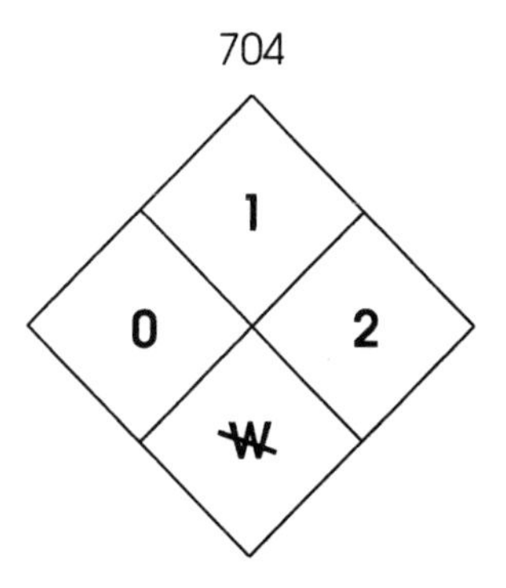

Notes

- Do not use water directly on actively burning magnesium, because an explosion may result. Water can be applied carefully on combustible material that is involved in the fire.
- Water must be used in huge volumes on large magnesium fires in order to be effective. Anticipate violent reactions if water is used. This procedure is not recommended.

Calcium

CALCIUM (Ca)
UN 1401
Dangerous
When Wet

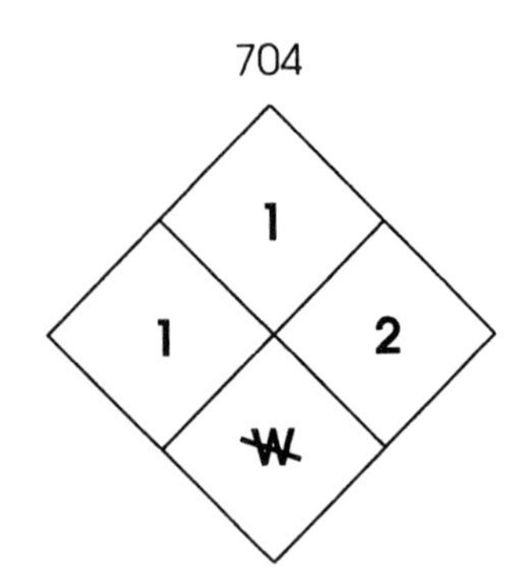

Calcium metal is a silvery, soft metal that oxidizes to grayish-white when exposed to air. Calcium reacts with water to form calcium hydroxide and hydrogen. The heat of such a reaction may ignite the hydrogen, the calcium itself, and nearby combustibles.

Calcium is used in the manufacture of polyester fibers and vacuum tubes. It is also used extensively in metallurgy as a deoxidizer for various metals and as an alloying agent.

Fire Extinguishing Guidelines

- Do not use water, carbon dioxide, or halogenated extinguishing agents.
- Use graphite, soda ash, or suitable dry powder.

ALUMINUM GROUP OF METALS

A summary of the physical and chemical properties of the aluminum group of metals is found in Figure 7-5.

Aluminum

Aluminum may be in the form of filings, powder, paste, or solid. Finely divided aluminum is more hazardous than large solid pieces. Vaporizing liquids, including halogenated extinguishing agents, should not be used, as they may react violently with the burning metal. Explosions are possible wherever aluminum powder or dust accumulates.

Most people are commonly aware of the many uses of aluminum, from aluminum foil to door frames. Aluminum is also used in the manufacture of printing inks, pharmaceuticals, construction materials, cans, containers, electrical equipment, metal alloys, and paints.

Physical and Chemical Properties of the Aluminum Group of Metals

Metal	Symbol	Melting Point		Boiling Point		Reactive With			Distribution
		°F	°C	°F	°C	O_2	H_2O	Halogen	
Aluminum	Al	1,220	660	4,442	2,450	Low	Low	Yes	Common
Gallium	Ga	86	30	4,357	2,403	Low	Low	Yes	Rare
Indium	In	315	157	3,776	2,080	Low	Low	Yes	Rare
Thallium	Tl	579	304	2,655	1,457	Low	Low	Yes	Rare

Source: *Sax's*

Figure 7-5

Fire Extinguishing Guidelines

- Control aluminum paste or slurry fires with carbon dioxide. Immediately smother with dry sand.
- Apply sodium chloride based powders, soda ash, or lime to small aluminum fires. Do not use water or halogenated extinguishing agents.
- Smother dry aluminum filings or powder fires in dry sand or other dry material recommended by the aluminum processor or supplier.
- Avoid spreading the burning material.
- Dry sand can be used to ring and isolate the fire.
- Move container from fire area if it can be done without risk.
- If aluminum powder burns alone, it will form an oxygen-excluding crust that will eventually extinguish the fire.
- For large fires, flood fire area with water from a distance.
- For massive fire in cargo area, use unmanned hose lines. If this is not possible, withdraw from the area and let the fire burn.
- Cool containers that are exposed to flames until well after the fire is out.

ALUMINUM (Al)
UN 1396
Dangerous
When Wet

704

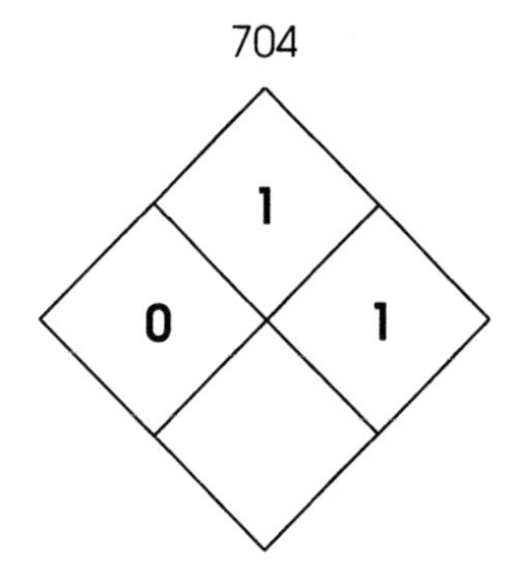

TITANIUM GROUP OF METALS

A summary of the physical properties of the titanium group of metals is found in Figure 7-6.

Titanium

Titanium is usually found as a dark gray, uncrystallized powder or as a white, lustrous metal. Titanium will burn in atmospheres of carbon dioxide, nitrogen, or air. Titanium is in its most hazardous form as finely divided metal, and it may ignite or explode spontaneously. It is also highly explosive in the molten form when mixed with water.

TITANIUM (Ti)
UN 2546
Spontaneously
Combustible

Titanium is as strong as steel but 45% lighter. For this reason, it is used extensively in the manufacture of missiles and aircraft. It is also very resistant to corrosion and is therefore used in ships and underwater machinery.

Physical Properties of the Titanium Group of Metals

Metal	Symbol	Melting Point °F	Melting Point °C	Boiling Point °F	Boiling Point °C
Titanium	Ti	3,051	1,677	5,931	3,277
Zirconium	Zr	3,366	1,852	6,471	3,577
Hafnium	Hf	4,041	2,227	8,316	4,602

Figure 7-6

Source: *Sax's*

Notes

Fire Extinguishing Guidelines

- Do not use water, carbon dioxide, or halogenated extinguishing agents.
- Allow the fire to burn itself out, if possible.
- Use graphite, soda ash, or dry powders on small fires.
- Argon and helium gases may be useful if applied in airtight enclosures.

Zirconium and Hafnium

ZIRCONIUM (Zr)
UN 1932
Spontaneously
Combustible

HAFNIUM (Hf)
UN 2545
Spontaneously
Combustible

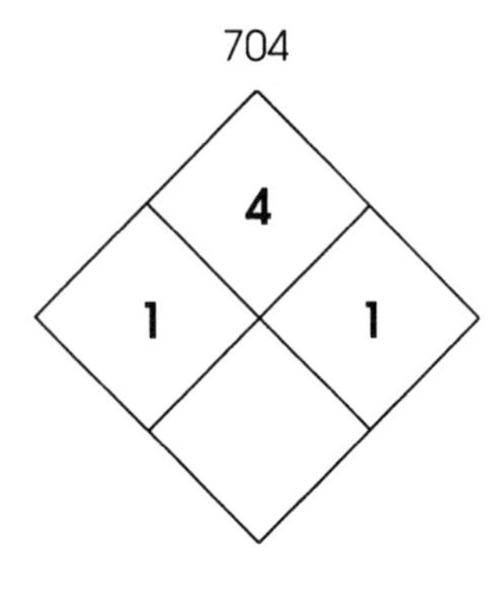

Both zirconium and hafnium are used as solid metal, crystals, or gray powder. Titanium/zirconium alloys are more pyrophoric and have a lower melting point than the pure constituent metals. Zirconium and hafnium are relatively stable under water at temperatures up to 122°F (50°C). At high temperatures, the dry powder form of these metals may combine explosively with oxygen, nitrogen, phosphorus, sulfur, halogens, and other non-metals. Dry powders have low ignition temperatures, burn intensely hot, and are difficult to extinguish. If water is mixed with zirconium or hafnium, the mixture should be at least 25% water by weight. Severe explosions are possible when less water is used to store these metals, primarily due to the production of hydrogen gas.

Zirconium is used for shielding nuclear reactors and for removing oxygen from molten steel. It is also used as a tanning agent, as a polishing powder for lenses, and as an ingredient in special welding fluxes. Hafnium is used in control rods in water cooled nuclear reactors, in light bulb filaments, and in the production of special glasses.

Fire Extinguishing Guidelines

- Do not use water, because water usually increases the burning rate.
- Do not use foam, carbon dioxide, or halogenated extinguishing agents.
- Allow the fire to burn itself out, if possible.
- Use sodium chloride based powders on small fires.
- Smother with dry sand, salt, graphite, or a ground mineral carbonate, such as dolomite.

ZINC GROUP OF METALS

Notes

A summary of the physical and chemical properties of the zinc group of metals is found in Figure 7-7.

Zinc

Zinc may heat spontaneously with air and moisture. The heat produced can ignite nearby combustible material. Zinc is not water soluble. Hydrogen is produced if zinc dust comes into contact with acids or alkali hydroxides.

In metallurgy, zinc is used with copper to form the common alloy brass and to galvanize iron. Zinc is also used in dry-cell batteries, in building materials, and as a paint pigment.

ZINC (Zn)
UN 1436
Dangerous When Wet
Spontaneously
Combustible
704

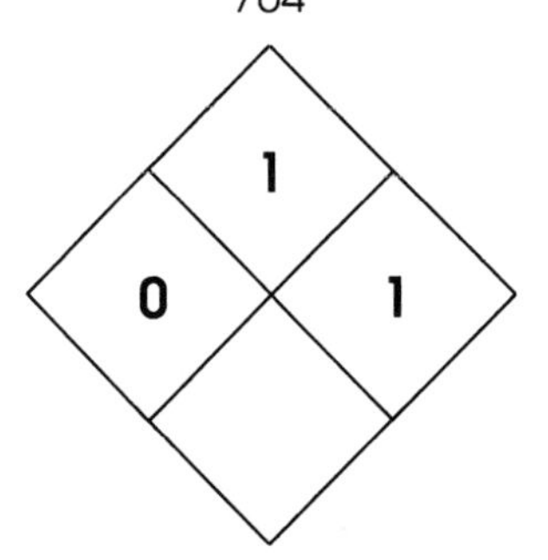

Fire Extinguishing Guidelines

- Do not use water, because zinc can react violently.
- Do not allow water to get inside the container.
- Do not use foam, carbon dioxide, or halogenated extinguishing agents.
- For small fires, use dry chemicals, soda ash, sodium carbonate, graphite based powders, lime, or dry sand.
- Move container from fire area if it can be done without risk.
- For large fires, withdraw and let the product burn, if possible.
- Use water only for large fires, using unmanned hose monitors.

Cadmium

Cadmium is a grayish-white powder or bluish-white soft metal used for metal coating. It reacts with oxidizing agents and releases hydrogen. It will ignite in dust form. Cadmium is a human carcinogen and should be regarded as extremely toxic. When heated, it oxidizes and emits dense brown poisonous fumes. Inhaling high concentrations of the fumes will cause pulmonary edema and can be fatal.

CADMIUM (Cd)
UN 2570
Poison
Keep Away
From Food

Fire Extinguishing Guidelines

- Use graphite, soda ash, and other dry powders.

Physical and Chemical Properties of the Zinc Group of Metals

Metal	Symbol	Melting Point °F	Melting Point °C	Boiling Point °F	Boiling Point °C	Reactivity	Toxicity	Distribution
Zinc	Zn	788	420	1,666	908	High	—	Galvanized iron
Cadmium	Cd	610	321	1,413	767	High	High	Electroplating
Mercury	Hg	-38	-39	675	357	Low	High	Instruments

Source: *Sax's*

Figure 7-7

Notes

MERCURY (Hg)
UN 2909
Corrosive

Mercury

At room temperature, mercury is an odorless, silver-white, extremely heavy, liquid metal. Solid mercury is tin-white. The health hazards of mercury are more predominant than the fire hazards. Mercury is highly toxic by skin absorption and inhalation of fumes or vapors. Mercury poisoning is most often associated with releases of mercury compounds into rivers and streams by certain industries. Mercury compounds are retained by the fish and wildlife that inhabit these areas. Because of this, mercury has been transmitted to other animals, or even humans who consume contaminated wildlife.

In the chemical industry, mercury compounds are often used as catalysts. Mercury is used in dentistry, in pharmaceuticals, in batteries, and, of course, in various calibrated instruments, such as thermometers and hydrometers.

Fire Extinguishing Guidelines

- For small fires, use dry chemicals, carbon dioxide, water, foam, or halogenated extinguishing agents.
- Move container from fire area if it can be done without risk.
- Use fog streams or standard firefighting foam on large fires, but expect violent reactions.

RADIOACTIVE PYROPHORIC METALS

Although radioactive materials will be covered in Chapter 10, the dual hazards of radioactive metals make it important to note these materials here. Radioactive incidents are best handled by radiological experts. The decision to attack a radioactive fire must be based upon sound knowledge and a clear understanding of the hazards involved. Common radioactive pyrophoric metals include:

- uranium
- plutonium
- thorium

EXTINGUISHING AGENTS FOR METAL FIRES

Figure 7-8 is a Guide to Extinguishing Agents for Fighting Metal Fires. The following key applies to this chart:

- **YES** means that the extinguishing substance is effective in most cases. The extinguishing agents mentioned should be used on small fires only. Do not expect any chemical agent to extinguish large metal fires.
- **NO** means that the extinguishing substance has undesirable or hazardous properties.
- **UNKNOWN** means that the extinguishing substance has not been used enough to determine whether it is desirable or not.

A GUIDE TO EXTINGUISHING AGENTS FOR FIGHTING METAL FIRES

METAL	EXTINGUISHING AGENT					
	Magnesium Oxide Powder	Sodium Chloride Based Powder (e.g., Met-L-X®)	Carbon Dioxide	Water	Foam	Explosive Hazard
Aluminum	Unknown	Yes	Yes, slurries only	Yes, slurries only	No	Moderate with water and aluminum dust
Hafnium	Unknown	Yes	No	No	No	Moderate with water and hafnium dust
Lithium	Yes	No	No	No	No	Extreme with water and in closed areas
Magnesium	Yes	Yes	No	Yes, but only for large quantities of solid non-burning metal	No	Extreme with water and in closed areas
Potassium	Unknown	Yes	No	No	No	Extreme with water and in closed areas
Sodium	Unknown	Yes	No	No	No	Extreme with water and in closed areas
Titanium	Unknown	Yes	No	Yes, but only for large quantities of solid non-burning metal	No	Moderate with water and in finely divided form
Zirconium	Unknown	Yes	Yes, but only for large quantities of solid metal	Yes, but only for large quantities of solid non-burning metal	No	Moderate with water and in finely divided form

Figure 7-8

COMMON NON-METALLIC FLAMMABLE SOLIDS

Camphor

CAMPHOR
UN 2717
Flammable Solid

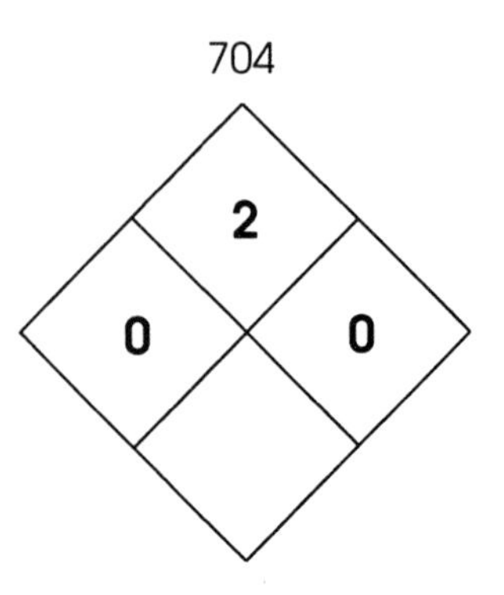

Camphor is a white or colorless crystalline or granular substance used in medicines, toothpastes, embalming fluid, and the manufacture of cellulose-nitrate plastics and insecticides. The vapors, which can be irritating, are similar to those of naphthalene.

At normal temperature, camphor undergoes sublimation, which is the process of a solid's changing directly into a gas (vapor) without going through the liquid state. Materials that sublimate can be ignited in the same way combustible liquids are ignited. These materials usually have a flash point that is above 150°F (66°C). Once ignited, the materials melt and flow like a flammable liquid, and the vapors produced can be toxic or narcotic.

Fire Extinguishing Guidelines

- Use carbon dioxide, dry chemical, or foam.

Note: Do not wear contact lenses when working with this chemical.

White Phosphorus

PHOSPHORUS (P)
UN 1381
Spontaneously
Combustible
Poison
(Environmentally
Hazardous Substance)

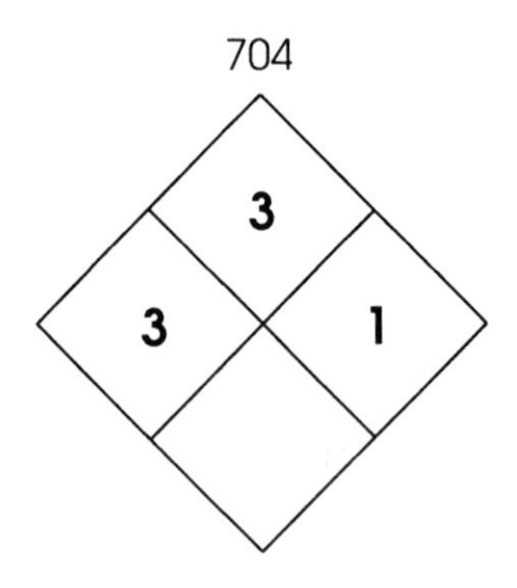

White or yellow phosphorus is a white to yellow-colored waxy solid with a pungent odor that is usually found in stick form. It ignites spontaneously with air at or above 86°F (30°C), or at room temperature when finely divided. It can explode when mixed with oxidizing materials. It is blanketed with an inert gas during shipment.

White phosphorus severely burns the skin and, if ingested, is toxic even in small quantities. It is also irritating to the eyes, throat and lungs. White phosphorus is heavier than and insoluble in water. It weighs 14.5 pounds per gallon. White or yellow phosphorus is used in poisons and incendiary bombs.

Fire Extinguishing Guidelines

- Deluge with water, taking care not to scatter the material.
- When the phosphorus has solidified and the fire is extinguished, cover the product with wet sand or dirt until it can be disposed of.

White Phosphorus Fire

Notes

Red Phosphorus

Red phosphorus is a reddish-brown powder used in manufacturing phosphoric acid, fertilizers, insecticides, and matches. It is much less reactive than white phosphorus; however, it may explode when mixed with oxidizing materials. Substances involved in red phosphorus fires will explode if the non-metal is mixed with or is contaminated by sufficient amounts of combustible materials.

Burning red phosphorus fumes are very irritating and slightly toxic in small concentrations. At high temperatures, red phosphorus can change into the more dangerous white phosphorus.

RED PHOSPHORUS
UN 1338
Flammable Solid
(Environmentally
Hazardous Substance)

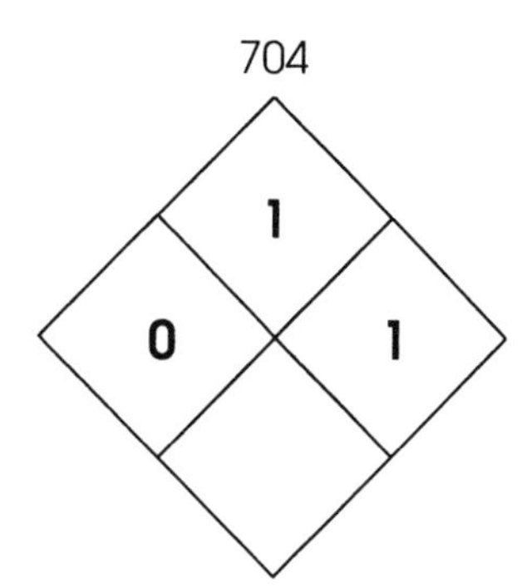

Fire Extinguishing Guidelines

- Use foam or dry chemicals.
- If the decision is made to flood the material with water, be aware that water can react with hot red phosphorus to form toxic phosphine.
- When the fire is extinguished, cover the product with wet sand or dirt.
- Use extreme caution during clean-up, since reignition may occur.

NAPHTHALENE
UN 1334
Flammable Solid
(Environmentally
Hazardous Substance)

704

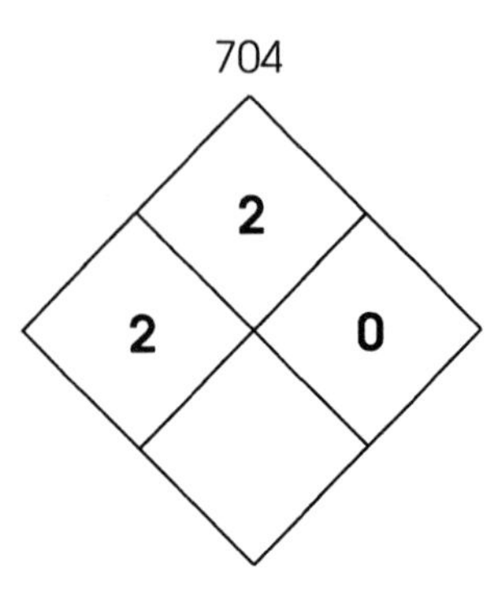

Naphthalene

Naphthalene occurs as volatile white crystals or flakes that smell like mothballs. It is used in manufacturing dyes, resins, fungicides, explosives, and, of course, mothballs. It is heavier than water and is soluble in water. Although naphthalene is difficult to ignite, it will burn. Naphthalene is toxic, and both the liquid and vapor irritate the skin, eyes, and respiratory tract.

Fire Extinguishing Guidelines

- Use water, foam, carbon dioxide, or dry chemical.
- Foam or direct water spray on molten naphthalene may cause extensive foaming.
- Use water spray to cool the burning material below its flash point.

SULFUR (S) - Non-Fire
UN 1350
Flammable Solid
(Environmentally
Hazardous Substance)

704

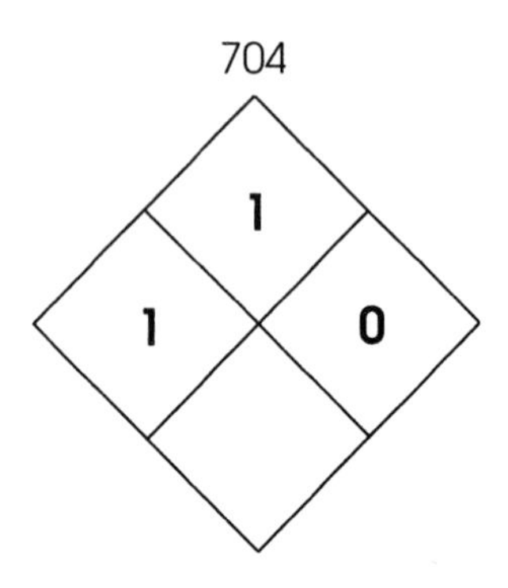

Sulfur

Sulfur is a pale yellow solid powder or yellow crystalline substance that ignites readily. Its dust or vapors form explosive mixtures with air and oxidizing materials. Sulfur dioxide is one of the toxic products produced when sulfur burns. Sulfur's flash point is 405°F (207°C), its ignition temperature is 450°F (232°C), and its melting point is 246°F (119°C). Sulfur is commonly encountered in molten form in haz-mat incidents.

The major use of sulfur is in the production of sulfuric acid. Sulfur is also used in fungicides and in the production of matches, fertilizers, pharmaceuticals, pesticides, rubber, explosives, dyes, and chemicals.

Fire Extinguishing Guidelines

- Use water spray, avoiding straight streams, which can scatter molten sulfur and dust.
- Use sand or additional sulfur to extinguish small fires.

SULFUR (S) - Fire

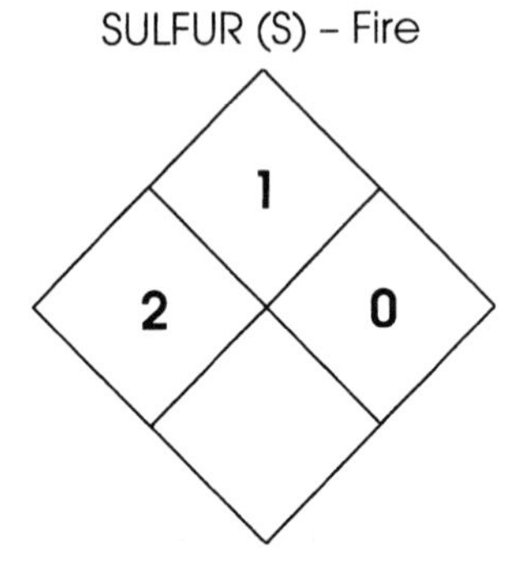

CLASS 4 PLACARD AND LABEL REQUIREMENTS

SUBSTANCE	VEHICLE PLACARD	PACKAGE LABEL	QUANTITY	SPECIAL REQUIREMENTS
Flammable Solid	**Flammable Solid** (black text on white background with seven vertical red stripes). FLAMMABLE SOLID 4 FLAMMABLE SOLID 4 Domestic US	**Flammable Solid** (black text on white background with seven vertical red stripes). FLAMMABLE SOLID FLAMMABLE SOLID Domestic US	Placard any quantity by rail; 1,001 lbs (454 kg) or more by truck.	Must not be loaded with some explosives and poisons. Must be segregated from other poisons and all corrosives.
Spontan-eously Combustible	**Spontaneously Combustible** (black text; the background must be red in the lower half and white in the upper half). SPONTANEOUSLY COMBUSTIBLE 4	**Spontaneously Combustible** (black text; the background must be red in the lower half and white in the upper half).	Placard any quantity by rail; 1,001 lbs (454 kg) or more by truck.	Must not be loaded with corrosives and some explosives and poisons. Must be segregated from other explosives and poisons.
Dangerous When Wet	**Dangerous When Wet** (white text on blue background) or **Flam-mable Solid** (black text on white with seven vertical red stripes in lower portion and white text on blue background in upper portion). DANGEROUS WHEN WET 4 FLAMMABLE SOLID 4 Domestic US	**Dangerous When Wet** (white text on blue background) or **Flam-mable Solid** (black text on white with seven vertical red stripes in lower portion and white text on blue background in upper portion). DANGEROUS WHEN WET 4 FLAMMABLE SOLID 4 Domestic US	Placard all quantities by rail or highway.	Must not be loaded with some explosives and poisons. Must be segregated from other poisons and all corrosives.

Note: If a subsidiary hazard is present, the subsidiary placard or label must **not** display a hazard class number. Placards identified "Domestic US" may be used in U.S. domestic highway transportation through October 1, 2001. Labels identified "Domestic US" may be seen on packages filled before October 1, 1991.

Notes

SHIPPING CONTAINERS

Flammable solids are transported and stored in a variety of containers. They are shipped in containers as small as mailing tubes and as large as rail box cars. Fiberboard tubes similar to those used for explosive materials, wooden boxes, and cardboard boxes are frequently used. Cardboard boxes are commonly used for highway and railroad fusees.

Medium amounts of flammable solids are shipped in metal and plastic pails and in 55 gallon drums. Some are also shipped in molten state, e.g., sulfur and aluminum. Because many of these materials are reactive, the following precautions are necessary to ensure safe transport:

- Lids must be secured tightly to reduce contamination.
- Some products may be immersed in a water bath.
- Some materials may be transported as a slurry mixture of flammable solid and water to increase stability.

Emergency response personnel must recognize the dangers that exist when a container leaks, when water evaporates, or when a lid is damaged, allowing the inerting (covering) gas or liquid to be released. In such cases, the material may spontaneously ignite when exposed to air.

Metal Cans of Calcium Carbide

Notes

Left: **Cardboard boxes of fusees.**
Below: **Flammable solids come in a variety of small containers.**
Bottom: **Rail transport of calcium carbide.**

Notes

KEY POINTS TO REVIEW

Flammable solids are very different from one another. The broad definitions for flammable solids include sludges, pastes, crystalline metals, and substances resembling thick liquids. Many of these materials become extremely dangerous if water is applied. Flammable solids are reactive, they can be extremely toxic, and some are radioactive. Remember the following key points:

- Flammable solids are regulated by the Department of Transportation, but combustible dusts are not.
- Some flammable solids can spontaneously ignite through slow oxidation or by pyrophoric ignition.
- Flammable solids can be either metallic or non-metallic.
- Be aware of the potential for explosive and violent reactions.
- Storage containers are designed to ensure safe shipment of these materials.

RESPONDING TO FLAMMABLE SOLID EMERGENCIES

Proper analysis of a haz-mat incident is the key to successful mitigation, and this is especially true with flammable solid emergencies. A flammable solid emergency can test even the most experienced haz-mat responder. Fortunately, accidents involving flammable solids have been relatively rare.

Successfully handling a flammable solid incident depends upon:

- the type of product
- the amount of product
- the availability of the proper extinguishing agent
- the emergency responder's ability to access technical specialists

There have been several transportation related incidents involving flammable solids. In most cases, emergency responders have failed to recognize or understand the chemical characteristics of the product. As a result, some responders and civilians have suffered severe respiratory exposure. This shows the need for responders to gain a working knowledge of the flammable solids likely to be involved in an emergency situation.

HAZ-MAT — HAZARD IDENTIFICATION

As in any haz-mat incident, if a flammable solid is involved in an emergency, it must be specifically identified before response personnel take action. Unlike many other hazard classes, flammable solids range widely in chemical characteristics and associated hazards. Mitigation techniques that work well for one flammable solid can be disastrous if used on another. Use the following steps in hazard identification:

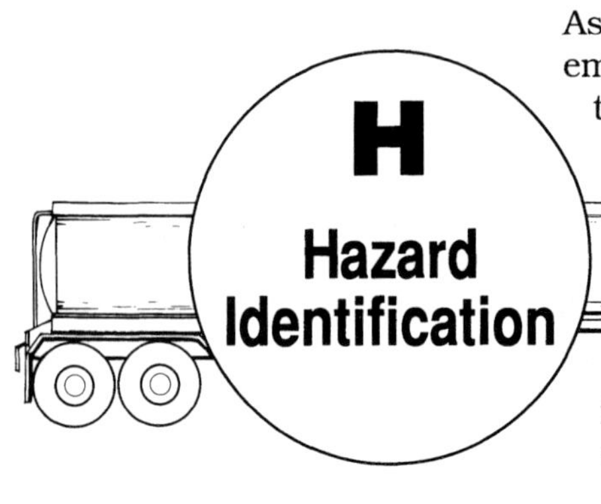

- ***Preplanning*** for a flammable solid incident is critical, especially for fixed facilities. Know where flammable solids are used in the community. Know the type and amount of product and the storage method used. Don't forget the acetylene generator at the local welding shop. These generators use calcium carbide, a dangerous when wet material, to create acetylene gas for welding operations.

- ***Location*** of the incident should indicate what materials are involved. This information, combined with an effective preplanning program and a good hazard assessment, will aid in successful mitigation.
- ***Markings or identification***, such as placards, labels, tank shapes, and stenciling, may aid in identifying the product.
- ***Chemical characteristics*** of flammable solid materials range from explosive reactions with air and water to production of large, billowing clouds of toxic gas and by-products. Know the specific characteristics of the product before taking action! Use informational resources such as CHEMTREC/CANUTEC, Material Safety Data Sheets, and on-line data banks to determine the hazards of the materials involved.

Once a flammable solid has been identified, evaluate the associated hazards.

- Locate the leak or fire.
- Assess how the material was released from its container.
- Find out the extent and severity of injuries.
- Determine the position of the container.
- Determine the integrity of the container.
- Find out if the container is exposed to fire or if it is on fire.
- Identify the color of the smoke, if there is a fire.
- Assess if there is an environmental hazard.

***Top:* Acetylene generator at welding shop—site for preplanning.**
***Right:* Acetylene generator to convert calcium carbide to gas.**

Notes

HAZ-MAT — ACTION PLAN

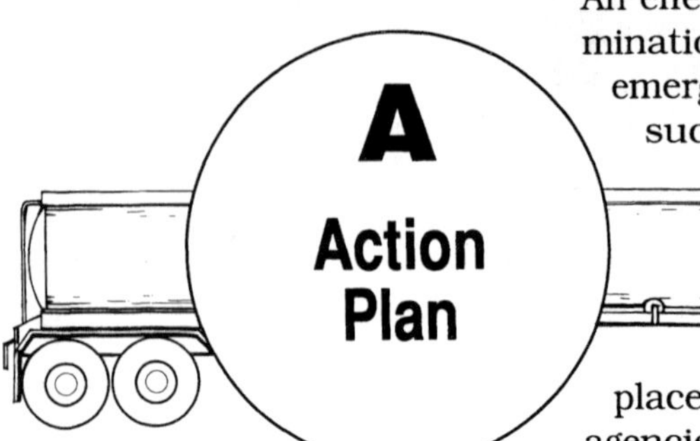

An effective action plan is based on sound information and the determination of associated hazards. Action plans for a flammable solid emergency should be based on specific information about the product, such as the amount of product, container type, and the most appropriate mitigation methods.

When large quantities of product are involved, extinguishment may not be the primary focus of the action plan. Efforts should be directed toward public protective action (i.e., evacuation or protecting in place), establishment of control zones, and notification of proper agencies, distributors, and technical specialists.

When responding to a flammable solid incident with **NO LEAK or FIRE**, consider the following factors:

- Determine the size of the container.
- Determine the quantity of product. The amount of product will dictate what resources are needed, the extent of public protective actions, and control zones.
- Verify the damage to the container. Will the container fail or develop leaks?
- Establish whether the container is airtight, or if it contains water or an inert substance to stabilize the product.
- Determine whether the inert substance is leaking.
- Anticipate possible leaks or fire.
- Set up evacuation/isolation corridors and control zones early.
- Isolate and contain the product until it can be safely removed.
- Prevent ignition from outside sources.

If there is a **LEAK but NO FIRE:**

- Establish a hot zone.
- Initiate public protective actions.
- Isolate, contain, and/or confine the product.
- Keep the product dry if it is a water reactive solid.
- Keep the product covered with sand, dirt, or other appropriate material if it is spontaneously combustible.
- Prevent ignition from outside sources.
- Know what resources you will need in case of fire, especially if large quantities of product are involved.

If there is a **LEAK and FIRE:**

- Establish a hot zone to minimize personnel exposure.
- Initiate protective actions according to the type and amount of material involved.
- Determine the best method of extinguishment, considering the type of material and size of the fire.

Notes

- Cover small fires with sand, dirt, or water spray. Keep the product wet, if appropriate.
- Have adequate water reserves. If water is appropriate, a large flammable solid fire will require large amounts of water.
- Use unmanned hose lines.
- Use only as much water as is needed for fire control. Extra water may scatter material.
- Be prepared to contain runoff if the runoff is contaminated.
- Protect all exposures.
- Let the product burn itself out, if this is a viable option.

HAZ-MAT — ZONING

Zoning requirements depend upon the product, the amount of product, and the severity of the incident. Check the table of initial isolation and protective action distances in the DOT *Emergency Response Guidebook* or follow local protocol for the proper zone distances. In the absence of available guidelines for specific flammable solids, consider the following recommendations:

Note: For the purposes of this text, a small spill is defined as one where the container size is equal to a 55 gallon drum or smaller. A large spill is defined as a larger container size.

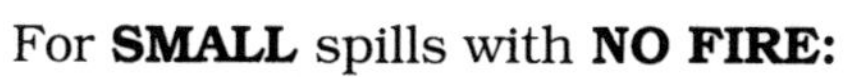

For **SMALL** spills with **NO FIRE:**

- Establish a protective action distance of 200 feet.
- Establish a minimum hot zone of 150 feet.

For **SMALL** spills with **FIRE:**

- Set up a minimum protective action distance of 500 feet.
- Set up a minimum hot zone of 200 feet.
- Consider establishing a warm zone.

For **LARGE** spills with **NO FIRE:**

- Establish a minimum protective action distance of 500 feet.
- Establish a minimum hot zone of 300 feet.
- Consider establishing warm and cold zones.

For **LARGE** spills with **FIRE:**

- Establish an initial protective action distance of 1,000 feet.
- Anticipate downwind evacuation to be as much as 1 mile long and 1/2 mile wide.
- Establish a minimum hot zone of 500 feet.
- Establish warm and cold zones.

Notes

HAZ-MAT — MANAGING THE INCIDENT

M

Managing

Scene Management

The flammable solid emergency can be very difficult to handle. If the incident commander or first arriving officer quickly determines the degree of danger, then it will be easier to obtain needed resources to handle the incident safely.

As previously discussed, incidents which involve large amounts of flammable solids may require a no-attack strategy. For example, if a tank car of white phosphorus is severely damaged and on fire, it is unlikely that the fire can be quickly extinguished. In such an incident, efforts will be better spent in evacuating the area and isolating the product. The key to successful scene management is to:

- preplan the strategy
- preplan the tactical operational considerations
- preplan the anticipated resource needs
- use an incident management system

Rescue Considerations

Thermal burns and severe skin, eye, and respiratory irritation are the most common injuries resulting from flammable solid emergencies. Signs and symptoms of exposure to flammable solids may be delayed. Because fires can produce large volumes of toxic gases, there may be large numbers of victims at the site.

General Treatment for Flammable Solid Exposures

Products .... Red or white phosphorus, lithium, sodium.

Containers .... Hermetically sealed (airtight) cans, drums, and tank cars. Sometimes sealed in an inert gas or water to avoid air exposure.

Life Hazard .... Irritating to mucous membranes; burns skin and eyes. Fumes from burning materials are very irritating and may lead to pulmonary edema and respiratory arrest. Ingestion of some products may be lethal. Some products may cause cardiac arrhythmias.

Signs/Symptoms ..... Burns to all moist surfaces (mucous membranes) and abdominal pain. Respiratory distress and pulmonary edema. If the product has been ingested, the victim may have pain to teeth and jaw. Vomit and diarrhea may be dark and luminous.

Basic Life Support .... Wear proper protective equipment and remove patient from contaminated area. Administer oxygen and remove patient's clothing. Gently brush all visible product from skin. Decontaminate with copious amounts of soap and water. Irrigate eyes thoroughly and continue during transport as necessary. If water reactive products are embedded in the skin, water should not be used. The embedded products can be covered with a light oil and the patient transported for surgical debridement. If phosphorus is involved, immerse affected body part in water to prevent burns. If ingested, activated charcoal may be used (follow advice of Poison Control and local protocols). Do not orally administer anything if the patient has a decreased level of consciousness or decreased gag reflex. Cover burn areas with sterile dressings after decontamination.

Advanced Life Support .... Assure an adequate airway. Start IV D5W TKO. Monitor for cardiac arrhythmias and pulmonary edema and treat as necessary. If no respiratory distress or trauma is present, analgesics may be given for pain relief. Follow local protocols for all drug therapy and medical treatment.

Other Information .... If the product has been ingested, the patient should avoid consuming oils or fats, as these may increase absorption.

Notes

HAZ-M**A**T — ASSISTANCE

Flammable solid emergencies will require assistance from many of the persons and agencies listed below:

- ***Dispatcher*** can coordinate resource acquisition.
- ***Local law enforcement*** can assist in site control and evacuation.
- ***HMRT*** or other specialized response groups can provide technical expertise and on-site tactical functions.
- ***Technical specialists*** are needed to successfully handle large flammable solid incidents. Such expertise may include a local distributor, a manufacturer's response team or representative, or a local chemical specialist.
- ***Private contractors*** are often employed to clean up large flammable solid spills. Anticipate this need and coordinate efforts with the manufacturer to expedite clean-up once the incident is stabilized. The sooner a product is removed, the less environmental damage there will be.

HAZ-MAT — TERMINATION

Terminating a flammable solid emergency may take several days. Termination, therefore, should be as systematic as the mitigation efforts. The following are the activities performed during this phase of the incident:

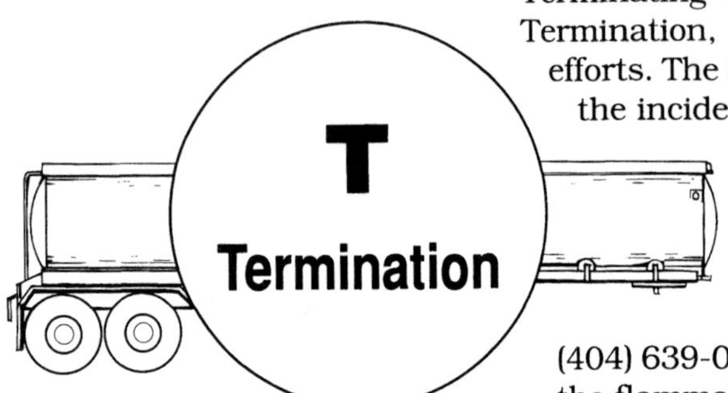

- Decontamination. Considering the diverse types of flammable solids, it is impossible to generalize about decontamination procedures. Be sure to contact the manufacturer or the Agency for Toxic Substances and Disease Registry (ATSDR) at (404) 639-0700 to obtain specific decontamination procedures for the flammable solids encountered.
- Rehabilitation
- Medical screening
- Exposure reporting
- Initial incident debriefing
- Post-incident analysis

KEY POINTS TO REVIEW

Accidents involving flammable solid materials have been relatively rare during the last 20 years. However, new technologies keep increasing the need for flammable solids, thereby increasing the potential number of emergencies involving these materials. Flammable solid emergencies often force emergency response personnel to attack the emergency indirectly. In many cases, the best strategy is to isolate the incident and let it burn itself out. However, remember the following key points:

- Understanding the physical characteristics of flammable solids is essential for developing action plans.
- Preplanning for the most likely incidents can help you deal successfully with an incident.
- Extinguishment may not be the primary focus during large flammable solid fires.
- Flammable solid fires can release extremely toxic gases and vapors.
- Be prepared for possible environmental impacts caused by flammable solids.
- Anticipate your needs prior to the incident.
- Use the HAZ-MAT checklist for flammable solid emergencies found in Appendix G.

Chapter 8

OXIDIZER EMERGENCIES

OBJECTIVES

After studying the material in this chapter, you will be able to:

- define oxidizers and organic peroxides
- state how to recognize the oxidizing potential of a substance
- list commonly used oxidizers and peroxides
- identify swimming pool chemicals and describe their hazards
- describe the hazards of oxidizers and organic peroxides
- identify and describe common oxidizing acids and oxidizing elements
- state the placarding and labeling requirements for oxidizers and organic peroxides
- describe common shipping containers used for oxidizers and organic peroxides
- describe control measures for an oxidizer emergency, using the acronym HAZ-MAT
- describe the effects of various extinguishing agents on oxidizers

UNDERSTANDING OXIDIZERS

INTRODUCTION

Oxidizing agents are unpredictable and can react violently without warning. Therefore, oxidizer emergencies can be some of the most dangerous that emergency response personnel face. As with any hazardous material, responders must be familiar with oxidizing agents and how they react.

The incident commander's action plan should reflect the fact that oxidizers are extremely unstable. If leaking oxidizing products become contaminated, they can be as dangerous as explosive materials. Oxidizer incidents should be approached in much the same way as an explosive emergency. In this chapter, you will gain a basic understanding of oxidizing materials, including:

Notes

- inorganic oxidizing materials
- organic peroxides
- oxidizing acids
- oxidizing elements

OXIDIZERS DEFINED

The Department of Transportation (DOT) defines an ***oxidizer*** as:

A material that may cause or enhance the combustion of other materials, generally by yielding oxygen (49 CFR 173.127(a)).

This definition states that *generally* oxidizers act by yielding oxygen. There are materials, however, that act as oxidizers even though they do not contain oxygen. For example, fluorine and chlorine support combustion, just like oxygen. In some cases, an oxidizing reaction can cause spontaneous ignition. All oxidizers release energy, and the faster energy is released, the greater the potential for harm. Refer to Appendix F for further classifications of oxidizers, organic peroxides, and unstable materials based on Article 80 of the Uniform Fire Code.

RECOGNIZING OXIDIZING POTENTIAL

To determine the amount of oxygen in a chemical, note the common prefixes and suffixes in the chemical name.

As a general rule, the more oxygen a substance contains, the greater the oxidizing potential. The amount of oxygen stored in a chemical molecule can generally be determined by looking at the common prefixes and suffixes in the chemical name. The common prefixes are "hypo-" and "per-." The common suffixes are "-ite" and "-ate." The greatest oxidizing potential exists when the chemical is in the "per-" state (O_4). Notice several examples:

Sodium **hypo**chlorite **Hypo**chlorous acid	NaCl**O** HCl**O**	The "hypo-" prefix indicates minimum oxygen content. In these examples, one oxygen atom is present in each molecule.
Sodium chlor**ite** Sodium nitr**ite**	$NaClO_2$ $NaNO_2$	The "-ite" suffix indicates medium oxygen content. In these examples, two oxygen atoms are present in each molecule.
Sodium chlor**ate** Sodium iod**ate**	$NaClO_3$ $NaIO_3$	The "-ate" suffix indicates medium oxygen content. In these examples, three oxygen atoms are present in each molecule.
Sodium **per**chlorate Potassium **per**manganate	$NaClO_4$ $KMnO_4$	The "per-" prefix indicates maximum oxygen content. In these examples, four oxygen atoms are present in each molecule.

COMMON INORGANIC OXIDIZING MATERIALS

Notes

Many oxidizers yield oxygen when exposed to heat or moisture. Others, such as chlorates and nitrates, produce toxic combustion products. A description of the common types of inorganic oxidizers follows.

Bromates

Bromates are powerful oxidizing agents. Most emit toxic bromine fumes when heated enough to decompose. They pose hazards similar to chlorates (see below), although they do not react as violently. Bromates are used in engraving and lithography. They are also used in sedatives and as corrosion inhibitors.

Chlorates

Chlorates are used in the manufacture of flares, gunpowder, and fireworks. Some chlorates (potassium chlorate, for example) are highly flammable when mixed with organic materials such as sugar and charcoal. Chlorates are very shock and friction sensitive.

Chlorites

Chlorites are bleaching agents that react like hypochlorites (see below) when exposed to fire. Chlorites require higher temperatures than hypochlorites to decompose, but the resulting explosion is significantly more violent. When mixed with acids, sodium chlorite decomposes to release chlorine gas.

Chromates

Chromates are used as additives in paint pigments. Although considered oxidizers, they do not present a significant fire hazard.

Hypochlorites

Hypochlorites, such as calcium hypochlorite, are used as bleaching agents and swimming pool disinfectants. Calcium hypochlorite releases oxygen and chlorine monoxide when involved in a fire. If exposed to air or moisture, hypochlorites can cause serious chemical burns.

Inorganic Peroxides

Inorganic peroxides, such as hydrogen peroxide, sodium peroxide, calcium peroxide, potassium peroxide, and strontium peroxide, decompose readily and yield oxygen in a fire or in contact with moisture. Inorganic peroxides are used as bleaching agents. They are also used in deodorants, germicides, and potassium superoxide.

Hydrogen peroxide is the most widely recognized inorganic peroxide because of its extensive use as an antiseptic. It is most commonly found as a 3% solution in water. Commercially it is available in 27.5%, 35%, 50%, and 70% concentrations. It is used to bleach textiles and leather and to synthesize other peroxides. In some industries, such as the aerospace industry, 90% concentrations of hydrogen peroxide can be found.

Notes

The hazards associated with hydrogen peroxide increase proportionally as the concentration is increased. An instantaneous reaction occurs when hydrogen peroxide solutions of 10% concentration are poured on grease. At 30% concentrations, an explosion is very likely. Hydrogen peroxide solutions of greater than 30% are also corrosive to the skin.

Alkaline earth metal peroxides (magnesium, calcium, and barium) are considered less reactive than alkali metal peroxides (sodium, lithium, and potassium). Sodium peroxide gives off corrosive by-products such as sodium hydroxide and hydrogen peroxide when it comes into contact with water. When potassium peroxide is mixed with combustibles, it can easily ignite by friction, heat, or contact with moisture. Prolonged exposure to fire may cause potassium peroxide to explode.

Nitrates

Nitrates yield oxygen when heated and release large amounts of toxic nitrogen oxide. Nitrates are used to make explosives, fertilizers, and some acids. Common nitrates include sodium nitrate, potassium nitrate, and ammonium nitrate.

Ammonium nitrate is a component of ammonium nitrate fuel oil (ANFO). When ammonium nitrate decomposes, it generates white smoke. If the burning rate increases, yellow smoke is generated. At about 700°F (371°C), brownish-red vapors are produced, indicating that toxic nitrogen oxides are being released and that a sudden detonation may occur. Ammonium nitrate can detonate if it is heated while confined or if it receives strong shocks, such as it would during an explosion. Ammonium nitrate was involved in a 1947 shipboard explosion in which the entire Texas City fire department and 500 civilians were killed. Another tragedy occurred in 1988, when six firefighters were killed in Kansas City in an explosion involving ammonium nitrate. This incident was discussed in Chapter 4.

Nitrites

Nitrites are normally less active oxidizing agents than nitrates. Nitrites contain one less oxygen atom per molecule, which makes them less reactive. Like nitrates, nitrites are used in explosives, fertilizers, and some acids.

Perborates

Perborates, such as sodium perborate, are used as disinfectants and bleaching agents. Perborates are soluble in water. They are non-combustible but will accelerate the rate at which combustible materials burn. Contact with liquid combustibles may result in spontaneous ignition.

Perchlorates

Perchlorates contain one more oxygen atom than chlorates and are powerful oxidizing agents. However, they are considered more stable than chlorates, chlorites, and hypochlorites. Perchlorates can explode if they are involved in fire. Common perchlorates include potassium perchlorate, sodium perchlorate, and ammonium perchlorate. Ammonium perchlorate is found in certain rocket propellants. Ammonium perchlorate does not readily burn, but it will if contaminated by combustible materials.

Permanganates

Friction can cause permanganates to ignite if they are mixed with combustibles. They will also ignite spontaneously if contaminated by corrosive liquids. For example, potassium permanganate, a disinfectant, is non-combustible, but it will accelerate burning. If permanganates contact liquid combustibles, the combustibles may spontaneously ignite. When in contact with sulfuric acid, permanganates may explode.

Swimming Pool Chemicals

One of the most common events that brings emergency responders in contact with oxidizers is a pool chemical incident. No hazard exists during normal use, such as when small amounts of pool chemicals are added to water. However, when water accidentally comes in contact with these materials, chlorine gas can be released. Accidental mixture of different pool chemicals, such as HTH and algicides or fungicides, prior to their use in water, can cause the materials to burst into flames.

There are many chemicals used in treating swimming pools, but most fall into two broad classes—calcium hypochlorite and chlorinated isocyanurates.

Notes

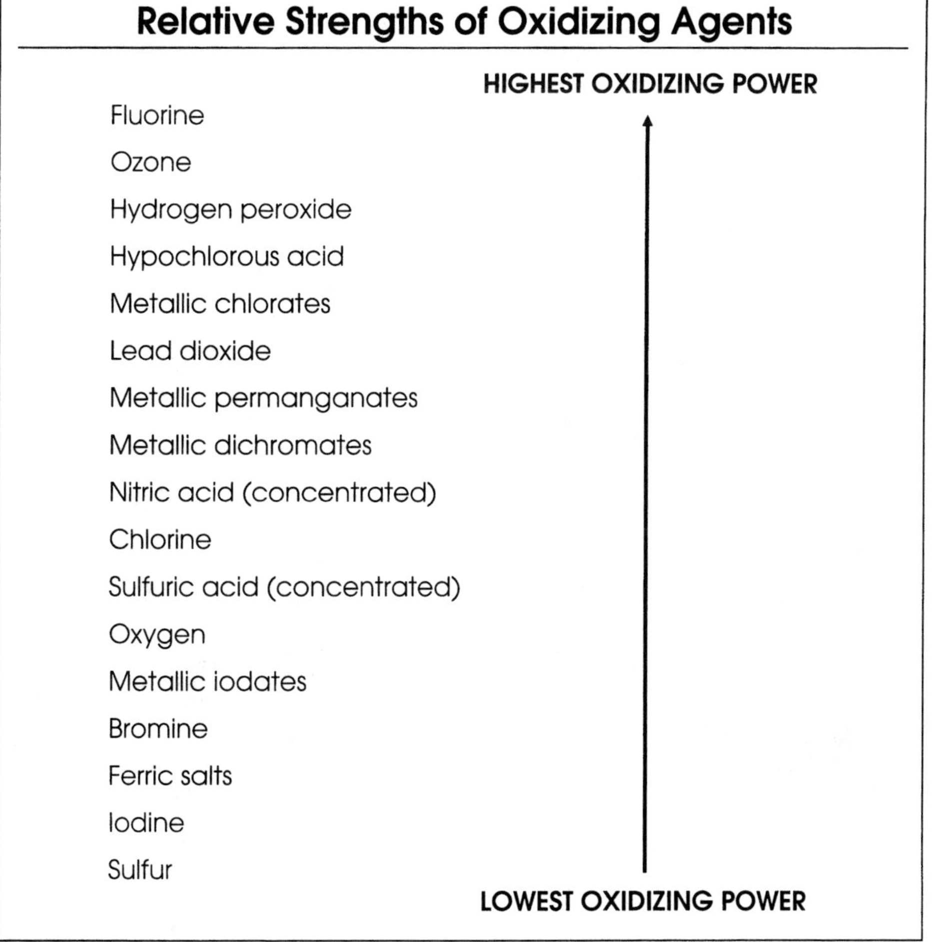

Source: Meyer

Figure 8-1

Notes

Calcium hypochlorite, commonly referred to as "cal hypo," is used to chlorinate pools. Many people refer to cal hypo as "HTH." Technically, HTH stands for high test hypochlorite, but it is also the trade name for a product line. Many chemicals have the initials HTH on the label, but they are not calcium hypochlorite. For example, Olin formulates calcium hypochlorite and a chlorinated isocyanurate both with the trademark of HTH™. In addition, many of these products are repackaged by independent wholesalers for retail sale. Always identify pool chemicals by their chemical name.

Chlorinated isocyanurates, commonly referred to as "isos," can be broken down into three categories:

- trichloroisocyanuric acid ("trichlor")
- sodium dichloroisocyanurate ("dichlor")
- potassium dichloroisocyanurate ("dichlor")

Calcium hypochlorite and chlorinated isocyanurates possess other hazards in addition to the oxidizing hazard. Look at these hazards closely:

- These substances are powerful chlorinating agents and yield enormous amounts of chlorine as well as oxygen. Often there is a number in the name of a pool chemical that indicates the available chlorine in the product.

CDB63	63% available chlorine
ACL56	56% available chlorine

- Calcium hypochlorite and chlorinated isocyanurates are not combustible, but they will decompose and release oxygen, chlorine, and toxic by-products at low temperatures.

Common Containers for Pool Chemicals

Notes

- Calcium hypochlorite and chlorinated isocyanurates are incompatible with each other. They are also incompatible with most other chemicals used in pool water treatment, such as algicides, clarifiers, pool conditioners, and tile cleaners.
- Some substances are incompatible with pool chemicals and other oxidizers. Following is a partial list of incompatible materials:
  - acetic acid
  - alcohols (methyl, ethyl, propyl, and higher alcohols)
  - aliphatic and aromatic unsaturated compounds
  - amines
  - ammonia and ammonium salts
  - ethers
  - floor sweeping compounds
  - glycerin
  - paints, oils, and greases
  - peroxides (hydrogen, sodium, calcium, etc.)
  - petroleum products (gasoline, kerosene, etc.)
  - phenols
  - quaternary ammonium compounds (such as algicides)
  - reducing agents (sulfides, sulfites, bisulfites, nitrates)
  - solvents (toluene, xylene, turpentine, etc.)

Contact the manufacturer for more detailed handling procedures when any substance contaminates a pool chemical. The chemical hotlines of the major companies that deal with these chemicals are listed in Figure 8-2.

Pool Chemical Information Sources

Company	Telephone Number	Type of Information
CHEMTREC	1-800-424-9300	Regarding calcium hypochlorite or any chlorinated isocyanurates in transit
Monsanto Chemical Company	1-800-325-1110	Regarding trichlor, sodium dichlor, or potassium dichlor
Olin Corporation Emergency Action Network (OCEAN)	1-800-OLIN-911 1-800-654-6911	Regarding calcium hypochlorite, trichlor, or sodium dichlor
PPG Industries	1-304-843-1300	Regarding calcium hypochlorite

Figure 8-2

Source: *Guidelines for Safe Handling and Storage of Calcium Hypochlorite and Chlorinated Isocyanurate Pool Chemicals*

Notes

HAZARDS OF OXIDIZERS

Reactions between oxidizing agents and combustible materials pose serious dangers and should always be considered. A basic understanding of these hazards is essential for emergency responders to correctly assess the risk to life and property during an incident. The primary hazards associated with oxidizers are:

- spontaneous ignition
- intensified combustion
- explosion
- production of toxic fumes

Spontaneous Ignition

Materials can spontaneously ignite by three methods:

- slow oxidation of a material
- heat and oxygen production by a water reactive oxidizer
- corrosive action of strong oxidizing acids

Slow oxidation is not usually a problem when dealing with the oxidizer hazard class. It is important to know, however, that slow oxidation by an oxidizing agent is an exothermic reaction. Therefore, if the heat produced by this reaction is contained by insulation or nearby combustibles, a fire is possible. Animal and vegetable products are very susceptible to the slow oxidation process.

When an inorganic peroxide is exposed to water, ***heat and oxygen are produced***. If fuel is introduced, the fire triangle is complete, and the possibility of fire exists. Fire is likely if combustibles are near the peroxide when a reaction occurs. If fire occurs, the peroxide will intensify the reaction by continuing to supply oxygen.

The ***corrosive action of any strong acid*** will generate heat. If the corrosive is also an oxidizer, the heat generated can ignite the exposed material. The fuel material usually must be fairly dry and in a finely divided form to ignite. However, depending on conditions and the oxidizer's strength, materials such as wood planking, tires, and brush may be affected. As the material is permeated by an oxidizing agent, there is a decrease in the material's ignition temperature. Remember that per-acids (perchloric, peracetic, and permanganic) are explosive at high concentrations or elevated temperatures.

Intensified Combustion

The most common hazard involving oxidizers is intensified combustion. The oxidizer increases the amount of oxidizing agent available to react with the fuel's free radicals. A free radical is a fragment of a molecule with an open or unsatisfied chemical bond. Of course, a free radical is very reactive and unstable. The oxidizer increases the rate of combustion and the amount of heat generated, because more fuel is being oxidized at a given moment than under normal conditions. In fact, for every 18°F (10°C) temperature increase in the reactants (the materials involved in the reaction), the reaction rate doubles.

Notes

Explosion

An explosion is possible whenever an oxidizer, especially a strong one, is mixed with a fuel and then exposed to heat, friction, shock, or pressure. Some strong oxidizers spontaneously explode if in contact with an organic material or other contaminant. Explosions can also occur when:

- heat and/or vapors over-pressurize a container
- there is an explosive decomposition (chemical breakdown) of the oxidizer
- strong liquid oxidizers or oxidizers in a molten state are contaminated by organic materials or water
- organic peroxides are exposed to minimal amounts of heat

Production of Toxic Fumes

Anticipate toxic vapors or toxic smoke in the event of an oxidizer incident. Inhaling the vapors (including oxidizers that are normally gases) and smoke of oxidizers causes serious damage to the respiratory system. Both vapors and smoke dissolve in the moist respiratory tract linings, usually producing corrosive liquids that damage lung tissue and result in pulmonary edema. In addition, smoke produced by an oxidizer is usually toxic.

The symptoms of toxic vapor inhalation, such as difficulty in breathing and pulmonary edema, may not appear immediately. The appearance of symptoms depends on the chemical involved and the severity of the exposure.

ORGANIC PEROXIDES DEFINED

The next division of chemical oxidizers is the organic peroxides. The DOT definition of an ***organic peroxide*** is:

> Any organic compound containing oxygen (O) in the bivalent –O–O– structure and which may be considered a derivative of hydrogen peroxide, where one or more of the hydrogen atoms have been replaced by organic radicals (49 CFR 173.128(a)).

Organic peroxides are found in reinforced plastic, plastic films, and synthetic rubber. They are used to bleach oils, waxes, syrups, fats, gums, and flours. They also are used in the textile, printing, and pharmaceutical industries.

Common Organic Peroxide Groups

- acetylene peroxides and hydroperoxides
- alkyl hydroperoxides
- alpha hydroperoxides
- alpha hydroperoxy ethers and ketones
- diacyl peroxides
- dialkyl peroxides and cycloperoxenes
- MEKP (methyl ethyl ketone peroxide)
- organometallic peroxides
- oxydialkyl peroxides
- peroxy acids and peroxy esters
- polymeric peroxides

Notes

"Organic peroxides are potential explosives never meant to be detonated."
– Dave Lesak

HAZARDS OF ORGANIC PEROXIDES

The hazardous properties of organic peroxides are related to their ability to decompose easily. Due to their molecular structure, organic peroxides are oxygen releasing agents. This property makes them strong and unstable oxidizers that create exothermic reactions.

While organic peroxides can act as oxidizers, many are flammable and explosive. Fires involving organic peroxides often result in an explosion. The violence of the explosion depends on the degree of confinement, the speed at which the organic peroxides are heated, and the specific organic peroxides involved.

Almost all organic peroxides have a temperature above which the decomposition process proceeds by itself. This is called ***self-accelerating decomposition temperature (SADT)*** and indicates the maximum safe storage temperature. See Figure 8-3 for the SADT of several organic peroxides. Some organic peroxides have such a low SADT that they must be shipped and stored under refrigeration to avoid high temperatures. This can be of great concern if the reliability of the refrigeration unit is in question.

For example, benzoyl peroxide strongly supports combustion. It can form explosive mixtures with finely divided combustible materials and will aid the burning of other combustible materials. The material may explode when exposed to fire or heat for extended periods.

Such was the case in 1962 in Norwich, Connecticut. Firefighters responded to a tractor trailer fire where 20,000 pounds of benzoyl peroxide and lauroyl peroxide were stored. A violent explosion occurred, killing four firefighters and seriously injuring two others.

Norwich, Connecticut, 1962 — Four Firefighters Killed

Notes

Other organic peroxides are equally reactive and dangerous. They pose multiple hazards and are extremely unpredictable. In defining the hazards of organic peroxides, the American Insurance Association has stated:

> Organic peroxides are vigorously reactive. Their thermal and shock sensitivity and their tendency to undergo autoaccelerative combustion make them potentially explosive. In some instances, they violently decompose to a point of detonation. Therefore, the problem of safety must be approached earnestly and intelligently by all concerned if these materials are to continue to find safe commercial acceptance.

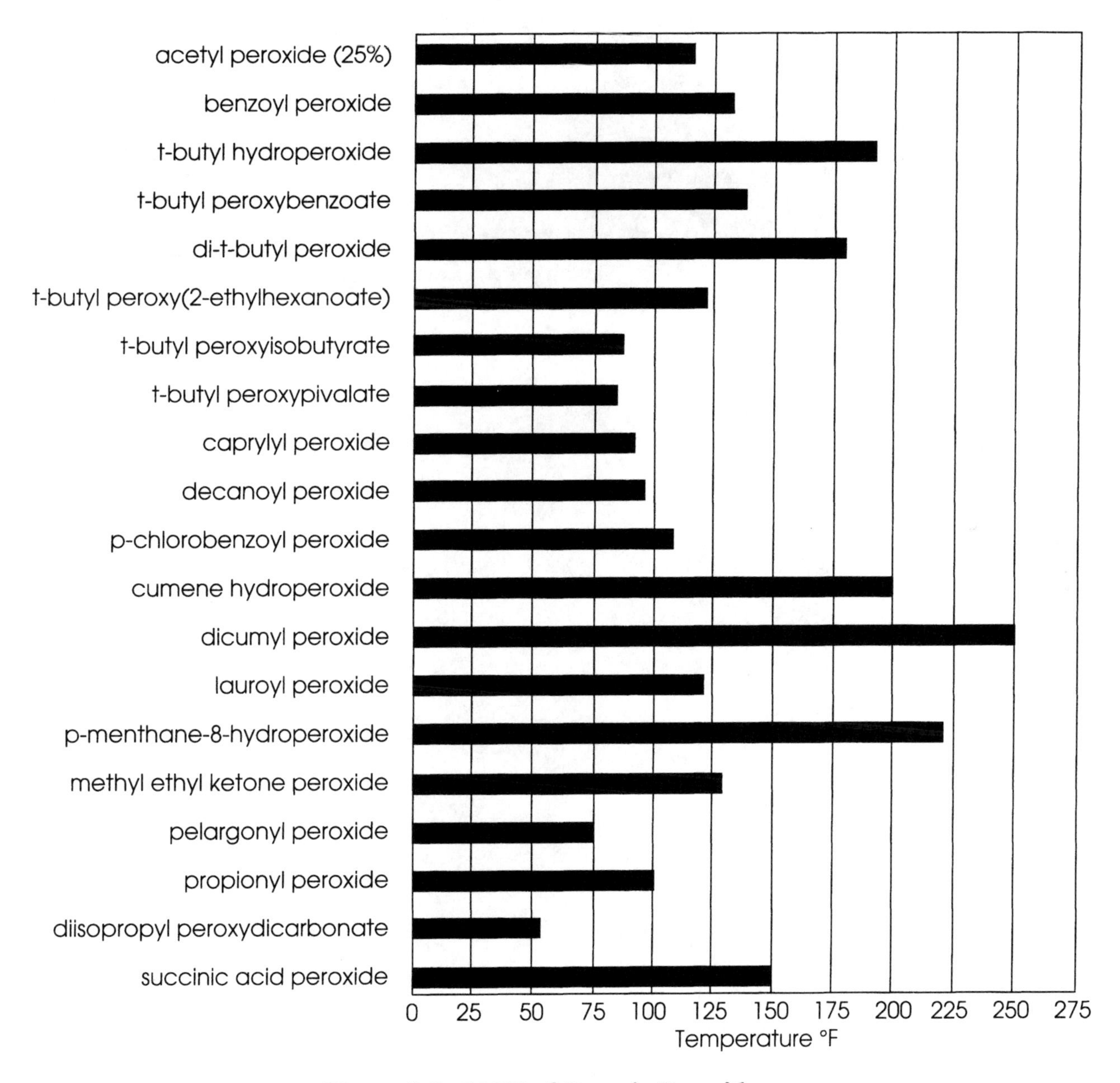

Figure 8-3. SADT of Organic Peroxides

Notes

Organic Peroxides in Shipment

Common Packaging for Organic Peroxides (note dual hazards)

Notes

Devastation of an Organic Peroxide Explosion

OXIDIZING ACIDS

Certain acids can act as oxidizing materials. Although acids will be covered extensively in Chapter 11, the dual hazards of these materials make it important to note a few oxidizing acids here. This group includes strong inorganic acids, such as concentrated nitric acid, hydrochloric acid, and many of the per-acids (perchloric, peracetic, permanganic, etc.). These materials are also water reactive and may become unstable and/or explosive when heated. The per-acids are powerful oxidizers and are explosive in high concentrations or at elevated temperatures. Consider two common per-acids, peracetic acid and perchloric acid:

- ***Peracetic acid*** is both a flammable liquid and an oxidizing agent. It is a clear, colorless, water soluble, corrosive liquid with an acrid odor. Peracetic acid may detonate if exposed to heat and shock. If spilled or leaking peracetic acid comes into contact with organic or combustible materials, it can spontaneously ignite or explode. It aids the burning of combustible materials.
- ***Perchloric acid*** is a clear, colorless, water soluble, oily liquid. In concentrations greater than 72%, perchloric acid can explode violently. Spilled or leaking perchloric acid can form explosive compositions if it mixes with combustible or organic materials. Perchloric acid is corrosive and will accelerate the burning of combustible materials. Even with concentrations of less than 72%, perchloric acid may violently explode or its container may rupture, if exposure to heat is prolonged.

Peracetic Acid
NA 2131
704

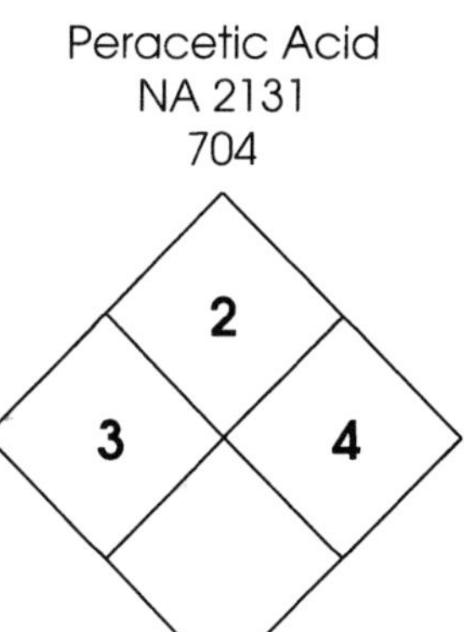

Perchloric Acid
UN 1873
704

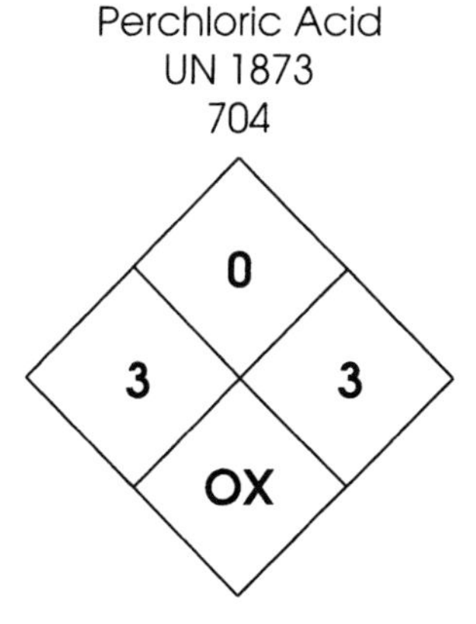

Notes

OXIDIZING ELEMENTS

The last group of oxidizers we will consider is the oxidizing elements. **Oxygen, fluorine**, and **chlorine** are three common oxidizing elements. At normal temperature and pressure, all three elements are gases. Oxygen, however, is the only one found as a gas in nature. Fluorine and chlorine are such strong oxidizers that they generally react with other materials to form more stable compounds. All three gases may be encountered in their pure state in transportation accidents.

Oxygen

OXYGEN (O_2)
(Compressed)
UN 1072

704

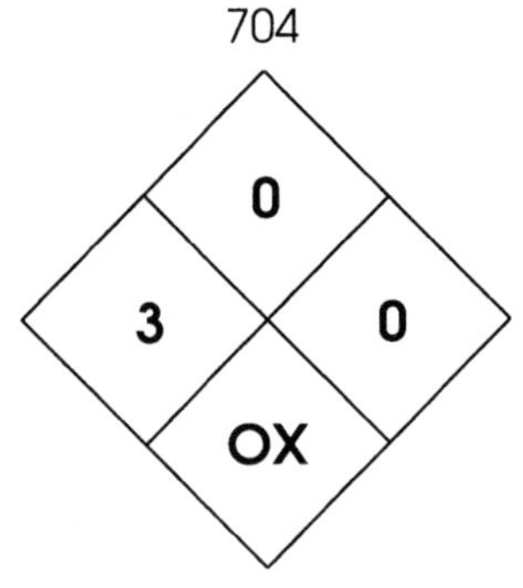

Oxygen is a colorless, odorless, tasteless gas. Elevated temperature may cause oxygen cylinders to explode. Oxygen is a strong oxidizer, although it is categorized as a non-flammable gas. Many combustibles, when subjected to an atmosphere of 100% oxygen, burn furiously. Figure 8-5 shows how oxygen affects the flammable range of hydrogen, methane, and ammonia. Liquid oxygen, when spilled onto petroleum based surfaces such as asphalt, can combine with other substances to form a shock sensitive explosive. Exposure to high concentrations of oxygen can confuse the respiratory system's awareness of carbon dioxide and oxygen levels in the blood, thus producing respiratory arrest.

Physical Properties of Oxygen

Specific Gravity	1.14	
Freezing Point	-361°F	-218°C
Boiling Point	-297°F	-183°C
Critical Temperature	-180°F	-118°C
Critical Pressure	235 psi	
Vapor Density (Air = 1.00)	1.105	
Gas to Liquid Ratio	875 to 1	

Source: Meyer

Figure 8-4

Oxygen's Effect on Flammable Limits

	Flammable Limits	
	In Air	In Oxygen
Hydrogen	4 - 74%	2.3 - 91%
Methane	5 - 15%	5.4 - 59%
Ammonia	16 - 25%	14 - 80%

Figure 8-5

Fluorine

Fluorine is a pale yellow gas with a pungent, irritating odor. It is incompatible with water, nitric acid, and most oxidizable materials. DOT requires "Poison Gas" (primary hazard) and "Oxidizer" (secondary hazard) labels when fluorine is transported.

Fluorine is not combustible, but it is a very strong oxidizer. Elevated temperature may cause fluorine cylinders to rupture. Fluorine will attack some forms of glass, plastic, rubber, and coatings.

As fluorine decomposes, toxic gases are produced, including hydrogen fluoride. Fluorine is highly toxic and can cause penetrating burns on contact. Continued inhalation exposure can cause serious edema and death within minutes. Due to fluorine's reactivity and toxic properties, do not try to extinguish a fluorine fire, but take appropriate public protective actions.

FLUORINE (F_2)
UN 1045

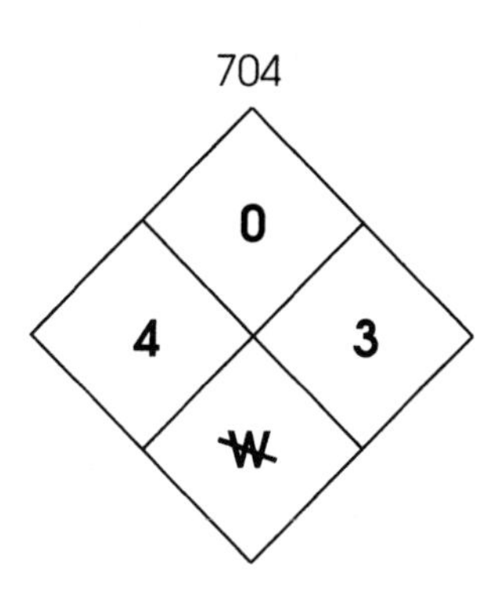

Physical Properties of Fluorine

Specific Gravity	1.1 at boiling point	
Freezing Point	-360°F	-218°C
Boiling Point	-305°F	-187°C
Critical Temperature	-200°F	-125°C
Vapor Density (Air = 1.0)	1.3	
Gas to Liquid Ratio	965 to 1	

Figure 8-6

Source: *Sax's*, Meyer

Chlorine

Chlorine is an amber or greenish yellow gas with a characteristic irritating odor. Chlorine is incompatible with combustible substances and finely divided metals. It is not combustible, but it is a strong oxidizer. Elevated temperatures may cause chlorine cylinders to burst. Contact with combustible substances such as gasoline, petroleum products, turpentine, acetylene, hydrogen, ammonia, sulfur, and finely divided metals may cause fires and explosions. Chlorine attacks some forms of plastic, rubber, and coatings. Like fluorine, chlorine is a toxic material, and prolonged exposure can cause pulmonary edema and death.

CHLORINE (Cl_2)
UN 1017

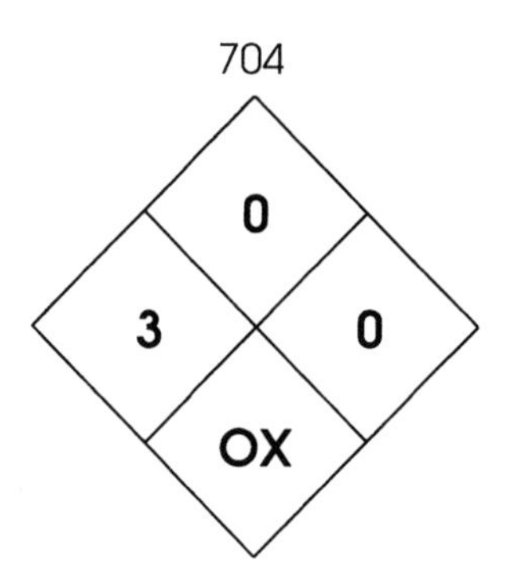

Physical Properties of Chlorine

Specific Gravity at 0°C	1.47 (liquid)	
Freezing Point	-150°F	-101°C
Boiling Point	-30°F	-34°C
Vapor Density	2.5	
Water Solubility Rating	0.7	
Gas to Liquid Ratio	458 to 1	

Figure 8-7

Source: *Sax's*, Meyer

CLASS 5 PLACARD AND LABEL REQUIREMENTS

SUBSTANCE	VEHICLE PLACARD	PACKAGE LABEL	QUANTITY	SPECIAL REQUIREMENTS
Oxidizer	**Oxidizer** (black letters on yellow background). OXIDIZER 5.1	**Oxidizer** (black letters on yellow background). If a subsidiary hazard is present, the subsidiary label must **not** display a hazard class number. OXIDIZER 5.1	Placard any quantity by rail; 1,001 lbs (454 kg) or more by truck.	Should not be loaded with some poisons and most explosives. Should be segregated (at least 4 ft away) from flammable liquids, some poisons, and corrosives. If combined load and 5,000 lbs (2,268 kg) picked up at one facility, then an **Oxidizer** placard is needed in addition to other placards.
Organic Peroxides	**Organic Peroxide** (black letters on yellow background). ORGANIC PEROXIDE 5.2	**Organic Peroxide** (black letters on yellow background). If a subsidiary hazard is present, the subsidiary label must **not** display a hazard class number. ORGANIC PEROXIDE 5.2	Placard any quantity by rail; 1,001 lbs (454 kg) or more by truck.	Should not be loaded with some poisons and most explosives. Should be segregated from some poisons and all corrosives. If combined load and 5,000 lbs (2,268 kg) picked up at one facility, then an **Organic Peroxide** placard is needed in addition to other placards.

Note: Some materials that have oxidation potential may not bear an "Oxidizer" or "Organic Peroxide" placard. Oxygen will have a "Non-Flammable Gas" or "Oxygen" placard (see page 141 for oxygen placard guidelines). Fluorine and chlorine are placarded as "Poison Gas" (see page 140 for chlorine placarding guidelines), but DOT assigns fluorine the secondary hazard of "Oxidizer."

Figure 8-8

SHIPPING CONTAINERS

Notes

Oxidizers and organic peroxides can be found anywhere, from a chemistry lab to a dry bulk hauler to a residential garage. One of the most common shipping packages for oxidizing materials is a multi-layered, plastic-lined, paper bag. Bagged material, such as ammonium nitrate, can range in size from 15 to 100 pounds.

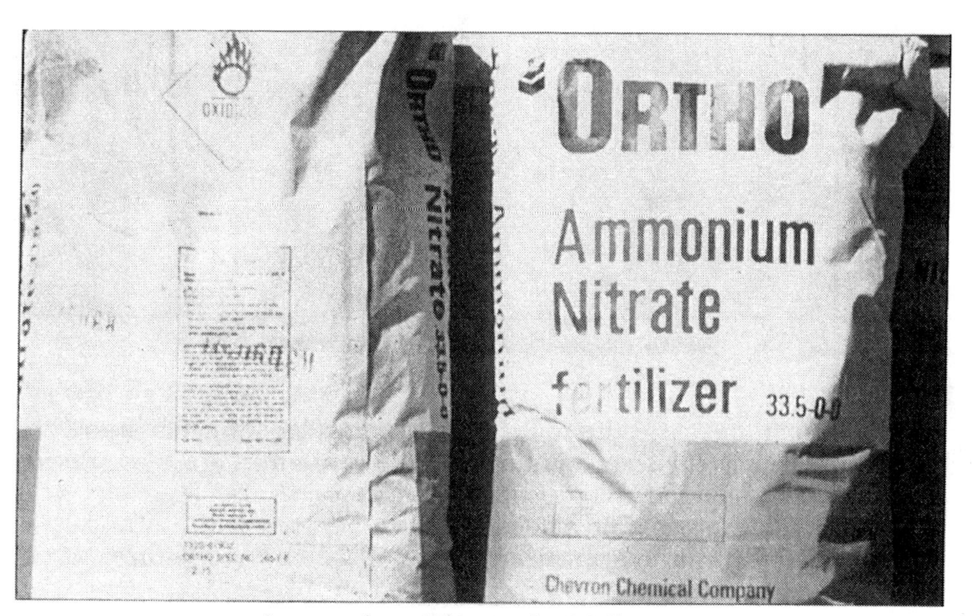

Ammonium Nitrate in 80 lb Bags

Bulk Transport of Ammonium Nitrate

Notes

Metal Casks Containing Oxidizers

Small quantities of oxidizers or organic peroxides are often stored in metal cans or polyethylene containers. These containers are then placed inside wooden or fiberboard boxes for shipment.

Oxidizers and organic peroxides are also shipped in 30 to 50 gallon fiberboard or plastic drums or carboys, or in 55 gallon metal drums. Metal casks are also used.

Agricultural oxidizing products are often transported as liquid slurry in heavy plastic tanks ranging in size from 150 to 5,000 gallons. These tanks must be placarded and labeled according to DOT regulations when moved on highways.

Bulk liquid or slurry oxidizing products are often transported in stainless steel tank trucks. Dry bulk granular commodities are shipped in dry bulk flowable tankers. These trucks are easily identified by their "W" or "V" silhouette.

Some organic peroxides can be transported by highway tank truck, but only by a special permit and exemption from the DOT. Although it is rare for peroxides to be transported in this way, responders should be aware of the possibility.

"W" Shaped Dry Bulk Hauler

Notes

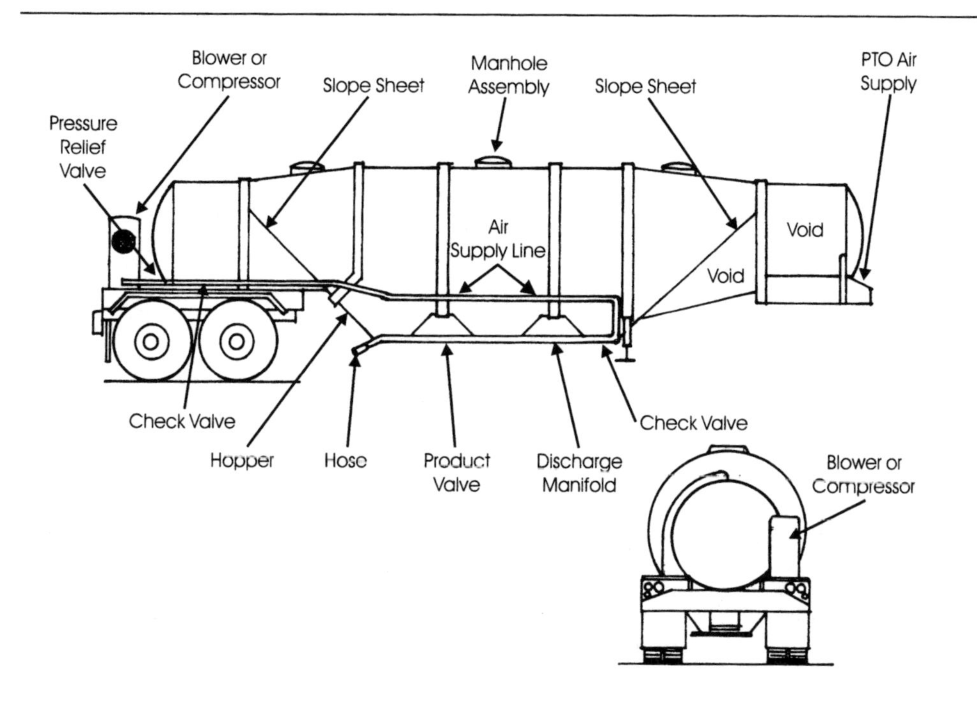

Figure 8-9. "W" Shaped Dry Bulk Hauler

"V" Shaped Dry Bulk Hauler (single dump)

Notes

Oxidizers transported by rail can easily be mistaken for other products.

Current federal regulations prohibit the transport of bulk organic peroxides by rail tank car. Bulk oxidizers, however, can be transported by rail car in slurry form or in box cars in bagged dry form.

KEY POINTS TO REVIEW

Oxidizers and organic peroxides may pose a variety of hazards. An oxidizing agent emergency can be as devastating and dangerous as an explosive incident. Therefore, responders must have a clear understanding of common oxidizing materials and their physical characteristics. Remember, oxidizers can be explosive, toxic, reactive, and generally undesirable to be around. Understanding the following points is critical for safe operations:

- Oxidizers may cause and will enhance the combustion of other materials.
- The more oxygen a substance contains, the greater the oxidizing potential.
- A swimming pool chemical incident may be one of the most common oxidizer incidents that emergency responders face.
- Oxidizers pose four primary hazards:
  - spontaneous ignition
  - intensified combustion
  - explosion
  - production of toxic fumes
- Organic peroxides are vigorously reactive and potentially explosive.
- Some organic peroxides must be shipped and stored under refrigeration.
- Oxidizing acids and oxidizing elements also possess oxidizing potential.

RESPONDING TO OXIDIZER EMERGENCIES

Notes

Responding to an oxidizing material emergency can be both challenging and dangerous. After studying the first part of this chapter, you should have a basic understanding of these two critical concepts:

- Oxidizers and organic peroxides support and intensify combustion, and a few may burn.
- Oxidizers and organic peroxides can be unpredictable and unstable, and they may react violently without warning.

Emergency response personnel must recognize that oxidizing materials can be extremely unstable and that, under certain circumstances, these materials can be as dangerous as explosives. The following considerations should be kept in mind when dealing with oxidizer emergencies.

HAZ-MAT — HAZARD IDENTIFICATION

H
Hazard Identification

At times, identifying the presence of oxidizing materials is not easy, because oxidizers can be found in all kinds of containers. As in any emergency, however, the product and its associated hazards must be identified in order for responders to develop a safe and effective action plan. Identification assessment should include the following points:

- ***Preplanning*** should determine what hazardous products are used at a site, how much of each product is used, and where each product is stored. Be familiar with fixed site facilities that use oxidizing materials.

Be familiar with fixed site facilities that use oxidizers.

Notes

- ***Location of the incident*** may indicate the type of material, the amount of material, and the container involved. Be familiar with common transport routes for bulk fertilizer shipments. These routes are possible haz-mat incident sites.
- ***Markings or identification***, such as placards, labels, and tank shapes, will also aid in identifying oxidizing materials.
- ***Senses.*** Look for bright, flare type combustion as well as white-gray or red-brown smoke. When reacting with another material, many oxidizers emit pungent odors. However, **never** use smell to identify an oxidizer, because the fumes are toxic.
- ***Chemical characteristics*** should be taken into consideration. Among other characteristics, oxidizing agents spontaneously ignite, and they intensify the combustion process.

Once oxidizing materials have been identified, determine the possible hazards:

- Locate the leak or fire.
- Assess the cause of the incident.
- Identify the type of container involved.
- Determine if the container is exposed to fire.
- Assess if there is an environmental hazard.

HAZ-MAT — ACTION PLAN

Extinguishing Agents

Responders should be aware of the effects that various common extinguishing agents have on oxidizers.

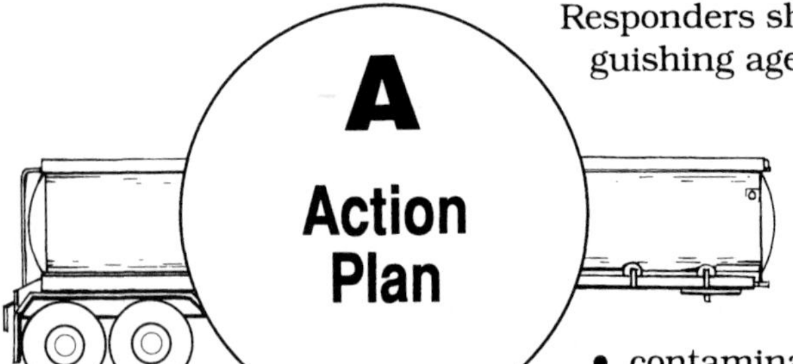

WATER APPLICATION

Water is the most widely used extinguishing agent, because it is the most effective and the most available. However, there are some hazards involved when using water during oxidizer incidents, including:

- contaminated runoff
- impregnation of combustibles
- possible reactions with water and other chemicals

There is always a potential for pollution of waterways and soil. As runoff from fire extinguishing efforts evaporates or drains off, an oxidizing residue is left on the soil or impregnates combustible material. This contamination can be dangerous. If an organic peroxide is involved, the combustible material becomes primed for spontaneous ignition or explosion as it dries.

During the overhaul phase of an oxidizer incident, it is important to note that oxidizing residues left on top of the soil may mix with other chemicals, such as leaking motor oil, and a violent reaction can occur. Unfortunately, these reactions can occur long after the incident. Therefore, when water is applied to a fire involving an oxidizer, give special consideration to the environment and subsequent clean-up.

Notes

FOAM AND CARBON DIOXIDE

Foam is ineffective in oxidizer emergencies, because it excludes only atmospheric oxygen. Atmospheric oxygen is not required for this type of combustion. Traditional firefighting foams break down quickly and are of no use during an oxidizer spill.

Carbon dioxide functions by disrupting the fire's chain reaction, but usually it is not effective in fighting an oxidizer fire. The use of foams or carbon dioxide on oxidizers and organic peroxides is not recommended, because a potential for reaction exists.

Some manufacturers produce foams that are effective for use on selected acids and bases, some of which are also oxidizers. The rate of breakdown for these foams is slower than for traditional foams. Be aware that these vapor suppressing foam agents are **not** fire extinguishing agents and should never be used when a fire is involved in the material or on water reactive materials.

DRY CHEMICAL EXTINGUISHING AGENTS

Dry chemical extinguishing agents are also ineffective. They interrupt the chemical chain reaction, but they are overcome by the oxidizer in all but the smallest of fires.

Special Considerations

- Always identify the product before determining the action plan.
- Don't forget that oxidizing materials can explode.
- If an incident is static, take time to determine the **very best** action plan.

Action Plan Considerations

If an oxidizer incident involves a container with **NO LEAK or FIRE**, consider the following factors:

- Determine the type and size of the shipping container. The container size will indicate what resources are needed and the extent of public protective actions.
- Find out how much product is in the container.
- Assess the damage to the container. Will the container fail or develop leaks?
- Isolate and prevent ignition by removing combustibles and eliminating outside ignition sources.
- Establish appropriate control zones if a leak is possible.

If there is a **LEAK but NO FIRE:**

- Establish control zones.
- Protect in place and/or evacuate the immediate area.
- Isolate and prevent ignition by removing combustibles and eliminating ignition sources.
- Determine if rescue can be performed safely.
- Control streets and request crowd control.
- Find out if the product is water soluble or water reactive.
- Prevent environmental contamination with control procedures.

Notes

- Determine how to minimize the environmental damage if the product has entered a storm sewer or waterway.

If there is a **LEAK and FIRE:**

- Search the available reference materials for technical information.
- Establish control zones.
- Protect in place and/or order an immediate evacuation according to the type and amount of material involved.
- Protect exposures.
- Anticipate the need for large volumes of water. The rate of combustion doubles for every 18°F (10°C) rise in temperature. A great amount of heat will be generated if the fire's temperature increases even a few hundred degrees.
- Use unmanned streams in hazardous areas to avoid endangering personnel.
- Anticipate the need for specialized response capabilities and additional resources. Call for assistance early.
- Runoff may need to be contained in order to prevent environmental problems. Anticipate clean-up operations.
- Recognize the fact that little can be done to rapidly extinguish or control an incident where large amounts of oxidizing agents are on fire. Develop action plans accordingly.

HAZ-MAT — ZONING

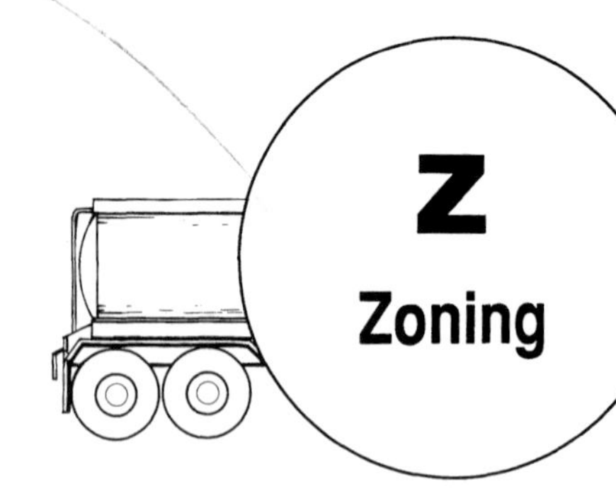

Develop control zones early in the incident. The distances should be based on the type and amount of oxidizing agent involved and the incident location. **Oxidizing materials can be as dangerous as explosives.** Check the table of initial isolation and protective action distances in the *Emergency Response Guidebook* or follow local protocol for the proper zone distances. In the absence of available guidelines for specific oxidizers, consider the following recommendations:

For **SMALL** releases (55 gallon drum container size or smaller) with **NO FIRE:**

- Establish a protective action distance of 250 feet.
- Establish a minimum hot zone of 100 feet.

For **SMALL** releases with **FIRE:**

- Set up a minimum protective action distance of 500 feet.
- Set up a minimum hot zone of 200 feet.
- Consider establishing a warm zone.

For **LARGE** releases (larger than a 55 gallon drum container) with **NO FIRE:**

- Establish a minimum protective action distance of 500 feet.
- Establish a minimum hot zone of 300 feet.
- Consider establishing warm and cold zones.

Notes

For **LARGE** releases with **FIRE:**

- Establish an initial protective action distance of 1,000 feet. Be prepared to expand to 2,500 feet if deflagration is possible.
- Anticipate downwind evacuation.
- Establish a minimum hot zone of 500 feet. Be prepared to evacuate all personnel to 2,500 feet if deflagration is possible.
- Establish warm and cold zones.

HAZ-MAT — MANAGING THE INCIDENT

Scene Management

Depending upon the quantity of product, the oxidizing emergency can place a significant burden on an agency's resources. Establish an incident management system early to promote successful mitigation.

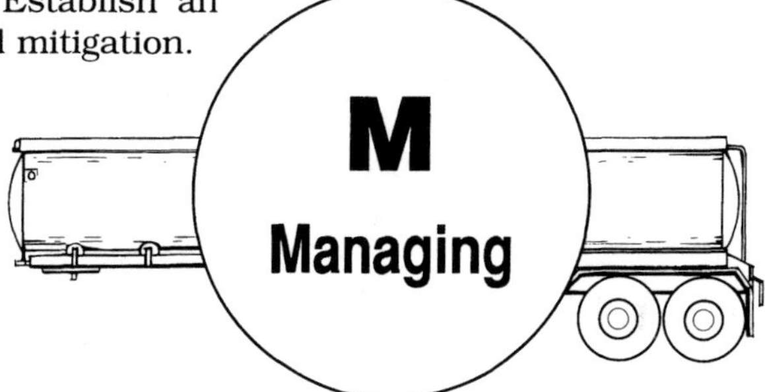

Oxidizing incidents can escalate quickly and require many agencies to work together. The incident commander must make an adequate assessment of the following:

- potential hazards
- rescue problems
- exposure problems
- tactical considerations such as water supply and environmental concerns
- immediate and future resource needs
- limitation of resources versus the size or complexity of the incident
- time vs. resource availability

In addition, several things are needed to bring an oxidizing agent fire under control. They are:

- a complete preplan for oxidizing incidents
- early resource identification
- short response time
- large supply of readily available water
- application of good fire ground tactics

Rescue Considerations

The decision to rescue victims at an oxidizing emergency should be based on considerations similar to those involved at an explosive incident. Determine the type of material involved before committing personnel. Do not attempt a rescue if the risk for emergency personnel is extreme and/or if the victims have no potential of survival.

If a rescue is made, treatment of exposed individuals will vary according to the product. Exposure to an oxidizing material may cause severe chemical burns and respiratory system injury. Because some products can be explosive in nature, victims may also suffer from multi-systems trauma and thermal burns. Due to the extremely toxic nature of many oxidizers, quick treatment is necessary to increase a victim's chance for survival.

Notes

General Treatment for Oxidizer Exposures

Products .... Ammonium nitrate, nitric acid, concentrated hydrogen peroxide.

Containers .... Liquids are contained in glass bottles, carboys, and aluminum and stainless steel drums. Solids are contained in layered paper bags, metal barrels and drums, tank cars, and tank trucks.

Life Hazard .... Many oxidizers are also corrosives (both acid and base) and will be covered in Chapter 11. Fumes are irritating to the eyes and mucous membranes. Damage to the respiratory tract, including laryngeal edema and pulmonary edema, may result from inhaling fumes or vapors. Upper airway obstruction due to edema is possible. Ingestion may result in burns to the gastrointestinal tract.

Signs/Symptoms ..... Chemical burns to all moist surfaces, possible respiratory distress, upper airway obstruction, and pulmonary edema. Signs of shock and increased heartbeat may be present. The patient may have a decreased level of consciousness due to hypoxia. If the product has been ingested, the patient may experience nausea, vomiting, and/or diarrhea. Eye damage ranging from irritation to severe burns may be present. The patient may exhibit skin irritation and chemical burns.

Basic Life Support .... Wear proper protective equipment and remove victim from contaminated area. Administer oxygen and remove patient's clothing. Brush or blot away any visible product and decontaminate with copious amounts of soap and water. Irrigate eyes thoroughly and continue during transport as necessary. If ingested, dilute with small quantities of water and give activated charcoal (follow advice of Poison Control and local protocols). Do not orally administer anything if the patient has a decreased level of consciousness or decreased gag reflex. Watch for shock and respiratory problems. Do not attempt to neutralize product on tissue. Cover skin burns with sterile dressings after decontamination.

Advanced Life Support .... Assure an adequate airway and assist ventilations as necessary. Start an IV LR (strict TKO). Monitor for cardiac arrhythmias/pulmonary edema and treat as necessary. Treat signs of hypovolemic shock with IV fluids as necessary, but watch for signs of fluid overload. Administer topical anesthetic to eyes for easier irrigation and to reduce pain. Follow local protocols for all drug therapy and medical treatment.

HAZ-MAT — ASSISTANCE

Even a small incident involving oxidizers may require assistance to mitigate the emergency safely. The following is a review of the key players needed during such an emergency.

- ***Dispatcher*** can assist in obtaining needed resources, technical expertise, or additional chemical product information.
- ***Law enforcement personnel*** can handle evacuation and site control.
- ***HMRT*** or specialized response group can provide technical expertise and equipment and can perform on-site tactical duties.
- ***Technical specialists***, such as manufacturers, can provide critical product information and can supply guidelines for containment and clean-up operations.
- ***Private clean-up contractors*** can remove product waste once the incident is safely under control.

HAZ-MAT — TERMINATION

Terminating an oxidizing incident is an important step toward ensuring personnel safety. Termination includes:

- decontamination
- rehabilitation
- medical screening
- post-incident analysis

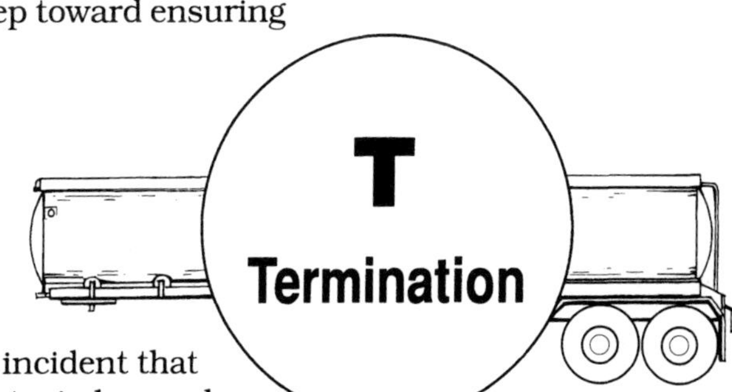

Decontamination

A thorough decontamination is mandatory after any incident that involves a hazardous chemical. Burning oxidizers emit toxic by-products that permeate clothing, tools, and equipment. Also, chlorinated chemicals such as swimming pool chemicals will break down fire retardant fabrics. Mild soap and water is a good decontamination solution for many oxidizers. However, consult reference materials or on-site technical specialists for the most effective solution. In some cases, disposal of clothing, tools, and equipment will be required. If disposal is required, be sure to follow proper hazardous waste regulations.

Oxidizers can contaminate some types of personal protective clothing, such as structural firefighting clothing worn by firefighters. If not adequately decontaminated, this clothing can burst into flames at the next structure fire.

Notes

Rehabilitation and Medical Screening

During this phase, personnel should rest and be evaluated as to their readiness to be reassigned. This phase should include:

- rest period of at least 20 minutes
- medical evaluation prior to reassignment
- exposure report
- initial incident debriefing

Post-Incident Analysis

Oxidizing emergencies are not common. Therefore, a post-incident analysis should be conducted with all involved agencies. This not only will provide insight into the problems encountered during the emergency, but may suggest different methods to use during the next oxidizing incident.

KEY POINTS TO REVIEW

Accidents involving oxidizing agents can challenge even the most experienced, well-equipped emergency response agency. Oxidizing materials have become common in our society, and they are being used in various industrial and residential applications. Oxidizing agents are similar to explosives in regard to their explosive potential. When oxidizers mix with other products, the oxidizing agent becomes even more unpredictable, volatile, and dangerous.

Emergency responders must understand the hazards of oxidizers. Determining the potential of the incident and making good tactical decisions whether to attack the incident or let it burn is based on the emergency responder's ability to assess the true dangers. Like explosives, many oxidizers have no "flashing lights" to warn of impending disaster. A "fight or flight" decision must be based on a knowledge of the material involved.

Two critical concepts regarding oxidizing agents are:

- Oxidizers support and intensify combustion.
- Oxidizers can be unpredictable and become unstable without warning.

Remember the following points when handling oxidizing agents:

- Preplanning is critical when preparing for an oxidizer emergency.
- Emergency responders must clearly understand the hazards of oxidizing materials.
- Be aware of agency limitations when mitigating a large oxidizer emergency.
- Determine the environmental dangers and special clean-up considerations.
- Never underestimate the potential of an oxidizing agent.

Chapter 9

POISON EMERGENCIES

OBJECTIVES

After studying the material in this chapter, you will be able to:

- define the types of poisonous materials
- list the types of effects of poison exposure
- identify common pesticide classes
- state the hazardous properties of solvents
- describe the various poisons used by the military
- describe the hazards present in clandestine drug laboratories
- state the placard and label requirements for poisonous materials
- explain how to read pesticide labels
- describe common containers for poisonous materials
- summarize procedures for responding to a poison incident, using the acronym HAZ-MAT

UNDERSTANDING POISONS

INTRODUCTION

Handling poisons presents a great challenge to emergency responders. Every time firefighters come to a fire scene, they are exposed to certain types of poisons, most of which are by-products of combustion. A few of the more common toxic by-products of combustion are carbon monoxide, hydrogen cyanide, and phosgene. Besides these poisons, however, there are countless manmade substances that emergency responders must deal with.

Improper response to an incident involving poisons can be deadly. Possible examples include the following:

- While working a traffic accident, a law enforcement officer unknowingly walks through a chemical. Several years later he contracts a rare form of cancer.
- Paramedics treat a pesticide poisoning victim, but they end up in the hospital themselves because of chemical exposure.

Notes

- A fire company responds to an industrial fire. A drum bursts, and several responders are covered with a thick, viscous substance. Within three years of the incident, each responder develops a form of cancer.

Poisons are "widow makers." It is imperative that responders understand the poison hazard class because of its extreme potential dangers. If you remember nothing else about poisons, remember the following:

- There is no second chance with some poisons. In some cases, a single breath or a few drops can be fatal.
- Death or chronic health problems can occur 10, 20, or more years after exposure.
- Many chemicals with other Department of Transportation (DOT) classifications are also poisonous.

DOT POISON CLASSIFICATIONS

To fully understand the potential dangers of poisons, it is necessary to examine the various classifications of these materials. DOT uses three classes to identify poisonous materials hazards:

- poison gas
- poisonous material
- infectious substance

Agricultural Supply Facility—Common Location for Poisons

Poison Gas

Notes

Poisonous gases are found in either a compressed or liquefied state. The DOT definition of a ***poison gas*** is:

> A material that is a gas at 68°F (20°C) or less and a pressure of 14.7 psi (101.3 kPa), which is known to be so toxic to humans as to pose a hazard to health during transportation, or is presumed to be toxic to humans based on tests conducted on laboratory animals (49 CFR 173.115(c)).

Poisonous Material

Poisonous materials are liquids, pastes, semi-solids, and solids that are known or presumed to be highly toxic to man. Most poisonous materials are agricultural chemicals, but a variety of plastics, fertilizers, and even rocket fuels are also toxic. Also included are some liquids, such as hydrocyanic acid, which were previously classed as Poison A and which are considered highly toxic. Prior to the October 1, 1991 revision of 49 CFR, DOT identified a class of poisons called irritant materials. These are now included in the poison classification. DOT defines a ***poisonous material*** as:

> A material, other than a gas, which is known to be so toxic to humans as to afford a hazard to health during transportation, or is presumed to be toxic to humans based on tests conducted on laboratory animals. A poisonous material can also be an irritating material, with properties similar to tear gas, which causes extreme irritation, especially in confined spaces (49 CFR 173.132(a)).

Infectious Substance

The DOT Hazard Class 6.2, Infectious Substances, includes three types of materials: infectious substances (synonymous with etiologic agents), diagnostic specimens, and biological products. DOT defines these substances as follows:

> An ***infectious substance*** is a viable microorganism, or its toxin, which causes or may cause disease in humans or animals. A ***diagnostic specimen*** is any human or animal material including, but not limited to, excreta, secreta, blood and its components, tissue, and tissue fluids, being shipped for purposes of diagnosis. A ***biological product*** is a material prepared and manufactured in accordance with the provisions of 9 CFR, which deals with biological products (49 CFR 173.134(a)).

Transportation accidents involving the three classifications of poisons are common. Injuries are also common in such incidents. Out of 569 injuries reported in 16,113 haz-mat incidents in 1994, 102 or nearly 18% were due to incidents involving poisonous materials, poisonous gases, and infectious substances.

Notes

OSHA CLASSIFICATIONS

The Occupational Safety and Health Administration (OSHA) defines chemicals by looking at the possible health effects on the worker or responder. OSHA uses six categories to classify health effects:

- corrosive
- irritant
- sensitizer
- toxic
- carcinogen
- highly toxic

We will discuss only the last five items here. Corrosives will be covered in Chapter 11. For more information, see 29 CFR 1910.1200, Appendix A.

Irritant

Irritants are chemicals that are not corrosive, but they cause a **reversible** inflammatory effect on living tissue by chemical action. Tear gas or similar chemicals in liquid or solid form are irritants that emit dangerous or intensely irritating fumes when exposed to air or fire.

Sensitizer

Sensitizers are chemicals that cause a substantial portion of exposed people to develop allergic reactions after repeated exposures. Some common sensitizers are hydroquinone, bromine, platinum, and isocyanates.

Toxic

A chemical is considered toxic if:

- It has a median lethal dose (LD_{50}) between 50 and 500 milligrams per kilogram of body weight when ingested. Aniline is a chemical that meets this criterion.
- It has an LD_{50} between 200 and 1,000 milligrams per kilogram of body weight when administered continuously over a 24 hour period to bare skin. Epichlorohydrin and acrylonitrile are examples of this type of chemical.
- It has a median lethal concentration (LC_{50}) between 200 and 2,000 parts per million of gas or vapor when inhaled continuously by laboratory rats over a two hour period. Ammonia, nitrogen dioxide, and boron trifluoride are examples of this type of chemical.

Carcinogen

A carcinogen is a material, such as benzene, vinyl chloride, beryllium, and carbon tetrachloride, that is capable of causing cancer in test animals and/or humans. A substance is considered to be a carcinogen if:

- It has been deemed a carcinogen or potential carcinogen by the International Agency for Research on Cancer (IARC).
- It is listed as a carcinogen or potential carcinogen in the *Annual Report on Carcinogens* published by the National Toxicology Program (NTP).
- It is regulated by OSHA as a carcinogen.

Highly Toxic

Notes

A chemical is considered highly toxic if:

- It has an LD_{50} of 50 milligrams or less per kilogram of body weight when ingested. Aldrin, ethylamine, and hydrogen cyanide are examples of substances that are highly toxic by ingestion.
- It has an LD_{50} of 200 milligrams or less per kilogram of body weight when administered continuously over a 24 hour period. Mustard gas is an example of this type of chemical.
- It has an LC_{50} of 200 parts per million or less of gas or vapor (or 2 milligrams per liter of mist, fume, or dust) when inhaled continuously by laboratory rats over a two hour period. Dimethyl-nitrosamine is an example of this type of chemical.

Measuring Toxicity and Exposure Values

Immediately Dangerous to Life or Health (IDLH) – Any atmospheric condition that poses an immediate threat to life, is likely to result in acute or immediate severe health effects, or would interfere with an individual's ability to escape from a dangerous atmosphere.

Median Lethal Concentration (LC_{50}) – The concentration of an inhaled substance that is necessary to kill 50% of test animals exposed to it within a specified time.

Median Lethal Dose (LD_{50}) – The dosage, administered by any route except inhalation, that is necessary to kill 50% of exposed animals in laboratory tests within a specified time.

Threshold Limit Value/Ceiling (TLV/C) – The maximum concentration that should not be exceeded, even instantaneously.

Threshold Limit Value/Short-Term Exposure Limit (TLV/STEL) – The maximum concentration to which workers can be exposed for up to 15 continuous minutes without suffering intolerable irritation, chronic or irreversible tissue change, or narcosis of sufficient degree to increase accident proneness, impair self-rescue, or reduce work efficiency.

Threshold Limit Value/Time-Weighted Average (TLV/TWA) – The maximum airborne concentration of a material to which average healthy workers may be repeatedly exposed for eight hours a day, 40 hours a week, without adverse effect.

Emergency Response Planning Guidelines (ERPG-2) – The maximum airborne concentration to which an average healthy individual can be exposed for up to one hour without experiencing irreversible or other serious health effects which could impair the person's ability to take protective action.

Notes

EFFECTS OF POISON EXPOSURE

As first noted in Chapter 2, the study of the adverse health effects of chemicals is called **toxicology**. Poisons interfere with the normal function of the body and may cause a number of negative health effects. The effects of exposure, however, can vary greatly. Exposure effects depend on the following:

- type of poison
- type of exposure (acute, chronic, or subacute)
- method of exposure
- general health and susceptibility of the person exposed

The two types of poison exposure effects that emergency responders should be most concerned with are:

- acute effects
- chronic effects

Acute Effects

The acute or immediate effects depend upon the amount (dose) and type of poison to which a person is exposed. An individual's body chemistry is also a factor. A poison's effect can vary greatly depending on the person's age and general health. A material that is toxic to one person may only mildly irritate another. Note the specific toxic and hazardous properties of formaldehyde and chlorine in Figures 9-1 and 9-2.

Toxic and Hazardous Properties of Formaldehyde

%	ppm	Hazards
100	1,000,000	
73	730,000	Flammable Range (7–73%)
50	500,000	
25	250,000	
7	70,000	Flash Point 122–185°F (50–85°C)
0.1	1,000	
0.025	250	LC_{50} (4 hours)
0.002	20	IDLH
0.0004	4	Discomfort in 10 minutes
0.0002	2	TWA/C, STEL
0.0001	1	Odor detectable
0	0	

Source: *Sax's*, CGA

Figure 9-1

Notes

Toxic and Hazardous Properties of Chlorine		
%	ppm	Hazards
100	1,000,000	
75	750,000	
50	500,000	
25	250,000	
0.1	1,000	Known fatal
0.0873	873	Lowest concentration known fatal to man
0.0293	293	LC_{50}
0.005	50	Potential irreversible injury
0.0015	15	Immediately painful irritation
0.001	10	IDLH
0.0001	1	STEL, TLV/C Odor detectable (.02-3.5 ppm)
0.00005	0.5	TWA
0	0	

Source: *Sax's*, NIOSH

Figure 9-2

Chronic Effects

There are three types of chronic or long-term effects:

- a carcinogenic effect, which results in some form of cancer
- a teratogenic effect, which results in birth defects in an unborn child
- a mutagenic effect, which results in a permanent gene and chromosome alteration that can be passed on to future generations

All three may emerge after the initial exposure. A one-time exposure can be aggravated by other materials, exposures, or habits. The cause of the long term effects is not fully understood, but in all three cases it is thought that a damaging chemical process occurs in the cells of the exposed tissue.

Notes

Local and Systemic Effects of Poisons

Once a person is exposed to a poison, the poison can act upon the body in several ways:

- The poison can have a localized effect. This means the chemical may affect only the exposed tissue.
- The poison can have a systemic or whole body effect. In this case, the poison has little to no effect on the exposed tissue. Instead, it enters the bloodstream and travels through the body until it finds susceptible tissue or until it is excreted.
- The poison may have a combined localized and systemic effect. In this case, the poison affects the exposed tissue and also enters the bloodstream.

The chemical and physical properties of selected poisons are listed in Figure 9-3. Note the hazard class and toxicity levels of the listed products.

Specific Effects of Infectious Substances

An infectious substance is a viable microorganism or its toxin that causes or may cause disease. The human body can defend itself against small quantities of these organisms. As the number of organisms and the number of toxins increase, however, the body's defenses are overwhelmed. The organisms travel through the body searching for susceptible tissue, where the toxins begin their damage. Infectious substances follow the same basic attack process as standard poisons. Many infectious substances are ranked among the most poisonous materials known to man, including the AIDS virus, botulism, and polio.

PHYSICAL AND CHEMICAL PROPERTIES OF SELECTED POISONS

PRODUCT	HAZARD CLASS	COLOR	ODOR	PHYSICAL STATE	VAPOR DENSITY	SPECIFIC GRAVITY	FLASH POINT	IGNITION TEMP.	FLAM. LIMITS	TLV	IDLH	LC_{50} or LD_{50}	LIFE HAZARD
Acetaldehyde	Flam. liquid	Colorless	Fruity	Liquid below 69°F/21°C	1.50	0.80	–36°F –38°C	347°F 175°C	4–57%	100 ppm	2,000 ppm	1,500 ppm (4 hrs)	Eye, skin, and lung irritant; narcotic effect from inhalation.
Acrolein	Poison, Flam. liquid	Colorless	Disagreeable	Liquid	1.94	0.84	< 0°F < –18°C	455°F 235°C	2.8–31%	0.1 ppm	2 ppm	66 ppm	Small amounts are highly poisonous.
Aniline	Poison	Colorless	Aromatic, amine-like	Liquid	3.22	1.02	158°F 70°C	1,139°F 615°C	1.3–11%	2 ppm	100 ppm	175 ppm	Skin absorption or inhalation causes anoxia due to the formation of methemoglobin.
Arsine	Poison gas, Flam. gas	Colorless	Garlic	Gas	2.66	—	—	Extreme flammability	—	0.05 ppm	3 ppm	—	Immediately dangerous to life and health.
Benzidine	Poison	White or slightly reddish	—	Powder crystal	—	1.25	—	Won't readily burn	—	—	—	214 mg/kg	Human carcinogen; exposure not permitted.
Benzotrifluoride	Flam. liquid	White	Aromatic	Liquid	5.04	1.20	54°F 12°C	—	—	—	—	10 g/kg	Moderately toxic in high concentrations.
Bromine	Corrosive, Poison	Red-brown liquid	Choking, irritating	Liquid	5.50	2.93	—	Not flammable	—	0.1 ppm	3 ppm	750 ppm	Both liquid and vapor cause severe burns. Highly toxic.
Chlorine trifluoride	Pois. gas, Oxidizer, Corrosive	Greenish yellow, almost colorless	Sweet, irritating	Gas	3.14	1.77	—	Not flammable	—	0.1 ppm	20 ppm	178 ppm	Extremely toxic and corrosive.
Chloropicrin	Poison	Colorless	Intense, penetrating	Liquid	6.69	1.65	—	Not flammable	—	0.1 ppm	2 ppm	1,600 mg/m^3	Very toxic. Short exposures may cause fatal lung damage.
Cyanogen	Poison gas, Flam. gas	Colorless	Pungent, penetrating, bitter almonds	Gas	1.80	0.87	—	—	6.6–32%	10 ppm	50 mg/m^3	350 ppm	Highly toxic when heated or when in contact with acids, water, or steam.
Dimethyl sulfate	Poison, Corrosive	Colorless	Faint onion-like	Liquid	4.35	1.33	182°F 83°C	370°F 188°C	—	0.1 ppm	7 ppm	45 mg/m^3	Toxic by inhalation, skin contact, or ingestion. Extremely irritating.
Ethylene dichloride	Flam. liquid, Poison	Colorless	Chloroform	Liquid	3.35	1.26	56°F 13°C	775°F 413°C	6.2–15.9%	10 ppm	50 ppm	1,000 ppm	Toxic by inhalation, skin contact, or ingestion.

Figure 9-3

Source: *Sax's*, Davis, NIOSH

PHYSICAL AND CHEMICAL PROPERTIES OF SELECTED POISONS (CONT.)

PRODUCT	HAZARD CLASS	COLOR	ODOR	PHYSICAL STATE	VAPOR DENSITY	SPECIFIC GRAVITY	FLASH POINT	IGNITION TEMP.	FLAM. LIMITS	TLV	IDLH	LC_{50} or LD_{50}	LIFE HAZARD
Fluorine	Poison gas, Oxidizer	Yellow	Pungent	Gas	1.70	1.14	—	Not flam-mable	—	1 ppm	25 ppm	150 ppm	Highly toxic gas causes severe burns to eyes, skin, and respiratory tract.
Hydrogen cyanide	Poison, Flam. liquid	Colorless	Bitter almonds	Liquid below 79°F/26°C	0.93	0.69	0°F -18°C	1,000°F 538°C	5.6–40%	10 ppm	50 ppm	323 ppm	A few breaths can cause death. Can be absorbed through the skin.
Hydrogen fluoride (hydrofluori c acid)	Corrosive, Poison	Colorless	—	Liquid below 67°F/19°C	0.70	1.26	—	Not flam-mable	—	3 ppm	30 ppm	342 ppm	Both liquid and gas states are irritating to eyes, skin, and respiratory tract.
Hydrogen sulfide	Poison gas, Flam. gas	Colorless	Rotten eggs	Gas	1.19	0.92	—	500°F 260°C	4.0–46%	10 ppm	100 ppm	444 ppm	Eye, skin, and respiratory tract irritant.
Methyl bromide	Poison gas	Colorless	Chloro-form-like vapor	Liquid or gas	3.27	1.70	—	998°F 537°C	13.5–14.5%	5 ppm	250 ppm	302 ppm	Toxic by inhalation, ingestion, or repeated exposure. Can cause severe burns.
Methyl isocyanate	Poison, Flam. liquid	Colorless	Sharp, causes tears	Liquid	2.00	0.96	0°F -18°C	995°F 535°C	5.3–26%	0.02 ppm	3 ppm	5.4 ppm	Highly toxic. Intense irritant to eyes, skin, and mucous membranes.
Nitrogen peroxide	Poison gas, Oxidizer	Colorless solid, yellow liquid, brown gas	Irritating	Liquid be-low 70°F/21°C, solid below 15°F/-9°C	1.49	1.45	—	May ignite other material	—	3 ppm	20 ppm	30 ppm	Eye, skin, and respiratory tract irritant.
Parathion	Poison	Yellow to dark	Garlic	Liquid	—	1.27	—	Not flam-mable	—	0.14 mg/m^3	10 mg/m^3	2 mg/kg	Very toxic. Can be fatal by inhalation, skin contact, or ingestion.
Phenol	Poison	Colorless to white	—	Crystals	3.24	1.07	175°F 79°C	1,319°F 715°C	1.8–8.6%	5 ppm	250 ppm	112 mg/kg	Severe tissue burns. Skin absorption or inhalation can cause death.
Phosgene	Poison gas, Corrosive	Colorless	Sweet, hay-like	Gas	3.40	1.37	—	Not flam-mable	—	0.1 ppm	2 ppm	3,200 mg/m^3	Death or delayed lung injury can result from inhalation.
Silver nitrate	Oxidizer	Colorless crystals	Odorless	Crystals melt at 414°F/212°C	—	4.35	—	Not flam-mable	—	0.01 mg/m^3	—	50 mg/kg	Absorbed through respira-tory and GI tract. Silver accumulates in elastic tissue and nervous system.

Source: *Sax's*, Davis, NIOSH

Figure 9-3 (cont.)

TARGET ORGANS OF SELECTED CHEMICALS

Notes

Respiratory (Lungs) .... Halogen and halogen acids, hydrogen sulfide, sulfur dioxide, phosgene, hydrogen cyanide, hydrogen chloride, hydrogen fluoride, hydrogen bromide, nitro compounds, hydrazine, arsine, phosphine, methyl mercaptan, solvent and fuel vapors and mists, chloropicrin, carbon monoxide, phenylamine, asbestos, coal dust, talc, acrolein, acrylonitrile, epichlorohydrin, styrene.

Hepatic (Liver) .... Vinyl chloride, aromatic hydrocarbons and many derivatives, chlorinated hydrocarbons, nitroethane, nitropropane and many other nitro compounds, picric acid, pentaborane, paraquat.

Nephritic (Kidneys) .... Mercury, calcium, carbon tetrachloride, halogenated hydrocarbons, nitro compounds, paraquat, pentaborane, picric acid.

Neurologic (Nervous System) ..... Organophosphates, carbon monoxide, mercury, halogenated hydrocarbons, chlorinated hydrocarbon insecticides and solvents, methyl mercaptan, pentaborane, styrene, tetraethyl lead, rotenone.

Hematic (Blood) .... Aromatics and many derivatives, nitrochlorobenzene, nitro compounds, chlorinated hydrocarbons, carbon monoxide, methyl mercaptan, aniline, anisidine, lead, methyl Cellosolve® (2-methoxyethanol), dichloromethane, nitric oxide, vinyl chloride, Warfarin®.

Skeletal (Bones) .... Fluorides, selenium, vinyl chloride.

Dermal (Skin) .... Arsenic, chromium, beryllium, other heavy metals, hexachloronaphthalene.

AGRICULTURAL CHEMICALS

Agricultural chemicals form the largest group of poisonous products. These substances were developed during the 19th century to control insects, weeds, and pests. They include a number of diverse products, many of which are able to injure or destroy living organisms. Agricultural chemicals may exist as liquids, dusts, granules, or compressed gases. Each physical state presents its own individual problems.

Four general classes of agricultural chemicals are covered in this chapter:

- insecticides for insect control
- fungicides for plant disease and organism control

Notes

- herbicides for weed control
- rodenticides for rodent control

Currently there are approximately 1,500 base chemicals that are combined in various forms to produce over 35,000 agricultural products. These products are so routinely used that at least one pesticide can be found in every building in the United States and Canada. Pesticides are so common that we sometimes forget that even in small amounts they can be very hazardous. Even short exposure can endanger human health and safety.

In addition to toxicity, pesticides generally exhibit one or more of the following characteristics: corrosiveness, flammability, ability to oxidize, or the mechanical dangers inherent with compressed gases. This diverse range of problems, coupled with common usage, is what makes the pesticide group so dangerous. Below are descriptions of some of the most widely used products.

Organophosphous Compound Insecticides (Organophosphates)

Compounds .... Abate®
DDVP (Vapona®)
Diazinon®
Dicapthon
Dimethoate (Cygon®)
Dursban®
Ethion
Fenthion (Baytex®)
Gardona
Malathion
Naled (Dibrom®)

Toxic by .... Ingestion
Inhalation
Skin absorption

Toxicity .... Moderate to extreme

Symptoms .... Onset: Immediate to hours.

Mild: Loss of appetite, headache, dizziness, weakness, anxiety, tremors of tongue and eyelids, excessive contraction of pupils, impairment of visual acuity.

Moderate: Nausea, salivation, lacrimation, abdominal cramps, vomiting, sweating, slow pulse, muscular tremors.

Severe: Diarrhea, pinpoint and non-reactive pupils, respiratory difficulty, pulmonary edema, cyanosis, loss of sphincter control, convulsions, coma, heart block.

Notes

Chlorinated Hydrocarbon Insecticides

Compounds .... Aldrin
Benzene hexachloride (BHC)
Chlordane
DDT
Dicofol (Kelthane®)
Dieldrin
Endrin
Kepone
Lidane (isomer of BHC)
Mirex
Thiodan
Toxaphene

Toxic by .... Ingestion
Inhalation
Skin absorption

Toxicity .... Slight to high

Symptoms .... Onset: Normally 20 minutes, but up to four hours. Symptoms include nausea, vomiting, restlessness, tremor, apprehension, convulsions, coma, respiratory failure, and death.

Carbamate Insecticides

Compounds .... Baygon®
Carbaryl (Sevin®)
Thiram
Vapam®
Zectran®

Toxic by .... Ingestion
Inhalation
Skin absorption

Toxicity .... Slight (carbaryl) to moderate (Baygon®)

Symptoms .... Onset: Usually several hours to days. Symptoms include constriction of pupils, salivation, profuse sweating, lassitude, lack of muscle coordination, nausea, vomiting, diarrhea, gastric pain, and tightness in chest.

Notes

Halogen Fumigants (Insecticides)

Compounds .... Methyl bromide
Sulfuryl fluoride

Toxic by .... Ingestion
Inhalation
Skin absorption

Toxicity .... Slight to high

Symptoms .... Onset: 4 to 12 hours following inhalation. Symptoms include dizziness, headache, loss of appetite, nausea, vomiting, abdominal pain, lassitude, weakness, slurring speech, staggering, mental confusion, mania, tremors, and convulsions.

Bromides cause rapid respiration, pulmonary edema, cyanosis, collapse, and death. Coma, absence of reflexes, and death due to respiratory or circulatory failure. Late manifestations may include bronchopneumonia, pulmonary edema, and respiratory failure. Methyl bromide may produce cutaneous blisters and kill via dermal exposure.

Arsenical Insecticides

Compounds .... Cacodylic (dimethylarsinic acid)
DSMA, MSMA (sodium methanearsonates)
Paris green
Sodium arsenite

Toxic by .... Ingestion
Inhalation

Toxicity .... Slight to high

Symptoms .... Onset: 30 minutes to many hours. Symptoms include vomiting, profuse painful diarrhea which becomes bloody later, pains in the esophagus, stomach, and bowel, dehydration, thirst, muscular cramps, cyanosis, weak pulse, cold extremities, headache, dizziness, vertigo, delirium or stupor, skin eruption, and convulsions. Two terminal signs: coma and general paralysis.

Ingestion: Chief signs are those of a violent gastroenteritis, burning esophageal pain, vomiting, watery or bloody diarrhea containing much mucus, later collapse, shock, marked weakness. Death generally due to circulatory failure.

Inhalation: May cause pulmonary edema, restlessness, dyspnea, cyanosis, and foamy sputum.

Notes

Cyanide Fumigants (Insecticides)

Compounds .... Hydrocyanic acid
Hydrogen cyanide
Organic bound cyanides (e.g., acrylonitrile)
Cyanogas®

Toxic by .... Ingestion
Inhalation

Toxicity .... Extreme

Symptoms .... Onset: Usually immediate—one of the fastest acting known poisons. A massive dose may cause unconsciousness and death without warning. In smaller doses, illness may last one or more hours.

Following ingestion, bitter, acrid, burning taste followed by constriction of membrane in throat. Other symptoms include salivation and nausea without vomiting. Anxiety, confusion, and dizziness. Respirations are variable—inspirations short, expirations prolonged. Odor of bitter almonds in breath and vomitus. Initial increase in blood pressure and slowing of heart, followed by rapid and irregular pulse, palpitation, and constriction of chest. Unconsciousness, convulsions, and death from respiratory failure.

Phosphine Fumigants (Insecticides)

Compounds .... Celphos
Delicia
Phostoxin (aluminum phosphide)

Toxic by .... Inhalation

Toxicity .... Extreme

Symptoms .... Onset: Usually immediate. Symptoms include nausea, vomiting, diarrhea, great thirst, headache, vertigo, pressure in chest, back pains, dyspnea, a feeling of coldness, and stupor or attacks of fainting. May develop cough with sputum of a green fluorescent color.

Chronic poisoning may be characterized by anemia, bronchitis, gastrointestinal disturbances, dental necroses, and disturbances of vision, speech, and motor functions.

Notes

Herbicides

Organic Acids and Derivatives

Compounds .... Dichlorophenoxyacetic acid (2,4-D)
Silvex (2-(2,4,5)-TP, a proprionic acid derivative)
Trichlorophenoxyacetic acid (2,4,5-T)

Toxic by .... Ingestion

Toxicity .... Mild to high

Symptoms .... Symptoms include weakness and lethargy, loss of appetite, diarrhea, muscle weakness, ventricular fibrillation, and/or cardiac arrest and death.

Urea

Compounds .... Betasan®
Bromacil
Hyvar-X®
Karmex® (diuron)
Telvar® (monuron)
Urab® (fenuron-TCA)
Urox®

Toxic by .... Ingestion

Toxicity .... Slight to moderate

Symptoms .... Handling may cause irritation of eyes, nose, throat, and skin. Ingestion may cause gastroenteritis.

Miscellaneous

Compounds .... Diquat
Endothall (dicarboxylic acid derivative)
Paraquat (quaternary ammonia derivative)

Toxic by .... Ingestion

Toxicity .... Slight to moderate

Symptoms .... Lethargy, convulsions, and coma

Coumarins, Indandiones (Rodenticides)

Compounds .... Diphacin®
Fumarin®
Pival®
Pivalyn®
PMP (valone)
Warfarin®

Toxic by .... Ingestion

Toxicity .... Slight (single dose) to high (multiple doses)

Symptoms .... After repeated ingestion for several days: Bleeding from nose, gums, and into conjunctiva, urine, and stool. Later symptoms include massive ecchymoses or hematoma of skin and joints, brain hemorrhage, shock, and death.

OTHER TOXIC MATERIALS

Notes

Organic Solvents

Responders often minimize the health hazards of organic solvents because flammability is often their first consideration. Many flammable liquids, such as methyl ethyl ketone (MEK), methanol, and acetone, are also extremely toxic and act as central nervous system depressants. The signs and symptoms of exposure to these chemicals are drowsiness, intoxication, or stupor.

These chemicals are not flammable until airborne concentrations reach the lower flammable limit (LFL). Long before this happens, some solvents can reach extremely toxic levels and can even cause unconsciousness. Refer to Figures 9-4 and 9-5 for the toxic and hazardous properties of MEK and gasoline.

Toxic and Hazardous Properties of MEK

%	ppm	Hazards
100	1,000,000	
11.5	115,000	Flammable Range (1.8–11.5%) Flash Point 22°F
1.8	18,000	
1.0	10,000	
0.3	3,000	IDLH
0.2	2,000	LC_{50}
0.1	1,000	
0.05	500	
0.035	350	Headache, eye irritation
0.03	300	STEL
0.02	200	TWA
0.0027	27	Range of odor detection (10–27 ppm)
0.001	10	
0	0	

Source: *Sax's*, NIOSH

Figure 9-4

Notes

Toxic and Hazardous Properties of Gasoline		
%	ppm	Hazards
100	1,000,000	
50	500,000	
7.1	71,000	Upper flammable limit
3.0	30,000	LC_{50}
1.6	16,000	Lowest concentration known fatal (5 minutes)
1.2	12,000	Lower flammable limit
1.0	10,000	Irritation (2 minutes)
0.2	2,000	Rapid dizziness, fainting
0.09	900	Lowest concentration causing serious symptoms
0.05	500	STEL
0.03	300	TWA
0.014	140	Eye irritation (8 hours)
0.000025	0.25	Odor detectable
0	0	

Source: *Sax's*

Figure 9-5

Military Poisons

The military has been interested in toxic chemicals for a long time. Chlorine, phosgene, and mustard gas were first used in the Battle of Ypres during World War I. At the beginning of the 21st century, chemical warfare is still a threat.

Most military gases are designed to bring about a "quick kill," or immediate incapacitation from eye injury and painful skin eruptions. Chemicals such as bromoacetone are strong irritants. Lewisite creates enormous skin blisters, like mustard gas, but it also acts as a systemic poison.

Nerve agents are another type of military war gas. Similar to organophosphate pesticides, many were originally developed for agricultural use. One extremely effective pesticide, tabun, was immediately introduced as a nerve agent. Nerve agents interfere with the chemical enzyme critical to muscle action. Common signs and symptoms of exposure are convulsions, muscle twitching, increased salivation, urination, constricted pupils, and abdominal cramps. See Figure 9-6 for examples of military markings for poisons.

Notes

CHEMICAL HAZARD

- Red figure indicates highly toxic chemical agents
- Yellow indicates harassing agents (riot control)
- White indicates phosphorus munitions

APPLY NO WATER

WEAR PROTECTIVE MASK OR BREATHING APPARATUS

Figure 9-6. U.S. Military Markings

Clandestine Drug Labs

Clandestine drug laboratories present unusual challenges for law enforcement and fire service personnel. These kinds of responses are increasing, and no part of the country is exempt. Drug Enforcement Agency (DEA) statistics show that the majority of clandestine laboratories produce three drugs: methamphetamine (82%), amphetamine (10%), and phencyclidine or PCP (2.5%). However, as new "designer drugs" are developed in an attempt to circumvent controlled substance laws, other drugs will be produced.

Illegal laboratories range from crude makeshift operations to highly sophisticated facilities, some of which are mobile. They can be set up anywhere and are often found in private residences, motel and hotel rooms, apartments, house trailers, houseboats, self-storage "mini-warehouses," and commercial establishments. Often these laboratories are hidden in houses or barns in remote rural areas. In many instances, they are discovered by fire service personnel responding to a building fire or explosion.

PRIMARY HAZARDS TO RESPONDERS

Clandestine drug laboratories present the same fire and chemical hazards found in legal drug manufacturing facilities. But what make clandestine operations more dangerous are the lack of fire and chemical release safeguards, untrained operators, unknown chemical compounds, booby traps, firearms and ammunition, improper storage containers, improper methods of chemical storage and use, and the total disregard for human safety by the operators.

A clandestine drug lab incident can simultaneously present a fire, a haz-mat incident, and a crime scene. Three primary hazards can be present:

- exposure to fire or explosion
- exposure to street drugs and related precursor chemicals
- weapons and booby traps

Notes

DEA Personnel at a Clandestine Drug Laboratory

Exposure to Fire and Chemical Hazards

Chemicals used to manufacture illicit drugs are numerous and varied. It is difficult to pinpoint which chemicals are used in any drug lab, because the chemicals depend upon the specific drug(s) being processed. Hazards present can include flammable and reactive chemicals, acids and caustics, strong oxidizers, and poisons.

Chemicals are often categorized according to the function they serve in the manufacturing process:

- ***Precursors*** are the raw materials, often a controlled chemical with serious health hazards. Examples include sodium and potassium cyanide.
- ***Reagents*** react chemically with the precursor but do not become part of the finished product. These often include chemicals that are flammable solids and water reactive, including lithium aluminum hydrate (LAH), magnesium metal turnings, sodium, potassium, and lithium metal.
- ***Solvents*** are used to dissolve, dilute, separate, and purify other chemicals. The most common solvents are toluene, methanol, ethyl ether, and acetonitrile.

Exposure to Street Drugs and Precursor Chemicals

Some of these materials have already been implicated in long-term disabilities among both users and law enforcement officers. There is some evidence, as in the case of PCP and its related precursors, of chemical and neurological disorders in children born to women who were exposed to PCP even before conception. This would include users, but also those who are occupationally exposed.

Weapons and Booby Traps

Anticipate the presence of weapons and booby traps at a suspected drug lab site. Examples of booby traps include trip wires anchored to an explosive device, light bulbs filled with gasoline inside refrigerators, leashed cobra snakes, and electrified fences, door mats, or door knobs.

RECOGNITION AND IDENTIFICATION

Due to the nature of illegal drug labs, most agencies have no prior knowledge of their existence or location. In many cases, emergency response personnel end up at these scenes because of a fire, EMS call, or related emergency. If an illegal lab is suspected, law enforcement personnel should be notified immediately, and responders should wait for their arrival.

Avoiding detection is a goal of all illegal labs. However, there are clues that should alert emergency responders to the presence of a lab. Look for heavily secured doors, bars on the windows, painted or covered windows, and taped door jambs and windows. Watch for unusual ventilation procedures. Vent pipes and large ventilation fans are often used to eliminate odors. Sometimes air conditioning is also used, even in the winter. Other external clues that should raise suspicions are the following:

- delivery of chemicals ranging from 5 gallon containers to 55 gallon drums in unusual locations, such as a private home
- presence of similar containers, both empty or full, in areas not usually associated with the use of chemicals
- presence of a "gray sludge" or residue outside or near a drain that may indicate the disposal of chemical by-products
- destruction of plant life around a structure or area, as a result of chemical dumping or emissions
- presence of large quantities of broken glass in the trash (flasks must be discarded after three or four batches)

If entry is made inside a structure, the presence of chemical hardware and equipment will be another indicator. Equipment may include Bunsen burners, heating mantles, beakers, glassware, and electric crockpots.

RESPONSE PROCEDURES

Local response protocol should follow DEA recommendations. Fire service involvement at a clandestine operation usually takes two forms:

- fire suppression operations at an actual fire
- stand-by operations to support police and/or hazardous materials response units who are dismantling an active laboratory

Notes

If an illegal lab is suspected, law enforcement personnel should be notified immediately.

Notes

Special tactical considerations are needed at a fire involving a clandestine drug lab. The incident commander should select a fire attack plan that will not expose responders, civilians, or the environment to chemical contamination. Tactical considerations include the following points:

- Treat the fire as a haz-mat incident. Position apparatus upwind, if possible.
- Be aware of the possibility of armed occupants.
- Evacuate and/or protect in place endangered exposures, as required.
- Limit the number of personnel and exposure times during rescue and firefighting operations.
- Have all personnel wear full protective clothing, including SCBA. Avoid standing in or breathing smoke, and avoid contact with all spilled chemicals.
- Avoid interior attacks. In most instances, the risk outweighs the benefit of extinguishment. Extinguishment should not be attempted unless personnel safety can be ensured.
- Use a minimum amount of water to avoid environmental damage and glassware breakage.
- Diking or other means to reduce contamination should be used to prevent chemicals and runoff water from contaminating the area.
- Avoid overhaul activities, because specialized chemical protective clothing is usually required.
- Do not touch or handle any items on the scene. Avoid equipment or chemical products, because they can be toxic and are considered evidence.
- Do not shut off power or water to equipment that could alter the cooking or cooling processes, as this action may cause an explosion.
- Consider as contaminated all fire suppression personnel who enter the hot zone and their clothing.
- Decontaminate all injured parties prior to transportation. Notify the medical facility of the incident specifics.
- Conduct a medical evaluation for all personnel involved in the incident.

PLACARDING AND IDENTIFICATION

Placards

All **poison gas** (Division 2.3) materials must be placarded, whether they are shipped by rail or truck, regardless of quantity. Certain **poison** (Division 6.1) materials that have a poison inhalation hazard must be placarded, whether shipped by rail or truck, regardless of quantity.

All other poison materials, including “Stow Away From Foodstuffs” materials, must be placarded when shipped by rail. Placarding is also required for 1,001 pounds (454 kg) or more gross weight shipped by truck.

If the load is a combined load and more than 5,000 pounds (2,268 kg) of a poison material is picked up at one facility, then the appropriate placard is required in addition to other placards.

Notes

"RESIDUE" PLACARD

After a poisonous material is unloaded from a rail car, an appropriate placard must appear with the word "Residue" displayed on the bottom half of the placard. When any packaging that originally required a haz-mat label is transported empty, it must have the label removed or completely covered. The residue placard is not applicable for highway transportation.

Labels

DOT requires one of the identification labels shown for chemicals that have adverse biological effects on humans. The labels warn of poison and have black lettering on a white background, together with the familiar skull and crossbones symbol. If the material is also a gas, it must be labeled as such. Poisonous materials that do not exceed certain toxicity levels may be labeled with the "Stow Away From Foodstuffs" label. These materials present a lower hazard than those with the "Poison" label. If a subsidiary hazard is present, the subsidiary label must **not** display a hazard class number.

INHALATION HAZARD

Materials that meet certain inhalation toxicity criteria must comply with other standards as regulated by DOT. The transport vessel must be placarded with the "Poison" or "Poison Gas" placards, as appropriate. In addition, the words "Inhalation Hazard" must clearly appear on the shipping papers and packages.

Markings for "Poison–Inhalation Hazard"

Notes

INFECTIOUS SUBSTANCES LABELING REQUIREMENTS

Every package that contains an infectious substance, a diagnostic specimen, or a biological product must exhibit an "Infectious Substance" label. Common microorganisms that are shipped as infectious substances include:

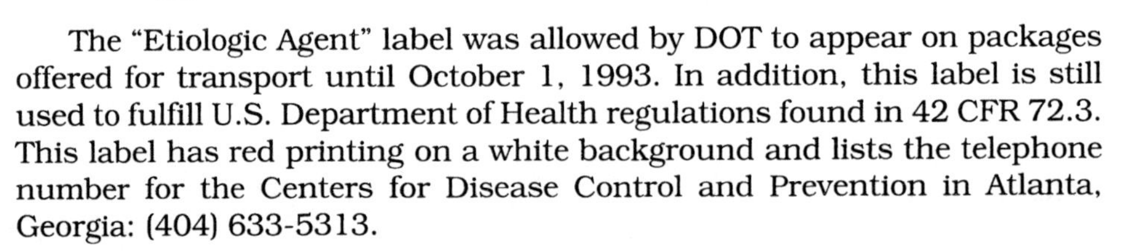

- anthrax
- anthrox
- blood and other potentially infectious materials
- botulism
- cholera
- encephalitis
- flu
- herpes
- HIV/AIDS
- molds
- polio
- rabies
- tetanus
- tuberculosis

The "Etiologic Agent" label was allowed by DOT to appear on packages offered for transport until October 1, 1993. In addition, this label is still used to fulfill U.S. Department of Health regulations found in 42 CFR 72.3. This label has red printing on a white background and lists the telephone number for the Centers for Disease Control and Prevention in Atlanta, Georgia: (404) 633-5313.

PESTICIDE LABELS

In contrast to labels required by DOT, pesticide labels are regulated by the Environmental Protection Agency (EPA). Emergency responders should scan pesticide labels for the following information:

- signal word
- EPA registration number
- product name
- first aid instructions or statement of treatment
- active ingredient statement

Signal Word

The signal word is one of the most important messages on the label. It alerts the user or responder to the material's toxicity. Manufacturers are required by law to use specified signal words on their products. Figure 9-7 lists these signal words and their meanings. See Figure 9-8 for the toxicity of selected pesticides.

Signal Words and Their Meaning					
Signal Word	Toxicity	Probable Oral Lethal Dose for 150 lb Person	Oral LD_{50} mg/kg	Dermal LD_{50} mg/kg	Inhalation LC_{50} mg/l
DANGER	Highly toxic	A taste to a teaspoon	Up to 50	Up to 200	0 to 0.2
WARNING	Moderately toxic	A teaspoon to a tablespoon	Over 50 to 500	Over 200 to 2,000	Over 0.2 to 2.0
CAUTION	Slightly toxic	An ounce to a pint	Over 500 to 5,000	Over 2,000 to 20,000	Over 2.0 to 20
CAUTION	Relatively non-toxic	A pint to a pound	Over 5,000	Over 20,000	Over 20
Note: All products must bear the statement, "Keep out of reach of children."					

Figure 9-7

Toxicity of Selected Pesticides			
Chemical	LD_{50} (mg/kg)	Chemical	LD_{50} (mg/kg)
Highly toxic to man		**Slightly toxic to man**	
Aldicarb (Temik®)	0.9	Cupravit®	750
Phorate (Forate®, Thimet®)	2.0	Kelthane®	800
Endrin	3.0	Malathion (cythion)	1,375
Azodrin®	8.0	Ronnel (Korlan®)	1,740
Demeton (Systox®)	8.0	Borax	2,660
Carbofuran (Furadan®)	11.0	Rermethrin (Ambush®, Pounce®)	4,000
Parathion (Folido®)	13.0	Benlate®	10,000
Dieldrin	50.0		
Moderately toxic to man			
Aldrin	67		
Paraquat	150		
Diazinon®	300		
Mirex	306		
Sevin®	500		

Figure 9-8

Notes

LD_{50} of Selected Chemicals

Chemical	LD_{50} (mg/kg)	Chemical	LD_{50} (mg/kg)
Highly toxic to man		**Slightly toxic to man**	
Strychnine	2.5	Chloroform	800
Sodium cyanide	6.4	Aspirin	1,000
Aflatoxin BI	7.0	Vanillin	1,580
Moderately toxic to man		Vitamin A	2,000
Nicotine	53	Sodium chloride (table salt)	3,000
Sodium nitrate	85	Ethyl alcohol	14,000
DDT	113	Sucrose (table sugar)	29,700
Phenobarbital, sodium salt	162		
Caffeine	192		
Copper sulfate	300		

Figure 9-9

EPA Registration Number

The EPA registration number is product specific. When contacting CHEMTREC/CANUTEC or product specialists, be sure to relay all hyphens, letters, and groupings exactly as they appear on the label. The EPA registration number will aid in identifying the product and its manufacturer. Then specific handling, treatment, and decontamination instructions can be obtained. Antidote information based on the type, length, and extent of exposure can also be obtained.

The EPA registration number appears as a two or three section number: 00000-0000-00.

- the first set of numbers identifies the manufacturer
- the second set identifies the specific product
- the third set indicates the crop on which it may be used
- a U.S. Department of Agriculture number may appear on products registered prior to 1970

Do not confuse the EPA **registration** number with the EPA **establishment** number. The EPA establishment number identifies only the plant or shop in which the product was compounded. Since several different products may come from the same plant, this information is of little value during an emergency.

Product Name

BRAND NAME

Each company has brand names for its products. The brand name is used in advertisements and is clearly seen on the label's front panel. It is the most identifiable name for the product.

Notes

COMMON NAME

Many pesticides have complex chemical names. Some are given another name, known as the common name, to make them easier to identify. For example, carbaryl is the common name for 1-naphthyl-N-methylcarbamate. A chemical made by more than one company is sold under several brand names, but all the products have the same common name or chemical name.

Statement of Treatment

The statement of practical treatment information found on the label is usually of most use to advanced life support personnel. In many cases, the statement may advise against the usual antidotes and list specific drugs to use for that particular pesticide. These differences must be relayed to the doctor, emergency room, or paramedic personnel.

Other important information may be found on the label, such as: "User precautions," "Use heavy duty natural rubber gloves only," "May penetrate rubber," or "Polyethylene provides a good barrier." Such statements inform emergency responders that normal protective equipment will not provide adequate protection. Such information is sometimes difficult to find and lacks details. However, it can warn emergency responders of inappropriate safety equipment.

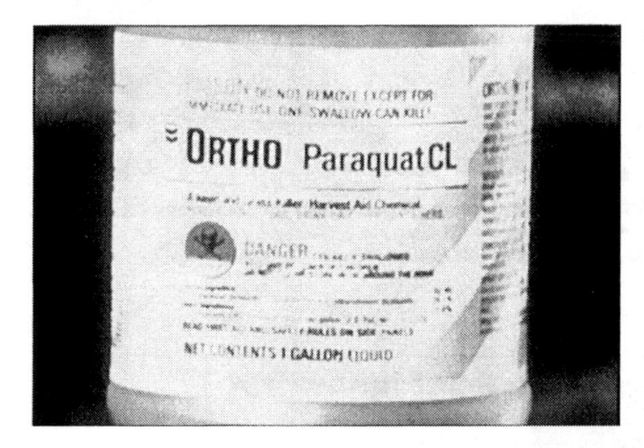

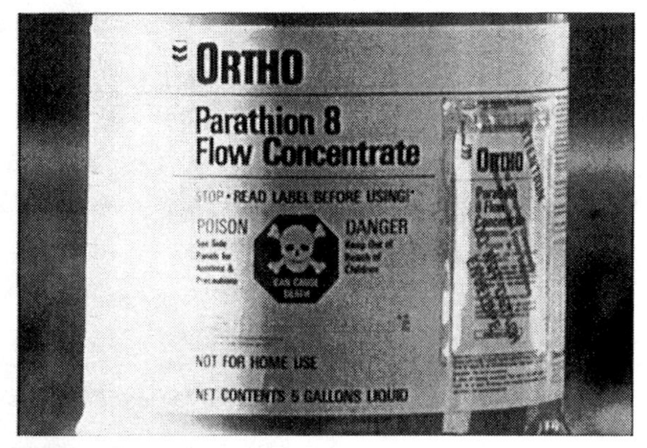

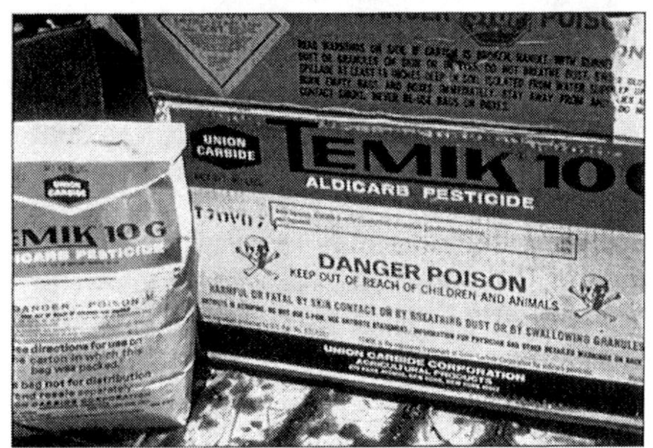

EPA Pesticide Labels

Notes

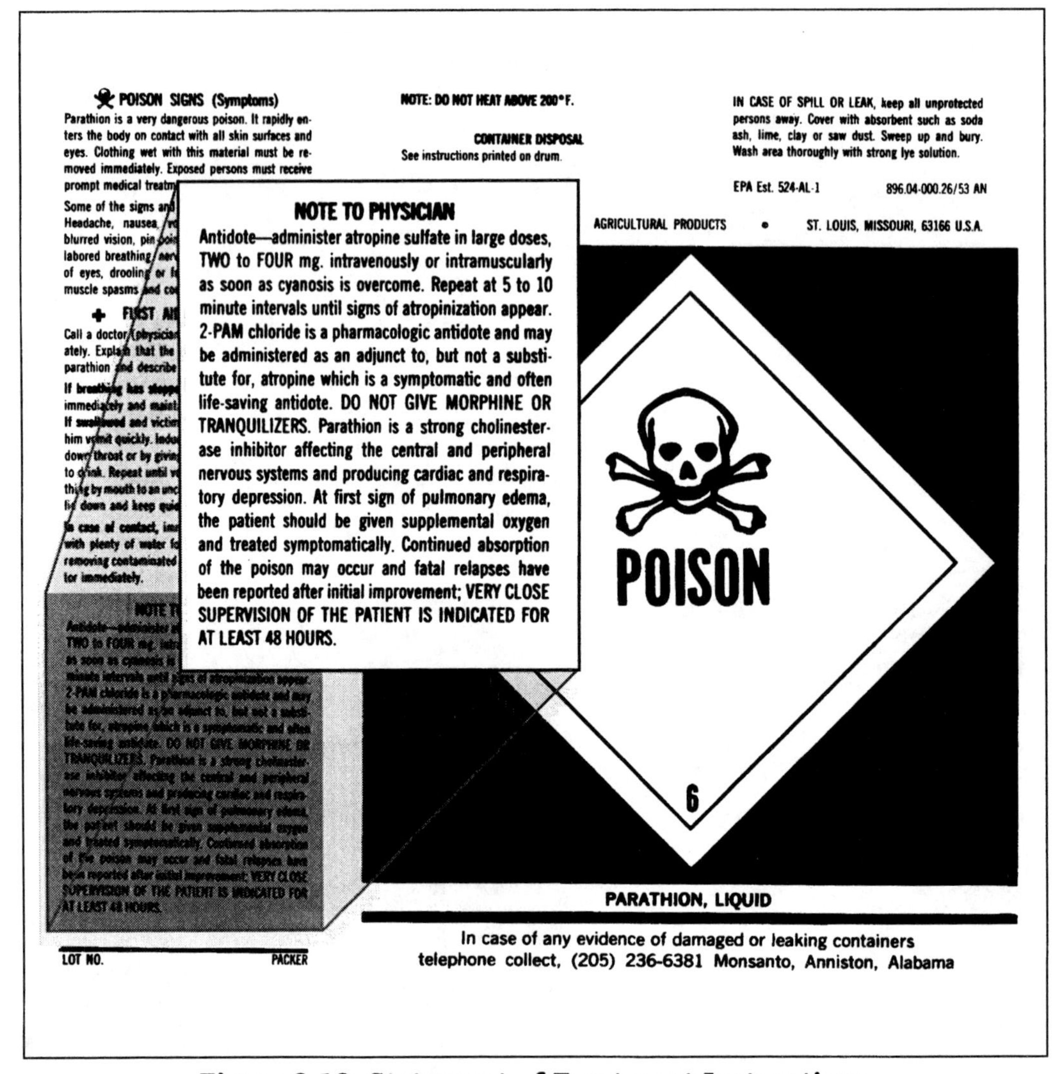

Figure 9-10. Statement of Treatment Instructions

Active Ingredient Statement

Every pesticide label must list the product's active ingredients. The active ingredients are listed clearly for quick identification. The amount of each active ingredient is listed as a percentage by weight or as pounds per gallon of concentrate.

Listing inert ingredients is not required, but the label must state what percentage of the contents the inert ingredients constitute. Responders should note that the "inert" ingredients may pose additional hazards, such as flammability.

Additional Pesticide Label Information

Notes

FORMULATION TYPE

Different types of pesticide formulations (such as liquids, wettable powders, and dusts) require different handling methods. The label states the formulation type of the package. A given pesticide may be available in more than one formulation.

NET CONTENTS

Note the net contents to determine the quantity of product in the container. It will be expressed in gallons, pints, pounds, quarts, or other units of measure.

MANUFACTURER NAME AND ADDRESS

A manufacturer or distributor is required by law to list the company's name and address on the label.

ESTABLISHMENT NUMBER

The establishment number identifies the factory in which the chemical was made. This number does not have to appear on the label, but will appear somewhere on each container. As previously noted, it should not be confused with the EPA registration number.

SYMBOLS

Symbols are an effective way to catch one's attention. That is why a skull and crossbones symbol is used on all highly toxic materials, in addition to the signal words DANGER and POISON. Pay attention to symbols on the label. They are a reminder that the contents can injure or kill.

PRECAUTIONARY STATEMENTS

Hazards to Humans (and Domestic Animals)

This statement warns of the ways in which the product may be poisonous to humans and animals. It also lists precautions to avoid being poisoned, such as specific protective equipment. If the product is extremely toxic, this section will inform physicians of proper treatment for poisoning.

Environmental Hazards

Pesticides are useful tools, but wrong or careless use can result in undesirable effects. To help avoid this, the label lists environmental precautions to follow, such as:

- "This product is highly toxic to bees exposed to direct treatment or to residues on crops."
- "Do not contaminate water when cleaning equipment or when disposing of wastes."
- "Do not apply where runoff is likely to occur."

Labels may contain broader warnings against harming birds, fish, and wildlife.

Notes

Use Classification Statement

Every pesticide label must state whether the contents are for general or restricted use. Restricted use indicates that the products can only be used by EPA certified contractors. General use products are available to all consumers. The EPA classes every product into one of these two uses. The classifications are based on:

- poisoning hazard
- pesticide use
- environmental effect

Physical and Chemical Hazards

This section of the label warns of any special fire, explosion, or chemical hazards posed by the product.

MISCELLANEOUS PESTICIDE LABEL INFORMATION

Other sections of the pesticide label include:

- general use statement
- restrictive use statement
- directions for use
- misuse statement
- reentry statement
- category for applicators
- storage and disposal directions

SHIPPING AND STORAGE CONTAINERS

Poison container types depend on the type of poison and its physical form (gas, liquid, solid). The transport container for a poison gas is quite different from the container used for an infectious substance. Poison gas containers can be pressure type vessels and must not rupture at a temperature of 130°F (54°C). **The containers may or may not be equipped with pressure relief valves because of the product's extreme toxicity.** A BLEVE is possible in the event of fire. Typical poison gas pressure vessels are found on rail tank cars, highway tank trucks, or cylinders that appear similar to standard compressed gas cylinders.

Other poisons are found in a variety of containers, ranging from aerosol cans to paper bags to large tanks. These materials range from pesticides to tear gas, and they can be either solids or liquids (including pastes and semi-solids).

Poisonous products and pesticides packaged for consumer use are often found in tin cans 1 gallon or less in size, in glass or plastic jars, or in cardboard tubes. They are packaged and shipped in cardboard boxes in lots of six to 24.

Bulk agricultural poisons are shipped in 2-1/2 to 5 gallon metal or plastic pails, in 15 to 100 pound multi-layered paper bags and fiberboard drums, and in every other conceivable type of container! Large amounts of dry poison material commonly are shipped in highway dry bulk carriers, recognized by their "V" or "W" tank silhouette. Bulk agricultural poisons can also be shipped in rail box cars and tank cars.

Notes

Hydrocyanic Acid Tank Car—Poison Inhalation Hazard

Acrolein Container—Note Dual Placards

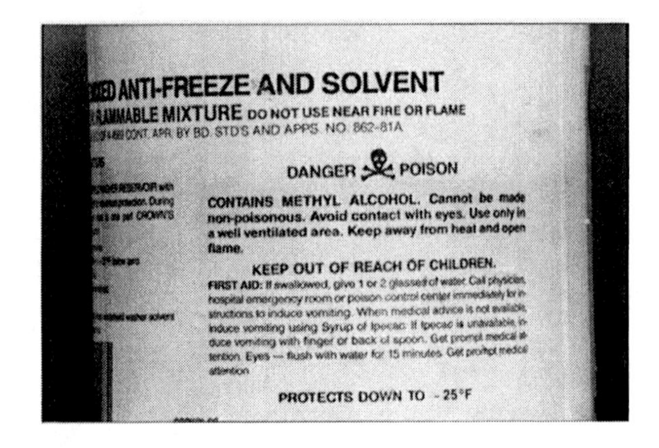

Poisons can be found at the grocery and hardware store!

Bulk Agricultural Poisons in 5 Gallon Cans

Notes

Agricultural Poisons in Multi-Layered Bags

Agricultural Products in 55 Gallon Drums

"W" shaped dry bulk haulers can carry bulk pesticides.

Insecticide at agricultural chemical facility. Be aware that fixed site containers may not be placarded.

Herbicides are often packaged in cases and 2-1/2 gallon plastic jugs.

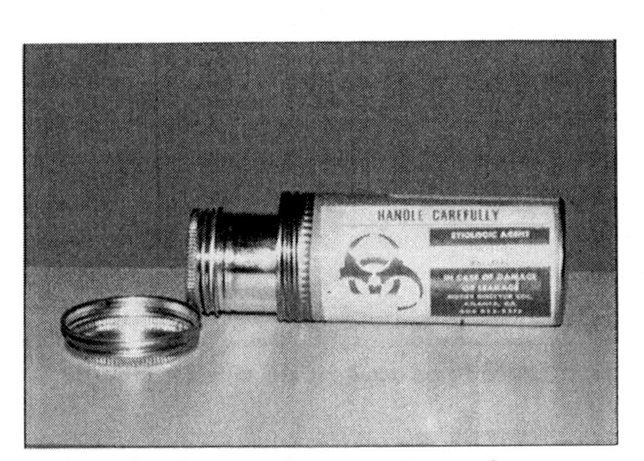

Infectious Substance Container

Infectious Substances

Infectious substances are found in a variety of containers, but most come in small metal screw-top tubes that are placed inside cardboard mailing tubes. The containers usually hold only a few ounces of agent. Normally, only 4 liters, or about 1 gallon, of infectious substance may be transported at a time. Most shipments contain only a few ounces per package in a petri dish or test tube. In addition to mailing tubes, cardboard and wooden boxes are allowed. Small glass jars or vials are shipped in the boxes. Mailing envelopes can also be used. Infectious substances may be transported by taxi cabs from hospitals or doctors' offices to a lab within the same city, or they may be sent through the mail.

KEY QUESTIONS TO REVIEW

Notes

Be able to answer the following questions:

- What are three things to remember about poisons?
- What are the DOT and OSHA poison classifications?
- Exposure effects are dependent upon what four things?
- What are the four kinds of agricultural chemicals?
- What placards are required for a poison gas shipment?
- What other placards are required for a poisonous material?

RESPONDING TO POISON EMERGENCIES

Fires and releases involving pesticides, poisons, and infectious substances can have permanent and disastrous effects. These materials are often referred to as "widow makers" because of the great threat they pose to emergency responders. Therefore, when mitigating a poison emergency, be aware that:

- smoke and vapors given off during a release and/or fire can be extremely toxic
- water runoff from fire operations can be contaminated and toxic
- toxic residues are common after a poison incident
- poison containers exposed to heat can pressurize, explode, and contaminate large areas
- extensive decontamination will be required

HAZ-MAT — HAZARD IDENTIFICATION

With poison emergencies, as with any other hazardous material incident, responders should first identify the hazard. It is easy to overlook the presence of poisonous materials. However, it is critical to identify what poisons are involved, if any, before taking action. A poor initial assessment can lead to **fatal** mistakes.

H

Hazard Identification

Remember that most poisons do not possess spectacular physical characteristics, unlike some other hazardous materials. When responders arrive at the scene of a chlorine leak, a magnesium fire, or a flammable liquid fire, it may be apparent what type of material is involved and what the material's general properties are. In contrast, a poison incident can disguise itself as a "normal" vehicle accident, with no physical indicators. After successfully mitigating the fire hazards and treating the injured, responders may discover later that the "fuel" they contacted was really a poisonous substance!

Let's review the basic steps of hazard identification:

- ***Preplanning*** is imperative for a safe, effective response. Learn the type and amount of product and the storage method used.

Notes

Chlorine Storage at Hot Tub Store—Site for Preplanning

- The ***location of the incident*** is very important, as it can dictate the way an incident should be handled. Location is also important when assessing life hazards, water supply, and environmental problems.
- ***Markings and identification***, such as placards, pesticide labels, and container shapes, are very helpful in identifying poisons.
- ***Noises***, such as hissing from valves or piping, help identify a leak that may be a poison.
- Many poisonous products have pungent, irritating ***smells***. Smell, however, is a very dangerous method of identification and should never be used. Some poison gases are colorless and odorless, while many other poisons have an odor threshold limit (the point at which a product can be detected by smell) that is equal to or close to the toxic level.

Once the presence of a poison is confirmed, the next step is to evaluate the associated hazards.

- Locate the leak or fire.
- Assess the cause of the incident.
- Determine the location, position, and condition of the containers.
- Determine if the container is exposed to fire.
- Assess if there is an environmental hazard.
- Determine the quantity of the material.

HAZ-MAT — ACTION PLAN

Notes

In an accident involving poisons where there is **NO LEAK or FIRE**, consider the following:

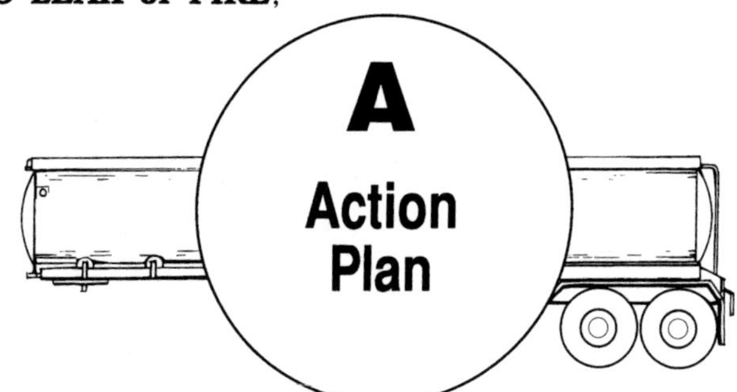

- Determine the size of the container.
- Establish appropriate control zones.
- Find out how much product is involved.
- Assess the damage to the container.
- Anticipate a course of action in case a leak or fire develops.
- Isolate the product.
- Eliminate all sources of ignition.
- Cover the container to prevent product contamination.

If there is a **LEAK but NO FIRE:**

- Approach safely from uphill and upwind directions.
- Establish control zones.
- Evacuate the area immediately.
- Reference and select the appropriate personnel protective equipment.
- Consider location of HVAC, air flows and vents, and drains if indoors.
- Isolate and contain the product.
- Eliminate all sources of ignition.
- Cover the product, if possible, with sand, dirt, or other appropriate material to reduce vapors and prevent contamination from rain and other sources.
- Anticipate the need for additional resources.
- Use technical specialists such as CHEMTREC/CANUTEC, local/state health department, private clean-up contractor, etc.
- Follow appropriate decontamination procedures for exposed emergency and civilian personnel.

If there is a **LEAK and FIRE:**

- Approach safely from uphill and upwind directions.
- Establish control zones.
- Protect in place or evacuate according to the incident size and weather conditions.
- Call for technical specialists such as CHEMTREC/CANUTEC, manufacturer, state/local health department, private clean-up contractor, etc.
- Anticipate additional resource needs.
- Decontaminate personnel and equipment. In some cases, equipment will have to be disposed of as directed by hazardous waste regulations.
- **Use as little water as possible.**

Notes

Are you ready for a poison emergency?

This last important point is different from usual firefighting practices. Runoff from a large gallon per minute flow may be contaminated with toxic products that can spread over large areas.

In addition, water will cool the burning poison, lowering the combustion temperature. Higher temperatures are desirable because they may cause the poison to break into less toxic compounds. For example, most organic pesticides are completely incinerated in two seconds at 1,800°F (982°C). The pesticides are converted to water, carbon dioxide, and hydrochloric acid. The 1,800°F is an ambient temperature not usually reached when burning ordinary combustibles. Cooler temperatures may also produce intermediate compounds of unknown properties that may be more toxic than the original compounds.

Steam produced by adding water to the fire can carry toxic combustion products into the air. As this toxic steam drifts to cooler air away from the fire, the airborne materials condense and deposit toxic fallout over a large area.

If water must be applied, use a fog spray. It allows more control than with straight streams. Straight streams break bags and bottles that can add fuel to the fire and increase the chance of contamination.

Control runoff water, as it is most likely to be contaminated. Build dikes to prevent the flow to lakes, streams, sewers, etc. Inform appropriate agencies if the water enters a drainage system. Contaminated runoff may be the most serious outcome of a fire involving poisons.

Think twice before using water on a poison fire!

HAZ-MAT — ZONING

It is critical to establish control zones during a poison emergency to maintain control over the incident and to minimize chemical exposure to personnel. Zoning requirements depend on the product, the amount, and the severity of the incident. Check the table of initial isolation and protective action distances in the DOT *Emergency Response Guidebook*, or follow local protocol for the proper zone distances. In the absence of available guidelines for specific poisons, consider the following minimum recommendations:

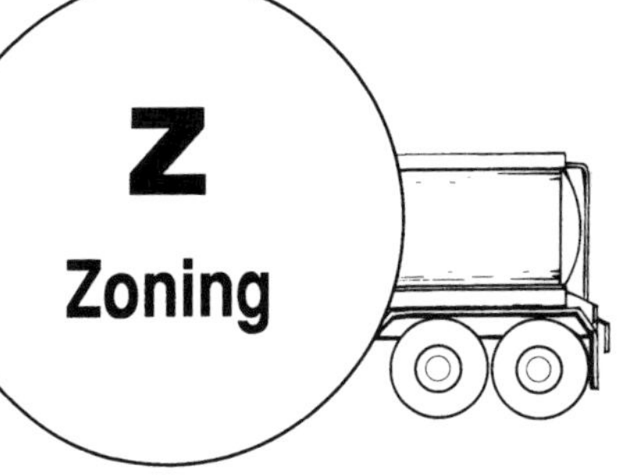

For **SMALL** spills (those involving containers the size of a 55 gallon drum or smaller) with **NO FIRE:**

- Establish a protective action distance of 250 feet.
- Establish a hot zone of 100 feet.

For **SMALL** spills with **FIRE:**

- Establish a protective action distance of 500 feet.
- Establish a hot zone of 250 feet.

Notes

For **LARGE** spills (those involving a container larger than a 55 gallon drum) with **NO FIRE:**

- Establish a protective action distance of 500 feet.
- Establish a hot zone of 250 feet.

For **LARGE** spills with **FIRE:**

- Establish an initial protective action distance of 1,000 feet.
- Anticipate establishing a downwind evacuation corridor of several thousand feet.
- Establish a hot zone of at least 500 feet.
- Establish warm and cold zones.

HAZ-MAT — MANAGING THE INCIDENT

Scene Management

A poison emergency can be difficult to manage. Responders may be dealing simultaneously with a large fire, toxic runoff, several victims, and an evacuation problem. With a correct, timely assessment of the incident, the incident commander can determine the resources needed. In many fire-related poison emergencies, a no-attack strategy may minimize the runoff and related environmental hazards. Use of the incident command system and control zones is necessary for incident management.

To ensure successful management of a poison emergency:

- preplan for known hazards
- preplan strategies
- preplan tactical operational considerations
- anticipate needed resources
- use an incident management system

Rescue Considerations

Chemical exposure by inhalation or skin absorption is the most probable health risk presented by a poison emergency. Depending on the product, health problems may not be immediately apparent due to the delayed effect of many of these chemicals.

It is important that emergency personnel do not become exposed during rescue operations or patient care. In most cases, victims should be decontaminated prior to medical treatment and transport to the hospital. Failure to do so may endanger emergency personnel, or transport equipment may become contaminated and therefore unusable.

Notes

Make sure personnel do not become exposed to poisons.

General Treatment for Poison Exposures

Products .... Arsenic, cyanide, organophosphates, carbamates, contact insecticides, herbicides, pesticides.

Containers .... Range from glass bottles, cans, and drums to heavy metal containers. Some farming products may be in layered paper bags.

Life Hazard .... Different poisons have varying effects on the body. Some present immediate symptoms, while others have delayed effects. Poisons may be absorbed through the skin, inhaled, or ingested. Generalized symptoms are discussed here, but it must be remembered that symptoms may vary markedly from product to product. Identification of the specific poison and the route of exposure is vital to determine proper treatment.

Signs/Symptoms ..... Respiratory distress, pulmonary edema, and possible respiratory arrest. Shock and cardiac arrhythmias may occur. Products may cause decreased level of consciousness, coma, or seizures. Nausea, vomiting, and abdominal pain may result. Eye irritation and skin burns are possible. Certain products, such as organophosphates, have specific signs and symptoms. These products cause SLUDGE syndrome (salivation, lacrimation, urination, defecation, GI pain, emesis). Products that cause a methemoglobinemia (aniline, nitrites, nitrates) produce signs of hypoxia and cyanosis that do not respond to oxygen.

Notes

Cyanides and hydrogen sulfide interfere with cellular respirations.

Basic Life Support .... Wear proper protective equipment and remove patient from contaminated area. Administer oxygen and remove patient's clothing. Brush or blot away any visible product and decontaminate with copious amounts of soap and water. Irrigate eyes thoroughly and continue during transport as necessary. If ingested, dilute with small quantities of water and give activated charcoal (follow advice of Poison Control and local protocols). Do not orally administer anything if the patient has a decreased level of consciousness or decreased gag reflex. Watch for signs of shock, pulmonary edema, and decreased level of consciousness/seizures and treat as necessary.

Advanced Life Support .... Assure an adequate airway and assist ventilations as necessary. Start IV D5W TKO if blood pressure is stable. Use replacement fluids such as LR if signs of hypovolemic shock or dehydration appear. Watch for signs of fluid overload. Monitor for cardiac arrhythmias/pulmonary edema and treat as necessary. Seizures may require treatment with an anticonvulsant. Administer topical anesthetic to eyes for easier irrigation and to reduce pain. Follow local protocols for all drug therapy and medical treatment.

Other Information .... Each poison has different actions and produces different signs and symptoms. Treat each according to its effect on the patient. Antidotes (i.e., cyanide antidote kit, atropine, 2 PAM) are available for a small number of products. The poison may be mixed with other products, resulting in a multiple product exposure. Pesticides are often mixed in a hydrocarbon solvent, resulting in exposure signs and symptoms of both products. Be cautious, and do not allow responders to become victims.

General Treatment for Infectious Substance Exposures

Notes

Examples .... Live bacteria and viruses, diagnostic specimens.

Containers .... Usually range from petri dishes to test tubes.

Life Hazard .... May be considered the most toxic substances handled. They enter the body along the same routes as other poisonous materials. Many of these substances are under study and may result in a disease about which little is known and for which there may be no cure.

Signs/Symptoms ..... Each disease process has its own specific symptoms. Effects are delayed.

Basic Life Support .... Wear proper protective equipment and remove patient from contaminated area. Administer oxygen and remove patient's clothing. Brush or blot away any visible product and decontaminate with copious amounts of soap and water. Irrigate eyes thoroughly and continue during transport as necessary. Assure that patient receives proper medical intervention. It is essential that the specific problem be identified so proper medical monitoring can be instituted.

Advanced Life Support .... Assure an adequate airway and assist ventilations as necessary. Start IV D_5W TKO. Administer topical anesthetic to eyes for easier irrigation and to reduce pain. Start symptomatic and supportive management. Treat any traumatic injuries that may have occurred during the incident and transport for further medical evaluation. Follow local protocols for all drug therapy and medical treatment.

HAZ-MAT — ASSISTANCE

Many resources may be needed to bring a poison emergency to a safe and successful end. The dispatcher, local law enforcement personnel, and specialized response teams will coordinate resources, evacuation, and site control and will perform on-site tactical functions.

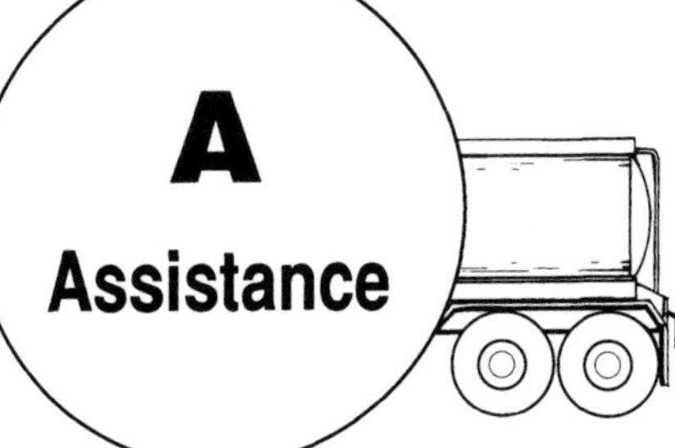

Technical specialists, such as CHEMTREC/CANUTEC, the toxic substance and disease registry for poison control centers, manufacturer's response teams, and chemists, will be valuable during a poison emergency. Be sure that the appropriate specialists are on file and that procedures for contacting them are in place.

Private contractors also may be needed. Arrange for technical assistance with the shipper or manufacturer before an incident.

Notes

Technical specialists may be needed for clean-up.

HAZ-MAT — TERMINATION

It may take several days to several months to completely terminate a poison emergency. The initial phases of termination are critical. Below is a review of termination activities.

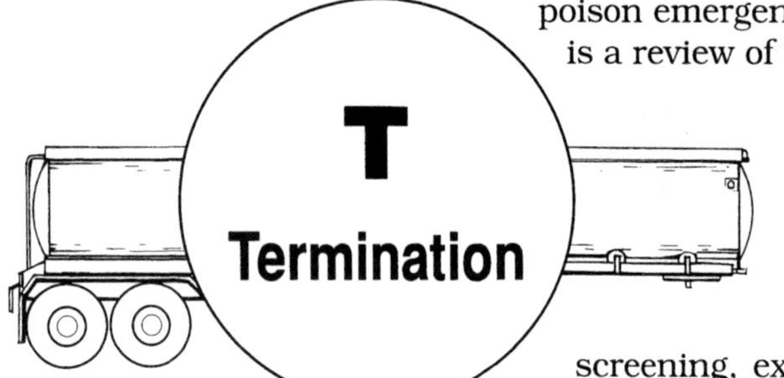

- ***Decontamination*** is critical during a poison incident. Personnel and equipment should be thoroughly cleaned according to the manufacturer's recommendations for the specific product involved.
- ***Personnel rehabilitation***, including medical screening, exposure reporting, and initial incident debriefing is essential. This is the time to evaluate the physical and mental health of responders, before they are assigned to other duties. Remember that many poison symptoms do not become apparent for several hours. This is also a good time to assess potential risks for personnel and to take action to minimize those risks.
- A ***post-incident analysis*** should be conducted with all public and private agencies to evaluate the methods of response. Use this as an opportunity to learn from past mistakes.

KEY POINTS TO REVIEW

Notes

Poisonous materials incidents are quite common for emergency responders, ranging from a residential garage fire to the crash of a crop dusting plane. Poisons are found virtually everywhere throughout any community. Be sure to understand the potential of a poison incident and how to mitigate the incident safely. Failure to recognize the presence, toxicity, and/or the effects of poisonous materials can be deadly.

Remember these key points:

- Understand the physical and chemical characteristics of poisons and how poisons can act on the body.
- Preplan for incidents that are likely to be encountered in your community.
- Recognize the environmental considerations of a poison emergency.
- Anticipate and assess resource needs.
- Remember that poisons are deadly. With many, you get only one chance.

Chapter 10

RADIOACTIVE EMERGENCIES

OBJECTIVES

After studying the material in this chapter, you will be able to:

- define radiation terms
- state the physical properties of radioactive materials
- identify methods of measuring radioactivity
- describe the types of radiation exposure
- state how to minimize radiation exposure
- state the placard and label requirements for radioactive materials
- describe the shipping containers used for radioactive materials
- identify special nuclear materials and their hazards
- identify and describe radioactive pyrophoric metals
- describe the multiple hazards that may be present at a radioactive incident
- list and describe the types of radiation monitoring equipment
- describe proper response considerations at a radioactive incident, using the acronym HAZ-MAT

UNDERSTANDING RADIOACTIVE MATERIALS

INTRODUCTION

The use of radioactive materials has developed and increased during the last 40 years, and these materials are now a vital part of modern society. Radioactive materials are used in medical diagnostic tests, in medical therapy, in medical and industrial research, in weapons production, in consumer product testing and fabrication, and in the generation of electric power.

Notes

Emergency responders will find radioactive materials at construction sites, laboratory facilities, and medical centers. In addition, remember that radioactive materials have to be transported from production sites to places of use and disposal. They may be transported by air, rail, highway, or water. Special handling procedures range from practically none to extreme, depending upon the amount and nature of the material.

Many tons of radioactive materials are transported annually. Also, the number of small users has grown, making radioactive materials commonplace in any community. This prevalence of radioactive materials presents a potential danger to lives and property and raises the risks for emergency response personnel.

Incidents involving radioactive materials can be grouped into five general categories, based on the location of the incident:

- isotope production facilities
- non-destructive testing (industrial radiography)
- nuclear reactor sites
- radioactive materials in transport
- radioisotopes in medical and research facilities

In 1994 there were 10 transportation incidents involving radioactive materials. This figure represented 0.06% of the 16,074 total incidents reported to the Department of Transportation.

DOT DEFINITION

The Department of Transportation defines a ***radiological material*** as:

> Any material that spontaneously emits ionizing radiation and has a specific activity greater than 0.002 microcuries per gram (49 CFR 173.403(y)).

In the discussion that follows, we will see that ionizing radiation includes alpha, beta, gamma, X ray, and neutron radiation. In order for a substance to be classed as a radioactive material for transportation, it must disintegrate at a rate greater than 0.002 microcuries per gram (see Units of Measurement on page 367).

PHYSICAL PROPERTIES OF RADIOACTIVE MATERIALS

Understanding radioactivity is based on a fundamental knowledge of the atomic structure of matter. Radioactive materials may be found as gases, liquids, or solids. Several key terms are listed on the next page. Refer to Appendix A for the definitions of other terms.

Definitions of Radiation Terms

Notes

Element – A substance that cannot be broken down into simpler substances by chemical means. There are 105 known elements, each with its own specific characteristics.

Atom – The simplest unit into which an element can be divided and still retain the specific properties of the element.

Molecule – A stable combination of two or more atoms. It is the smallest structural unit that displays the physical and chemical properties of a compound.

Atomic particle – A particle that makes up an atom. All atoms except hydrogen are made up of electrons, protons, and neutrons. Hydrogen has only an electron and a proton. Protons and neutrons make up the nucleus of the atom, while electrons surround the nucleus. Most atoms are stable. Those that aren't can become stable by emitting radiation.

Radiation – Visible light, radio waves, sound waves, or ionizing radiation. In this text, only ionizing radiation will be discussed.

Ionizing radiation – Radiation which changes the physical state of the atoms it strikes, causing them to become electrically charged or "ionized." In some circumstances, the presence of such ions can disrupt normal biological processes. Ionizing radiation may therefore present a health hazard to man. Atoms that emit ionizing radiation are said to be radioactive. Such atoms are actually going through a process of decay or disintegration. Radioactive materials are not only produced for use in nuclear reactors, but are also found naturally on earth. Ionizing radiation that occurs naturally is called ***natural background radiation***. The three most common types of ionizing radiation are alpha particles, beta particles, and gamma rays.

- ***Alpha particles***, designated by the Greek letter α, are the least penetrating type of ionizing radiation because they are relatively heavy. They do not penetrate the skin and can be stopped by a piece of thin paper or clothing. Alpha particles travel only about two inches (51 mm) in air. A great health hazard exists, however, when alpha emitting materials are inhaled or swallowed or when they enter the body through a wound.
- ***Beta particles***, designated by the Greek letter β^- or β^+, are small particles ejected from the nucleus of a radioactive element. Beta particles are more penetrating than alpha particles, having the capacity to penetrate skin and clothing. Aluminum foil will shield out low energy beta particles. Beta particles can travel several yards in air. They can injure the body both externally and internally.

Notes

- ***Gamma rays***, designated by the Greek letter γ, are electromagnetic radiation of high energy emitted from the nucleus of radioactive atoms. These rays are extremely penetrating and travel close to the speed of light. They can travel through the air and deeply penetrate or pass through human tissue. Several inches or millimeters of dense material, such as lead, give only partial protection from gamma rays (see Figure 10-1). Because gamma rays can invade the body, they are sometimes called "penetrating radiation." Like beta particles, gamma rays can damage both external and internal body tissues.
- ***X rays*** are a familiar form of ionizing electromagnetic radiation. They can penetrate human tissue. X rays can be harmful depending on the dose.
- ***Neutrons*** are another type of ionizing radiation. They are produced by only a few elements during radioactive decay, fission, or nuclear reaction. Neutron radiation exposure is rare, but it can severely damage the body externally and internally.

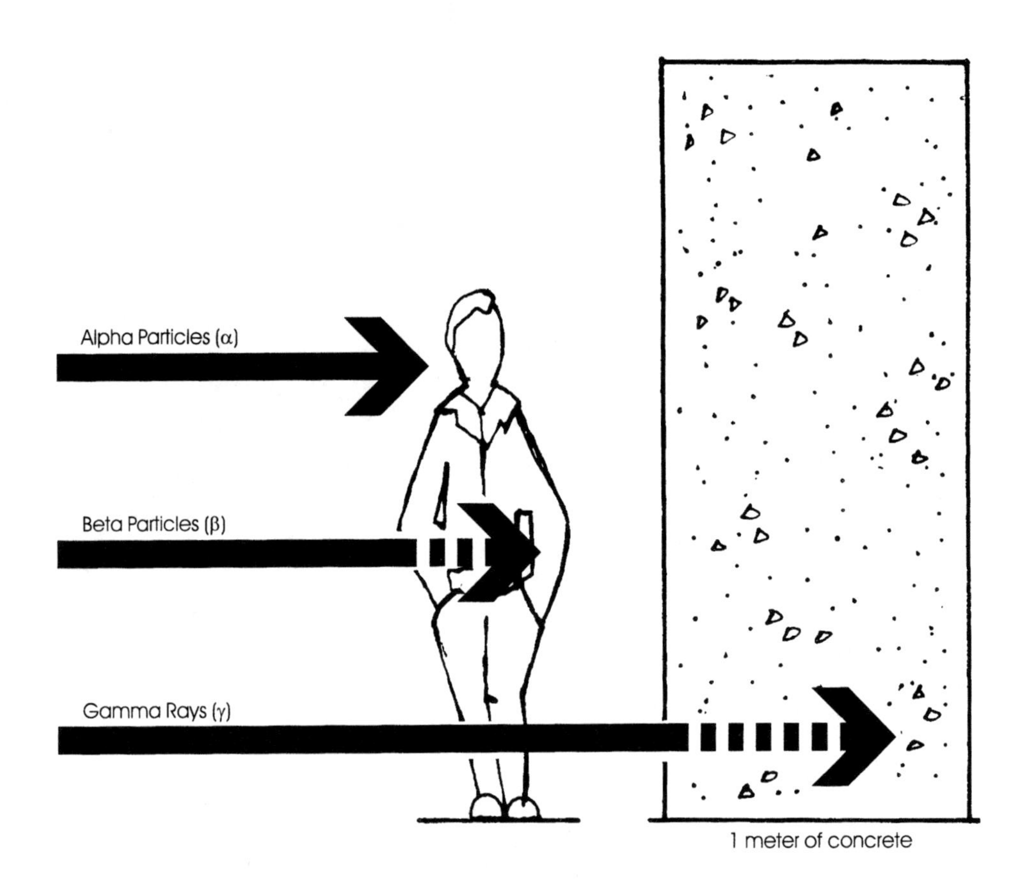

Figure 10-1. Penetrating Power of Radiation

UNITS OF MEASUREMENT

Notes

Emergency responders must understand radioactivity measurement. The following terms will be discussed:

- activity and half-life
- exposure
- dose
- dose equivalent

Activity and Half-Life

The ***curie (Ci)*** is the basic unit of measurement for activity. It describes the amount of radioactivity in a material sample. The ***specific activity*** of a radioactive material is the amount of activity per unit weight of the substance.

Because the curie expresses a relatively large amount of radioactivity, activity is often expressed as some fraction of a curie. For example, a millicurie (mCi) equals one thousandth of a curie, and a microcurie (μCi) equals one millionth of a curie.

Half-life is the time it takes for the activity of a radioactive element to lose one-half of its radioactive intensity through radioactive decay. The half-life of known radioactive elements ranges from fractions of a second to millions of years. Emergency responders should note the half-life of a material, but half-life usually is not a crucial factor of incident mitigation. The more active a material is, the shorter its half-life. Note that the curie and half-life are **not** measures of biological hazard. The half-life values of several common radioactive elements are listed in Figure 10-2.

Half-Life of Common Radioactive Materials

Element	Half-Life
Uranium-235	713,000,000 years
Carbon-14	5,580 years
Radium-226	1,620 years
Strontium-90	38 years
Cobalt-60	5.3 years
Iodine-131	8 days
Sodium-24	15 hours
Polonium-212	Less than 1/1,000,000 second

Source: *Hawley's*

Figure 10-2

Notes

Exposure

Exposure is a measurement that indicates the amount of ionization produced by a radioactive source. The ***roentgen (R)*** is the basic unit of measurement for exposure to gamma radiation, but it may also be used to measure alpha and beta radiation. This measurement is not always an accurate indicator of what the exposure will do to the body.

Note: One roentgen (R) equals 1,000 milliroentgens (mR). The exposure rate may also be expressed as milliroentgens per hour (mR/hr).

Dose

Dose is the accumulated quantity of ionizing radiation. It measures how much energy is absorbed per gram of absorbing material. The measurement unit of dose is the ***radiation absorbed dose*** or ***rad***. The rad measures the quantity of radiation absorbed by the body upon exposure to a substance with a given activity. The absorbed dose in tissue is about 1 rad when the exposure in air is 1 roentgen.

Dose Equivalent

Dose equivalent is a quantity used in radiation safety. The measurement unit of dose equivalent is the ***roentgen equivalent man*** or ***rem***. It quantifies radiation by its estimated biological effect in man (see Figure 10-3). Rem measures effects of radiation that are caused by the equivalent absorption of 1 roentgen of radiation. In emergency situations, use the following equivalents:

$$1 \text{ roentgen} = 1 \text{ rad} = 1 \text{ rem}$$

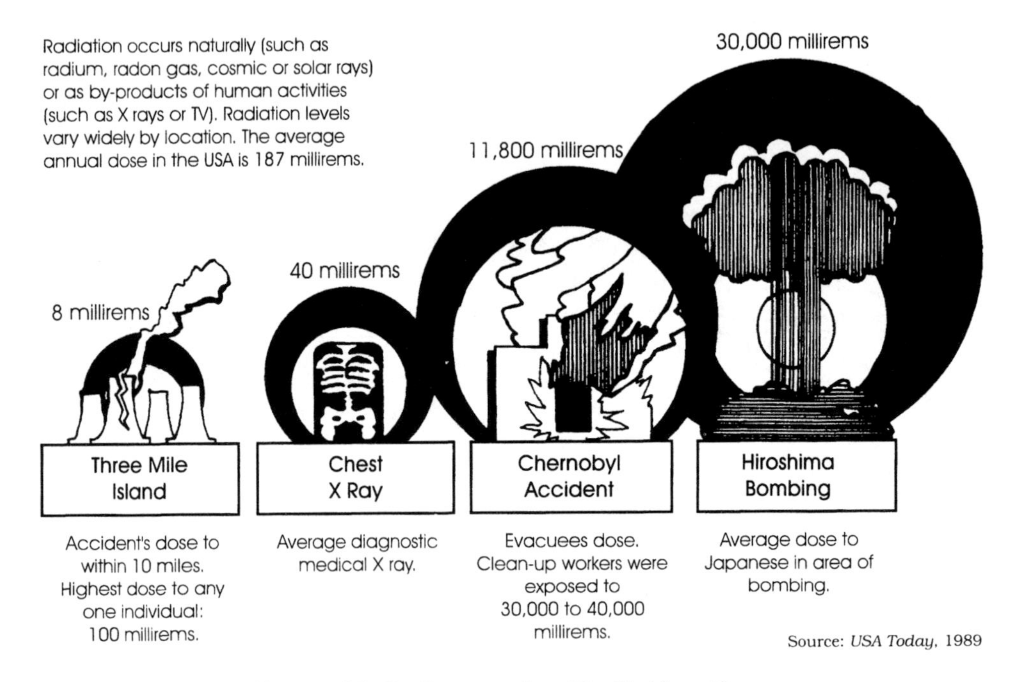

Figure 10-3. Comparing Radiation Doses

According to the Nuclear Regulatory Commission (NRC), the maximum dose equivalent limits are as follows:

- general public – 100 millirems (mrem) per year
- over 18 years old – 5 rems per year
- one time exposure, for emergency or industry worker – 25 rems

The effects of various levels of exposure are listed in Figure 10-4.

Effects of Radiation Exposure

Roentgens (R)	Milliroentgens (mR)	Effect of Single Dose Over Whole Body
<20	<20,000	Clinically not detectable
25	25,000	Maximum limit during one emergency exposure
50	50,000	Blood changes, but no illness expected
100	100,000	Blood changes, nausea, fatigue can occur
200	200,000	10% fatal, genetic complications probable
300	300,000	20% fatal, 100% illness
450	450,000	50% fatal, high percentage of sterility
650	650,000	95% fatal
1,000	1,000,000	Survival unlikely

Figure 10-4

TYPES OF RADIATION EXPOSURE

An accident involving radioactive materials can result in one or more of the following types of exposure:

- external or internal contamination by radioactive materials, such as alpha and beta particles
- external or internal radiation by radioactive materials, such as a gamma source
- combined radiation injury

Severity of injury depends upon the type of radiation source, the intensity of the source, and the amount of time a person is exposed. Because victims of radiation exposure rarely show immediate signs or symptoms of radiation, emergency responders should presume victims have experienced radiation injury until proven otherwise.

Notes

Presume a radiation injury exists until proven otherwise.

Notes

Contamination

Contamination can occur when radioactive vapors, liquids, or solids are released into the environment, thereby coming in direct contact with people either externally and/or internally. **Externally**, the skin and clothing can become contaminated. **Internal contamination** occurs when radioactive materials enter the body through the lungs, digestive tract, skin, or wounds and are subsequently metabolized.

For example, the protective clothing of responders at a release and fire involving a radioactive material will be externally contaminated if personnel walk through the ash or water runoff. Other people could be internally contaminated if they breathe the smoke. Responders should try to avoid contamination, but if contamination takes place, an immediate and thorough decontamination is required.

Radiation

External radiation or irradiation occurs when all or part of the body is exposed to penetrating radiation from an external source. The severity of a radiation induced injury depends upon the amount of radiation an individual receives. Beta radiation can penetrate skin and clothing. Gamma radiation can pass through the body, as happens with ordinary chest X rays.

External radiation differs from contamination in that actual contact with the radioactive material need not occur for the person to be irradiated. An individual is not "radioactive" after external radiation by a gamma source. If the person is contaminated by alpha or beta particles, however, radiation may be detected. Consider again the scenario in which radioactive materials are involved in a release and fire. Responders who are upwind and uphill of the incident may be exposed to radiation but not contamination.

Internal radiation or ***incorporation*** is the depositing of radioactive materials into body cells, tissues, and target organs such as bone, liver, thyroid, or kidney. Internal radiation occurs when radioactive materials find their way into the body by an individual's choice, as in a medical test, or by accident. Internal radiation is caused by internal contamination. In general, radioactive materials distribute throughout the body according to their chemical properties (e.g., radium to bone, iodine to thyroid).

Combined

Contamination and radiation can occur in combination and be complicated by physical injury or trauma.

MINIMIZING RADIATION EXPOSURE

At a radioactive incident, the first priority of emergency response personnel should be to avoid contamination and to minimize radiation exposure for themselves and for incident victims. There are three radiation protection factors that alter the radiation dose. They are:

- time
- distance
- shielding

Notes

Time

Time is an important protection factor against radiation. The less time spent in a radiation field, the less radiation exposure accumulated. Many radiation monitoring devices measure exposure in milliroentgens per hour (mR/hr). If the exposure rate is constant, an exposure rate of 60 mR/hr means that for each minute spent in a radiation field, a person will be exposed to 1 mR.

Distance

Distance also reduces exposure by means of the ***inverse square law***. The general concept of the inverse square law is that **the farther a person is from the radiation source, the lower the radiation dose.**

By measuring the radiation exposure rate at a given distance and then doubling the distance, the radiation intensity is decreased by a factor of four. For example, a dose that measures 8 mR/hr two feet from the source would measure only 2 mR/hr at four feet. Conversely, when the distance is reduced by half, for example, from two feet to one foot, the exposure rate increases from 8 mR/hr to 32 mR/hr (see Figure 10-5).

The inverse square law applies only for point sources such as those used in radiography. In situations in which radioactive materials are released and scattered or in which radioactive fallout is present, the inverse square law does not apply. Radiation exposure, however, is always reduced by moving away from the radioactive source. The key is to stay as far away from the radioactive source as possible.

Shielding

Shielding refers to the ability of a dense material to stop radiation. The denser a material, the greater its ability to stop radiation. Sometimes a vehicle, a mound of soil, or a piece of heavy equipment is available to shield emergency responders and reduce exposure. However, in many emergency situations shielding is often limited to lightweight protective clothing, such as gloves, shoe covers, structural firefighting clothing, coats and jackets, or surgical clothing. Such clothing is sufficient to stop all alpha and some beta radiation, but it does not stop penetrating gamma radiation (see Figure 10-6).

Minimizing Radiation Injury Summary

- Time – Spend as little time as possible in the immediate danger area.
- Distance – Stay as far away from danger as possible.
- Shielding – Wear appropriate protective clothing and stay behind barriers.

Notes

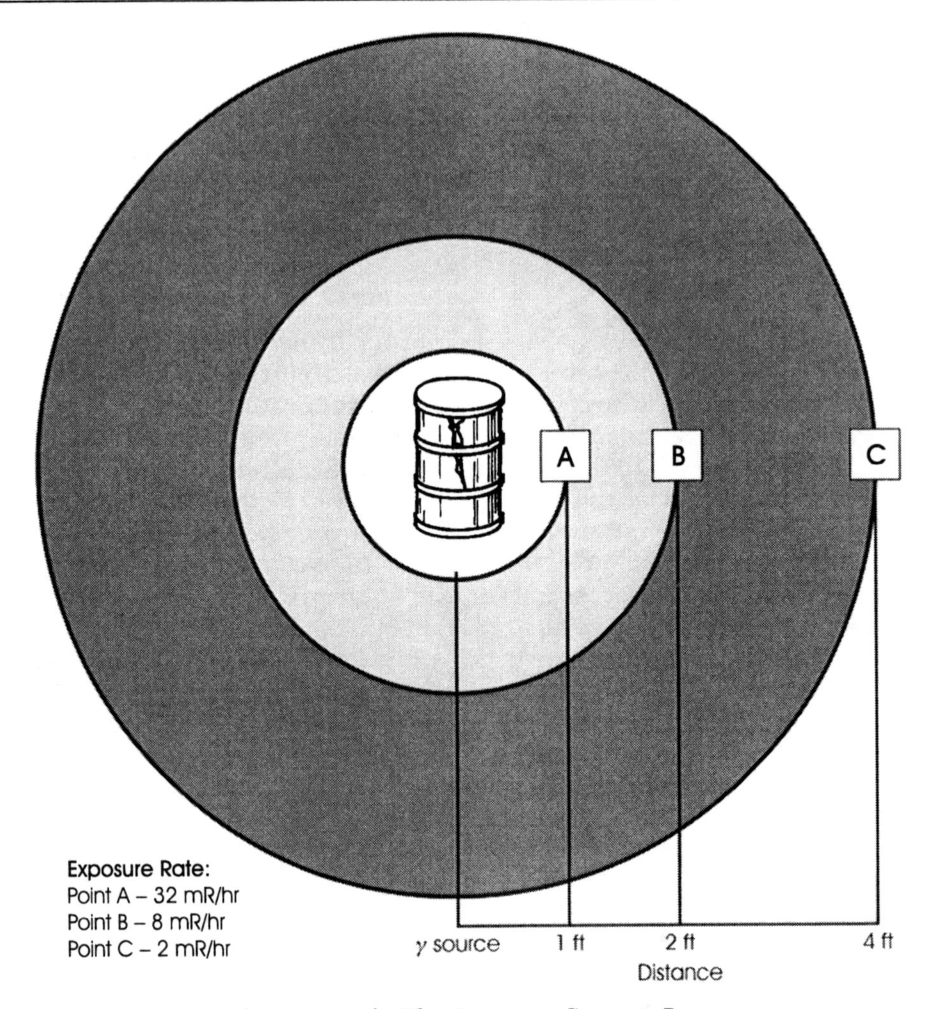

Figure 10-5. The Inverse Square Law

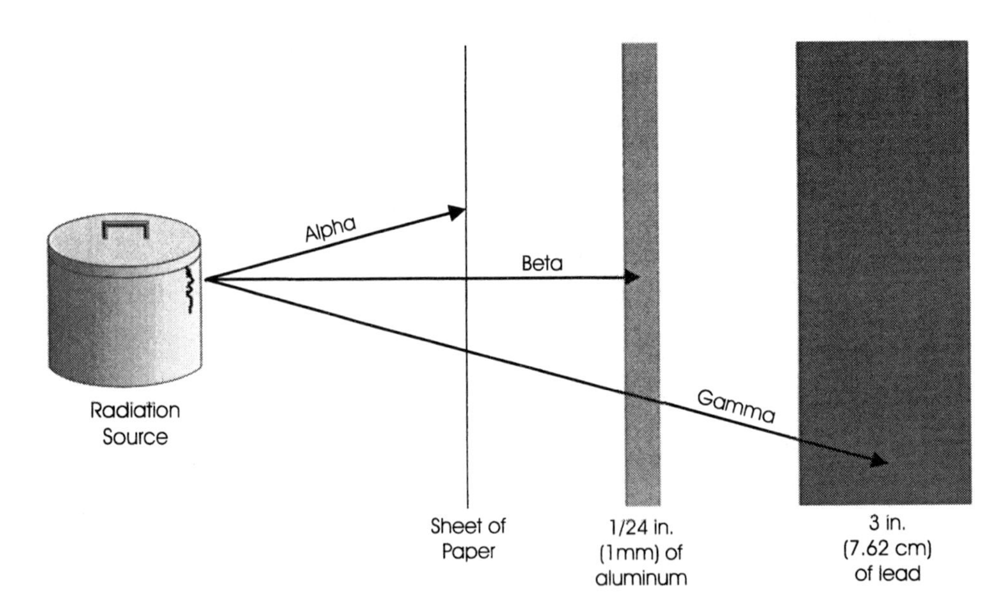

Figure 10-6. Radioactive Shielding

PLACARD AND LABEL REQUIREMENTS

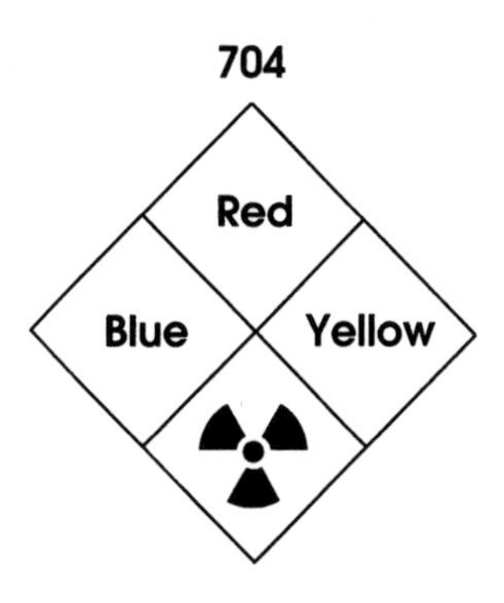

Radioactive materials are identified by the distinctive warning labels and placards displayed on containers, packages, and transport carriers. One distinctive marking is the black or magenta "propeller" on a yellow background. Fixed storage facilities may also display a 704 placard or other markings, indicating a radioactive material's presence. These facilities may have warning symbols at strategic locations to identify radiation hazards.

Labels

All packages of radioactive material must bear two identifying warning labels on opposite sides of the outer package. The two exceptions to this rule are packages that contain limited quantities or low specific activity (LSA) materials that present no hazard.

There are three different labels for packages of radioactive materials. The correct label is usually determined by the external radiation level or by the type and quantity (curies) of radionuclide (contents). Package labels must specify the radionuclide and quantity. In addition, Yellow II and Yellow III labels indicate the transport index. The ***transport index*** is equal to the maximum radiation level (in millirem per hour) three feet from the package, or the degree of nuclear safety control required for packages containing material that has the capability of undergoing fission (fissile material). See 49 CFR 173.403(bb) for additional information.

RADIOACTIVE WHITE I

A Radioactive White I label is a white diamond shaped label with black printing and a propeller symbol. One red vertical bar appears in the bottom half of the diamond. The criteria for use of this label are a maximum accumulated dose rate of less than 0.5 millirem per hour (mrem/hr) at the surface of the package and a zero dose rate at three feet from its surface. The white background indicates a minimal radiation hazard. **In the event of container failure, much higher levels of radiation may be present because of loss of shielding.** See 49 CFR 172.403(c) for further details.

RADIOACTIVE YELLOW II

A Radioactive Yellow II label is yellow in the upper half of the diamond symbol and white in the lower half. The lettering, symbol, and border are again black. There are two red vertical bars in the bottom half of the diamond. The label is employed when the radiation level is from 0.5 mrem/hr up to 50 mrem/hr at the surface and 1 mrem/hr or less at three feet from the package.

RADIOACTIVE YELLOW III

A Radioactive Yellow III label is identical to the Radioactive Yellow II label, except it has three vertical red bars rather than two. This label is used when the surface radiation is above 50 mrem/hr up to 200 mrem/hr, and the dose rate three feet from the surface of the package is above 1 mrem/hr up to 10 mrem/hr. This label is essential on all packages carrying fissile Class III radioactivities, regardless of their dose rate. If a package bears this label, the rail or highway vehicle in which it is carried must be placarded, regardless of the amount being carried.

Notes

Radioactive Package Label Requirements

Label	Package Radiation Level
Radioactive White I	Almost no radiation; 0.5 mrem/hr maximum on surface.
Radioactive Yellow II	Low radiation levels; above 0.5 mrem/hr to 50 mrem/hr maximum on surface, 1 mrem/hr maximum at three feet.
Radioactive Yellow III	Higher radiation levels; above 50 mrem/hr to 200 mrem/hr maximum on surface, above 1 mrem/hr to 10 mrem/hr maximum at three feet. Also required for fissile Class III or large quantity shipments regardless of radiation level.

Note: If the dose rate exceeds 200 mrem/hr, or 10 mrem/hr at three feet from the surface of the package, the radioactive material may not be offered for transport in anything but exclusive use vehicles.

Figure 10-7

"RESIDUE" LABEL

Reusable shipping containers are frequently transported empty but may be internally contaminated. When in transit, these containers must bear the "Residue" label. Anticipate that residual materials may be present in empty containers. Such residues should be handled with care.

Placards

DOT regulations require that radioactive placards be displayed on vehicles transporting one or more packages bearing Radioactive Yellow III labels, even if in Type A packages (see below). Low specific activity (LSA) radioactive materials transported as full loads or in exclusive use vehicles must be placarded. LSA packages do not require placards, but when they are shipped in full loads, the carrier must be marked with "Radioactive Materials LSA." All four sides of the carrier must be placarded.

In addition, radioactive materials that have other hazardous properties may be required to display other identifying labels and placards. For example, uranium hexafluoride fissile that contains more than 1.0% U^{235}, and uranium hexafluoride, an LSA material that contains 1.0% or less of U^{235}, are required to be labeled "Radioactive" and "Corrosive."

SHIPPING CONTAINERS

Notes

There are two packaging classifications for radioactive materials: Type A and Type B. Curie limits for each package type are based on the relative hazards of the specific radioactive materials being transported.

Type A Containers

All packaging is designed to prevent radiation release and damage to the radiation shielding. Type A packaging is designed to withstand the stress of transit under normal non-accident conditions. Because Type A packages contain small quantities or LSA materials, accidents generally will not result in serious radiation hazards. Therefore, such packaging must withstand only moderate stress, such as heat, cold, reduced air pressure, vibration, impact, water, drop, penetration, and compression. Low level radioactive shipping containers can be fiberboard or cardboard boxes, wooden boxes, or steel drums. In addition, there are special Type A containers for transporting large quantities of LSA material or low level radioactive waste.

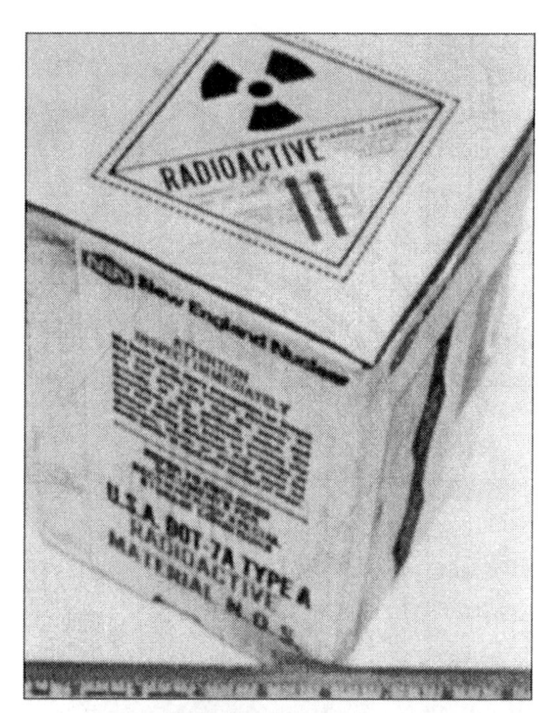

Right: **Type A container for medical isotopes.**
Below and bottom right: **Type A storage drums.**

Notes

Type A Container for Uranium Hexafluoride

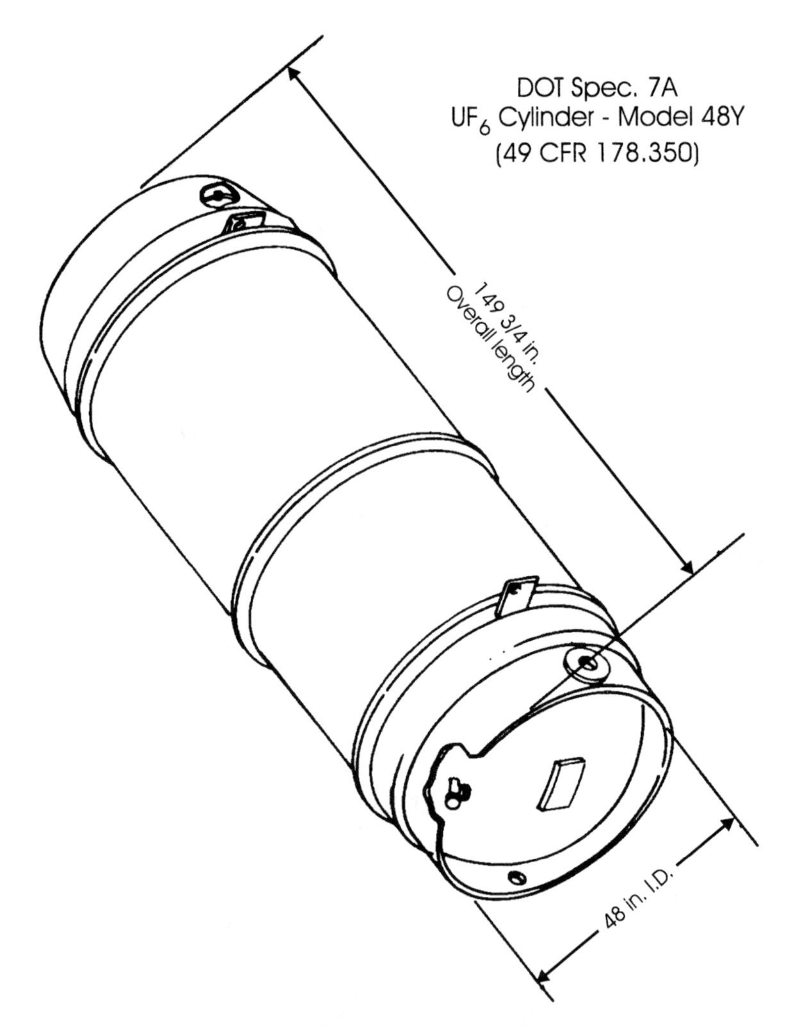

Figure 10-8. Diagram of Type A Container Shown in Photo

Notes

Low Level Contaminated Material

Low Level Radioactive Waste

Notes

Type B Containers

Type B packages are designed for transport of high level radiation sources. Since the potential hazard from damage to this type of container is greater than for a Type A container, there are additional structural design requirements. Type B packaging is designed to withstand serious accident damage tests with a limited loss of shielding and no leakage of radioactivity. The accident damage tests for Type B containers include:

- 30 foot free drop onto an unyielding surface
- puncture test that involves a free drop of 40 inches onto a 6 inch diameter steel pin
- thermal exposure at 1,475°F (802°C) for 30 minutes
- water immersion for eight hours for all fissile materials

Actual accident tests with spent fuel casks were conducted by Sandia Laboratories. The accident scenarios were as follows:

- crash of a tractor-trailer rig carrying a spent nuclear fuel cask into a massive concrete barrier at 84 miles per hour
- high speed (81 mph) impact of a locomotive into a tractor-trailer mounted with a fuel cask
- impact at 81 mph of a railroad car carrying a spent nuclear fuel cask, followed by exposure to a fire

These full scale tests showed that the spent fuel casks are extremely rugged and capable of surviving the worst conceivable accidents. Modern casks are constructed to more rigid requirements and can be expected to survive equally well.

Spent fuel cask prior to test

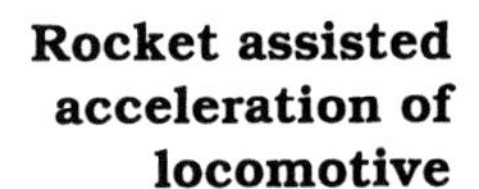

Rocket assisted acceleration of locomotive

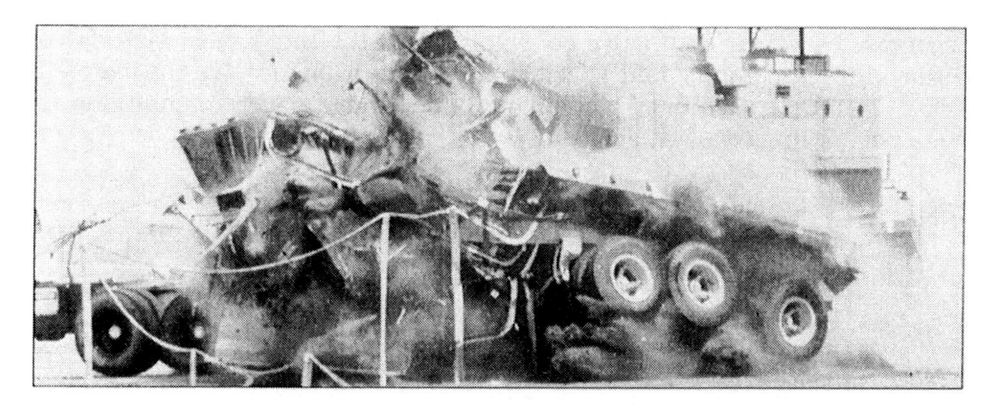

Impact of locomotive and tractor-trailer rig

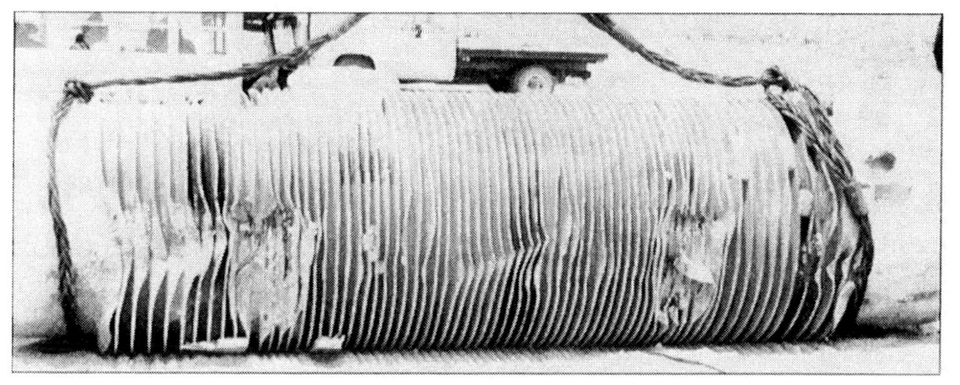

Spent fuel cask after test

Quantities of radioactive material that exceed the limits for small Type B packages are termed large quantities. These shipments are limited to a few high-integrity Type B packages designed with the approval of the DOT, the Department of Energy (DOE), and the Nuclear Regulatory Commission (NRC). Stronger radioactive sources or hazardous waste shipping containers can be lead, steel, or concrete casks. Type B containers are easily

Large Container for Type B Materials

Notes

recognized by their distinctive construction. If the radioactive material is fissile, the shipper must take special precautions to ensure that the material will not undergo a nuclear chain reaction under normal conditions of transport or in case of an accident.

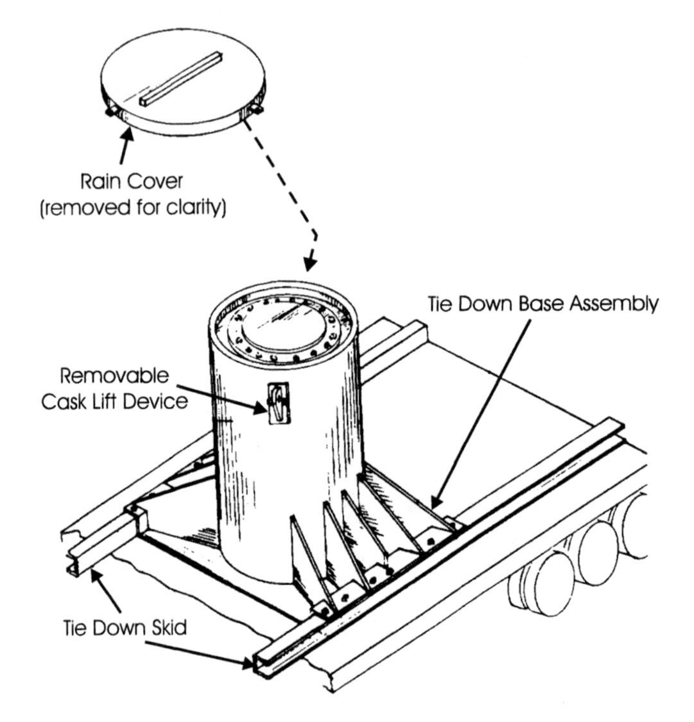

Figure 10-9. Outer View of a Type B Large Container

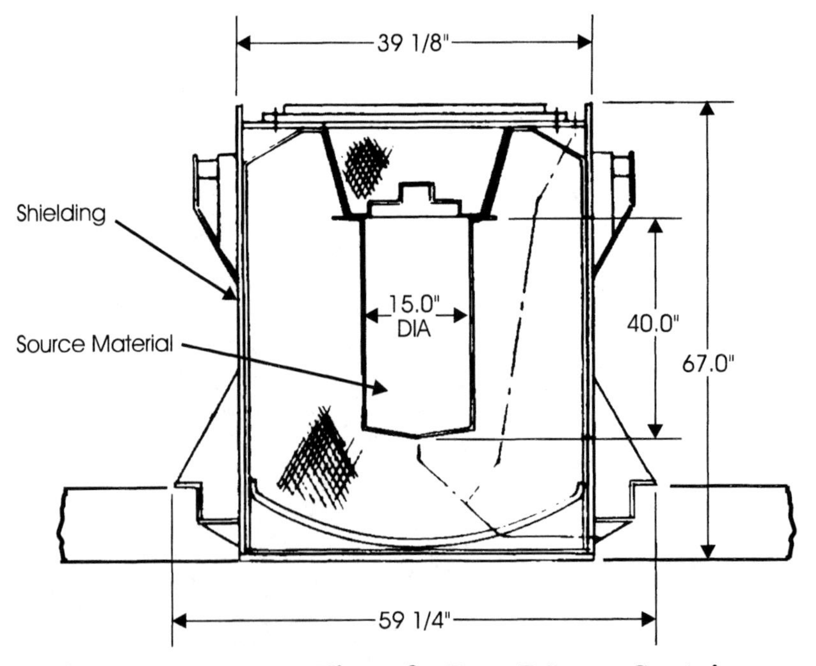

Figure 10-10. Inner View of a Type B Large Container

FIXED SITE STORAGE VESSELS

Notes

Emergency responders should be aware of the locations in the community that use and store radioactive materials. Get to know local radioactive material storage sites in order to increase safety in the event of an emergency operation. Don't forget university campuses, hospitals, medical research facilities, and companies that use industrial radiography equipment.

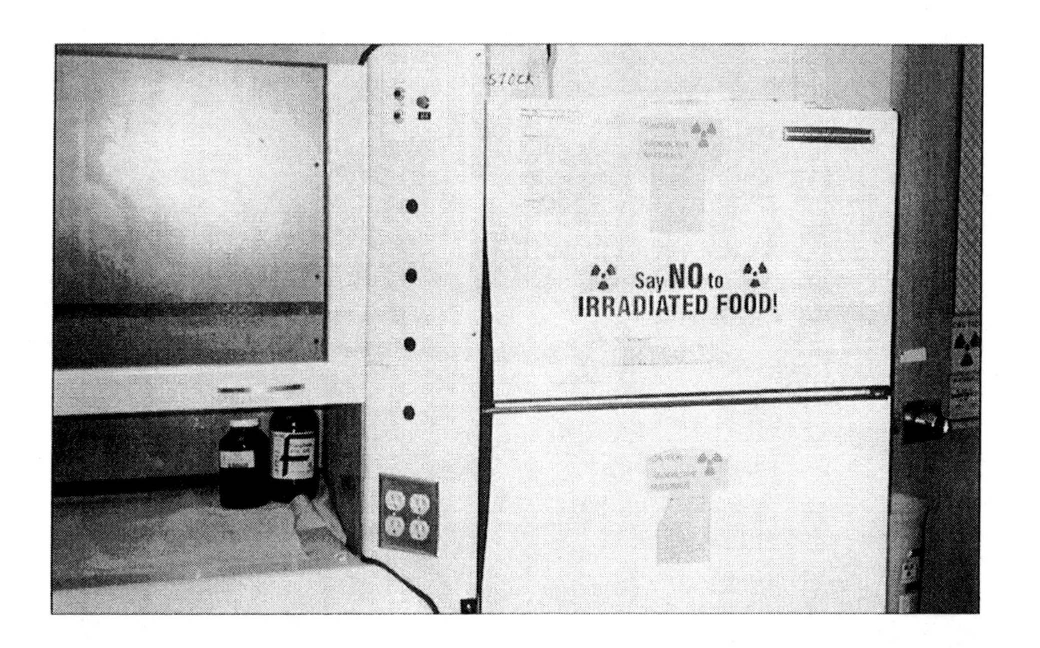

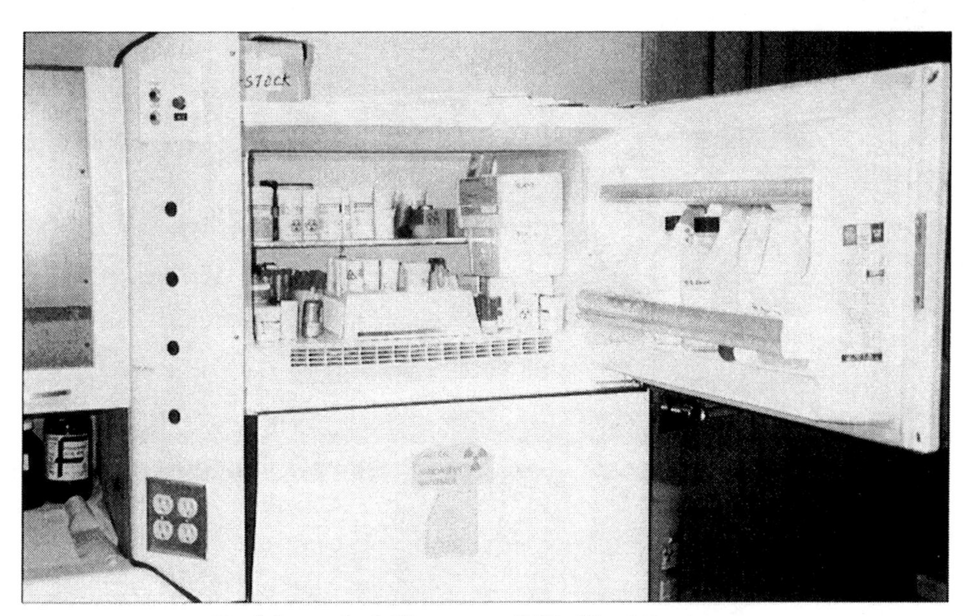

Refrigerated Storage of Medical Isotopes

Industrial Radiography Equipment

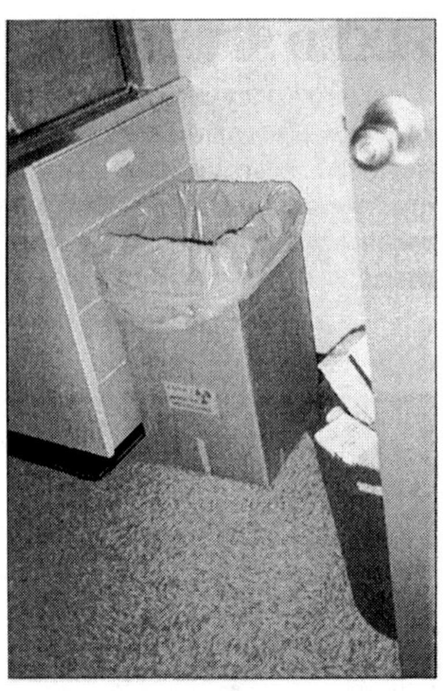

Waste Receptacle for LSA Waste With Short Half-Life

Fixed Site Storage Using Lead Bricks for Shielding

Notes

Drum Storage Using Concrete Bricks for Shielding

A gamma source storage site at a large university that uses a remote retrieval system.

Notes

SPECIAL NUCLEAR MATERIALS

Special nuclear materials are used in manufacturing reactor fuel and nuclear weapons. Reactor fuels (both new and spent) and nuclear weapons components are commonly transported. Special nuclear materials include:

- deuterium
- enriched lithium
- enriched uranium-235
- neptunium-237
- plutonium
- tritium
- uranium-233
- other materials so classified by the Department of Energy

Most special nuclear materials present a greater radiological hazard than source materials because of a higher specific activity and radiotoxicity. Many times, special nuclear materials are transported with armed escorts in separate vehicles. If there is an accident during transport, the escort provides valuable assistance and information about potential hazards. During transport of military classified weapons, the escort may isolate the accident site, provide no information to responders, and keep non-military personnel away from the site. Use discretion in such cases, as military escorts are armed, highly trained, and under strict orders to maintain site security.

SOURCE MATERIALS

The most commonly transported source materials are uranium ore, concentrates and refined products such as U_3O_8 (commonly known as yellow cake), and unenriched uranium hexafluoride (UF_6). Source material includes:

- natural uranium
- thorium

Breached Drums Leaking Yellow Cake

- other material specified by the Department of Energy as material that can be used in the production of special nuclear materials

Notes

NUCLEAR WEAPONS

Nuclear weapons pose an extra threat because they consist of uranium highly enriched with uranium-235 or plutonium, high explosives, and detonating devices. A nuclear weapon's two most hazardous components are the conventional non-nuclear high explosives and plutonium.

Conventional high explosives constitute the primary hazard of accidents involving nuclear weapons.

Most nuclear weapons contain various amounts of conventional high explosives. These high explosives constitute the primary hazard of accidents involving nuclear weapons. Such accidents or fires should be treated like similar conventional high explosives accidents. **If a nuclear weapon is on fire, the high explosives may ignite and detonate.**

The major radiological danger is the release of plutonium. Plutonium is a heavy metal which looks like stainless steel when first processed but rapidly oxidizes to a characteristic brownish-black color. When involved in a fire, metallic plutonium may produce radioactive oxide particles that pose serious hazards if inhaled or ingested. If the high explosive component of a nuclear weapon detonates, it may pulverize plutonium into very small particles and contaminate a large geographic area. Because plutonium emits alpha particles, it does not pose a radiation hazard outside the body. As stated earlier, alpha particles have a short range and cannot penetrate the skin. However, alpha particle contamination can occur if plutonium is inhaled or otherwise absorbed into the body, such as through ingestion or shrapnel wounds.

RADIOACTIVE PYROPHORIC METALS

Uranium

URANIUM (U)
UN 2979
Radioactive
Spontaneously
Combustible

Uranium is a naturally occurring radioactive element. It is found in isotopic forms, including U^{238} (the most abundant), U^{235}, U^{234}, and U^{233}. These isotopes primarily emit alpha radiation and, occasionally, some gamma and beta rays. Uranium exposure can cause subsequent health risks.

Finely divided uranium may ignite spontaneously at temperatures less than 572°F (300°C). It can be stored in an inert atmosphere such as helium. Massive amounts of uranium (1 kilogram or more) are difficult to ignite and usually are not a pyrophoric hazard. Beware of "critical mass" problems with enriched uranium, such as U^{235}. ***Critical mass*** is the smallest mass of fissionable material that can support chain reactions. For U^{235} the critical mass is about 33 pounds.

Uranium is a source of fissionable isotopes, plutonium, and radium salts. It is also used as a catalyst in gas manufacturing. Its melting point is approximately 2,066°F (1,130°C).

Fire Extinguishing Guidelines

- Contact local, state, or DOE radiological response teams.
- Carbon dioxide, soda acid, and foam fire extinguishers are not recommended.

Notes

- Do not flood large fires with water because of the danger of spreading the radioactivity.
- Use sodium chloride based powders and dry magnesium oxide powder.
- Barium, graphite chips, and asbestos blankets can be used to extinguish uranium fires.
- Totally immerse burning uranium in water only if the evolved hydrogen gas can be adequately dissipated.
- Let the fire burn, if possible.

Plutonium

PLUTONIUM (Pu)
UN 9170
Radioactive
Spontaneously
Combustible

Plutonium is an artificially produced radioactive element. Several isotopes exist, including Pu^{238} and Pu^{239}. Pu^{238} and Pu^{239} emit alpha radiation and some gamma radiation. If plutonium enters the body, particularly the lungs, it can cause serious health problems.

Finely divided plutonium turnings, filings, and powders are a greater pyrophoric hazard than large amounts of 500 grams or more. Plutonium may spontaneously ignite in air. Burning plutonium explodes in the presence of halogenated hydrocarbons. Plutonium is used in nuclear reactors. Its melting point is approximately 3,353°F (1,845°C). The critical mass of Pu^{239} is about 10 pounds.

Fire Extinguishing Guidelines

- Contact local, state, or DOE radiological response teams.
- Lead powder, iron powder, copper powder, foam, plutonium dioxide, and halogenated extinguishing agents are not considered effective.
- Do not apply water directly to burning plutonium metal. Use water to control associated combustible fire (non-plutonium).
- Use sodium chloride based powders and dry magnesium oxide powder.

Thorium

THORIUM (Th)
UN 2975
Radioactive
Spontaneously
Combustible

Thorium is a naturally occurring radioactive element. Thorium-232 is a common isotope. It primarily emits alpha and gamma radiation. Thorium's pyrophoric properties are similar to those of uranium. The danger of thorium and other radioactive metals is the possibility of internal radiation due to inhalation, ingestion, or skin absorption. Thorium is used to manufacture incandescent mantles, photoelectric cells, X ray tubes, sunlamps, and certain alloys. Its melting point is approximately 3,353°F (1,845°C).

Fire Extinguishing Guidelines

- Contact local, state, or DOE radiological response teams.
- Carbon dioxide and foam fire extinguishers are not recommended.
- Use sodium chloride powders, graphite, soda ash, and magnesium oxide.
- Flood large fires with water or let burn, if possible.

A GUIDE TO EXTINGUISHING AGENTS FOR FIGHTING RADIOACTIVE METAL FIRES

Radioactive Metal	Magnesium Oxide Powder	Sodium Chloride Based Powder (e.g., Met-L-X®)	Carbon Dioxide	Water	Foam	Explosive Hazard
Plutonium*	Yes	Yes	Yes, but only for large quantities of solid metal	No	No	Moderate with water in closed areas
Thorium	Yes	Yes	Yes, but only for large quantities of solid metal	Yes, but only for large quantities of solid, non-burning metal	No	Moderate with water in closed areas; powder may ignite spontaneously in air
Uranium*	Yes	Yes, but only for large quantities of solid, cool, non-burning metal	Yes, but only for large quantities of solid, cool, non-burning metal	No	No	Moderate with water in closed areas

KEY: **YES** means that the extinguishing substance is effective in most cases. The chemicals mentioned above should be used on small fires only. Do not expect any chemical agent to extinguish large metal fires.

NO means that the extinguishing substance has undesirable or hazardous properties.

* Beware of "critical mass" problems with enriched uranium (U^{235}) and plutonium (Pu^{239}). Critical mass is the smallest mass of fissionable material that can support chain reactions.

Figure 10-11

Notes

A radioactive substance may also be corrosive, flammable, or toxic.

MULTIPLE HAZARD SITUATIONS

Other hazardous substances may be present at a radioactive incident besides radioactive material. It is quite possible that these other substances may pose greater hazards to emergency response personnel than the radioactive material. The hazards may come through direct exposure or through the interaction of the materials with each other and/or with the radioactive material. In addition, the radioactive substance may also be corrosive, flammable, toxic, or a member of any of the other hazard classes. Emergency service personnel must make note of other hazardous materials present at the incident and must take necessary precautions during response operations. Removing uncontaminated and undamaged packages of radioactive materials will greatly reduce the risk of fire or chemical contamination.

Several multiple hazard situations deserve special mention. One is the combination of radioactive materials with **flammables** or **explosives**. Flammables will intensify a fire and will cause more radioactive product to be released. Radioactive materials are not permitted to be shipped with Division 1.1, 1.2, or 1.3 explosives, but they may be shipped with lower hazard explosives. During a fire involving radioactives and explosives, establish an evacuation zone of at least 2,500 feet and clear personnel from the area.

Many radioactive materials are shipped as a **corrosive** material. Of particular note is uranium hexafluoride (UF_6). Breached containers of this compound can release highly toxic hydrogen fluoride gas. This toxic gas is as hazardous as the uranium radiation. Emergency personnel should know that fumes, smoke, and irritating or noxious odors can be very dangerous. Shipments of uranium hexafluoride must bear both "Radioactive" and "Corrosive" labels and placards.

As previously discussed, pyrophoric metals such as uranium, thorium, and plutonium may **spontaneously ignite** and burn when they are in the form of finely divided metal. The residue will release airborne radioactive material. Responders should watch for smoke from packages of these materials.

MONITORING RADIATION INCIDENTS

The complete and accurate evaluation of a radiological accident requires appropriate monitoring equipment. Radiation monitoring devices that are most likely to be available to emergency responders are those provided for emergency management. These include two survey meters, the CD V-700 Geiger-Muller (GM) counter and the CD V-715 ionization chamber. Also included are the CD V-138 and CD V-742 pocket chamber dosimeters with the CD V-750 charging unit. These instruments detect and measure gamma radiation. The CD V-700 can also detect some types of beta radiation.

Be aware of the limitations of using emergency management issued survey meters.

Be aware of the limitations of using emergency management issued survey meters:

- These monitors were designed for nuclear attack emergencies, which pose radiological hazards much different from those of transportation or fixed site accidents.

Notes

- **None of the emergency management instruments can detect or measure alpha or low energy beta radiation.** As a result, emergency responders often do not have devices to monitor the types of radiation they may encounter. It is best to supplement these instruments with units that measure alpha and low level beta activity.
- Many emergency management meters have not been properly maintained. The calibration and batteries of **all** meters must be checked monthly.

CD V-700 GM Survey Meter

The CD V-700 GM survey meter is a low range (0.5–50 mR/hr) instrument for roughly measuring gamma ray exposure rates and detecting beta radiation. The probe has a closable shield to discriminate between gamma and beta radiation. When the shield is open, both beta and gamma radiation are detected. When the shield is closed, the beta radiation is blocked and only gamma radiation is detected. Headphones permit rapid surface monitoring without having to watch the meter face.

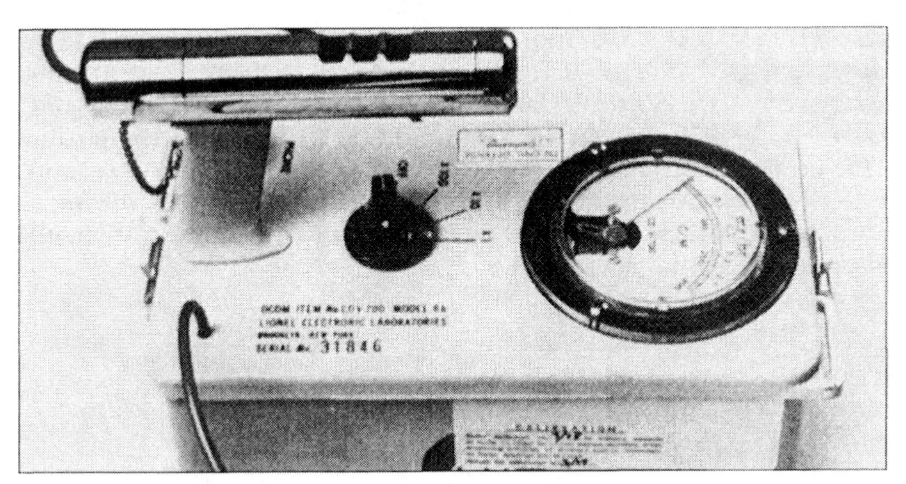

CD V-700 GM Low Range Survey Meter

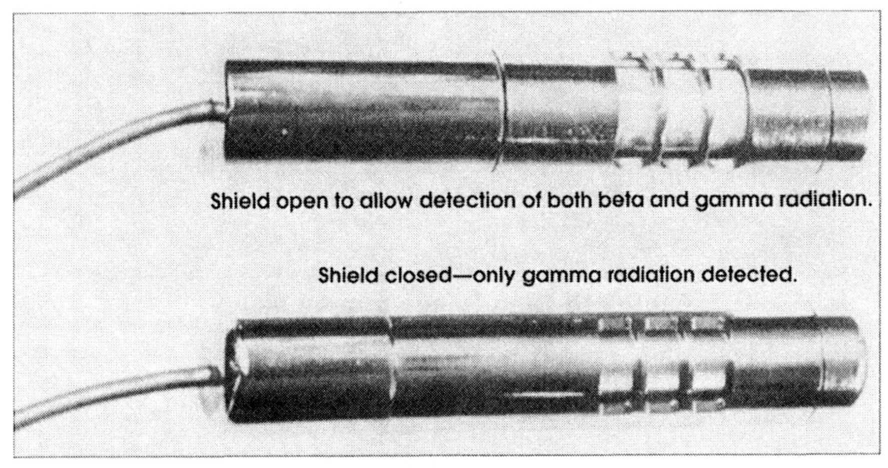

CD V-700 Probe

Notes

Caution: The CD V-700 may "saturate" at radiation exposure rates above 1 R/hr, and the meter reading will not be accurate. A higher range instrument should be used whenever radiation exposure levels exceed 50 mR/hr. Fifty mR/hr is the full scale reading on the least sensitive scale.

CD V-715 Survey Meter

The CD V-715, an ionization chamber survey meter, is a high range instrument for measuring gamma ray exposure rates. It cannot detect beta rays. This meter has two controls. The first is the selector switch which has seven positions:

- circuit check
- off
- zero
- x100 (times 100)
- x10
- x1
- x0.1

At the x1 position, the measured exposure rate is read directly from the meter. At the x0.1, x10, and x100 positions, the meter readings are multiplied by a factor of 0.1, 10, or 100, respectively. The maximum measurements are 500 mR/hr at the x0.1 position, 5 R/hr at the x1 position, 50 R/hr at the x10 position, and 500 R/hr at the x100 position.

The second control on the meter is used to adjust the meter reading to zero during the operational check and to adjust for zero drift during long periods of operation. Without proper zero adjustment, the instrument readings may have large errors.

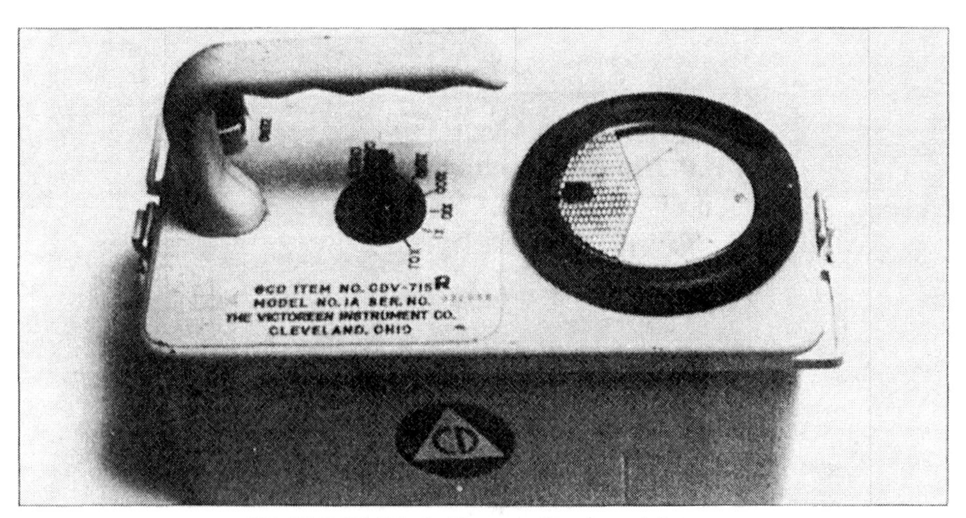

CD V-715 High Range Survey Meter

Pocket Chamber Dosimeters

Notes

The CD V-138 and CD V-742 pocket ionization chamber dosimeters monitor the total gamma radiation to which the wearer is exposed. The CD V-138 monitors relatively low levels of exposure and has a maximum scale reading of 200 mR. The CD V-742 ranges up to 200 R (200,000 mR) and monitors high levels of exposure. Both dosimeters should be worn by emergency response personnel. Before each use, the chambers must be reset or zeroed, using the CD V-750 dosimeter charger. After this initial setting, the wearer can read the dosimeter by looking through it from the pocket clip end. Total radiation exposure is indicated wherever the cross hairs lie on the numbered scale. **Both the CD V-138 and CD V-742 should be charged to zero before each use and every three months to prevent false readings.**

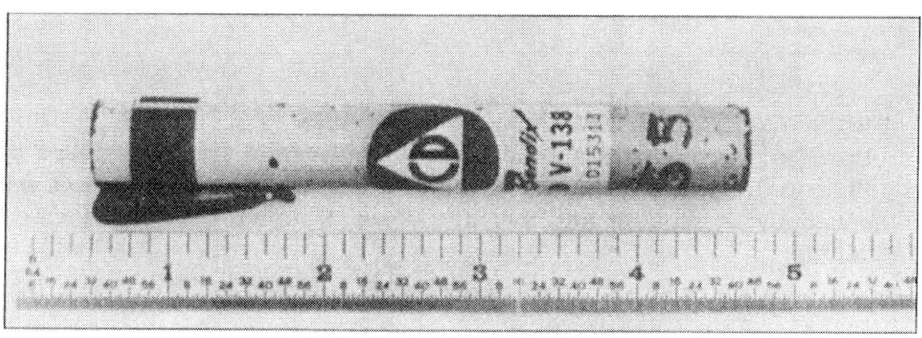

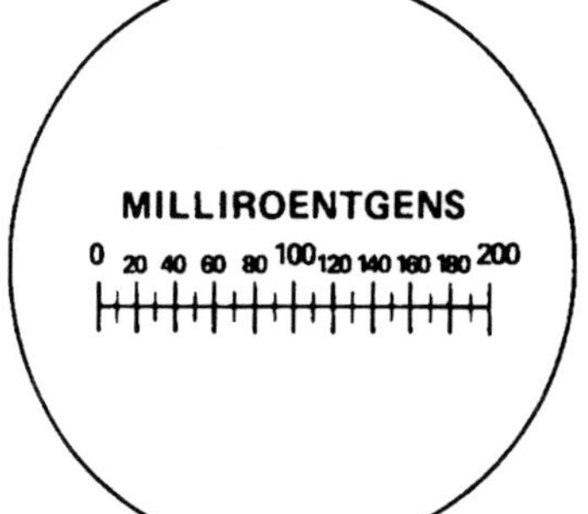

CD V-138 Pocket Dosimeter and Scale

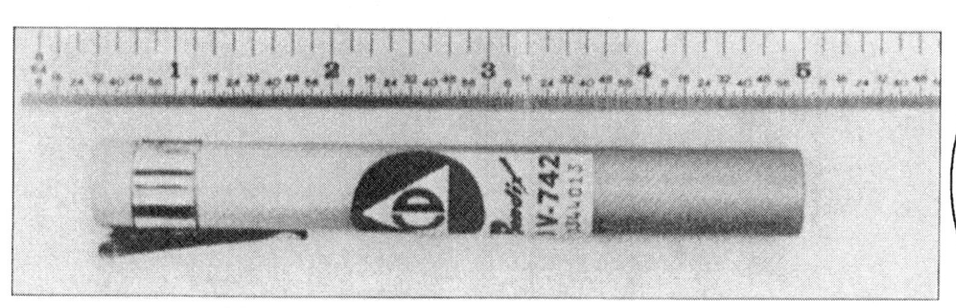

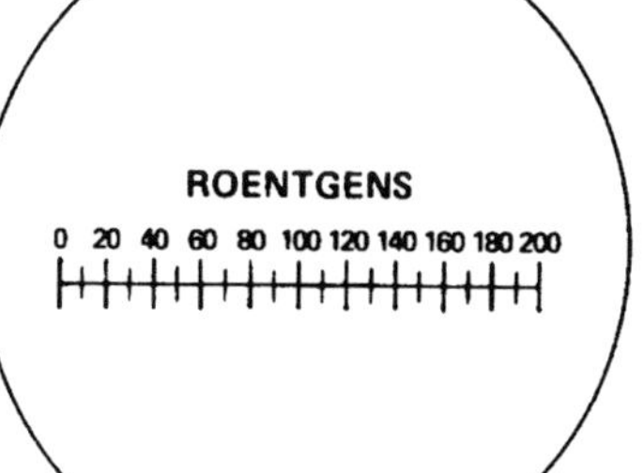

CD V-742 Pocket Dosimeter and Scale

Notes

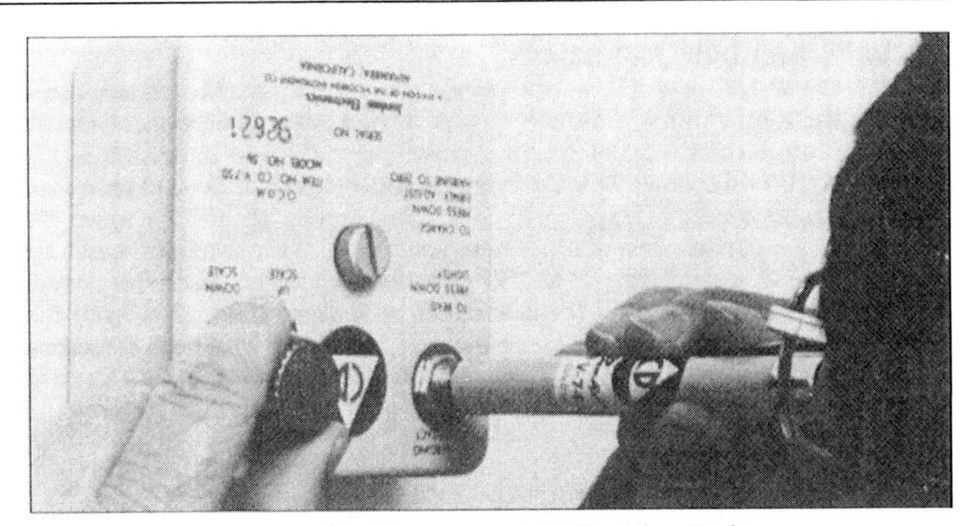

CD V-750 Charger and Calibration Unit

Commercial Radiation Detection Devices

Emergency personnel should always understand the capabilities of monitoring equipment. Several commercial radiation detection devices are also available to monitor and measure doses of alpha, beta, and gamma radiation.

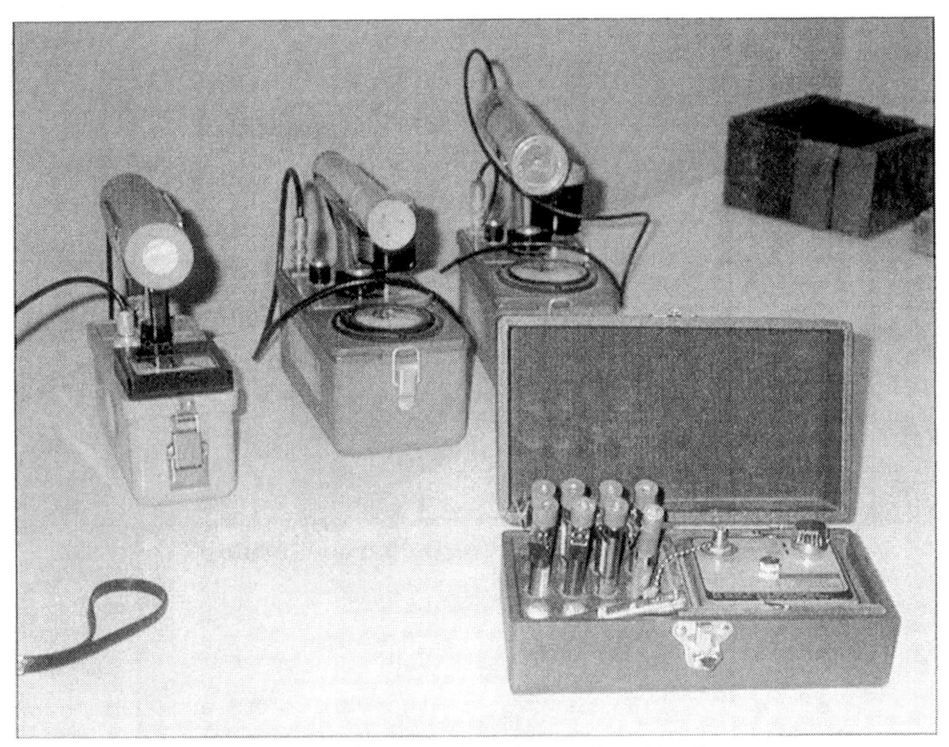

Commercial Radiation Detection Devices

Notes

Be familiar with the agencies in your area that may be a resource during a low level radiation emergency. These agencies may have trained personnel, low level monitoring devices, and emergency plans to deal with this kind of emergency. Possible resource agencies include:

- hospitals
- local companies
- private contractors
- manufacturers of the products
- state agencies
- state police
- universities
- laboratories

KEY POINTS TO REVIEW

- Radioactive materials are a vital part of modern society.
- The most common types of ionizing radiation are alpha, beta, and gamma.
- Understand the physical properties of radioactive materials.
- Understand the measurement units for radioactive materials.
- Be aware of the injuries caused by radioactive exposure.
- Know the three ways to minimize exposure.
- Understand placard and label requirements.
- Be able to recognize the common radioactive shipping containers.
- Understand radiation monitoring and its limitations.

RESPONDING TO RADIOACTIVE MATERIAL EMERGENCIES

Responding to a radioactive emergency presents unique challenges. The first challenge is recognizing and identifying the radioactive source, as radioactive materials do not have easily distinguishable physical and chemical characteristics. Therefore, the danger is often invisible, in the form of airborne particulates and gamma radiation.

The second challenge is monitoring. Different devices monitor different types of radiation. Know the capabilities and limitations of monitoring equipment. Successful mitigation of a radioactive emergency depends upon:

- recognizing the hazard
- monitoring the hazard
- the amount of radiation
- the experience level and knowledge of the responder

Notes

HAZ-MAT — HAZARD IDENTIFICATION

Preplanning

Radiation accidents can occur at fixed sites and during transport. Examples of possible accident sites include:

- biological firms
- construction sites
- hospitals
- industrial labs
- industrial plants
- medical offices
- military installations
- nuclear power plants
- transport routes—land, sea, air
- universities

Emergency responders must identify locations where radioactive materials are used and stored. Be aware of the types and quantities of materials; fire, accident, or health hazard potential; and evacuation requirements.

Use placards and labels to identify radioactives.

The same information should be compiled for commonly transported radioactive materials. The most commonly used transportation routes for radioactive materials should be identified.

In an emergency situation, use placards, labels, fixed site facility markings, and container shapes to help identify radioactive materials. Once the material is identified, consider its hazards and answer the following questions:

- Where is the leak/fire?
- What caused the accident?
- Are there any injuries?
- Are the radioactives exposed to fire?
- Can the hazards be accurately identified with the available monitoring equipment?

HAZ-MAT — ACTION PLAN

Notes

An effective action plan is based on accurate information. The action plan for a radioactive emergency must be based on the particular product, the amount of product involved, the amount and type of radiation, and the type of container.

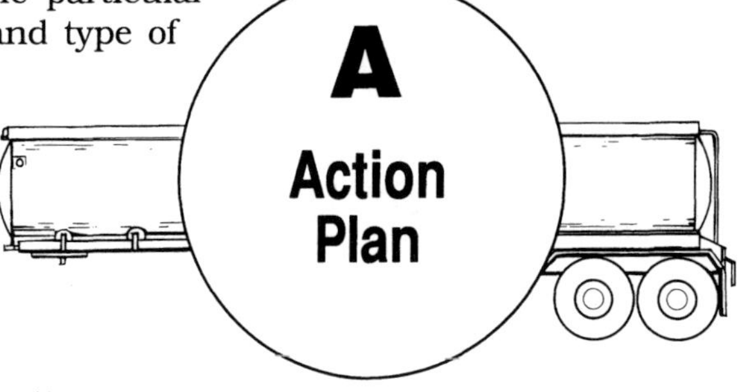

If there is **NO LEAK or FIRE** during an accident involving radioactive materials, proceed as follows:

- Approach uphill and upwind.
- Appropriately monitor with personal dosimeters and radiation detection meters to determine radiation hazards.
- Establish appropriate control and public protective action zones.
- Isolate the incident and surrounding area.
- Request radiological specialists.
- Avoid the material except when performing rescue functions.
- Use "TDS" (time, distance, shielding) to the maximum advantage.
- Proper time management can reduce radiation exposure. Remember: total dose = dose rate x exposure time.
- Always wear full protective equipment, including SCBA.
- Prepare for thorough decontamination of personnel and equipment when alpha and beta sources are involved.

If there is a **LEAK and FIRE**, proceed as above, plus:

- Immediately establish control zones and evacuate to the appropriate distance.
- Extinguish the fire if there is no risk for emergency personnel. If there is a risk, let the fire burn.
- Contain contaminated runoff.
- Prepare to extensively decontaminate personnel and equipment.

HAZ-MAT — ZONING

Zoning is based on the type and amount of product and the severity of the incident. Use appropriate monitoring equipment, check local protocols and other available information sources, and use the following recommendations for appropriate zone distances.

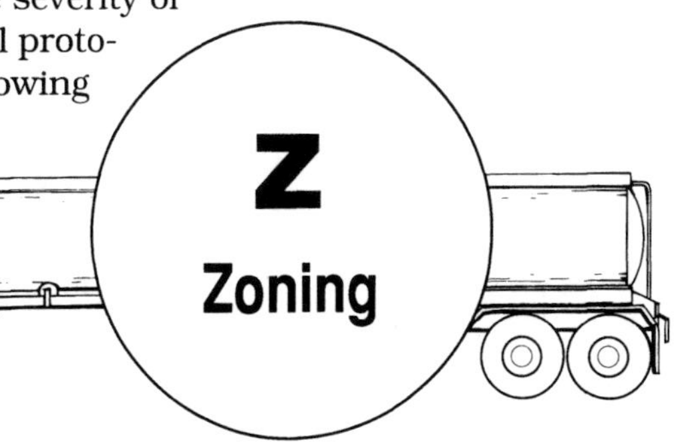

For incidents involving isolated **low specific activity materials:**

- Establish a public protective action distance of 500 feet.
- Establish a hot zone of at least 300 feet or at a distance that limits exposure to 2 mR/hr.
- Establish a warm zone of at least 500 feet.
- Consider establishing a cold zone.

Notes

For incidents involving **hot source materials** or when fire is involved:

- Establish a public protective action distance of 1,500 feet.
- Establish a hot zone of at least 500 feet or at a distance that limits exposure to 2 mR/hr.
- Establish a warm zone of 1,000 feet or at a distance that limits exposure to 1 mR/hr. Be prepared to expand zones if monitoring indicates larger zones are needed.
- Establish an appropriate cold zone.

For incidents involving **nuclear fuels, waste fuels, or weapons which are on fire:**

- Establish a public protective action distance of at least 3,000 feet, especially downwind.
- Establish a hot zone at a distance that limits exposure to 2 mR/hr.
- Do not enter the site. **The risk is too great.**

HAZ-MAT — MANAGING THE INCIDENT

Scene Management

The most effective way to manage a radioactive emergency is to have monitoring equipment that adequately assesses the hazards and to know how to use it. The development of a sound incident management plan is also important. In many cases, adequate equipment and expertise are not available to the first due officer. Therefore, it is essential to:

- preplan strategy for radioactive emergencies
- anticipate needed resources, such as monitoring equipment, and have these resources available
- preplan tactical operational considerations
- use an incident command system

The following are basic radioactive incident considerations:

- assess radiation exposure
- limit exposure time for responders
- keep a safe distance from the incident
- use appropriate shielding techniques
- perform rescue size-up
- recognize multiple hazards during size-up
- confine or isolate the product
- extinguish the fire or allow it to burn
- monitor surroundings with appropriate equipment
- anticipate needed resources
- decontaminate

Preplanning is the key to a successfully managed radioactive emergency. Anticipate the hazards and preplan basic strategy, tactical considerations, and needed resources.

Rescue Considerations

The question of rescue always arises when dealing with radioactive emergencies. If rescue is attempted, emergency personnel should never be exposed to a total dose of more than 25 roentgens. Twenty-five roentgens is the accepted recommended exposure limit for any single life-threatening emergency. In fact, some agencies do not allow personnel to receive more than 5 roentgens per year. Therefore, it is essential to determine the radiation level and estimate the time needed for rescue efforts. It is recommended that personnel be rotated so that no individual receives an unacceptable dose. Remember:

dose rate x exposure time = total dose

Note the following example:

20 R/hr	x	1/4 hr	=	5 R
(dose rate)		(exposure time)		(total dose)

A successful rescue may require multiple teams, each working short periods. Working in such a manner will reduce individual exposure. Always be prudent and remember the three factors that help reduce exposure:

- time
- distance
- shielding

Where possible, use bullhorns or vehicle public address systems to coach people out of the hazard area. Exposing rescue personnel to high levels of radiation is not acceptable.

Trauma is the most likely injury encountered during a radioactive emergency. Unless the source is extremely radioactive, emergency responders probably will not notice the signs and symptoms of radiation sickness. Nevertheless, it is important to know the general care and treatment for radiation exposure.

Notes

During rescue, emergency personnel should never be exposed to a total dose of more than 25 roentgens.

General Treatment for Radiation Exposure

Products .... Cobalt-60, iridium-192, cesium-137, radium-226, uranium-235, radioactive waste.

Containers .... Range from small lead capsules used in the medical field to large lead "pigs" used in transporting waste.

Life Hazard .... Trauma from the incident will most likely be the most life threatening condition. Radiation can pose health problems from external irradiation or exposure, from internal and/or external contamination, and from incorporation into cells or target organs. ***Note:*** Incorporation cannot occur without contamination. Exposure may cause severe gastrointestinal or neurological damage. Patients may also experience a loss of bone marrow function. Products may act as carcinogens. In most cases, the effects of the radiation exposure are delayed for days or longer. Concentrate on the immediate health concerns of the patient.

Notes

Lead Pigs Containing Radioactive Waste

Signs/Symptoms ..... Nausea, vomiting, high fever, and swelling of the nose, mouth, and throat. The patient may be short of breath and have a decreased level of consciousness. The heartbeat may be abnormally fast. Certain products may cause eye irritation and skin burns. Severity of exposure effects depends on the patient's prior health and the degree of exposure.

Basic Life Support .... Wear proper protective clothing and remove patient from the contaminated area. Administer oxygen. In cases of gamma exposure, no further decontamination is necessary. In cases of small quantity releases resulting in alpha or beta contamination, the patient should be decontaminated by knowledgeable individuals in order to reduce the chance of internalization of the contamination. The best practice may be to carefully remove and isolate the patient's clothing and package the patient in a manner that will prevent the spread of contamination. The patient can be decontaminated after arrival and assessment at the emergency department. If high levels of radioactive contamination or other chemical contaminants are present, brush or blot away any visible product and decontaminate with copious amounts of soap and water. Irrigate eyes thoroughly and continue during transport as necessary. Treat injuries as in normal situations.

Notes

Advanced Life Support .... Assure an adequate airway and assist ventilations as necessary. Monitor cardiac rhythm and treat as needed. Assess patient for other injuries and/or medical problems and treat as necessary.

Other Information .... Simple protection and normal treatment procedures are the best care for radiation accident victims. Most symptoms from radioactive product exposure are delayed. An accurate history of the incident and exposure is vital.

HAZ-MAT — ASSISTANCE

Radioactive emergencies usually require the following assistance:

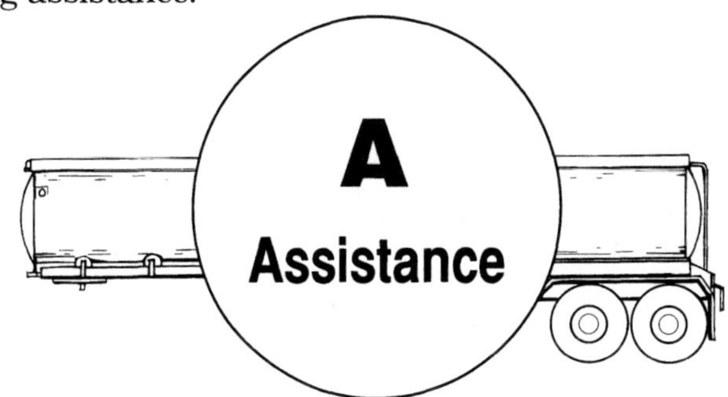

- trained dispatchers
- law enforcement
- specialized response teams
- technical specialists, such as:
  - Nuclear Regulatory Commission (NRC), telephone (301) 415-7000
  - Department of Energy (DOE)
  - health department
  - hazardous materials response team
  - manufacturer of the product
  - private clean-up firms
  - private individuals
  - hospital staff
  - university

Regional coordinating offices for radiological assistance have been set up by the DOE. These regional offices provide advice and assistance for radiological incidents. The DOE has divided the United States into eight regions, each with a Regional Coordinating Office (see Figure 10-12).

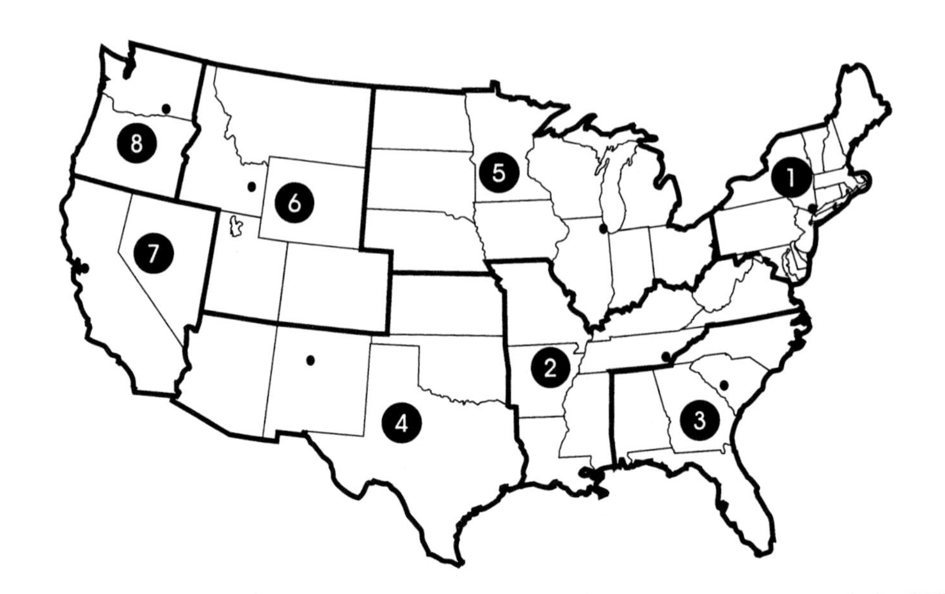

U.S. Department of Energy Regional Coordinating Offices for Radiological Assistance			
	Regional Coordinating Office	**Address**	**Telephone**
1	Brookhaven National Laboratory	Upton, NY 11973	(516) 282-2200
2	Oak Ridge Operations Office	P.O. Box 2001 Oak Ridge, TN 37831	(615) 576-1005 or (615) 525-7885
3	Savannah River Operations Office	P.O. Box 616 Aiken, SC 29808	(803) 725-3333
4	Albuquerque Operations Office	P.O. Box 5400 Albuquerque, NM 87185	(505) 845-4667
5	Chicago Operations Office	9800 S. Cass Avenue Argonne, IL 60439	(708) 252-4800 or (708) 252-5731
6	Idaho Warning Communications Center	785 DOE Place Idaho Falls, ID 83402	(208) 526-1515
7	Oakland Operations Office	1301 Clay Street Oakland, CA 94612	(510) 637-1794
8	Richland Operations Office	P.O. Box 550 Richland, WA 99352	(509) 376-8519

Figure 10-12

HAZ-MAT — TERMINATION

Successfully terminating a radioactive incident is critical to ensuring the long term health of personnel.

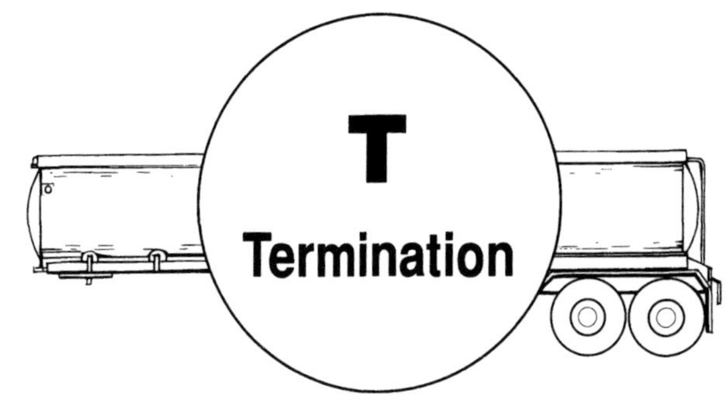

Decontamination

Final decontamination should include cleaning and monitoring all personnel and equipment. Be sure that no one takes any material home. Mild soap and water is a good decontamination solution for these materials. However, consult reference materials, product specialists, and the Agency for Toxic Substances and Disease Registry for the recommended solution.

Rehabilitation and Medical Screening

Use this phase of the incident to evaluate personnel. The evaluation should include:

- rest period
- medical evaluation
- exposure reporting
- initial incident debriefing
- final contamination monitoring

Post-Incident Analysis

Successful mitigation of radioactive emergencies requires interaction among several agencies. The post-incident analysis provides each agency the opportunity to evaluate its own performance. The lessons learned from the analysis can make the next radioactive emergency safer for all involved.

KEY POINTS TO REVIEW

- Understand the physical and chemical characteristics of radioactive materials.
- Preplan the types of incidents likely to be encountered.
- Recognize the limitations of the available monitoring equipment.
- Develop a resource needs list.
- Know how to acquire the needed resources.
- Consult the HAZ-MAT checklist for radioactive materials found in Appendix G.

Chapter 11

CORROSIVE EMERGENCIES

OBJECTIVES

After studying the material in this chapter, you will be able to:

- define corrosives
- describe the physical and chemical properties of corrosive materials
- describe the hazards of acids and bases
- list common acids and bases
- state the placard and label requirements for corrosive materials
- describe common corrosive containers
- describe response strategies for corrosive emergencies, using the acronym HAZ-MAT

UNDERSTANDING CORROSIVES

INTRODUCTION

Emergencies involving corrosive materials are the second most frequent haz-mat incident that responders face. Only flammable liquid emergencies are more common. The types and hazards of corrosive materials are many and varied. Corrosive materials are shipped and handled in all types of containers, including rail tank cars, trailer on flatcars (TOFC), container on flatcars (COFC), tank trucks, box trailers, vans, ships, and barges. Container sizes range from 1 quart glass bottles to large mobile and fixed tanks.

Corrosives are commonly used in metal cleaning and plating, soaps, textiles, paper, pulp, flashlight batteries, car batteries, disinfectants, mining, and many other chemical processes. With such widespread use, it is important to understand the corrosive hazard class.

Notes

DEFINITIONS

The Department of Transportation (DOT) defines a ***corrosive material*** as:

> Any liquid or solid that causes visible destruction or irreversible alterations in human skin tissue at the site of contact, or a liquid that has a severe corrosion rate on steel or aluminum (49 CFR 173.136(a)).

Emergency responders should remember that any material that destroys steel also can destroy living tissue. There are two categories of materials capable of destroying tissue and metal: **acids** and **bases**. Bases are also referred to as alkali or caustic materials. The DOT, however, does not distinguish between acids and bases and includes both under the corrosive heading.

PHYSICAL AND CHEMICAL PROPERTIES OF CORROSIVES

Acids and bases are completely opposite in character, yet both are defined as corrosive. In fact, acids and bases are so different that they can neutralize each other. In 1814, a French chemist named Guy-Lussac concluded that although acids and bases exhibit different chemical behaviors, it was necessary to define them in relation to each other. This concept is still used to understand these two chemical types.

The word acid is derived from the Latin, *acidus*. It means tart or sour and refers to the taste of acetic acid in vinegar or citric acid in citrus fruits. Very simply, ***acids*** are compounds that form hydrogen ions (H^+) when dissolved in water. To be chemically exact, they actually form hydronium ions (H_3O^+).

Alkali is an Arabic word that refers to the ashes of the saltwort and glasswort plants, and their aqueous solutions. As we have seen, alkaline materials are chemically referred to as bases. ***Bases*** are compounds that form hydroxide ions (OH^-) when dissolved in water. The chemical characteristics of such solutions can be called either alkaline, caustic, or basic.

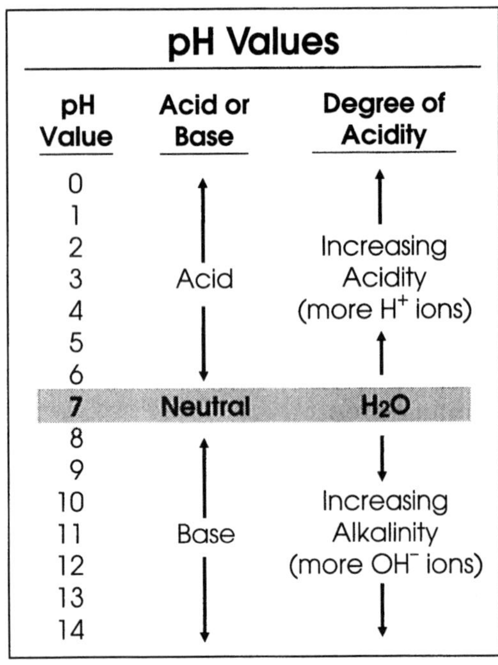

Figure 11-1

pH Value

Emergency responders can determine whether a chemical is an acid or a base, and its relative strength, by using the pH scale. pH describes the acidity or alkalinity of a solution. Solutions having a pH from 0 to 7 are called acids, while solutions having a pH from 7 to 14 are called bases (see Figure 11-1). Pure water has a pH value of 7 since it is neither an acid or a base. The pH scale is a logarithmic scale, which means that a one unit change in pH is a 10-fold change in acidity or alkalinity. Adequately trained personnel can measure pH values by using pH meters, papers, and strips. Figure 11-2 gives the pH values for several common substances and products.

Notes

Approximate pH of Common Substances and Products			
Substance	**pH**	**Product***	**pH**
Human blood	7.4	Potassium hydroxide	13.0
Milk	6.6	Sodium hydroxide	13.0
Tomatoes	4.2	Trisodium phosphate	12.0
Apples	3.1	Sodium carbonate	11.6
Carbonated water	3.0	Ammonium hydroxide	11.1
Vinegar	2.8	Borax	9.2
Lemons	2.3	Sodium bicarbonate	8.4
Ginger ale	2.0 - 4.0	Pure water	7.0
		Boric acid	5.2
		Alum	3.2
		Acetic acid	2.9
		Citric acid	2.2
		Phosphoric acid	1.5
		Sulfuric acid	1.2
		Hydrochloric acid	1.1

* pH value depends on the concentration in water.

Figure 11-2

Strength and Concentration

It is also important to understand the difference between strength and concentration. The two are often misunderstood or confused. ***Strength*** is the percentage of ionization that occurs when a material is mixed with water. It is expressed as weak or strong (see Figure 11-3).

Concentration is the amount of material that is mixed with a specific volume of water. In other words, it is a ratio of material to water. It is expressed in percent of weight or volume. For example, commercial concentrations can vary from 38% hydrochloric acid to 98% sulfuric acid.

When reading a label or shipping papers, be careful not to confuse concentration with strength. It is possible to have a weak acid in high concentration. For example, in low concentrations boric acid is weak enough to be used as an external eye wash. However, boric acid can be found in concentrations of up to 85%. Not only are strength and concentration important in determining the hazards of a chemical, but they are also important for developing plans to contain or neutralize the product.

Notes

Relative Strengths of Acids and Bases

Acids (Decreasing Acid Strength ↓)	
Perchloric acid	$HClO_4$
Sulfuric acid	H_2SO_4
Hydrochloric acid	HCl
Nitric acid	HNO_3
Phosphoric acid	H_3PO_4
Nitrous acid	HNO_2
Hydrofluoric acid	HF
Acetic acid	CH_3COOH
Carbonic acid	H_2CO_3
Hydrocyanic acid	HCN
Boric acid	H_3BO_3

Bases (Decreasing Base Strength ↓)	
Amines	NH_2^-
Hydroxides	OH^-
Carbonates	$CO_3^=$
Ammonia	NH_3

Figure 11-3

Neutralizing Acids and Bases

An acid is neutralized by a base and vice-versa. Hydrogen ions and hydroxide ions chemically react to form neutral water:

$$H^+ + OH^- \rightarrow H_2O + \text{heat}$$

There are three general ways to neutralize acids or bases:

- A strong acid plus a strong base produces a more neutral solution.
- A strong acid plus a weak base usually produces a weaker acid, but can produce a base.
- A weak acid plus a strong base usually produces a weaker base, but can produce an acid.

The product must be positively identified before any action is taken to neutralize a corrosive material. Acids can be neutralized by mixing them with bases such as slaked lime (calcium hydroxide) or soda ash (soda carbonate). A combined base and acid may create a neutral substance (pH = 7), but the reaction may also produce considerable heat and toxic fumes. The violence of the reaction is largely dependent upon the strength and concentration of the materials. Sometimes such a reaction occurs so fast that it appears to be explosive. In any case, the materials will splatter if mixing conditions are not carefully controlled. Be aware that the term "neutralization" does not necessarily mean safe. The degree of neutralization depends on the concentration and quantity of the reacting chemicals.

The term "neutralization" does not necessarily mean safe.

Dilution is usually an acceptable way to reduce the pH level of **small** corrosive spills of one gallon or less. However, most corrosive products require large amounts of water to dilute them to the point where they will be safe to discard. A significant amount of hazardous waste will be generated if dilution is not successful.

HAZARDS OF ACIDS AND BASES

Notes

There are five basic hazards associated with corrosive materials:

- skin or tissue damage
- vapor inhalation
- reactivity with other materials
- flammability
- instability and toxicity

Skin or Tissue Damage

A corrosive's greatest hazard is its ability to destroy living tissues (referred to as chemical burns). Tissue damage begins the instant the chemical agent comes into contact with the tissue. Some of these materials, particularly strong acids and bases, are so corrosive that even brief exposure can severely damage the skin. With other corrosive materials, victims may not realize they have been exposed because the damage doesn't reveal itself until hours later. By that time the burn will be severe and medical treatment will be difficult. Tissue damage will continue until the chemical reaction is stopped. There are four ways to stop the damaging chemical action of a corrosive:

Corrosives destroy human tissue.

- flush or rinse with water
- physically remove the contaminant
- neutralize
- dilute

Of these possibilities, flushing is usually the emergency responder's best choice because the other methods are more difficult to accomplish. Physically removing the corrosive is difficult with liquids and finely divided solids, and this process also tends to miss small portions of the material. Neutralizing should be avoided because it produces heat, and heat may add thermal burns to the victim's chemical burns. Dilution is acceptable if water is in short supply.

Flushing incorporates two of the other methods. It physically removes any material on the tissue and dilutes it as well. To be most effective, flushing must continue for a minimum of 15 minutes. This applies to the eyes as well as to other parts of the body. Of course, flushing is **not** recommended if the corrosive is water reactive.

Flushing is not acceptable when the victim has contacted any dry corrosive, because when mixed with water the dry corrosive creates a corrosive liquid. Brush any dry corrosive from the skin and clothing and flush only if the area of contact can be continuously flooded with a large amount of water.

Vapor/Inhalation Hazard

There are several corrosive materials capable of forming massive vapor clouds. Such situations are difficult for even the best staffed and equipped emergency response agency. In many cases, large scale public protective actions will be necessary. The presence of a vapor cloud increases the possibility of casualties. The most common injuries will include damage to mucous membranes, such as the eyes, throat, and airway. Moist tissues

Notes

(areas of the body most susceptible to sweating), such as armpits, groin, and lower back, also may be affected.

While most corrosive vapor clouds are corrosive, irritating, or toxic, the majority are also water soluble. Water fog streams, therefore, can help knock down, control, and disperse the vapor cloud. However, water must not come in contact with the product because **many vapor producing corrosives are also water reactive**. If contact is made, the amount of vapor and the severity of the incident usually increase.

Reactivity With Other Materials

During any haz-mat incident, materials may mix, forming dangerous reactions. Corrosives are no different! They may react with the shipping container, the metal on the truck or rail car, or the rain puddle in the street. Several complications can occur when corrosives mix with other materials.

CORROSIVES AND METAL

Corrosives cause metal destruction.

When a corrosive material leaks, one possible reaction is metal destruction. This is of particular concern when the exposed metal happens to be the container in which the corrosive is stored or other containers nearby. In such a situation, slight stress can prompt container failure. The stress may be as slight as the material's weight or the disturbance of someone attempting to plug the leak. In either case, it is difficult to determine the level of destruction.

Always use caution when attempting to patch or over-pack a leaky corrosive container. Be aware that hydrogen gas can be produced simultaneously with metal destruction. Hydrogen is an odorless, colorless gas. It is lighter than air and has a flammable range of 4% to 75% by volume in air. Hydrogen gas is especially dangerous if it is produced inside a building, because it is not readily detected and poses a major hazard. Large amounts of hydrogen gas can be trapped in high locations in an enclosure. If an ignition source is present, a tremendous explosion is possible. If hydrogen gas is produced inside a container, it can cause the container to over-pressurize and fail. If the hydrogen is outside and not otherwise confined, most of it will rise and be diluted by the air, reducing the potential of an explosion.

CORROSIVES AND WATER

Acid into water – never water into acid.

Some corrosive materials are water reactive and should never come in contact with water. Generally, though, acid may be poured into water, but **never pour water into acid**. When water comes in contact with acid, it causes the rapid generation of heat (exothermic reaction), turning the water into steam. Expect one or more of the following results when water is mixed improperly with acid:

- violent reaction
- heat generation
- vapor production
- over-pressurization of the container (caused by one of the first three reactions when inside a closed container)

Notes

Mixing water improperly with acid can result in a violent reaction, heat generation, and/or vapor production.

CORROSIVES AND COMBUSTIBLE MATERIALS

During a corrosives incident, it is possible that the corrosive will become mixed with a flammable or combustible material. There are many possible reactions, depending upon the type and quantity of the materials. When a strong acid or base is mixed with a flammable or combustible liquid, heat may be generated in a manner similar to when water is added to an acid. Because of the temperature increase, there is also an increase in vapor. The liquid may be heated enough to reach its flash point or go above it. If enough vapor is produced and if an ignition source is present, a fire will result. The heat generated may also spontaneously ignite liquids with low ignition temperatures.

Some corrosives are also strong oxidizing agents. If an oxidizing material comes in contact with a combustible liquid or solid, the combustible may spontaneously ignite. After ignition, the corrosive will react like an oxidizer—it will greatly intensify the rate of combustion. If the corrosive happens to be nitric acid and the combustible is a cellulose material (wood, paper, cotton, or any plant material), the chemical reaction may produce nitrocellulose. Nitrocellulose is a highly flammable material that can explode. In addition, when nitrocellulose burns it produces toxic gases.

CORROSIVES AND POISONS

If corrosives mix with poisons, the primary concern is the toxic vapors produced by the decomposing poison. The poison vapors may be far more toxic than the corrosive vapors. Also, many poisons are hydrocarbon based, thereby creating fire dangers similar to those outlined earlier.

Flammability

Acids and bases are classified into two basic groups: organic and inorganic. An organic acid or base contains carbon within its compound and is likely to be flammable. An inorganic acid or base has no carbon and is not flammable. Certain inorganic acids and bases, however, can act as oxidizing agents and are capable of igniting combustible materials.

Inorganic corrosives can ignite combustible materials.

Notes

Instability and Toxicity

Some corrosives exhibit instability and toxicity. These materials, particularly the per-acids (perchloric, perfluoric, peracetic) and the organic acids (hydrocyanic, acrylic, picric, methacrylic), can explode, form polymers, and decompose. They are also poisonous. For example, when picric acid becomes crystallized, it can be so sensitive that even the slightest movement can cause detonation. Acrylic and methacrylic acids are monomers (materials that undergo planned chain reactions to form plastics) used in the manufacturing of acrylic plastics. If exposed to heat or sudden shock, they can polymerize. Polymerization is a chemical chain reaction that, when not controlled, closely resembles an explosion. In other words, polymerization produces heat, light, fragments, and shock waves (when the material is in a container) that can easily be mistaken for an explosion.

It is also important to note that some weak acids and bases are hazardous not because of their strength or concentration, but because of their toxicity. For example, hydrocyanic acid and hydrogen sulfide are extremely toxic and flammable, yet they are weak acids.

COMMON INORGANIC ACIDS

Sulfuric Acid

Sulfuric acid is the most widely used industrial chemical in the United States. Sulfuric acid can be found in vehicle batteries, metal plating facilities, and steel mills. It is used to manufacture other acids and many other substances. It has been said that the consumption of sulfuric acid is a measuring gauge for the economic status of a nation. Sulfuric acid is often called battery acid and dipping acid. Another name for sulfuric acid is **oil of vitriol**.

Sulfuric Acid
H_2SO_4

704

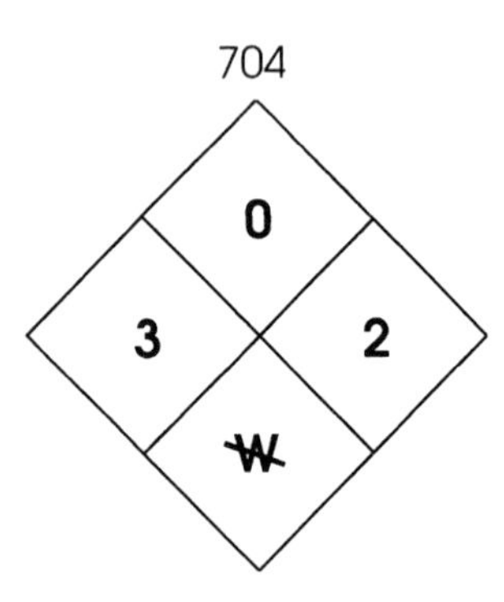

Sulfuric acid is not flammable, but it does pose health risks to emergency response personnel. Primarily these risks include severe chemical burns if the liquid is splashed onto tissue and burns to mucous membranes if the vapor is inhaled. Sulfuric acid differs from most other inorganic acids in that it has a great affinity for water. This means that the acid can even attract water from other compounds. Even small amounts of water that come into contact with sulfuric acid can splatter and boil violently. Sulfuric acid also acts as an oxidizer.

Physical Properties of Concentrated Sulfuric Acid

Concentration in Water	98.33%
Specific Gravity	1.83
Boiling Point	554°F (290°C)
Melting Point	50°F (10°C)
Vapor Pressure at 68°F (20°C)	Less than .001 mm Hg
Water Solubility	Soluble

Source: *Sax's*, Davis

Figure 11-4

Notes

Fuming sulfuric acid ($H_2S_2O_7$) is a form of sulfuric acid that reacts even more violently with water. It is sometimes referred to as "super concentrated" sulfuric acid or **oleum**. Fuming sulfuric acid is also an oxidizer, but it is far more hazardous than sulfuric acid. Violent explosions result when it reacts with water. It is such a strong dehydrating and oxidizing agent that it has been known to ignite combustibles. The viscous liquid has a pungent, choking odor.

Remember these three facts about sulfuric acid:

- Concentrated sulfuric acid produces much heat when dissolved in water.
- Concentrated sulfuric acid has a strong affinity for water.
- Sulfuric acid can react with other substances, creating hazardous consequences. For example, when mixed with sodium, a dilute aqueous solution of sulfuric acid can react explosively. Oxidizing agents containing chlorine and oxygen react with concentrated sulfuric acid to produce an unstable chlorine oxide.

Sulfuric Acid Storage at Manufacturing Facilities

Notes

Sulfuric Acid Storage at Manufacturing Facilities

Hydrochloric Acid

Hydrochloric Acid
HCl

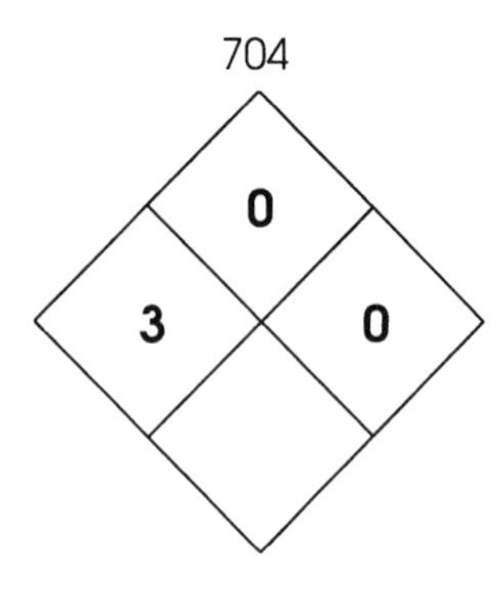

Hydrochloric acid (HCl) and hydrofluoric acid (HF) are also referred to as **halogen acids**. They are non-flammable acids that are corrosive, toxic, and extremely irritating when the vapors are inhaled. Other halogen acids are hydrobromic acid (HBr) and hydriodic acid (HI).

Technical grade hydrochloric acid, normally yellow in color, is called **muriatic acid**. Hydrochloric acid is used commercially to synthesize a variety of chlorides, such as zinc chloride used by plumbers as a soldering flux. Masons use hydrochloric acid to clean mortar from brick surfaces. Iron is cleaned in hydrochloric acid before it is galvanized. Hydrochloric acid is commonly used in the home. It is found in toilet bowl cleaners and other household cleaning fluids. During a hydrochloric acid spill or leak, hydrogen chloride vapor is released into the air. The permissible exposure limit (PEL) of hydrochloric acid is only 5 ppm, and the immediate danger to life or health level (IDLH) is 100 ppm. All hydrochloric acid spills should be taken seriously.

Notes

Physical Properties of Concentrated Hydrochloric Acid	
Concentration in Water	36 - 38%
Specific Gravity	1.19
Boiling Point	-121°F (-85°C)
Melting Point	-174°F (-114°C)
Vapor Pressure at 64°F (18°C)	4 atm
Water Solubility	Soluble

Figure 11-5

Source: *Sax's*, Davis

Inhalation Effects of Hydrogen Chloride

Concentration in Air (ppm)	Signs and Symptoms
1 - 5	Limit of odor detection
5	Permissible exposure limit (PEL)
5 - 10	Mild irritation of mucous membranes
35	Irritation of throat on short exposure
50 - 100	Barely tolerable
100	Immediate danger to life and health (IDLH)
1,000	Danger of lung edema and respiratory failure after short exposure

Figure 11-6

Hydrofluoric Acid

Hydrofluoric acid is a colorless, fuming liquid with a penetrating, pungent odor. It is used in water fluoridation, in petroleum refining, in hardening cement, in ceramics, as a wood preservative, and in etching and frosting glass. Hydrofluoric acid is also widely used in the manufacture of computer chips.

Hydrofluoric Acid
HF

704

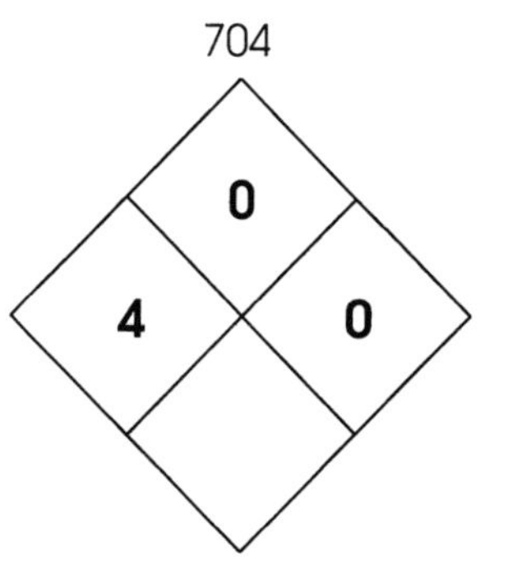

Gaseous hydrogen fluoride is formed when concentrated sulfuric acid and a fluoride salt are heated. When hydrogen fluoride is dissolved in water, it forms hydrofluoric acid.

Hydrofluoric acid is a strong acid, and it has some characteristics not shared by other common acids. For example, hydrogen fluoride dissolves glass and calcium silicate, a common component of glass. It also poses serious health hazards, having the highest health hazard rating of all inorganic acids. Inhaling hydrofluoric acid fumes causes pulmonary edema, and a slight exposure to the liquid can cause deep and severe burns. Hydrofluoric acid has been called a "bone seeker" because it can be absorbed through the skin, causing bone and bone marrow damage.

Notes

Physical Properties of Concentrated Hydrofluoric Acid	
Concentration in Water	The concentrated solutions usually contain 38–70% HF, but commercial strengths up to 100% HF are available.
Specific Gravity	1.26
Boiling Point	68°F (20°C)
Melting Point	-117°F (-83°C)
Vapor Pressure at 36°F (2.5°C)	400 mm Hg
Water Solubility	Soluble

Source: *Sax's, Hawley's*

Figure 11-7

Nitric Acid
HNO_3

704

	0	
3		0
	OX	

Nitric Acid

Nitric acid, also called **aqua fortis**, is a non-flammable, colorless to brown liquid. Next to sulfuric acid, nitric acid is the second most important industrial acid. In the United States alone, more than 8 million tons of nitric acid are produced annually. Nitric acid is principally used in the manufacture of ammonium nitrate fertilizers, explosives, and nitrated organic compounds. Several common explosives, such as nitroglycerin and trinitrotoluene (TNT), are made with nitric acid.

Spilled or leaking nitric acid (68% concentration) is easily identified by the reddish fumes of the highly toxic nitrogen dioxide produced. In fact, this form of nitric acid is called **red fuming nitric acid**. Visual identification of nitric acid in lesser concentrations may not be as easy, but these concentrations are no less dangerous because nitrogen dioxide vapor is still present.

Physical Properties of Concentrated Nitric Acid	
Concentration in Water	68 – 70%
Specific Gravity	1.50
Boiling Point	187°F (86°C)
Melting Point	-44°F (-42°C)
Vapor Pressure at 68°F (20°C)	
White Fuming Nitric Acid	62 mm Hg
Red Fuming Nitric Acid	103 mm Hg
Water Solubility	Soluble

Source: *Sax's*, Davis

Figure 11-8

Notes

Concentrated Nitric Acid
Transported by Rail

During fires or accidents involving nitric acid, a number of nitrogen oxides may be present, including nitrous oxide (N_2O), nitrogen trioxide (N_2O_3), and the highly toxic nitrogen dioxide (NO_2). When nitrogen dioxide is inhaled, it converts to nitric acid in the lungs. It can cause pulmonary damage that usually is not apparent until four or more hours after exposure.

Concentrated nitric acid is a strong oxidizer and can react explosively with metallic powders and natural organic compounds, such as turpentine and carbides. It can also spontaneously ignite wood, sawdust, and other cellulose materials. Nitric acid can severely burn the skin. The skin turns yellow within minutes after contact.

Nitric acid mixtures with less than 50% nitric acid are labeled "Corrosive." Mixtures with over 50% nitric acid are labeled "Corrosive" and "Oxidizer." Fuming nitric acid bears the "Poison" label in addition to "Corrosive" and "Oxidizer" labels.

Perchloric Acid

Perchloric Acid
$HClO_4$

704

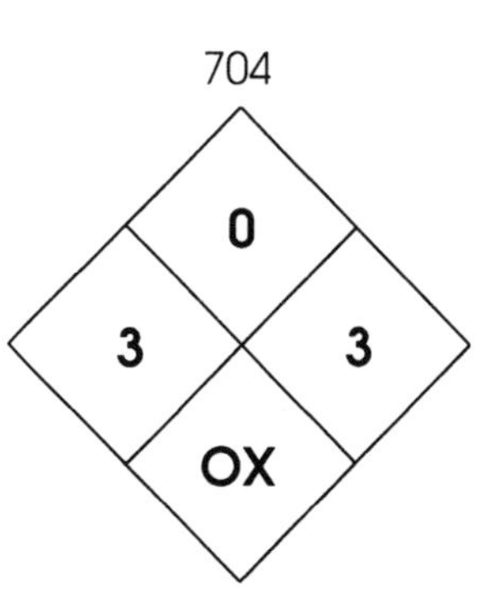

Perchloric acid forms when potassium perchlorate is treated with sulfuric acid. It is used extensively by electroplating and chemical industries. One pound reagent bottles of 60% to 70% perchloric acid solution are often found in laboratories.

Because perchloric acid is extremely reactive, it has been labeled one of the "bad actors." Pure liquid perchloric acid decomposes and explodes spontaneously. Water, however, slows down the decomposition of the oily, colorless liquid.

Spontaneous ignition can occur when perchloric acid is mixed with paper, wood, or cotton. As perchloric acid warms, it can spontaneously explode. One drop of anhydrous perchloric acid on paper has been known to detonate. When heated under atmospheric pressure to near 198°F (92°C), it will explode and decompose to oxygen, water, and chlorine. Reactions

Notes

Physical Properties of Concentrated Perchloric Acid	
Concentration in Water	65–70%
Specific Gravity	1.77
Boiling Point	66°F (19°C)
Melting Point	–170°F (–112°C)
Water Solubility	Very soluble

Source: *Sax's, Hawley's*

Figure 11-9

between perchloric acid and common organic compounds often result in explosions. Organic acids, alcohols, ethers, and other organic compounds violently decompose or explode when mixed with perchloric acid.

Phosphoric Acid
H_3PO_4

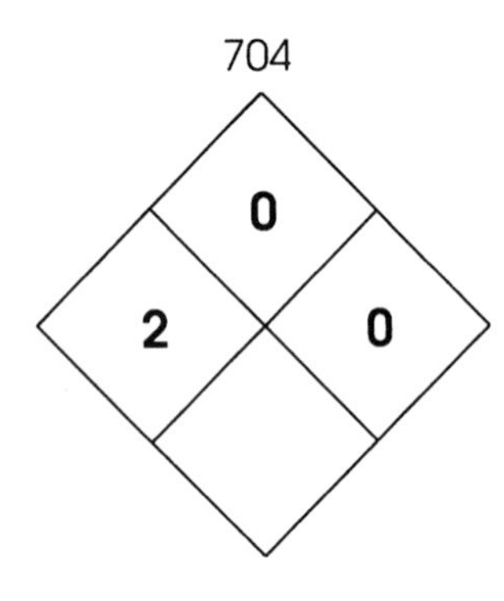

Phosphoric Acid

Phosphoric acid is a clear, colorless liquid. Among other things, it is used in the manufacture of fertilizers and detergents and in food processing. Phosphoric acid is formed when phosphate rock is treated with sulfuric acid.

Phosphoric acid is considered a moderate to strong acid. It is not as hazardous as other acids discussed. It does, however, slowly corrode metals at ordinary temperatures. Phosphoric acid is a poor oxidizing agent. However, this acid can cause serious skin burns.

Physical Properties of Concentrated Phosphoric Acid	
Concentration in Water	85% generally
Specific Gravity	1.86
Boiling Point	500°F (260°C)
Melting Point	108°F (42°C)
Vapor Pressure at 68°F (20°C)	0.0285 mm Hg
Water Solubility	Very soluble

Source: *Sax's*, Davis

Figure 11-10

COMMON ORGANIC ACIDS

Notes

The acids previously discussed are inorganic acids. The following acids are some common organic acids. Organic acids do burn, although because of their relatively high ignition temperatures, flammability is generally not a major hazard. As with most carbon-containing compounds, the flammability of organic acids varies.

Acetic Acid

The most common organic acid is acetic acid. It is responsible for the sharp odor and sour taste of vinegar. Vinegar contains from 3% to 6% acetic acid. Acetic acid is found in two forms: glacial or concentrated, and diluted. Glacial acetic acid is usually greater than 99.8% pure, while diluted solutions are found in varying concentrations.

Acetic acid is prepared commercially in several ways. The most common procedure involves acetylene as a starting material. Acetic acid solutions are generally purified by freezing. The unfrozen portion of the solution that contains the impurities is discarded.

In industry, acetic acid is used to synthesize other organic materials, such as acetone, esters, perfumes, dyes, plastics, and pharmaceuticals. Cellulose acetate and polyvinyl acetate are produced with acetic acid. The acid also is often employed in the paper and dye industries.

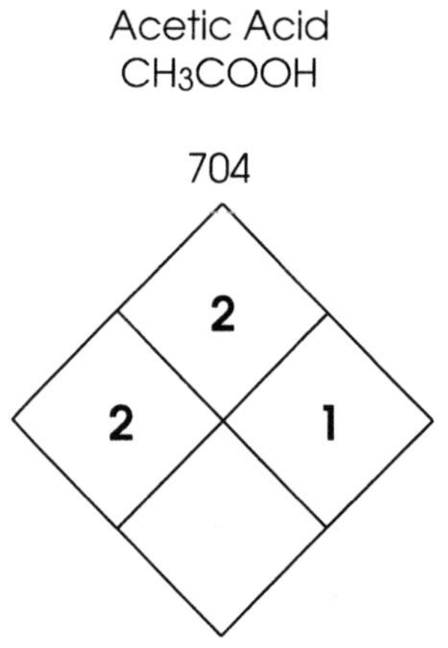

Physical Properties of Concentrated Acetic Acid

Concentration in Water	99 - 100%
Specific Gravity	1.05
Boiling Point	244°F (118°C)
Melting Point	61°F (17°C)
Vapor Pressure at 68°F (20°C)	11 mm Hg
Water Solubility	Soluble
Flash Point	109°F (43°C)
Autoignition Temperature	869°F (465°C)
Flammable Range	5.4 - 16%

Source: *Sax's*

Figure 11-11

Formic Acid

Formic acid is found naturally in red ants and pine needles. It is an acute irritant and is toxic if inhaled in even small doses. Formic acid is widely used for cleaning animal hides and in the dye industry.

Formic Acid
HCOOH

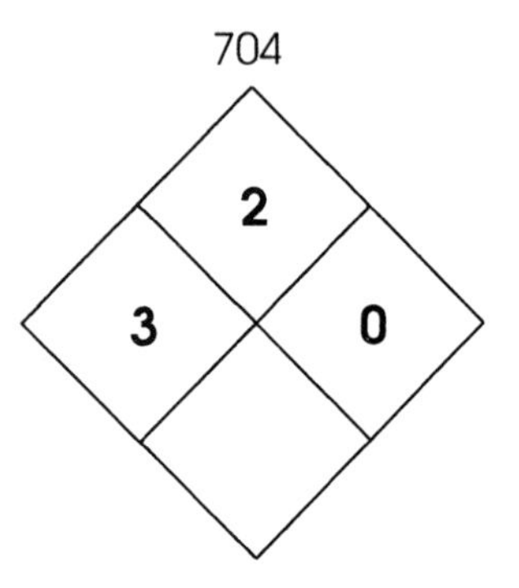

Physical Properties of Concentrated Formic Acid

Concentration in Water	85%, 90%
Specific Gravity	1.23
Boiling Point	213°F (101°C)
Melting Point	47°F (8°C)
Vapor Pressure at 68°F (20°C)	23 - 33 mm Hg
Water Solubility	Soluble
Flash Point	156°F (69°C)
Autoignition Temperature	1,114°F (601°C)
Flammable Range	18 - 57%

Source: *Sax's*, Davis

Figure 11-12

Hydrocyanic Acid

Hydrocyanic Acid
HCN

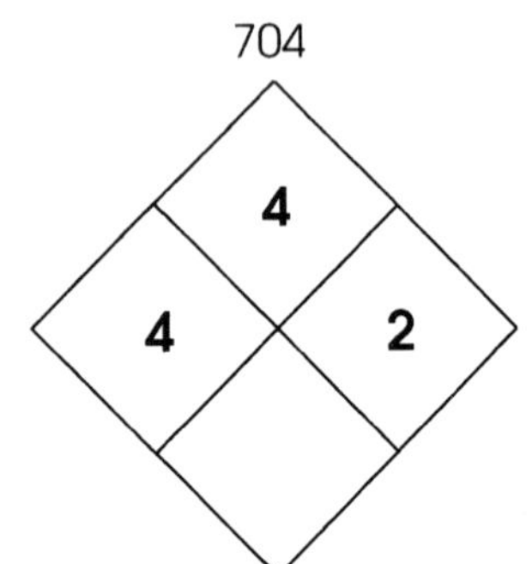

Hydrocyanic acid is a well recognized chemical, mainly because it is so toxic. It is also called **prussic acid**, and in the gaseous form it is called **hydrogen cyanide**. Hydrocyanic acid is a weak acid, but it is classified by DOT as a poison/inhalation hazard substance. It is also flammable and environmentally hazardous. Hydrocyanic acid has the highest health hazard and flammable rating of all organic acids. It is found in solutions of up to 96% concentrated HCN or as a liquefied compressed gas. The liquefied gas is transported under its own vapor pressure and is stabilized to avoid polymerization.

Physical Properties of Concentrated Hydrocyanic Acid

Concentration in Water	Up to 96%
Specific Gravity	0.69
Boiling Point	79°F (26°C)
Melting Point	9°F (-13°C)
Vapor Pressure at 50°F (10°C)	400 mm Hg
Water Solubility	Soluble
Flash Point	0°F (-18°C)
Autoignition Temperature	1,000°F (538°C)
Flammable Range	5.6 - 40%

Source: *Sax's*

Figure 11-13

PHYSICAL PROPERTIES OF SELECTED ACIDS

Product	Formula	Strength	Organic/ Inorganic	Oxidizing Agent	Reducing Agent	Toxic	Water Reactive	Flash Point °F	Flash Point °C	Specific Gravity	Vapor Density	Boiling Point* °F	Boiling Point* °C	Melting Point* °F	Melting Point* °C
Perchloric acid	$HClO_4$	Strong	Inorganic	Yes	No	Yes	No	—	—	1.77	—	66	19	-170	-112
Sulfuric acid	H_2SO_4	Strong	Inorganic	Yes	No	Yes	Yes	—	—	1.83	—	554	290	50	10
Hydrochloric acid	HCl	Strong	Inorganic	No	No	Yes	No	—	—	1.19	1.3	-121	-85	-174	-114
Nitric acid	HNO_3	Strong	Inorganic	Yes	No	Yes	No	—	—	1.50	—	187	86	-44	-42
Phosphoric acid	H_3PO_4	Strong	Inorganic	No	No	Yes	No	—	—	1.86	—	500	260	108	42
Hydrofluoric acid	HF	Weak	Inorganic	No	No	Yes	No	—	—	1.26	0.7	68	20	-117	-88
Acetic acid	CH_3COOH	Weak	Organic	No	No	Yes	No	109	43	1.05	2.1	244	118	61	16
Formic acid	$HCOOH$	Weak	Organic	No	No	Yes	No	156	69	1.23	1.6	213	101	47	8
Hydrocyanic acid	HCN	Weak	Organic	No	No	Yes	No	0	-18	0.69	0.9	79	26	9	-13
PHYSICAL PROPERTIES OF SELECTED BASE GROUPS															
Methylamine (all organic amines)	Varies by specific compound	High	Organic	No	Yes	Yes	Water soluble	—	—	—	1.1–2.0	—	—	—	—
Aniline	$C_6H_5NH_2$	High	Organic	No	Yes	Yes	Yes	158	70	1.02	3.2	364	184	21	-6
All metal hydroxides	Varies	High	Inorganic	No	No	Yes	Yes	—	—	—	—	—	—	—	—
All metal carbonates	Varies	Low	Inorganic	No	No	No	No	—	—	—	—	—	—	—	—
Ammonia (anhydrous)	NH_3	Low	Inorganic	No	Yes	Yes	Water soluble	—	—	0.77	0.60	-28	-33	-108	-78

Figure 11-14

* The boiling point and melting point of a solution depend on the concentration.

Notes

COMMON INORGANIC BASES

The most common bases are sodium hydroxide (lye), potassium hydroxide (potash), and calcium hydroxide (slaked lime). As discussed previously, DOT labels lye and potash as corrosive materials.

Like acids, bases are divided into two groups: inorganic and organic. The inorganic bases are non-flammable. The organic bases, which contain carbon and hydrogen, are flammable. The relative strength of common bases is shown in Figure 11-14.

The following reactions are common to all bases:

- reaction with acids, sometimes in a violent manner
- reaction with some metals to produce hydrogen and a metallic salt
- reaction with body tissue, proteins, and fats

Sodium Hydroxide

Sodium Hydroxide
NaOH

704

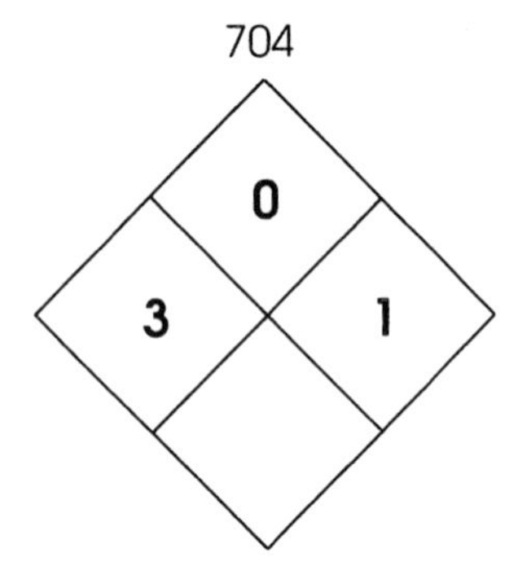

The most common corrosive base is sodium hydroxide, commonly called **lye** or **caustic soda**. Sodium hydroxide is used in pulp and paper manufacturing, in textile processing, in petroleum refining, in chemical and soap manufacturing, in medicine, in etching and electroplating, and in chemical laboratories.

Sodium hydroxide is found as white deliquescent flakes, sticks, beads, or lumps, or as a solid mass. A deliquescent substance is one that absorbs moisture on exposure to air. When in liquid form, sodium hydroxide usually is dissolved in either water or alcohol at concentrations of 50% to 75% sodium hydroxide. Considerable heat is generated when sodium hydroxide dissolves in water, making it possible to initiate some fires.

Sodium hydroxide is moderately toxic and can severely burn skin tissue. The mist, dust, and solutions of sodium hydroxide cause severe injury to the eyes, mucous membranes, and skin. Solutions of 25% to 50% on the skin cause irritation in about three minutes. Concentrated solutions will cause severe burns, even during short exposures.

Physical Properties of Sodium Hydroxide	
Specific Gravity	2.12
Boiling Point	2,534°F (1,390°C)
Melting Point	605°F (318°C)
Vapor Pressure at 68°F (20°C)	Essentially zero
Water Solubility	42g/100g of water

Source: *Sax's*

Figure 11-15

Potassium Hydroxide

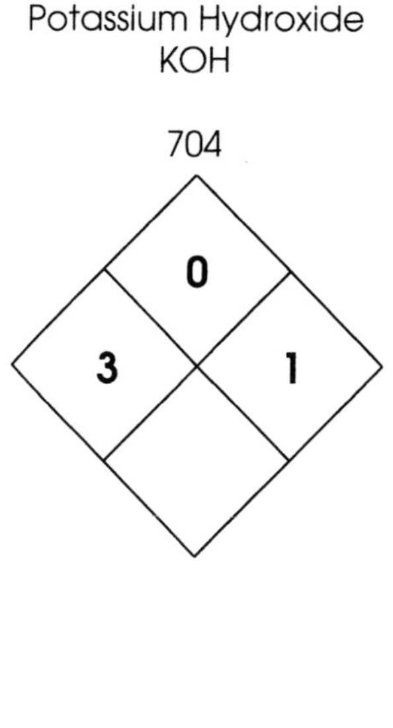

Potassium hydroxide, also known as **caustic potash** or **potash lye**, is used in the manufacture of soap, bleaches, medicine, and other potassium salt substances. Potassium hydroxide poses hazards similar to those of sodium hydroxide. Both sodium and potassium hydroxides can have extreme exothermic reactions with water. Potassium hydroxide is considered a stronger base than sodium hydroxide, but the difference is of little significance in light of the severe health hazards of each.

Physical Properties of Potassium Hydroxide	
Specific Gravity	2.04
Boiling Point	2,408°F (1,320°C)
Melting Point	680°F (360°C)
Vapor Pressure at 68°F (20°C)	Essentially zero
Water Solubility	107g/100g of water

Source: *Sax's*

Figure 11-16

COMMON ORGANIC BASES

Organic bases are highly toxic and flammable, although these properties vary according to the substance. Organic bases are derived from organic functional groups called amines. All amines are highly toxic when inhaled, ingested, or absorbed through the skin. They are either flammable or combustible.

Aniline

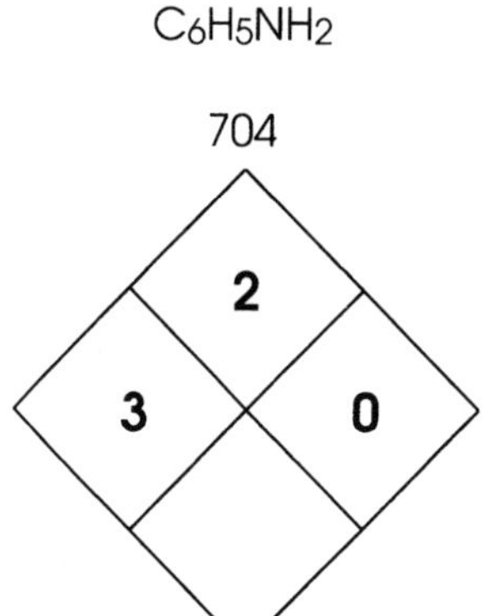

Aniline, also called **aminobenzene** and **phenylamine**, is a colorless to dark brown liquid with an odor similar to ammonia. It produces a burning taste when inhaled. It is used to manufacture dyes, photographic chemicals, agriculture chemicals, drugs, and other chemical products.

Like all organic amines, aniline is highly toxic. It is classified as a poison material. It is highly toxic when absorbed through the skin. Toxic nitrogen oxides are produced during combustion of aniline.

Notes

Physical Properties of Aniline

Specific Gravity	1.02
Boiling Point	364°F (184°C)
Melting Point	21°F (-6°C)
Vapor Pressure at 68°F (20°C)	0.6 mm Hg
Water Solubility	Slight
Flash Point	158°F (70°C)
Autoignition Temperature	1,139°F (615°C)
Flammable Range	1.3 - 11%

Source: *Sax's*, Davis

Figure 11-17

Diethylamine
$(C_2H_5)_2NH$

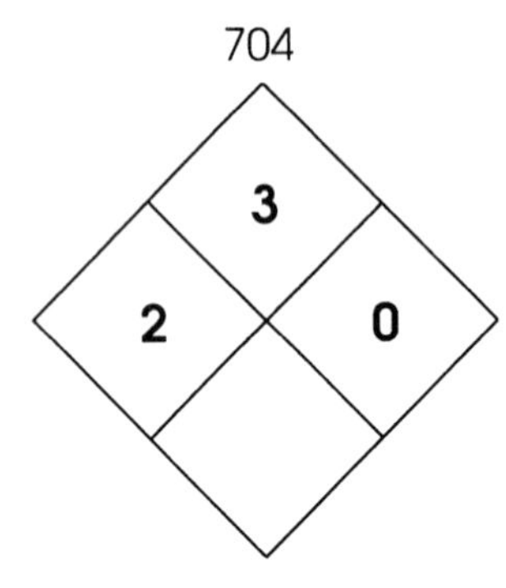

Diethylamine

Diethylamine is used in dyes, pharmaceuticals, and synthetic processing. It is a corrosion inhibitor on some metals. Diethylamine is a colorless liquid with an ammonia-like odor. It also has a fishy odor, a common characteristic of aliphatic amines.

The liquid is irritating and burns tissue. The vapor is toxic and damages lung tissue. It is moderately toxic when ingested. Diethylamine poses an extreme fire hazard, because it has a flash point below 0°F (–18°C) and a flammable range of 1.8–10.1%. Toxic nitrogen oxides are released during combustion.

Physical Properties of Diethylamine

Specific Gravity	0.71
Boiling Point	132°F (56°C)
Melting Point	-38°F (-39°C)
Vapor Pressure at 68°F (20°C)	195 mm Hg
Water Solubility	Soluble
Flash Point	-0.4°F (-18°C)
Autoignition Temperature	594°F (312°C)
Flammable Range	1.8 - 10.1%

Source: *Sax's*, Davis

Figure 11-18

Notes

Physical Properties of Other Common Organic Bases		
Name	Diethylenetriamine	Methylamines
Synonyms	Diaminodiethylamine	Mono-, di-, trimethylamine
Physical State	Liquid	Liquid
Water Solubility	Soluble	Soluble
Flash Point	215°F (102°C)	0° to 32°F (–18° to 0°C)
Flammable Range	1 - 10%	Mono: 5.0 – 20.8% Di: 2.8 - 14.4% Tri: 2.0 - 11.6%
Vapor Density	3.5	1.1 - 2.0
Health Hazard	Toxic. Contact with tissue or eyes causes burns.	Liquid causes burns. Vapor irritating. Toxic liquefied gas.

Source: *Sax's*, Davis

Figure 11-19

PLACARD AND LABEL REQUIREMENTS

As mentioned in Chapter 9, some materials require the "Poison/Inhalation Hazard" markings as well as the "Corrosive" label. Examples include phosphorus trichloride and chlorosulfonic acid. Some corrosive materials may also be labeled as "Oxidizer." "Corrosive" placards also will be displayed with the radioactive materials uranium hexafluoride fissile and uranium hexafluoride LSA. The DOT placard and label requirements for **corrosive materials** are outlined in Figure 11-20.

CLASS 8 PLACARD AND LABEL REQUIREMENTS			
VEHICLE PLACARD	**PACKAGE LABEL**	**QUANTITY**	**SPECIAL REQUIREMENTS**
Corrosive (white letters on black background and black drawing on white background).  CORROSIVE 8	**Corrosive** (white letters on black background and black drawing on white background).  CORROSIVE 8	Placard any quantity by rail; 1,001 lbs (454 kg) or more by truck.	If 5,000 lbs (2,268 kg) picked up at one facility, then a **Corrosive** placard is needed in addition to other placards.

Figure 11-20

Notes

SHIPPING CONTAINERS

Shipping containers for acids and bases range from small glass jugs to the MC312/DOT412 liquid carrier. Emergency responders can gain an important advantage in incident management if they quickly identify an acid or base by its container. A discussion of the most common corrosive shipping containers follows.

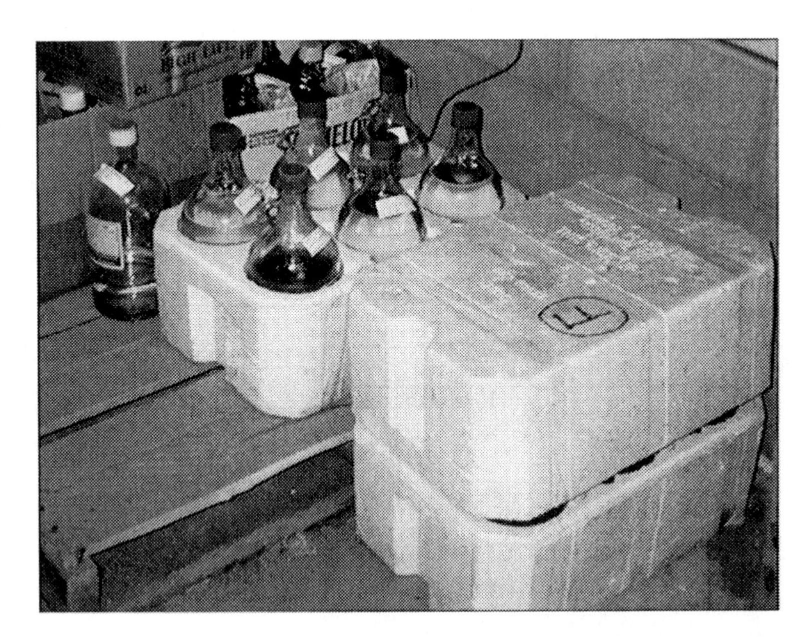

The glass jug size ranges from 1/2 gallon to 5 gallons. For shipping, these jugs are usually found encased in polystyrene or Styrofoam packing. Plastic jugs also are used, but usually for acids such as muriatic acid.

***Left:* Gallon glass containers encased in Styrofoam packing. *Below:* Corrosives in reinforced carboys.**

Several corrosives are found as a dry solid or powder. They are transported in 55 gallon fiberboard or metal drums, small glass containers packed in cardboard, and in multi-layered paper bags. Many liquid corrosives are shipped in 55 gallon plastic drums.

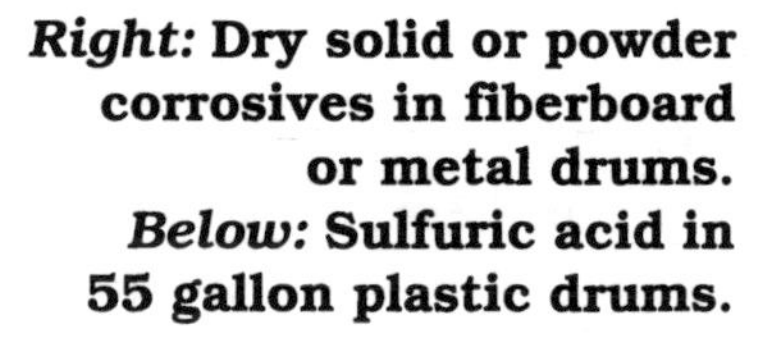

***Right:* Dry solid or powder corrosives in fiberboard or metal drums.**
***Below:* Sulfuric acid in 55 gallon plastic drums.**

Notes

MC312/DOT412 Corrosive Liquid Carrier

Bulk corrosive liquid tank trucks and tank cars are common containers for acid and base materials. Large corrosive carriers usually can be identified by external rings or the smaller capacity. Because corrosives are very dense and can weigh up to three times as much as an equal volume of water, tank construction is designed for strength and small volumes. Mild corrosives can be found in the MC307/DOT407 tank truck (see Chapter 6).

The MC312/DOT412 may have visible reinforcing rings or an insulated double shell in which the reinforcing rings are not visible. The dome/cover valve assembly provides overturn and splash protection. Most assemblies are made of stainless steel and may require a plastic, rubber, glass, or wax liner depending upon the product. Tank trucks that haul stronger corrosives have air pressure top unloading. Many of the tanks have a 35 psi working pressure. If the unloading pipe extends externally to the bottom of the tank, it is either valved at both the top and end of the tank or else valved at the top and capped at the end. ***Note:*** The MC312/DOT412 is not typically bottom valved but uses an eduction system as noted above.

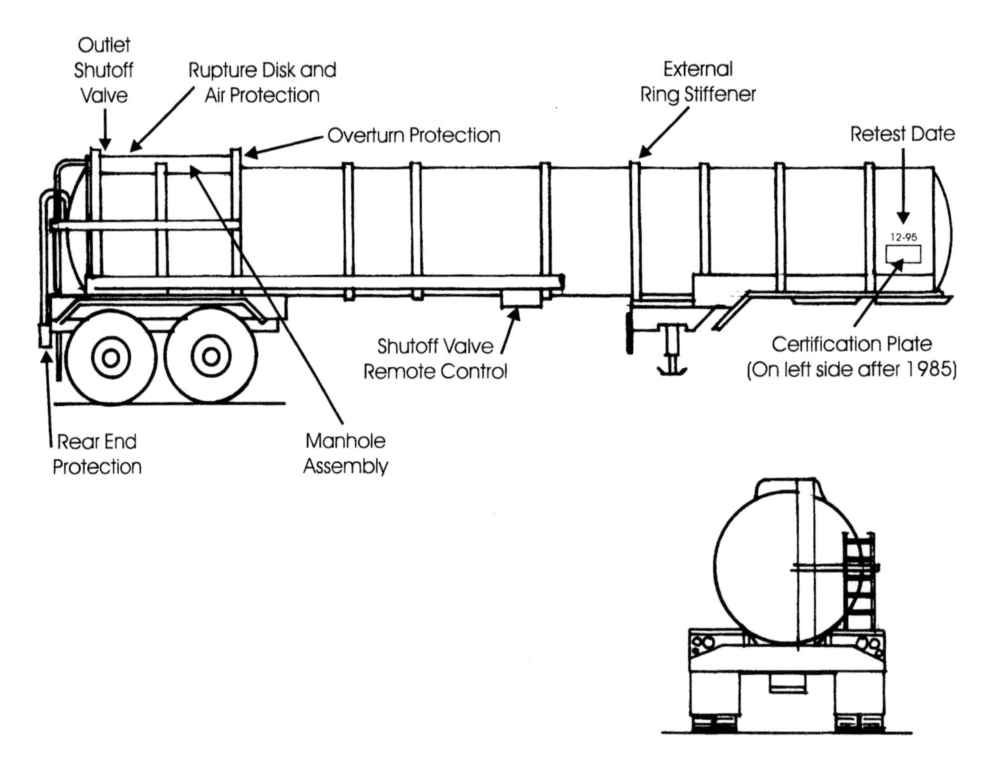

Figure 11-21. MC312/DOT412 Corrosive Liquid Carrier

Adapted with permission from *Hazardous Materials for First Responders*, © 1988, Board of Regents, Oklahoma State University.

Notes

MC312/DOT412 Corrosive Liquid Carrier

Rail Transport

Rail car capacities range from 4,000 to 24,000 gallons. The safety devices on both rail and tank cars are similar to those found on other liquid carriers.

Rail Tank Carrying a Corrosive Liquid

Notes

KEY QUESTIONS TO REVIEW

- What is an acid?
- What is a base?
- What is the difference between strength and concentration?
- What are the hazards associated with corrosives?
- What are the common acids and bases?
- What are the placarding and labeling requirements for corrosives?
- What are the common shipping containers for corrosives?

RESPONDING TO CORROSIVE EMERGENCIES

Responding to corrosive emergencies presents unique challenges. The vapors and fumes of corrosive materials can usually penetrate structural firefighting clothing. If combustibles are contaminated by corrosives or if organic corrosives are near combustibles, there is a potential fire and explosion hazard. Many burning corrosives produce highly toxic fumes. Serious burns are possible if flesh is exposed to acidic or caustic materials. Haz-mat technicians will be needed to determine whether to neutralize or dilute the corrosive material. In addition, they will need to know if water can be applied safely without prompting a reaction. Since corrosives are so common, responders are likely to encounter a corrosive emergency.

HAZ-MAT — HAZARD IDENTIFICATION

During an incident, it is imperative to identify the corrosive materials involved. Responders should ***preplan*** for such emergencies. Know where corrosives are used in the community, how much is used, and the common storage methods. You should also know the common transportation routes over which corrosives are moved and whether sufficient water supplies are available along the way. The life hazards present at fixed site facilities and along major thoroughfares should also be identified.

When an emergency occurs, an incident's ***location*** may indicate what products are involved, the quantity, and what type container or vessel is used.

Always note ***markings or identifying features*** such as placards, labels, and tank shapes. Corrosive containers such as Styrofoam packing crates, carboys, and the MC312/DOT412 are easily distinguished.

Listen for ***noises***, such as a hissing sound caused by increased internal pressure in the leaking tank. Pungent ***odors*** can also indicate a corrosive release, but be aware that these fumes may be toxic. Many acids produce pungent odors when released. ***Fuming*** orange or reddish brown vapor clouds or ***discolored materials*** are also clues that corrosives are present.

Notes

Once you know that corrosives are present, identify the exact product and its hazards:

- Determine the location of the leak and/or fire:
  - building
  - roadway
  - fixed site facility
  - tractor trailer
  - confined space
  - open field
- Determine the cause of the incident:
  - traffic accident
  - equipment failure
  - container failure
  - chemical reaction
  - fire
  - train accident
  - accidental spill
- Identify injuries:
  - type of injuries
  - number of injuries
  - severity
- Determine the chemical/physical potential of the product:
  - potential for vapor or fume production
  - strength and concentration
  - degree of toxicity
  - reaction with other materials
  - reaction with water
  - procedures for safely neutralizing or diluting the product
  - degree of environmental impact

HAZ-MAT — ACTION PLAN

When developing an action plan, the incident commander must assess the agency's training and ability to control the incident. Determine if appropriate equipment is available. Do not attempt to control a leak if properly trained personnel and equipment are not on hand. Don't forget to initiate proper decontamination procedures if needed.

In a transportation incident, the container may need to be uprighted. Because corrosives are often much heavier than equal volumes of water, it may be necessary to off-load the container before attempting to upright it. Evaluate the need to transfer the product from the container. This procedure is usually done by on-site technicians or private contractors under the supervision of emergency response personnel.

Notes

Off-Loading an Overturned Bulk Container

Off-Loading a Leaking Rail Car Containing Nitric Acid

Notes

If there is **NO LEAK or FIRE** during a corrosives accident:

- Determine the type and amount of material.
- Establish appropriate control zones.
- Identify resources needed to control the incident.
- Assess container damage.
- Develop a containment plan if a leak is possible.

If there is a **LEAK but NO FIRE:**

- Maintain control zones and initiate appropriate public protective actions.
- Contain the material to avoid contaminating other materials and to reduce environmental damage.
- Control the leak if properly trained personnel and equipment are on hand.
- Evaluate the need to transfer the product to another container after the leak is stopped.

If there is a **LEAK and FIRE:**

- Reevaluate control zones and increase if necessary.
- Order immediate, appropriate public protective actions.
- Contain the material.
- Protect exposures.

Leaking Nitric Acid Being Contained by Ditching and Diking

Notes

- For a leak and fire involving an **inorganic acid**, remember:
  - Exposed organic material can become contaminated and oxidize to the point of ignition.
  - Smoke can be toxic and heavily saturated with corrosive gases.
  - Some inorganic acids are oxidizing agents.
  - Some acids and bases may react violently with water and spread the hazard.
- For a leak and fire involving an **organic corrosive**, remember:
  - Many organic corrosives are combustible.
  - Smoke may be highly toxic and corrosive.
  - The best choice may be to let the fire burn itself out.
  - Anticipate the resources needed for decontamination and clean-up operations.

HAZ-MAT — ZONING

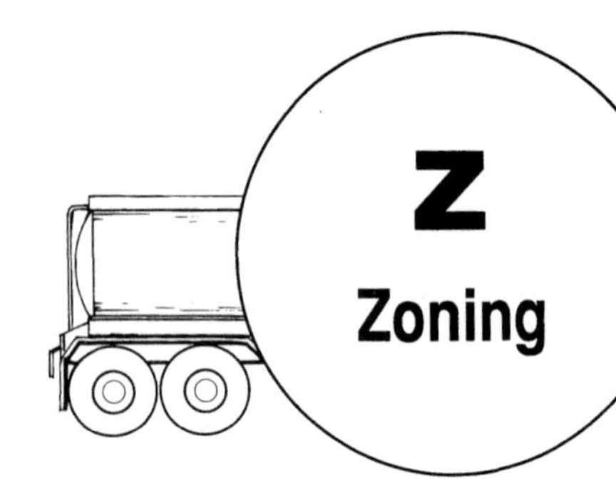

It is extremely important to isolate the corrosives incident and establish entries and exits for personnel. Once again, such action begins with the first arriving emergency responder. Be sure to check appropriate information resources or local protocol for the proper zone distances. Adjust the following recommended distances according to the specific material involved. Zoning for an acetic acid spill would be completely different from zoning for a fuming nitric acid spill.

For **SMALL** spills (those involving a container no larger than a 55 gallon drum) from drums, containers, and tanks with **NO FIRE:**

- Evacuate an area of 250 feet, if necessary.
- Establish a minimum hot zone of 150 feet.

For **SMALL** spills with **FIRE:**

- Evacuate an area of 350 feet.
- Establish a minimum hot zone of 200 feet.

For **LARGE** spills (those involving containers larger than a 55 gallon drum) or fuming liquid incidents with **NO FIRE:**

- Establish an evacuation area of 1,000 feet.
- Establish a minimum hot zone of 500 feet.
- Establish warm and cold zones.

For **LARGE** spills or fuming liquid incidents with **FIRE:**

- Establish an initial evacuation area of 1,000 feet.
- Anticipate extensive downwind evacuation.
- Establish a minimum hot zone of 500 feet.
- Establish warm and cold zones.

HAZ-MAT — MANAGING THE INCIDENT

Notes

Scene Management

Corrosive emergencies can be managed effectively when emergency responders are prepared. The following are key factors for successful corrosive emergency management:

- know hazards in the area
- preplan tactical operational considerations
- preplan resources
- quickly define the severity of the incident
- establish control zones
- use an incident management system

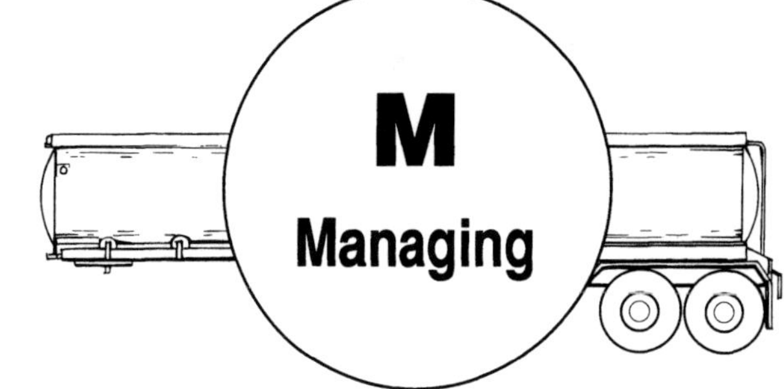

Rescue Considerations

Rescue operations are dangerous for emergency responders when corrosive materials are involved. Structural firefighting clothing will not provide sufficient protection when dealing with heavy concentrations of liquid and/or vapor. Therefore, it is critical to completely assess hazards before committing to a rescue operation.

The most common injuries encountered during a corrosive emergency are burns on the skin, mucous membranes, and respiratory tract. Because these products are so toxic, other life-threatening complications may arise. For example, exposure to hydrocyanic acid requires immediate treatment with a cyanide antidote kit. The following is an outline for general care and treatment of corrosive-related injuries:

General Treatment for Corrosive Material Exposures

Products .... Hydrochloric acid, sulfuric acid, sodium hydroxide, potassium hydroxide, potassium hypochlorite.

Containers .... Glass bottles, plastic carboys, plastic-lined cans and drums, lined tank cars and tank trucks.

Life Hazard .... Irritation to tissues resulting in severe respiratory injury. Upper airway burns and edema as well as lower airway burns and pulmonary edema may result. Shock and severe tissue burns are possible. Ingestion may lead to gastric burns and perforation.

Signs/Symptoms ..... Shortness of breath, upper airway obstruction, and signs of pulmonary edema. Tachycardia and signs of shock with decreased level of consciousness. Nausea, vomiting, and abdominal pain may result from ingestion. Severe skin and eye burns may result.

Notes

Basic Life Support .... Wear proper protective equipment and remove patient from contaminated area. Administer oxygen and remove patient's clothing. Brush or blot away any visible product and decontaminate with copious amounts of soap and water. Irrigate eyes thoroughly and continue during transport as necessary. Do not induce vomiting. If ingested, dilute with small quantities of water (follow advice of Poison Control and local protocols). Do not orally administer anything if the patient has a decreased level of consciousness or decreased gag reflex. Watch for signs of shock and respiratory problems. Do not attempt to neutralize product on the skin. Cover skin burns with sterile dressings after decontamination.

Advanced Life Support .... Assure an adequate airway and assist ventilations as necessary. Start an IV LR (strict TKO). Monitor for cardiac arrhythmias/pulmonary edema and treat as necessary. Treat signs of hypovolemic shock with IV fluids as necessary, but watch for signs of fluid overload. Administer topical anesthetic to eyes for easier irrigation and to reduce pain. Follow local protocols for all drug therapy and medical treatment.

Other Information .... Avoid all contact with product. There is a danger of an exothermic reaction if the product is neutralized. Some products are extremely toxic and may require specific antidote therapy (i.e., hydrocyanic acid and hydrofluoric acid). Always follow local medical protocols for antidote and drug usage.

HAZ-MAT — ASSISTANCE

To successfully conclude a corrosive emergency may require additional resources. Use the following resources:

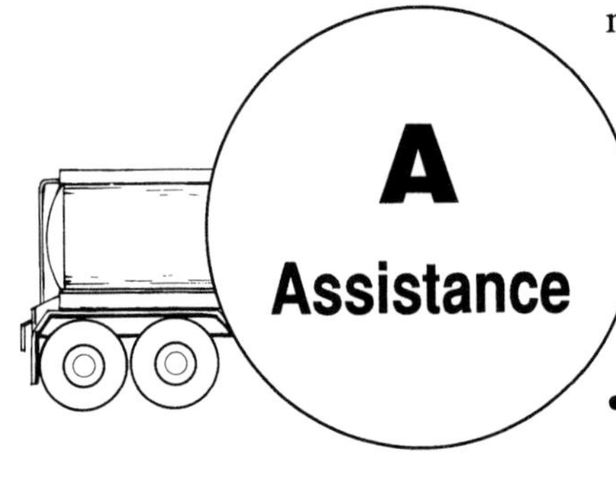

- ***Dispatcher*** can coordinate resource acquisition.
- ***Local law enforcement*** can provide site control.
- ***Specialized response teams*** can furnish technical expertise and equipment.
- ***Private technical advisors*** and ***clean-up operators*** are warranted.
- ***Technical specialists*** may be critical to safely mitigate a corrosive emergency.
- ***Local distributors, manufacturers, and/or chemical experts*** can provide important information.
- ***Private contractors*** can perform tasks such as operating vacuum trucks, providing absorbent pads, and neutralizing spilled product during a corrosive incident.

HAZ-MAT — TERMINATION

Notes

Terminating a corrosive emergency can be very time consuming. Extensive on-site clean-up and decontamination are common. In addition, if equipment such as emergency vehicles or SCBA are exposed to concentrated corrosive vapors, they may become unsafe to operate and may need to be rebuilt. As previously discussed, termination should be as systematic as actions taken during the actual emergency. Termination consists of:

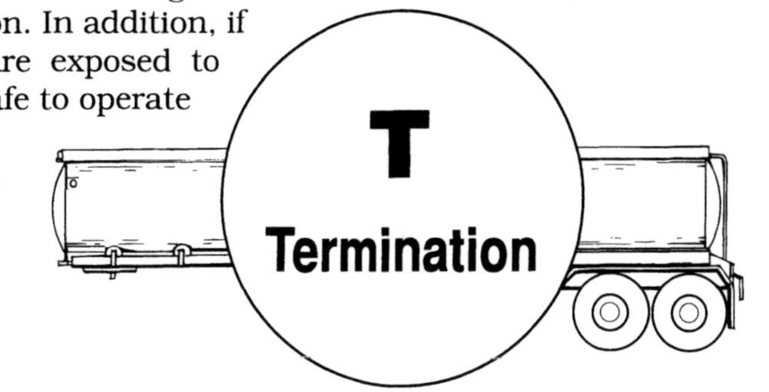

- decontamination
- rehabilitation
- medical screening
- exposure reporting
- initial incident debriefing
- post-incident analysis

KEY POINTS TO REVIEW

Corrosive materials accidents are common, yet response agencies often fail to recognize the hazards associated with them. Emergency responders should have a clear understanding of the corrosive hazard class and should be prepared for such emergencies. If an agency is not prepared, personal education and training should continue until all personnel:

- understand the chemical and physical characteristics of corrosive materials
- can follow the developed response strategies for corrosive materials
- have preplanned for known corrosive hazards to assist in successful incident mitigation
- are able to use the HAZ-MAT checklist for corrosive materials found in Appendix G

Chapter 12

MISCELLANEOUS HAZARDOUS MATERIAL EMERGENCIES

OBJECTIVES

After studying the material in this chapter, you will be able to:

- define miscellaneous hazardous materials
- define other regulated materials (ORMs)
- define hazardous wastes
- describe the use of the "Dangerous" placard
- explain how to identify international shipments of hazardous materials
- describe other hazardous substances likely to be encountered
- summarize incident response considerations for miscellaneous hazardous material emergencies

UNDERSTANDING MISCELLANEOUS HAZARDOUS MATERIALS

INTRODUCTION

An incident involving a miscellaneous chemical can be as routine as a broken pesticide bottle at the local hardware store or misuse of a corrosive household cleaning agent. It can also involve a major spill, such as a spill from a rail car containing molten sulfur. Never underestimate the potential risk of a miscellaneous chemical emergency, even when the material is only slightly hazardous or is found in small quantities.

Notes

MISCELLANEOUS HAZARDOUS CHEMICALS DEFINED

Review the definition for ***hazardous materials*** found in Chapter 1:

> A material or materials *accidentally released* from the original container and *used in a manner not originally intended.* Hazardous materials include materials that are *unintentionally contaminated or mixed with other chemicals* or involve some outside reactive source such as heat, light, liquids, shock, or pressure (emphasis added).

There are many substances that can cause harm if they are *accidentally released* from the original container, *used in a manner not intended,* or *contaminated* with another substance. This chapter will cover several incident types that pose hazards to emergency responders:

- miscellaneous hazardous materials
- other regulated materials (ORMs)
- hazardous wastes
- hazardous materials shipped under the "Dangerous" placard
- international shipments
- elevated temperature products

A Miscellaneous Haz-mat Emergency in Disguise

MISCELLANEOUS HAZARDOUS MATERIALS

Notes

The Department of Transportation (DOT) defines a ***miscellaneous hazardous material*** (Hazard Class 9) as:

> A material which presents a hazard during transportation, but which does not meet the definition of any other hazard class. This class includes any material which has an anesthetic, noxious, or other similar property which could cause extreme annoyance or discomfort to a flight crew member so as to prevent the correct performance of assigned duties, or any material which meets the definition for an elevated temperature material, a hazardous substance, or a hazardous waste (49 CFR 173.140).

Shipping containers for miscellaneous hazardous materials are as varied as the materials being transported. These materials may be found in fiberboard or wooden boxes, wooden barrels, bags, metal drums, tank trucks, and tank cars. Selected materials found in this hazard class include:

- ammonium nitrate fertilizers with less than 70% ammonium nitrate
- carbon dioxide, solid (dry ice)
- cotton
- environmentally hazardous substances, liquid or solid, not otherwise specified (n.o.s.)
- fish material or fish scrap, stabilized
- lead sulfide
- life saving appliances
- lithium batteries
- other regulated substances, liquid or solid, n.o.s.
- polychlorinated biphenyls (PCBs)
- polyhalogenated biphenyls, liquid or solid
- polystyrene beads, expandable, evolving flammable vapor
- sulfur, molten
- vehicles, self-propelled, including internal combustion engines and electric storage batteries
- zinc dithionite or zinc hydrosulfite

OTHER REGULATED MATERIALS (ORMs)

Other regulated materials (ORMs) are regulated because they may pose an unreasonable risk to health, safety, or property when transported. They do not, however, meet the DOT definition of a hazardous material. Prior to the October 1, 1991 revision of 49 CFR, ORMs were divided into five categories—ORM-A, B, C, D, and E. Under the new regulations, many ORMs A, B, C, and E were incorporated into the miscellaneous hazardous material classification (Class 9), while others were included in their respective hazard class. **The only ORM classification remaining after October 1, 1993, is ORM-D.** However, responders need to be familiar with the definitions of

Notes

ORM-A, B, C, and E for a long time to come, because these materials may be stored in warehouses and other locations into the next century.

ORM–A

ORM-A materials have an anesthetic, irritating, noxious, toxic, or other similar property. If involved in a leak during transport, they can cause extreme discomfort to the passengers and crew of transport vehicles. Remember, ORM classifications A, B, C, and E may still be encountered in storage locations.

Examples of ORM-A materials include:

- bromochloromethane
- camphene
- carbon tetrachloride
- chloroform
- ethylene dibromide
- naphthalene
- solid carbon dioxide
- trichloroethylene

ORM–B

Leaking ORM-B materials can damage transport vehicles or vessels. These materials can corrode aluminum, posing a serious danger for aircraft. Most of these materials are now classified as corrosive materials under 49 CFR as revised October 1, 1991.

Examples of ORM-B materials include:

- calcium oxide (unslaked lime, quicklime)
- copper chloride
- ferric chloride
- metallic mercury

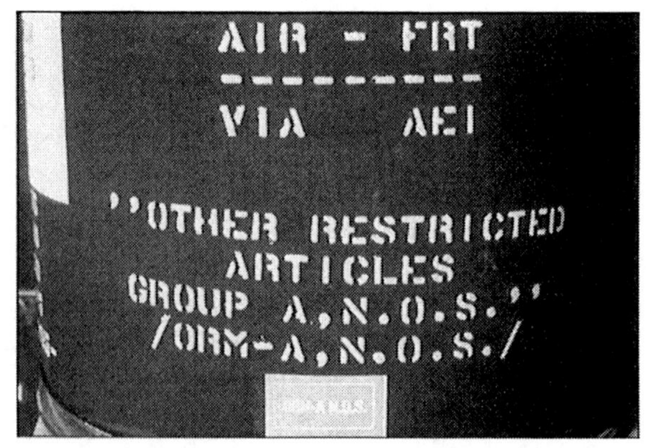

55 Gallon Drum Containing an ORM-A

ORM–C

Notes

ORM-C materials are considered unsuitable for transport unless properly identified and prepared. This group is specifically named in DOT regulations, and almost all are regulated for water transportation only.

Examples of ORM-C materials include:

- battery parts
- calcium cyanide (not hydrated)
- castor beans
- copra
- cotton batting
- excelsior
- fish meal
- fish scrap
- hay or straw
- metal boring
- oakum
- oiled materials
- pesticides
- petroleum code
- rosin
- rubber curing compound
- sulfur

ORM–D

ORM-D materials are packaged consumer commodities found on store shelves. They include such products as hair spray, lighter fluids, camping fuel, pesticides, charcoal briquettes, cleaning products, and hundreds of other store shelf commodities. They pose a limited hazard during transportation because of the form, quantity, or packaging (see 49 CFR 173.144). All non-bulk packaging containing an ORM-D must be marked on at least one side or end with the ORM-D designation. No hazard specific labels are required for this class of materials unless transported by air. Placards are not required.

The gross weight of an ORM-D case, including all the inner packages, may not exceed 66 pounds (30 kg). Although there are weight restrictions on individual cases, there are no restrictions on the number of cases that can be shipped. Therefore, emergency responders could encounter an unplacarded tractor trailer completely filled with ORM-D materials.

Examples of ORM-D materials include:

- Compressed gas in containers (normally spray cans) with a water capacity of 4 ounces (120 ml) or less. Compressed gas in material solutions, in limited quantities and internal pressures, must be non-flammable, non-poisonous, and non-corrosive.

Common ORM-D Containers

- Flammable liquids in inner packages not over 1 gallon (4 l). The allowable volume of the inner package may be less, depending on the flash point and boiling point of the specific liquid. Alcoholic beverages in packages of 1 gallon (4 l) or less are exempted.
- Flammable solids in inner packages not over 11 pounds (5 kg). The allowable volume of the inner package may be less, depending on the reactivity of the specific material. Spontaneously combustible materials may not be classified as ORM-D. Charcoal briquettes in packages not more than 66 pounds (30 kg) may be classed as ORM-D.
- Oxidizers in inner packages not over 1 gallon (4 l) for liquids and 11 pounds (5 kg) for solids. The allowable volume of the inner package may be less, depending on the reactivity of the specific material.
- Organic peroxides in inner packages not over 1 ounce (30 ml) for liquids and 1.1 ounces (30 g) for solids.
- Poisonous materials in inner packages not over 8 ounces (250 ml) for liquids and 8.8 ounces (250 g) for solids. Poisonous gases and many poisonous liquids and solids may not be classified as an ORM-D, based on the toxicity of the specific material.
- Corrosive materials in inner packages not over 1 gallon (4 l) for liquids and 11 pounds (5 kg) for solids. The allowable volume of the inner package may be less, depending on the corrosivity of the specific material. Some corrosive materials may not be classified as ORM-D.
- Miscellaneous hazardous materials in inner packages not over 1 gallon (4 l) for liquids and 11 pounds (5 kg) for solids.

ORM–E

Notes

Most of these materials are now classified according to their respective hazard class under 49 CFR. ORM-E materials are either hazardous wastes or hazardous substances as defined by the Environmental Protection Agency (EPA).

During the last two decades the recognition and control of hazardous waste has become a serious problem for most industrialized nations. A recent *USA Today* study of the EPA's toxic release inventory data stated that over 7 billion pounds of toxic products are released into the air, water, and land. Another 3.3 billion pounds are shipped for treatment, reclamation, and disposal. Much of that 3.3 billion pounds is classed as hazardous waste.

HAZARDOUS WASTE

Hazardous waste is one of the most commonly encountered miscellaneous commodities. The EPA defines ***hazardous waste*** as:

> A material that possesses properties, such as flammability, corrosivity, reactivity, or toxicity, that could pose dangers to human health or the environment if discarded.

Hazardous waste may be stored until enough material is gathered to warrant disposal. There are many different types of waste, including by-products of chemical reactions, biomedical waste, and contaminated water runoff or soil. The waste label is often combined with the standard hazard class placard or a blank white marking with the four-digit United Nations identification number. When dealing with any waste, identify the substance and its major ingredients. Be aware that hazardous wastes may not be placarded or labeled like other transported commodities, especially during storage.

Hazardous Wastes in Common Containers

Notes

UNDERSTANDING THE "DANGEROUS" CLASSIFICATION

The DOT "Dangerous" placard can be seen every day on over the road traffic. When this placard is displayed, emergency responders should recognize that there are two or more dangerous chemicals involved. There are several requirements for multiple commodities shipped with this placard.

The "Dangerous" placard may never be used when there is any quantity of the materials listed in Figure 12-1.

Materials That May Not Be Shipped Under the "Dangerous" Placard

Placard Name	Hazard Class or Division Number
Explosives 1.1 (formerly Class A explosives)	1.1
Explosives 1.2 (formerly Class A or B explosives)	1.2
Explosives 1.3 (formerly Class B explosives)	1.3
Poison Gas	2.3
Dangerous When Wet	4.3
Poison (Inhalation Hazard only)	6.1
Radioactive (Yellow III label)	7

Figure 12-1

A vehicle containing 1,001 pounds (454 kg) or more (gross weight) of two or more of the substances listed in Figure 12-2 may use the "Dangerous" placard in place of the separate placards usually required for each material. For example, a "Dangerous" placard may be used on a shipment of 1,500 pounds of flammable liquid and 1,200 pounds of oxidizers. However, the individual placards are required if the vehicle is carrying more than 5,000 pounds (2,268 kg) of any one material picked up from a single facility. This regulation does not apply for portable tanks, tank cars, or cargo tanks.

The "Dangerous" placard should never be taken lightly during a haz-mat emergency. The hazards present when products combine may be much greater than the hazards for each product shipped separately. For example, 1,500 pounds of burning gasoline is more dangerous in the presence of an oxidizer. Beware of the multiple hazards of an incident involving the "Dangerous" placard.

Notes

Materials That May Be Shipped Under the "Dangerous" Placard

Placard Name	Hazard Class or Division Number
Explosives 1.4 (formerly Class C explosives)	1.4
Explosives 1.5 (formerly Blasting Agents)	1.5
Explosives 1.6	1.6
Flammable Gas	2.1
Non-Flammable Gas	2.2
Flammable Liquid	3
Combustible Liquid	n/a
Flammable Solid	4.1
Spontaneously Combustible	4.2
Oxidizer	5.1
Organic Peroxide	5.2
Poison (excluding Inhalation Hazard material)	6.1
Stow Away From Foodstuffs (Packing Group III only)	6.1
Corrosive	8
Miscellaneous Hazardous Material	9

Figure 12-2

INTERNATIONAL SHIPMENTS

Hazardous materials shipped across international boundaries may have markings that are significantly different from domestic markings, especially those used prior to 49 CFR as revised October 1, 1991. For example, Canada and many other countries have adopted wordless placards and labels. Canada and the United States, however, have reciprocity regarding the use of wordless and worded placards and labels. Many nations now use the United Nations publication, *Recommendations for the Transport of Dangerous Goods*, when promulgating regulations. Several examples of Canadian and international placards and labels appear in Figure 12-3.

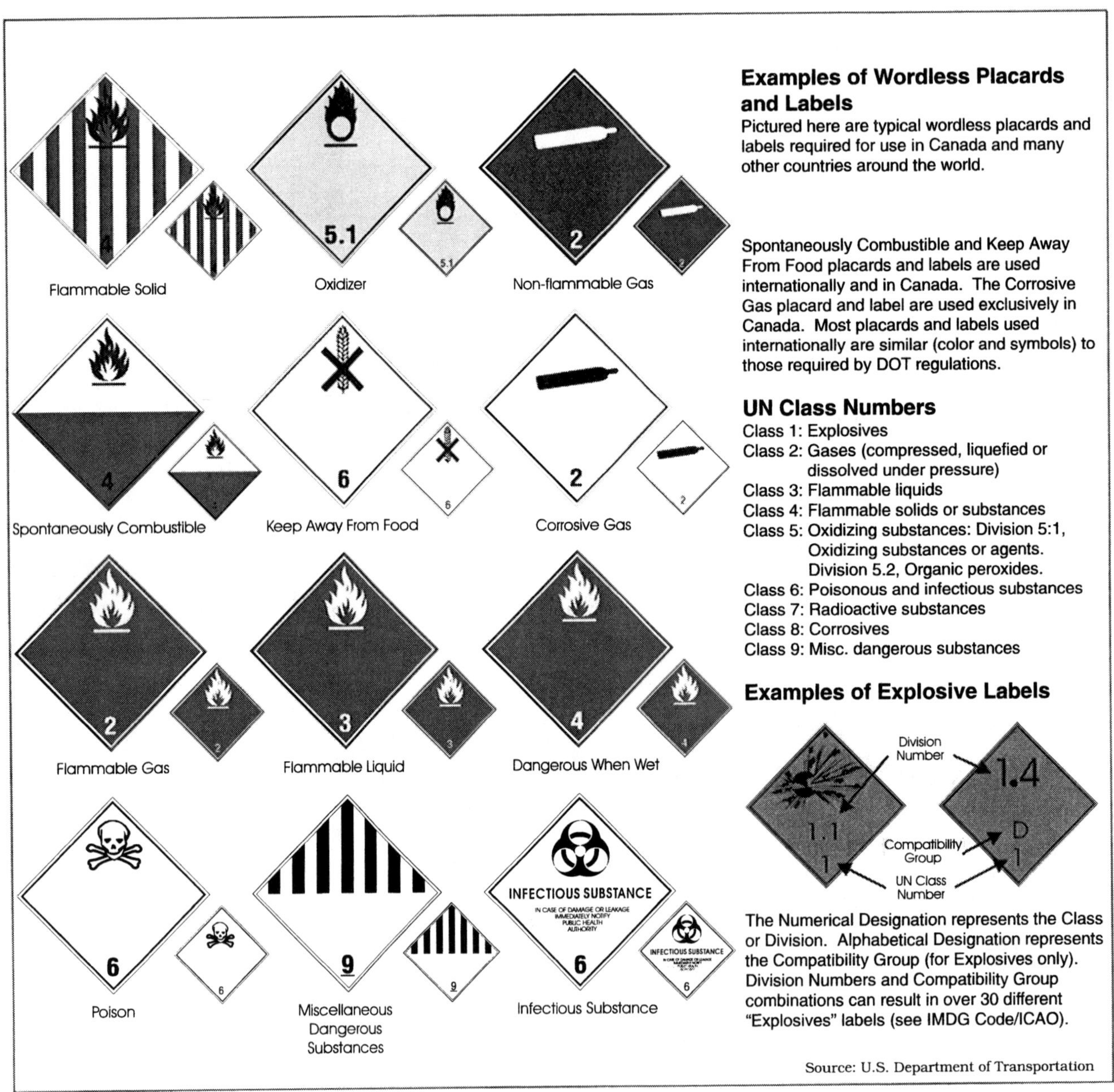

Figure 12-3. Canadian and International Placards and Labels

ELEVATED TEMPERATURE MATERIALS

Notes

Many substances usually seen as solids are transported in the molten state. These substances include molten salts used for cleaning steel, high flash point products that pour more easily when hot, and chemicals such as molten sulfur. Emergency responders must know where these products are used in their community and must be able to identify the containers.

The "Miscellaneous Hazardous Material" placard may be found on **some** of these products. In addition, current DOT regulations require that bulk containers of these materials be marked on each side and each end with the word "Hot." Molten aluminum and molten sulfur must be marked "Molten Aluminum" and "Molten Sulfur," respectively.

Molten Steel Release

Molten Sulfur Release in Benicia, California

Notes

These materials may pose a serious threat to responders due to the elevated temperature and/or the chemical hazards of the product. Specifically identify the product and the associated hazards before initiating emergency response procedures.

POLYCHLORINATED BIPHENYLS (PCBs)

Polychlorinated biphenyls (PCBs) are a class of chemicals found inside capacitors and transformers. The fluid is clear to amber in color and has the consistency of oil. PCBs from a capacitor that has exploded may be black.

Common Uses

In the United States, the EPA banned the production and manufacture of PCBs in 1979. However, electrical transformers manufactured prior to this time used PCBs as insulating fluids. They were also commonly used in many other types of electrical equipment, such as circuit breakers, lighting ballasts, and capacitors. In addition, they were used in heat transfer systems and hydraulic systems.

A typical transformer may contain 200 gallons of fluid with a PCB concentration of 50% to 60%. However, some transformers hold as much as 1,000 gallons of liquid PCBs.

Over the past 20 years, electrical capacitors have remained very similar in appearance, but electrical design changes make it difficult to pinpoint precisely how much PCB material is in each unit. A good average is approximately 1-1/2 gallons per unit. When a capacitor unit ruptures, as much as 1-1/2 gallons of fluid may be released. If a strong wind is blowing, the fluid may be splattered over a large area.

Utility facilities, such as pole top capacitors or facility storage containers, will usually be labeled if they contain PCBs. If not, PCBs can be identified by such registered trade names as Aroclor®, Pyranol®, Inerteen®, Chlorextol®, No-Flamol®, Abestol®, or the familiar askarel.

PCB Health Hazards

PCBs are known carcinogens. They may also cause a number of other adverse health problems, such as birth defects, liver damage, acne, impotence, and death. PCBs are relatively stable, so once they enter the body they will remain there for a long time. They tend to settle in the liver, fat cells, and other target organs.

Smoke from PCB fires may expose people to substantial health risks. The incomplete combustion of PCBs may form other toxic compounds, such as dioxin. A PCB transformer fire can release some of the most toxic by-products that responders will ever encounter. Realizing this, the EPA now requires the registration of PCB-containing transformers with the appropriate local fire department. Responders should know where such transformers are located in their jurisdiction. The EPA also requires the isolation of these transformers from flammable and combustible materials, as well as from locations near ventilation equipment and heating and air conditioning ductwork.

PCB Response Considerations

Notes

- Avoid any inhalation of soot and smoke if an electrical transformer catches fire.
- Avoid all contact with PCB fluids.
- Contact utility crews, who are usually trained and equipped to perform the required clean-up and disposal of any contaminated material.
- Isolate the area and wait for utility employees to take charge of the spill.
- Do not flush the area with water.
- Dike and contain all liquids to avoid human contact and spreading to other locations.
- Do not rely on the sense of smell to detect harmful PCB vapors. During a PCB leak or spill, vapors will not be detected even in an enclosed or poorly ventilated area.
- Do not allow a PCB spill to enter a water drainage system.

Electric Transformer Containing Polychlorinated Biphenyls

Notes

RESPONDING TO MISCELLANEOUS HAZ-MAT EMERGENCIES

Treat every incident as a haz-mat incident until you are 100% certain it is not.

The hazards of a miscellaneous chemical emergency are sometimes overlooked because there may be only small amounts of the material. That small amount could be a bottle of malathion dropped in the local grocery store. The incident might also involve an unplacarded truck on fire, containing 10 vehicle batteries, 50 pounds of Sevin®, and a case of Diazinon®. All of these incidents are haz-mat incidents. Remember to treat every incident as a haz-mat incident until you are 100% certain it is not.

Dealing with a miscellaneous chemical emergency is no different from any other haz-mat emergency. A systematic approach guarantees a safe and effective response. Briefly review the HAZ-MAT acronym:

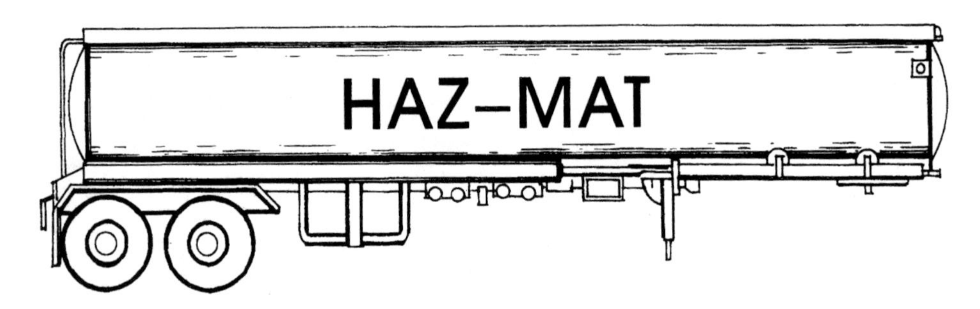

H – Hazard Identification
Recognize and identify the hazardous material.

A – Action Plan
Evaluate the situation, determine what action needs to be taken, estimate immediate and long term needs, and delegate command.

Z – Zoning
Control the risk by establishing a hot zone (exclusion), a warm zone (contamination reduction), and a cold zone.

M – Managing the Incident
Establish the necessary incident command structure.

A – Assistance
Determine the need for additional resources, such as more fire companies, haz-mat teams, technical assistance, or private contractors.

T – Termination
Determine what is needed to clean up, decontaminate, administer physical exams, rehabilitate, and conduct a post-incident analysis.

Whatever system you use, be it HAZ-MAT, Benner's D.E.C.I.D.E. Process, or a response system designed by your agency, **the important thing is to use it on every incident**.

EPILOGUE

Notes

In no way does this book provide all the information needed to mitigate every haz-mat emergency safely. It is our hope, however, that this text has challenged you in some way. For many, this material may have been simply a review, while for others it may have been a revelation.

The job of emergency responders has changed dramatically during the last 20 years. The saying, "Yesterday I couldn't spell hazardous materials responder; today I am one," is true for many responders across the United States and Canada. However, there is light at the end of the tunnel. More and more regulations are designed to reduce the risks involving hazardous materials in our communities. Right-to-know laws and development of training and equipment standards will improve safety for emergency responders.

Emergency responders have come a long way in a short period of time. Use this text as a foundation upon which to build. As stated earlier: Preparation for haz-mat incidents is much like opening doors down an endless hallway. Each opened door can provide more insight into handling the next incident. Never stop opening those doors, and be careful out there!

Appendix A

GLOSSARY OF COMMON HAZ-MAT TERMS

A

Absolute pressure – True pressure, which equals atmospheric plus gauge pressure. It is abbreviated psia.

Absolute zero – Lowest point on the Kelvin scale, at which there is total absence of heat.

Absorbed dose – An accumulated quantity of ionizing radiation. It is a measure of the energy absorbed per gram of absorbing material. The unit of absorbed dose is the radiation absorbed dose (rad). See also Dose; Rad.

Absorbent material – Material used to soak up liquid hazardous materials. Examples: sawdust, clays, charcoal, polyolefin-type fibers.

Absorption – (1) A common method of spill control in which a liquid hazardous material is absorbed or "picked up." (2) Taking in of toxic materials by oral, dermal, or inhalation exposure.

Acid/corrosive – Material usually containing an H^+ ion which is capable of dehydrating other materials.

Active ingredient – Chemical with pesticidal action. Active ingredients are listed on a pesticide container as a percentage by weight or as pounds per gallon of concentrate. See Inert Ingredient.

Activity – The decay rate of radioactive material, expressed as the average number of nuclear disintegrations per second. See also Curie.

Acute poisoning – Poisoning by a single exposure to a toxic chemical.

AFFF (Aqueous Film Forming Foam) – A foam agent used for vapor and fire control for flammable liquids.

Air bill – Shipping paper prepared from a bill of lading which accompanies each piece of an air shipment.

Air lift axle – Single air-operated axle which, when lowered, will convert a vehicle into a multi-axle unit. This provides the vehicle with a greater load-carrying capacity.

Air reactive – Materials that will ignite at normal temperatures when exposed to air.

Air spring – Flexible, air-inflated chamber in which air pressure is controlled and varied to support the load and absorb road shock. Formerly called air bag.

Alcohol foam – Foam which blankets flammable liquids in the same manner as conventional foam, but it is intended for use with liquids which are soluble in water, such as alcohol and acetone. Must be applied more carefully than regular foam because the mechanical strength of the bubbles is less.

Alpha particle – A specific particle ejected spontaneously from the nucleus of some radioactive elements. It is identical to a helium nucleus (He), which has an atomic mass of 4 and an electrostatic charge of +2. It has low penetrating power and short range. The most energetic alpha particle will generally fail to penetrate the skin. The danger occurs when matter containing alpha-emitting radionuclides is introduced into the lungs or wounds. Symbol: α.

Anhydrous – Free from water, dry. Example: anhydrous ammonia.

Antidote – Material administered to an individual exposed to a poison to counteract the poison's toxic effects.

Area of refuge – A holding area within the hot zone where exposed or contaminated personnel are protected from further contact or exposure until they can be safely decontaminated or treated.

Asphyxiating materials – Substances that can cause death through displacing the oxygen in the air.

Assisting agencies – Outside agencies that provide supporting services at a haz-mat scene that are not within the responsibility or capability of the responding agency, but are requested by the incident commander. Such services would include, but are not limited to, road closures and detours, technical advice, sampling and monitoring capabilities, clean-up, off-loading, and disposal.

Atmospheric pressure – Pressure caused by the weight of air elevated above the earth's surface. At sea level it equals 14.7 pounds per square inch (psi).

Atom – The smallest particle of an element which cannot be divided or broken up by chemical means. It consists of a central core called the nucleus, which contains protons and neutrons. Electrons revolve in orbits around the nucleus. To be broken into smaller units, an atom must be subjected to nuclear reaction.

Autoignition – Process in which a material ignites without any apparent outside ignition source. The material's temperature is raised to its ignition temperature by heat transferred by radiation, convection, combustion, or some combination of all three.

Axle weight – Amount of weight transmitted to the ground by one axle or the combined weight of two axles in a tandem assembly.

B

B-end – The end of a railroad car where the handbrake is located. It is usually the initial reference point when communicating rail car damage.

Background radiation – Naturally occurring radiation. It includes cosmic rays and radiation from natural radioactive elements, both outside and inside the bodies of people and animals. It is also called natural radiation. Man-made sources of radioactivity contribute to total background radiation levels. Approximately 90% of background radiation from man-made sources is related to the use of ionizing radiation in medicine and dentistry.

Baffle – Intermediate partial bulkhead that reduces the surge effect in a partially loaded tank.

Base/caustic – A material usually containing a hydroxide ion (OH^-) and capable of hydrolyzing other materials.

Beta particle – A small particle ejected spontaneously from the nucleus of a radioactive element. It has the mass of an electron and has a charge of –1 or +1. It has medium penetrating power and a range of up to a few meters in air. Beta particles will penetrate only a fraction of an inch of skin tissue. Symbol: β^- or β^+.

BLEVE – Acronym for boiling liquid expanding vapor explosion.

Blow down valve – Manually operated valve used to quickly reduce tank pressure to atmosphere.

BMCS – The Bureau of Motor Carrier Safety, under the Federal Highway Administration of DOT. It is responsible for establishing regulations for in-service motor vehicles and their drivers.

Boiling point – Temperature at which a liquid's vapor pressure equals atmospheric pressure.

Boilover – Explosive expulsion of hot and flaming oil due to steam expansion.

Boom – A floating physical barrier serving as a continuous obstruction to the spread of a contaminant.

Boyle's Law – Law which states that when the temperature and mass of a gas are kept constant, the product of the pressure and volume is equal to a constant.

British thermal unit (BTU) – The heat necessary to raise the temperature of 1 pound of water by 1 degree Fahrenheit.
Buddy system – A system of organizing employees into work groups in such a manner that each employee of the work group is designated to be observed by at least one other employee in the work group.
Bulk packaging – Packaging that has an internal volume greater than 119 gallons (450 liters) for liquids, a capacity greater than 882 pounds (400 kg) for solids, or a water capacity greater than 1,000 pounds (453.6 kg) for gases.
Bulkhead – Structure used to protect against damage caused by shifting cargo and/or to separate loads.
Bung – Cap or screw used to cover the small opening in the top of a metal drum or barrel.

C

Carboy – Glass or plastic bottle used for the transportation of liquids, ranging in capacity to over 20 gallons.
Carcinogenic – Material capable of producing cancer in test animals and/or humans.
Cargo manifest – Shipping paper listing contents of transporting vehicle or vessel.
Cargo tank – Tank permanently mounted on a tank truck or tank trailer, which is used for the transportation of liquefied and compressed gases, liquids, and molten materials. Examples include MC306/DOT406, MC307/DOT407, MC312/DOT412, MC331, and MC338. May also be any bulk liquid or compressed gas packaging not permanently attached to a motor vehicle which (because of its size, construction, or attachment to the vehicle) can be loaded or unloaded without being removed from the vehicle.
CAS number – The Chemical Abstract Service number. Often used by state and local right-to-know regulations for tracking chemicals in the workplace and the community.
centi – Metric prefix for 0.01, abbreviated c.
Certification label – Label permanently affixed to a trailer's forward left side stating the vehicle conforms to all applicable Federal Motor Vehicle Safety Standards in effect on the date of original manufacture.
Charged particle – An ion; an elementary particle that carries a positive or negative electrical charge.
Charles' Law – Law which states that if the volume of a gas is kept constant and the temperature is increased, the pressure increases in direct proportion to the increase in absolute temperature.
Chemical protective clothing – Items made from chemical resistive materials, such as clothing, hood, boots, and gloves, that are designed and configured to protect the wearer from hazardous materials.
Chemical reaction – Process of producing a compound by mixing two or more compounds or elements, which causes transfer or rearrangement of shared electrons.
CHEMTREC – Chemical Transportation Emergency Center. Provides haz-mat emergency assistance by relaying information on the chemical(s) involved. CHEMTREC contacts the manufacturer or other expert for additional information and on-site assistance.
Chronic poisoning – Poisoning resulting from repeated exposure to sublethal doses over a period of time.
Clean-out fitting – Fitting on the top of a tank to facilitate washing the interior.
Clean-up – Incident scene activities to remove the hazardous material and contaminated debris, including dirt, waste, road surfaces, containers, vehicles, and extinguishment tools and materials. Includes reclaiming the scene to how it existed prior to the incident. Clean-up is not usually the function of emergency response agencies, but overseeing clean-up operations is the responsibility of the incident commander. Technical guidance for clean-up can be given by the HMRT officer.
COFC – Container on flatcar.
Cold zone – The control zone of a hazardous materials incident that contains the command post and such other support functions as are deemed necessary to control the incident.
Combination – Change that occurs when two chemicals are combined and result in a different chemical.
Combustible dust – Particulate material that when mixed in air will burn or explode.
Combustible liquid – See Hazardous Materials, Categories of.

Combustible metal – Metal that will burn.

Combustion – Rapid oxidation or chemical combination, usually accompanied by heat or light.

Command – To direct and delegate authoritatively through an organization that provides effective implementation of departmental control procedures.

Command post/location – Overall incident management base for commander. Position in a safe, strategic location. Agencies involved at the incident should provide liaison officers to the command post.

Command post/vehicle – Safely and strategically located vehicle that provides the incident commander with a facility for tactical planning. It includes such resources as multiple radio channels, resource and reference books, maps, reports, etc.

Common name of pesticide – Well-known name accepted by the Environmental Protection Agency to identify the active ingredients in a pesticide, which are listed under the active ingredient statement on the label.

Compatibility charts – Permeation and penetration data supplied by manufacturers of chemical protective clothing to indicate chemical resistance and breakthrough time of various garment materials as tested against a battery of chemicals. These test data should be in accordance with ASTM and NFPA standards.

Compound – Substance with unique chemical and physical properties that is composed of two or more elements that have chemically reacted.

Compressed gas – Material or mixture having container pressure exceeding 40 psi at 70°F (21°C), or an absolute pressure exceeding 104 psia at 130°F (54°C).

Compressed gas in solution – Non-liquefied gas dissolved in a solvent at high pressures.

Concentration – The percentage of an acid or base dissolved in water. Concentration is *not* the same as strength.

Condensation – Process of going from gaseous to liquid state.

Conduction – Heat transfer through the movement of atoms within a substance.

Confinement – Procedures taken to keep a material in a defined area.

Connection box – Compartment that contains fittings for trailer emergency, service brake connections, and electrical connector to which the lines from the towing vehicle may be connected. Formerly called junction box, light box, or bird box.

Consignee – Person receiving shipment.

Consist – Rail shipping paper similar to cargo manifest. Contains list of the cars in the train in order and may list those cars carrying hazardous materials and their location on the train.

Consumer commodity – A material that is packaged and distributed in a form intended or suitable for sale through retail sales agencies for consumption by individuals for purposes of personal care or household use. This term also includes drugs and medicines.

Container – Article of transport equipment of a permanent nature, strong enough for repeated use. It is specifically designed to facilitate the carriage of goods by one or more modes of transport without intermediate reloading. It is fitted with devices permitting handling, particularly transfer from one mode to another. The term container does not include vehicles. Also referred to as freight container, cargo container, or intermodal container.

Container chassis – Trailer chassis having a frame with locking devices for securing and transporting a container as a wheeled vehicle.

Container ship – Ship specially equipped to transport large freight containers in horizontal or vertical container cells. Containers usually are loaded and unloaded automatically by special cranes.

Container specification number – Number on a shipping container preceded by the initials DOT. This indicates the container was built according to federal specification.

Containment – All activities necessary to keep a material in its container.

Contaminant – Toxic material found as a residue in or on a person, animal, or object where it is not wanted.

Contamination – Process of transferring a toxic material from its source to people, animals, the environment, or equipment, that may act as a carrier.

Contamination, radioactive – Deposition of radioactive material in any place where it is not desired, particularly where it can be harmful.

Contamination reduction corridor – Area where decontamination procedures take place. It is usually located within the warm zone.

Control – Defensive or offensive procedures, techniques, and methods used in the mitigation of a hazardous materials incident, including containment, extinguishment, and confinement.

Control agent – Material used to contain or extinguish a hazardous material or vapor.

Control zones – Areas designated at hazardous materials incidents based upon safety and the degree of hazard.

Convection – Heat transfer from one place to another by actual motion of the hot material.

Coordination – Administration and management of several tasks so as to act together in a smooth manner. The act of bringing together in a uniform fashion the functions of several agencies.

Corrosive material (DOT) – See Hazardous Materials, Categories of.

Cosmic rays – High-energy particulate and electromagnetic radiations which originate outside the earth's atmosphere.

Covalent bond – Chemical bond in which atoms share electrons in order to form a molecule.

Critical pressure – Maximum pressure required to liquefy a gas that is at or below its critical temperature.

Critical temperature – Maximum temperature to which a liquid can be heated and still remain a liquid. Additional heating will cause the liquid to vaporize irrespective of pressure.

Criticality – A self-sustaining nuclear reaction that may occur with fissionable materials under special conditions. This type of reaction occurs in nuclear reactors and atomic weapons.

Critique – An element of incident termination which examines the overall effectiveness of the emergency response effort and develops recommendations for improving the organization's emergency response system.

Crossover line – Pipe installed in a tank piping system to allow unloading from either side of tank.

Cryogenic liquid – Liquid with a boiling point below −130°F (−90°C).

Cubic capacity – Useful internal load-carrying space usually expressed in cubic feet, cubic yards, or cubic meters. Also referred to as available cube or cube.

Curie – Basic measuring unit of radioactivity. One curie is equal to 37 billion disintegrations per second. Abbreviation: Ci.

D

Dam – Defensive confinement procedure consisting of constructing a dike or embankment to totally immobilize a flowing waterway contaminated with a liquid or solid hazardous substance.

Dangerous cargo manifest – Cargo manifest used on ships listing all hazardous materials on board and their location.

Dangerous goods – Term used for hazardous materials in Canadian transportation.

deca – Metric prefix for 10, abbreviated da.

Decay, radioactive – Disintegration of the nucleus of an unstable atom through spontaneous emission of charged particles, electromagnetic radiation, or both.

deci – Metric prefix for 0.1, abbreviated d.

Decomposition – Breaking down of a substance to a less complex form by the introduction of heat through the addition of neutralized chemicals or biodegradation.

Decontamination (contamination reduction) – Physical or chemical process of reducing and preventing the spread of contamination from persons and equipment used at a hazardous materials incident.

Decontamination corridor – Distinct area within the warm zone that functions as a protective buffer and bridge between the hot zone and the cold zone, where decontamination stations and personnel are located to conduct decontamination procedures.

Defensive tactics – Spill and fire control tactics that are less aggressive, in which certain areas may be "conceded" to the emergency, with response efforts directed toward limiting the overall size or spread of the problem.

Deflagration – Intense burning rate of some explosives (e.g., black powder).

Degradation – (1) Chemical action involving the molecular breakdown of a protective clothing material or equipment due to contact with a chemical. (2) Molecular breakdown of the spilled or released material to render it less hazardous during control operations.
Detonation – Extreme rapid self-propagating decomposition of an explosive, accompanied by a wave that passes along the body of the explosive, instantaneously converting the explosive into gas.
Dewar – Small insulated container (less than 25 gallons) used for temporary storage or handling of cryogenic liquids.
Dike – Defensive confinement procedure consisting of an embankment or ridge on ground used to control the movement of liquids, sludges, solids, or other materials.
Diluent – Liquid or solid material used to carry or dilute an active ingredient.
Dilution – Application of water to water-miscible hazardous materials to reduce to safe levels the hazard they represent. In decontamination applications, dilution is the use of water to flush a hazardous material from protective clothing and equipment. It is the most common method of decontamination.
Dip tube – Device installed for pressure unloading of product from the top of the tank.
Distillation – Process of going from liquid to gaseous to liquid state.
Diversion – Defensive confinement procedure to intentionally control the movement of a hazardous material into an area where it will pose less harm to the community and the environment.
Dome – Circular fixture on the top of a tank car containing valves and relief devices.
Dose – (1) The amount of substance ingested, absorbed, and/or inhaled during an exposure period. (2) General term denoting the amount of absorbed radiation or energy. If unqualified, it refers to absorbed dose. It must, however, be qualified for special purposes. For example, if used to represent exposure expressed in roentgens (R), it measures the total ionization that the radiation could produce in air. See also Absorbed Dose.
Dose rate – The absorbed dose delivered per unit time. It usually is expressed as rads per hour or in multiples or sub-multiples of this unit, such as millirads per hour. The dose rate is commonly used to indicate the hazard level. See also Dose; Rad.
Dosimeter – A small, pocket-sized ionization chamber used for monitoring personnel radiation exposure. It is charged before use. Then the accumulated radiation exposure is measured as the device discharges.
DOT – United States Department of Transportation.
Double – Trailer combination consisting of a truck tractor, semi-trailer, and full trailer coupled together. Formerly called double-trailer or double-bottom.
Dry chemical – Special fire extinguishing chemical (sodium or potassium bicarbonate or monosodium phosphate powder), usually available from a semi-fixed or portable extinguisher.
Dummy coupler – Fitting used to seal the opening in an air brake hose connection (gladhands) when the connection is not in use; a dust cap.

E

Edema – Swelling of tissues due to fluid accumulation.
Electromagnetic energy – Energy form of the same composition as light, transmitted by waves in small energy packets known as photons.
Electromagnetic radiation – A traveling wave motion that occurs when changing electric and magnetic fields. Familiar electromagnetic radiations range from those having short wavelengths, such as X rays and gamma rays, to those having relatively long wavelengths, such as radar and radio waves.
Element – A substance that cannot be broken into any other substances by chemical means.
Elevated temperature material – Material which (1) is a liquid and is at a temperature at or above 212°F (100°C) during transportation, (2) is a liquid with a flash point at or above 100°F (38°C) that is transported at or above its flash point, or (3) is a solid and is at a temperature at or above 464°F (240°C) during transportation.

Emergency decontamination – Physical process of immediately reducing the contamination of individuals in situations that are potentially life-threatening, without the formal establishment of a contamination reduction corridor.

Emergency response plan – Plan that establishes guidelines for handling hazardous materials incidents as required by 29 CFR 1910.120.

Emergency shutoff lever – Automatic or manually operated safety valve control that stops the flow of a liquid.

Emergency valve – Self-closing tank outlet valve.

Emergency valve operator – Device used to open and close emergency valves.

Emergency valve remote operator – Secondary closing means that is remote from tank discharge openings for operation in event of fire or other accident.

Emulsifiable concentrate – Material mixed with a solvent (usually petroleum), forming an emulsion when mixed with water for application.

Emulsion – Mixture in which one liquid is suspended as tiny drops in another liquid, such as oil in water.

Encapsulated source – A radionuclide sealed in containers such as tubes or needles. Also called a sealed source.

Endangered area – Actual or potential area of exposure from a hazardous material. Sometimes referred to as the engulfed area.

Endothermic – Absorbing heat during a reaction.

EPA registration number – Number appearing on pesticide label to identify individual pesticide product. May appear as "EPA Reg. No."

Etiologic agent – An infectious substance.

Evacuation – Public protective option which results in the removal of fixed facility personnel and the public from a threatened area to a safer location.

Exothermic – Giving off heat during a reaction.

Expansion ratio – Amount of gas produced by the evaporation of one volume of liquid at a given temperature.

Explosion – Sudden and rapid production of gas, heat, noise, and a shock wave, within a confined space. Sudden release of a large amount of energy in a destructive manner. It is the result of powders, mists, or gases undergoing instantaneous ignition, or liquids or solids undergoing sudden decomposition, or pressurized vessel undergoing overpressure rupture, with such a force as to generate tremendous heat, cause severe structural damage, occasionally generate a shock wave, and propel shrapnel.

Explosive (DOT) – See Hazardous Materials, Categories of.

Explosive limits – See Flammable Limits.

Exposure – Subjection of a person to a toxic substance or harmful physical agent through any route of entry (e.g., inhalation, ingestion, skin absorption, or direct contact).

Exposure, radioactive – Amount of ionization in air produced by X rays or gamma radiation. The unit is the roentgen (R). For practical purposes, 1 roentgen = 1 rad = 1 rem for X rays and gamma radiation. See also Rad; Rem; Roentgen.

Extremely flammable – Liquid with a flash point of 20°F (–7°C) or lower, determined by closed cup or Seta flash test.

F

FHWA – Federal Highway Administration. The DOT division concerned with highway construction and usage. Other similar divisions of DOT relate to air, rail, and water transportation.

Fifth wheel – Device used to connect a truck tractor or converter dolly to a semi-trailer to permit joining the units. It is composed of a lower part consisting of a trunnion, plate, and latching mechanism mounted on the truck tractor (or dolly), and a kingpin assembly mounted on the semi-trailer.

Fifth wheel pick-up ramp – Steel plate designed to lift the front end of a semi-trailer to facilitate engagement of the kingpin into the fifth wheel.

Fill opening – Opening on top of a tank used for filling the tank usually incorporated in a manhole cover.

First responders, awareness level – Individuals who are likely to witness or discover a hazardous substance release, who have been trained to initiate an emergency response sequence by notifying the proper authorities of the release. They would take no further action beyond notifying the authorities of the release.
First responders, operations level – Individuals who respond to releases or potential releases of hazardous substances as part of the initial response to the site for the purpose of protecting nearby persons, property, or the environment from the effects of the release. They are trained to respond in a defensive fashion without actually trying to stop the release. Their function is to contain the release from a safe distance, keep it from spreading, and prevent exposures.
Fissile material – Material that is capable of undergoing fission. Fissile material requires special precautions during transport to assure nuclear criticality safety. Fissile materials include plutonium-238, plutonium-239, plutonium-241, uranium-233, and uranium-235.
Fission – Nuclear transformation characterized by the splitting of an atomic nucleus into at least two other nuclei and the release of a relatively large amount of energy.
Flame impingement – Points where flames contact the surface of a container.
Flammable – Material that is easily ignited and burns with extreme rapidity.
Flammable compressed gas – Flammable material or mixture in a container with pressure exceeding 41 psia at 68°F (20°C).
Flammable gas – See Hazardous Materials, Categories of.
Flammable limits – Range of gas or vapor concentrations (percent by volume in air) that will burn or explode if an ignition source is present. Limiting concentrations are commonly called the *lower flammable limit* (LFL) and the *upper flammable limit* (UFL). Below the LFL the mixture is too lean to burn, and above the UFL it is too rich to burn.
Flammable liquid – See Hazardous Materials, Categories of.
Flammable range – Difference between the minimum and maximum volume percentages of the material in air that forms a flammable mixture.
Flammable solid (DOT) – See Hazardous Materials, Categories of.
Flash point – Minimum temperature at which a substance gives off flammable vapors, which will ignite upon contact with sparks or flame.
Flashing – Liquid-tight rail on top of a tank that contains water and spillage and directs it to suitable drains. May be combined with DOT overturn protection.
Flashing drain – Metal or plastic tube that drains water and spillage from flashing to the ground.
Forbidden – Prohibited from being offered or accepted for transportation.
Free radical – Fragment of a molecule with an open or unsatisfied chemical (covalent) bond. These fragments are very reactive and unstable in almost all cases.
Freezing point – Temperature in °C or °F at which a liquid solidifies.
Full protective clothing – Protective clothing worn primarily by firefighters which includes helmet, fire-retardant hood, coat, pants, boots, gloves, PASS device, and self-contained breathing apparatus designed for structural firefighting. It does not provide specialized chemical splash or vapor protection.
Fusible plug – Safety relief device in the form of a plug of low-melting metal, which blocks the safety relief device channel under normal conditions and is intended to yield or melt at a set temperature to permit the escape of gas.

Gamma rays, gamma radiation – High-energy electromagnetic radiation originating in atomic nuclei. It accompanies many nuclear reactions including fission, radioactive decay, and neutron capture. Gamma rays are identical to high-energy X rays except that X rays do not originate from atomic nuclei. They instead are produced in ways such as slowing down fast, high-energy electrons. Gamma rays are the most penetrating type of radiation and represent a major external hazard. Symbol: γ.
Gauge pressure – Pressure read on a gauge which does not take atmospheric pressure into account. The abbreviation for this pressure reading is psig.

Geiger counter, G-M meter – Instrument used to detect and measure radiation. The detecting element is a gas-filled chamber with an electrical discharge that will spread over the entire anode when triggered by a primary ionizing event.

Gladhands – Fittings for connection of air brake lines between vehicles, also called hose couplings, handshakes, or polarized couplers.

Gram – Weight of 1 cubic centimeter of water at 4°C.

Gram-calorie – Amount of heat necessary to raise the temperature of 1 gram of water 1 degree Centigrade, from 14.5°C to 15.5°C.

Gross decontamination – Initial phase of the decontamination process during which the amount of surface contaminant is significantly reduced. This phase may include mechanical removing and initial rinsing.

Gross weight – Weight of a trailer plus the weight of its entire contents.

H

ΔH – Change in heat. It usually refers to heat added to or given off by a reaction.

Half-life – Length of time it takes for a radioactive substance to lose one-half of its radioactive intensity.

Hazard class – Group of materials designated by the Department of Transportation that share a common major hazardous property (e.g., radioactivity, flammability).

Hazard sector – That function within an overall incident management system that deals with the mitigation of a hazardous materials incident.

Hazard sector officer – Person responsible for the management of the hazard sector.

Hazardous material – Material or substance in a quantity or form that when not properly controlled or contained may pose an unreasonable risk to health, safety, property, and the environment. It requires implementing special control procedures to supplement standard departmental procedures, and it may require the use of specialized equipment and reference material. For the purpose of this book, "hazardous material," "hazardous substance," "dangerous material," and "dangerous chemical" are synonymous.

Hazardous materials, categories of:

a. Explosive – Any substance, article, or device which is designed or is able to function by explosion, with extremely rapid release of gas and heat.

b. Flammable gas – Any material which is a gas at 68°F (20°C) or less and 14.7 psi of pressure and which (1) is ignitable when in a mixture of 13% or less by volume with air at atmospheric pressure, or (2) has a flammable range with air at atmospheric pressure of at least 12% regardless of the lower flammable limit.

c. Non-flammable, non-poisonous, compressed gas – Any material or mixture which exerts in the packaging an absolute pressure of at least 41 psia at 68°F (20°C) and does not meet the definition of a flammable gas or a poisonous gas. This category includes compressed gas, liquefied gas, pressurized cryogenic gas, and compressed gas in solution.

d. Gas poisonous by inhalation – A material which is a gas at 68°F (20°C) or less and a pressure of 14.7 psi which (1) is known to be so toxic to humans as to pose a hazard to health during transportation, or (2) is presumed to be toxic to humans based on tests conducted on laboratory animals.

e. Flammable liquid – A liquid having a flash point of not more than 141°F (61°C), or any liquid with a flash point at or above 100°F (38°C) that is transported at or above its flash point in a bulk packaging.

f. Combustible liquid – Any liquid that does not meet the definition of any other hazard class and that has a flash point above 100°F (38°C) and below 200°F (93°C), as determined by specific tests listed in 49 CFR 173.120(b)–(c).

g. Flammable solid – Any of the following three types of materials: (1) wetted explosives, (2) self-reactive materials that are liable to undergo heat-producing decomposition, or (3) readily combustible solids that may cause a fire through friction or that have a rapid burning rate as determined by specific tests listed in 49 CFR 173.142(a).

h. Spontaneously combustible material – A liquid or solid that can ignite after coming in contact with air, without an external ignition source, or that is liable to self-heat when in contact with air, without an energy supply.

i. Dangerous when wet material – A material that is liable to become spontaneously flammable or to give off flammable or toxic gas as a result of contact with water.

j. Oxidizer – A material that may cause or enhance the combustion of other materials, generally by yielding oxygen.

k. Organic peroxide – Any organic compound which may be considered a derivative of hydrogen peroxide, where one or more of the hydrogen atoms have been replaced by organic radicals.

l. Poisonous material – A material, other than a gas, which (1) is known to be so toxic to humans as to afford a hazard to health during transportation, (2) is presumed to be toxic to humans based on tests conducted on laboratory animals, or (3) is an irritating material with properties similar to tear gas, which causes extreme irritation, especially in confined spaces.

m. Infectious substance – A viable microorganism or its toxin or any other agent that causes or may cause disease in humans or animals.

n. Radioactive material – Any material that spontaneously emits ionizing radiation and that has a specific activity greater than 0.002 microcuries per gram.

o. Corrosive material – A liquid or solid that causes visible destruction or irreversible alterations in human skin tissue at the site of contact, or a liquid that has a severe corrosion rate on steel or aluminum.

p. Miscellaneous hazardous material – A material which presents a hazard during transportation but which does not meet the definition of any other hazard class. This class includes (1) any material which has an anesthetic, noxious, or other similar property which could cause extreme annoyance or discomfort to a flight crew member so as to prevent the correct performance of assigned duties, or (2) any material which meets the definition for an elevated temperature material, a hazardous substance, or a hazardous waste.

q. ORM-D (Other regulated material) – A material such as a consumer commodity which, although otherwise subject to the regulations, presents a limited hazard during transportation due to its form, quantity, and packaging.

Hazardous materials incident – Any spill, leak, rupture, fire, or accident that results in or has the potential to result in the loss or escape of hazardous material from its container.

Hazardous materials response team (HMRT) – An organized group of employees, designated by the employer, who are expected to perform work to handle and control actual or potential leaks or spills of hazardous substances requiring possible close approach to the substance.

Hazardous materials specialists – Individuals who respond and provide support to hazardous materials technicians. While their duties parallel those of technicians, they require a more detailed or specific knowledge of the various substances they may be called upon to contain.

Hazardous materials technicians – Individuals who respond to releases or potential releases of hazardous materials for the purpose of stopping the leak. They generally assume a more aggressive role in that they are able to approach the point of a release in order to plug, patch, or otherwise stop the release of a hazardous substance.

Haz-mat – Abbreviation for hazardous material.

Head – Front and rear closure of a tank shell.

Heat exhaustion – Mild form of shock caused when the circulatory system begins to fail as a result of the body's inadequate effort to give off excessive heat.

Heat of fusion – Quantity of heat that must be supplied to a material at its melting point to convert it completely to a liquid at the same temperature.

Heat of vaporization – Quantity of heat that must be supplied to a liquid at its boiling point to convert it completely to a gas at the same temperature.

Heat rash – Inflammation of the skin resulting from prolonged exposure to heat and humid air and often aggravated by chafing clothing.

Heat stroke – Severe and sometimes fatal condition resulting from the failure of the temperature regulating capacity of the body.

Heating tube – Tube installed inside a tank used to heat the contents. Also may be called fire tube.

hecto – Metric prefix for 100, abbreviated h.

Hemispherical head - Head that is half a sphere in shape. Used on MC331 high pressure tanks.
High temperature protective clothing - Protective clothing designed to protect the wearer against short-term high temperature exposures. It includes both proximity suits and fire entry suits.
Hopper - Sloping panels at the bottom of a tank that direct dry bulk solids to the outlet piping.
Hose tube - Housing used on tank and bulk commodity trailers for the storage of cargo-handling hoses. Also called hose troughs.
Hot zone - Area immediately surrounding a hazardous materials incident, which extends far enough to prevent adverse effects from hazardous materials releases to personnel outside the zone.
Hydrocarbons - Compounds primarily made up of hydrogen and carbon. Examples include LPG, gasoline, and fuel oils.
Hygroscopic - Absorbing and retaining moisture from the air.
Hypergolic reaction - Immediate spontaneous ignition when two or more materials are mixed.

I

ICC - Interstate Commerce Commission. An independent federal government agency in the executive branch, not affiliated with DOT, charged with administering acts of Congress affecting rates and routes for interstate commerce.
Ignition temperature - Minimum temperature to which a material must be heated to cause the formation of free radicals (chemical bond breakage). If radical formation occurs in the presence of an oxidizer, combustion results.
Impermeable - Cannot be penetrated by liquid or vapor.
Incident command - System of command and control designed to assure the smooth implementation of immediate and continued operational procedures until the incident has been contained or abated.
Incident commander (IC) - Representative of the agency responsible for overall direction and control of immediate on-scene mitigation functions.
Incipient fires - Fires in the beginning stages.
Inert ingredient - Material in a pesticide formula that has no pesticidal activity but could be flammable or combustible.
Infectious substance - See Hazardous Materials, Categories of.
Ingestion - Taking in of toxic materials through the mouth.
Inhalation - Taking in of toxic materials by breathing through the nose or mouth.
Inorganic chemistry - Chemistry dealing with compounds not containing carbon.
Inverse square law - Law which states that gamma radiation intensity is inversely proportional to the square of the distance from a point source.
Ion - Atomic particle, atom, or chemical radical bearing either a negative or positive electrical charge.
Ionic bond - Chemical bond in which atoms of different elements transfer or exchange electrons. As the electrons are exchanged, charged particles known as ions are formed.
Ionization - The process of separating a normally neutral atom or molecule into electrically charged components. The term also describes the degree or extent to which this separation occurs. Ionization removes an electron (a negative charge) from an atom or molecule, either directly or indirectly, and leaves a positively charged ion. The separated electron and ion are called an ion pair.
Ionizing radiation - Electromagnetic radiation (X ray and gamma ray radiation photons) or particulate radiation (electrons, positrons, protons, neutrons, and heavy particles) capable of producing ions by direct or secondary processes.
Irradiation - Exposure to ionizing radiation.
Isolation perimeter - Designated crowd control line surrounding the hazard control zones. It is always the line between the general public and the cold zone.
Isotopes - Forms of the same chemical element having identical chemical properties but differing in their atomic masses and nuclear properties.

J

Jacket – Metal cover that protects tank insulation.
Jackknife – Condition of truck tractor/semitrailer when their relative positions to each other form a 90° angle or less around the trailer kingpin.

K

kilo – Metric prefix for 1,000, abbreviated k.
Kingpin – Attaching pin on a semi-trailer that mates with and pivots within the lower coupler of a truck tractor or converter dolly while coupling the two units together.
Kingpin assembly – See Upper Coupler Assembly.

L

Labels – Diamond shaped markers 3.9 inches (100 mm) square that are required on individual shipping containers.
LD50 – Estimated lethal dosage to 50% of the population that was exposed.
LD5030 – Estimated lethal dosage to 50% of the population exposed in 30 days.
Leak – Release of a flammable, toxic, poisonous, or noxious liquid or gas that poses a threat to air or ground quality and health safety.
Limited quantity – Maximum amount of a hazardous material for which there is a specific labeling or packaging exception.
LIN – Liquid nitrogen.
Liquefied gas – Gas that is partially liquid at a temperature of 70°F (21°C).
Liquid splash protective clothing – Garment portion of a chemical protective clothing ensemble that is designed and configured to protect the wearer against chemical liquid splashes but not against chemical vapors or gases.
Liter – Volume of 1 kilogram of water at 4°C.
Lower flammable limit – See Flammable Limits.
LOX – Liquid oxygen.

M

Manhole – Openings with removable, lockable covers and large enough for a person to enter a tank trailer or dry bulk trailer.
Manifest – Shipping document that lists the commodities being transported on a vessel.
Manifold – Device used to join a number of discharge pipelines to a common outlet.
Medical monitoring – Ongoing, systematic evaluation of individuals at risk of suffering adverse effects of exposure to heat, stress, or hazardous materials as a result of working at a haz-mat emergency.
Medical surveillance – Comprehensive medical program for tracking the overall health of the program's participants (e.g., HMRT personnel, public safety responders, etc.). Medical surveillance programs consist of pre-employment screening, periodic medical examinations, emergency treatment provisions, non-emergency treatment, and record keeping and review.
mega – Metric prefix for 1,000,000, abbreviated M.
Melting point – Temperature in °F or °C at which a solid becomes a liquid.
Meter – Measure of length based on the spectrographic color line of the element krypton.
micro – Metric prefix for 10^{-6} or 0.000001, abbreviated μ.
milli – Metric prefix for 0.001, abbreviated m.
Miscible – Two or more liquids that can be mixed and will remain mixed under normal conditions.
Mixture – Combination of chemicals that contain two or more substances and do not lose their individual identities.
MLD – Estimated minimum lethal dose.
Molecule – Two or more atoms combined by either ionic or covalent interaction of electrons (bond). The smallest unit into which a compound may be broken.
Monitoring – Periodic or continuous determination of present ionizing radiation or radioactive contamination for health protection purposes. Also referred to as "surveying."
Monitoring equipment – Instruments and devices used to identify and quantify contaminants.
MPBB – Abbreviation for *maximum permissible body burden*. The maximum amount of a specific radionuclide that supposedly produces no adverse health effects.

Mutagenic – Capable of producing genetic changes in animals and/or humans, which are passed on to subsequent offspring.

N

Necrosis – Cell or tissue death due to disease or injury.

Neutralization – Process of neutralizing a hazardous materials liquid spill by applying another material to the spill which will react chemically with it to form a less harmful substance. Also refers to those materials which can be used to neutralize the effects of a corrosive material (e.g., acids and bases).

NFPA 704 – National Fire Protection Association Pamphlet 704. This publication describes a system for marking hazardous materials as to health hazard, flammability, and reactivity.

NHTSA – National Highway Traffic Safety Administration. A division of DOT responsible for establishing motor vehicle safety standards and regulations for *new* vehicles. Formerly called National Highway Safety Bureau (NHSB).

Non-flammable gas – See Hazardous Materials, Categories of.

Non-liquefied gas – Gas that is entirely gaseous at a temperature of 68°F (20°C).

NOS or NOIBN – Notations designating *not otherwise specified* or *not otherwise identified by name*, which appear on shipping papers when the materials conform to a haz-mat definition, but are not listed by generic name in the regulations (e.g., Flammable NOS).

Nucleus, atomic – The small, positively charged core of an atom. It is about 1/100,000 diameter of the atom, but it contains nearly all the atom's mass. All nuclei except hydrogen contain both protons and neutrons. The hydrogen nucleus consists of a single proton.

Nuclide – A general term referring to any nuclear species of the chemical elements. There are about 270 stable nuclides and about 1,250 radioactive nuclides (radionuclides).

O

Offensive tactics – Aggressive leak, spill, and fire control tactics designed to quickly control or mitigate the problem.

Oil and Hazardous Materials–Technical Assistance Data System (OHM–TADS) – Organization within the Environmental Protection Agency (EPA) that provides information on hazardous substances to emergency teams responding to spills.

Organic chemistry – Chemistry dealing with compounds containing carbon.

Organic peroxide – See Hazardous Materials, Categories of.

Organochlorine compounds (chlorinated hydrocarbons) – Synthetic organic pesticides that contain chlorine, carbon, and hydrogen and affect the central nervous system (e.g., DDT and endrin).

Organophosphates – Synthetic organic pesticides that contain carbon, hydrogen, and phosphorus and are toxic to humans because they prevent proper transmission of nerve impulses (e.g., parathion and malathion).

ORM-D – See Hazardous Materials, Categories of.

Outage – Space left in a vessel filled with a flammable liquid.

Outlet valve – Valve farthest downstream in a tank piping system to which the discharge hose is attached.

Outrigger – Structural load-carrying members attached to and extending outward from the main longitudinal frame members of a trailer.

Overpacking – Use of a specially constructed drum to overpack damaged or leaking containers of hazardous materials for shipment.

Overturn protection – Fittings on top of a tank in case of rollover. May be combined with flashing rail or flashing box.

Oxidizer – See Hazardous Materials, Categories of.

Oxidizing agent – Material that gains electrons from the fuel during combustion.

Oxidizing material – Substance that readily yields oxygen to stimulate the combustion of organic matter.

Oxygen deficient atmosphere – Atmosphere which contains an oxygen content less than 19.5% by volume at sea level.

P

Package markings – Descriptive name, instructions, cautions, weight, or specification marks required on the outside of haz-mat containers.

Packing group – Classification of hazardous materials based on the degree of danger represented by the material. There are three groups: Packing Group I indicates great danger, Packing Group II indicates medium danger, and Packing Group III indicates minor danger.

Patching/plugging – Use of chemically compatible patches and plugs to reduce or temporarily stop the flow of materials from small holes, rips, tears, or gashes in containers.

Penetration – Movement of a material through a suit's closures, such as zippers, buttonholes, seams, flaps, or other design features of chemical protective clothing, and through punctures, cuts, and tears.

Permeation – Chemical action involving the movement of chemical, on a molecular level, through intact material.

Personal protective equipment – Equipment provided to shield or isolate a person from the chemical, physical, and thermal hazards that may be encountered at a hazardous materials incident. It includes both personal protective clothing and respiratory protection.

Pesticide – Chemical used to destroy, prevent, or control any living organism considered to be a pest.

pH – Negative logarithm of the hydrogen ion concentration of a solution. Because it is a negative logarithmic relationship, the smaller the number of the pH, the greater the concentration of hydrogen (H^+) ions.

Physical properties – Properties of a material that relate to the physical states common to all substances: solid, liquid, or gas.

Pick-up plate – Sloped plate and structure of a trailer located forward of the kingpin. It is designed to facilitate engagement of the fifth wheel to the kingpin.

Piggyback transport – Type of shipping in which bulk containers from one mode, such as highway transportation, are placed on flatcars or container ships for transportation by another mode, such as rail or marine.

Pintle – Connecting device at the rear of a vehicle used to pull a trailer, with provision for easy coupling.

Placards – Diamond shaped markers 10.8 inches (273 mm) square that are required on the transporting vehicle, such as a truck or tank car or a freight container.

Poison – Substance which, even in small amounts, can produce injury to susceptible tissues by a chemical action.

Polarized couplers – Fittings for connecting air brake lines between vehicles. The service and emergency couplings are unilateral and will not mate with each other.

ppm – Parts per million.

Pressure – Force applied over a given area.

Protection in place – Directing fixed facility personnel and the general public to go inside a building or structure and to remain indoors until the danger from a hazardous materials release has passed.

Protective clothing – Equipment designed to protect the wearer from heat and/or hazardous materials contacting the skin or eyes.

psia – See Absolute Pressure.

Public information officer – Individual responsible for interface with the media or other appropriate agencies requiring information direct from the incident scene. This person is a member of the command staff.

Public protective actions – Strategy used by the incident commander to protect unexposed people from a hazardous materials release by evacuating or protecting in place.

Pumpoff line – Pipeline that runs from the tank discharge openings to the front of the trailer usually mounted on the tractor.

Pyrophoric – Capable of igniting spontaneously when exposed to dry or moist air at or below 130°F (54°C).

R

rad – Abbreviation for *radiation absorbed dose.* A rad is the unit of absorbed dose. It measures the energy imparted to matter by ionizing particles per unit mass of irradiated material at the place of interest. A rad is approximately equal to the absorbed dose in tissue when the exposure in air is 1 roentgen (R) of medium-voltage x-radiation.

Radiation – The energy propagated through space or a material medium such as waves. An example is electromagnetic energy waves. Radiation, or radiant energy, when unqualified, usually refers to electromagnetic radiation. Such radiation is classified according to frequency, such as Hertzian, infrared, visible (light), ultraviolet, X ray, and gamma ray. Alpha and beta radiation, or rays of mixed or unknown type, such as cosmic rays, also can be called radiation.

Radiation accident – Accident in which there is an unintended exposure to ionizing radiation or radioactive contaminants.

Radioactive material (DOT) – See Hazardous Materials, Categories of.

Radioactivity – Spontaneous emission of alpha, beta, or gamma rays from the nucleus of an unstable atom. As a result, the radioactive atom converts or decays into an atom of a different element that may or may not be radioactive. The final result of one or more stages of radioactive decay is usually a stable non-radioactive atom.

Radioisotopes – Artificially radioactive elements.

Radionuclide – See Nuclide.

Reactive – Capable of or tending to react chemically with other substances.

Reactivity – Degree of ability of one substance to undergo a chemical combination with another substance.

Reducing agent – Fuel that becomes chemically changed by the oxidizing process.

Reid vapor pressure – Equilibrium pressure exerted by vapor over liquid at 100°F (38°C), expressed in pounds per square inch absolute.

Relay emergency valve – Combination valve in an air brake system controlling brake application. It provides for automatic emergency brake application should the trailer become disconnected from the towing vehicle.

rem – Abbreviation for *roentgen equivalent man*. A special unit of radiation dose equivalent. The rem dose equivalent is numerically equal to the absorbed dose multiplied by the quality factor (Q), the distribution factor, and any necessary modifying factors.

Reportable quantity (RQ) – Designated amount of a hazardous substance that, if spilled or released, requires immediate notification to the National Response Center (NRC).

Respiratory protection – Equipment designed to protect the wearer from the inhalation of contaminants.

Retention – Defensive spill confinement method. It involves temporary containment of a hazardous material in an area where it can be absorbed, neutralized, or picked up for proper disposal.

Roentgen – Measure of exposure to radiation, abbreviated R. One roentgen is approximately 2.1 billion ionizations from X rays or gamma radiation per cubic centimeter of air.

Rotary gauge – Gauge for determining the liquid level in a pressurized tank.

Running lights – Marker, clearance, and identification lights required by regulations.

Rupture – Physical failure of a container that releases or threatens to release a hazardous material. Physical failure may be due to forces that cause punctures, creases, tears, corrosion, breakage, or collapse of the container.

Rupture disc – Safety device in the form of a metal disc that closes the relief channel under normal conditions. The disc bursts at a set pressure to permit the escape of gas.

S

Safety officer – Individual responsible for monitoring and assessing safety hazards and unsafe conditions and developing measures for ensuring personnel safety.

Sandshoe – Flat steel plate that serves as ground contact on the supports of a trailer. It is used instead of wheels, particularly where the ground surface is expected to be soft.

Sealed source – Radioactive source sealed in an impervious container. The container has sufficient mechanical strength to prevent contact with and dispersion of the radioactive material under the conditions of use and wear for which the container was designed. Generally used for radiography or radiation therapy.

Secondary contamination – Process by which a contaminant is carried out of the hot zone and contaminates people, animals, the environment, or equipment outside the hot zone.

Self-accelerating decomposition temperature (SADT) – Temperature above which decomposition of an unstable material proceeds by itself, independently of the external temperature.

Self-contained breathing apparatus (SCBA) – Positive pressure, self-contained breathing apparatus (SCBA) or combination SCBA/supplied air breathing apparatus for use in atmospheres that are immediately dangerous to life or health (IDLH).

Semi-trailer – Truck trailer with one or more axles designed to carry the weight of the front end, a substantial part of its own weight, and the weight of its load upon a truck tractor.

Sensitizer – Chemical that causes a substantial proportion of exposed people or animals to develop an allergic reaction in normal tissue after repeated exposure to the chemical.

Shear sections – Safety feature, such as a valve or joint, built onto a cargo tank designed to fail or break completely to prevent a failure or break of the tank itself.

Shipping package or warning label – Label affixed to a package of hazardous material to identify the package contents. Department of Transportation regulations establish the design and use requirements for these labels.

Shipping papers or shipping documents – Forms containing a description of the materials being transported. Such papers must accompany all packages of radioactive material.

Site management and control – Management and control of the physical site of a haz-mat incident. Includes initially establishing command, approach, and positioning; staging; establishing initial perimeter and hazard control zones; and implementing public protective actions.

Skin absorption – Introduction of a chemical or agent into the body through the skin.

Sliding fifth wheel – Fifth wheel assembly capable of being moved forward or backward on the truck tractor to vary load distribution on the tractor, and to adjust the overall length of combination.

Slope sheet – Panels located at each end of payload compartment that direct product by gravity to hoppers.

Soil contamination – Contamination of the ground area where a hazardous material spill or fire occurs, or where contaminated runoff water flows.

Solubility – Ability of a solid, liquid, gas, or vapor to dissolve in water or other specified medium. The ability of one material to blend uniformly with another, as a solid in a liquid, liquid in liquid, gas in liquid, or gas in gas.

Solution – Even dispersion (mixing) of molecules of two or more substances. The most commonly encountered solutions involve mixing of liquids and liquids, or solids and liquids.

Solvent – Liquid that will dissolve a substance to form a solution (e.g., water, petroleum distillate, xylene, or methanol).

Specialist employees – Employees who, in the course of their regular job duties, work with and are trained in the hazards of specific hazardous substances and who will be called upon to provide technical advice or assistance to the incident commander at a haz-mat incident.

Specific gravity – Ratio of the weight of a product's volume to the weight of an equal volume of water. In the case of liquids of limited solubility, the specific gravity will predict whether the product will sink or float on water. If the specific gravity is greater than 1.0, the product will sink. If the specific gravity is less than 1.0, the product will float.

Spill – Release of a liquid, powder, or solid form of a hazardous material out of its original container.

Spontaneous combustion – Process whereby heat is generated within a material by either a slow oxidation reaction or by microorganisms.

Stabilization – Incident scene activities directed to channel, restrict, and/or halt the spread of hazardous material; to control the flow of a hazardous material to an area of lesser hazard; to implement procedures to insure against ignition; or to control a fire in such a manner as to be safe. Safe fire control measures include controlled burns, flaring off, or extinguishing the fuel by consumption.

Staging – Management of committed and uncommitted emergency response resources (personnel and apparatus) to provide orderly deployment.

Staging area – Safe area established for temporary location of available resources closer to the incident site to reduce response time.

Standard Transportation Commodity Code (STCC) – Number designating categories of articles being shipped.

Strength – Amount of ionization that occurs when an acid or base is dissolved in a liquid.

Stress – State of tension put on or in a shipping container by internal chemical action, external mechanical damage, or external flames or heat.

Structural firefighting protective clothing – Protective clothing normally worn by firefighters during structural firefighting operations. Often called turnout or bunker gear.

Sublimation – Process of changing a solid directly to a gas without passing through the liquid phase.

Subsidiary hazard class – Classification indicating a hazard of a material other than the primary hazard assigned.

Surfactant – A chemical that lowers the surface tension of a liquid.

Sump – Low point of a tank at which the emergency valve or outlet valve is attached.

Survey instrument – Portable instrument used for detecting and measuring radiation under varied physical conditions. The term covers a wide range of devices.

Switch list – List of the train cars used by railroad crews in a yard when they are making up a train.

Systemic – Pertaining to the internal organs and structures of the human body.

T

Tandem – Two-axle suspension.

Target organ/tissue – Organ or tissue that is susceptible to a given poison.

Technical assistance – Personnel or printed materials which provide technical information on the handling of hazardous materials.

Temperature – Condition of an object that determines whether heat will flow to or from another object.

Teratogenic – Material that is capable of producing birth defects in animals and humans.

TIR – Transport International Routier, meaning international transport by road. A customs agreement that exists among many countries, principally European, permitting vehicles or containers, properly approved and certified, to be sealed under customs direction in one country and be transported across borders of member countries without reinspection until arrival at final destination. Once arrived, the seal is removed under customs supervision.

TLV – Threshold limit value. The estimated exposure value below which no ill health effects should occur to an individual.

TOFC – Trailer on flatcar. Also refered to as a piggyback.

Toxic – Materials that can be poisonous if inhaled, swallowed, or absorbed into the body through cuts or breaks in the skin.

Toxicology – Study of chemical or physical agents that produce adverse responses in the biologic systems with which they interact.

Transport index – Number placed on a radioactive materials package label that indicates the control required during transport. The transport index is the radiation level, in millirems per hour, at three feet from the accessible external package surface. For fissile Class II packages, it is an assigned value based on critical safety requirements for the package contents. Abbreviation: TI.

Transportation – Methods of transporting commodities and materials including highway, railroad, pipeline, waterborne vessels, aircraft, and other means.

Triaxle – Three-axle construction in which at least two of the axles are equally spaced at approximately 48 to 50 inches apart. The third axle may be spread or equally spaced.

Truck – Self-propelled vehicle carrying its load on its own wheels and primarily designed for transportation of property rather than passengers.

Truck tractor – Powered motor vehicle primarily designed for drawing semi-trailers and constructed so that it carries part of the trailer weight and load.

Truck trailer – Vehicle without motor power primarily designed for transportation of property rather than passengers; designed to be drawn by a truck or truck tractor.

TTL – Threshold toxic limit. The estimated exposure value below which no ill health effects should occur to an individual.

U

UN/NA identification number – Four-digit identification number assigned to a hazardous material by the Department of Transportation. On shipping documents it may be found with the prefix "UN" (United Nations) or "NA" (North American).

Unstable – Capable of rapidly undergoing chemical changes or decomposition.

Upper coupler assembly – Coupling device consisting of an upper coupler plate, reinforcement framing, and fifth wheel kingpin mounted on a semi-trailer. Formerly called upper fifth wheel assembly.

Upper flammable limit – See Flammable Limits.

V

Vapor density – This is actually a specific gravity rather than a true density because it equals the ratio of the weight of a vapor or gas (with no air present) compared to the weight of an equal volume of air at the same temperature and pressure. Values less than 1.0 indicate that the vapor or gas tends to rise, and values greater than 1.0 indicate that the vapor or gas tends to settle. However, temperature effects must be considered.

Vapor pressure – Equilibrium pressure of the saturated vapor above the liquid. It is measured in millimeters of mercury (760 mm Hg = 14.7 psia) at 68°F (20°C) unless another temperature is specified.

Vapor protective clothing – Garment portion of a chemical protective clothing ensemble that is designed and configured to protect the wearer against chemical vapors or gases.

Vaporization – Process of going from a liquid or a solid to a gaseous state.

Vehicle warning placard – Sign displayed on the outside of a haz-mat carrier indicating the type of cargo. The design and use of placards are specified by DOT regulations.

Viscosity – Flow resistance to a liquid. This characteristic increases or decreases with the temperature of a liquid.

W

Warm zone – Control zone at a hazardous materials incident site where personnel and equipment decontamination and hot zone support take place. It includes control points for the access corridor, helping to reduce the spread of contamination.

Water fog – Finely divided mist produced by either a high or low velocity fog nozzle used for knocking down flames and cooling hot surfaces.

Water pollution – Contamination of water by a pesticide or other unwanted material.

Water reactive material – Material that will decompose or react when exposed to moisture or water.

Water solubility – Ability of a liquid or solid to mix with or dissolve in water.

Waybill – Shipping paper prepared by the railroad from a bill of lading. It accompanies a shipment and is carried by the engineer or conductor.

X

X rays – Penetrating electromagnetic radiation with wavelengths that are shorter than those of visible light. They are usually produced by bombarding a metallic target with fast electrons in a high vacuum. It is customary to refer to photons originating in the nucleus as gamma rays, and to those originating in the extranuclear parts of the atom as X rays. These rays are sometimes called roentgen rays after their discoverer, W. C. Roentgen.

Appendix B

MATERIAL SAFETY DATA SHEETS (MSDS) AND SHIPPING PAPERS

MATERIAL SAFETY DATA SHEET

(Approved by U.S. Department of Labor "Essentially Similar to Form OSHA-20")

SECTION 1 **NAME & PRODUCT**

Chemical Name: Acrylonitrile		Catalog Number: AX0350
Trade Name & Synonyms: Vinyl Cyanide	CAS #107-13-1	Chemical Family: Nitrile
Formula: CH_2:CHCN		Formula Weight: 53.07

SECTION 2 **PHYSICAL DATA**

Boiling Point, 760 mm Hg (°C)	77.3°	Specific Gravity (H_2O = 1)	0.806
Melting Point (°C)	-83.5°	Solubility in H_2O, % by wt. at 20°C	7.35%
Vapor Pressure at 20°C	83 mm Hg	Appearance and Odor	Clear yellow liquid
Vapor Density (air = 1)	1.83		pungent odor
Percent Volatiles by Volume	100%	Evaporation Rate (Butyl Acetate = 1)	4.5

SECTION 3 **FIRE AND EXPLOSION HAZARD DATA**

Flash Point (test method) 30°F (TCC)	Flammable Limits	Lel 3%	Uel 17%	

Extinguishing Media: CO_2, alcohol foam, dry chemical. (water may be ineffective)

Special Hazards and Procedures: Wear self-contained breathing apparatus and protective clothing.

Unusual Fire and Explosion Hazards: Violent polymerization may occur in presence of concentrated alkali. Thermal decomposition produces highly toxic fumes.

SECTION 4 **REACTIVITY DATA**

Stable	XX	Conditions to Avoid: Polymerizes in the absence of oxygen or exposure to light.
Unstable		

Materials to Avoid

( ) Water (X) strong Acids (X) Bases ( ) Corrosives (X) Oxidizers

(X) Other (specify) Alkali, Br, NH_3 Cu and copper alloys, amines.

Hazardous Decomposition Products: HCN, NO_2, COx

SECTION 5 **SPILL OR LEAK PROCEDURES AND DISPOSAL**

Steps to be Taken in Case Material is Released or Spilled: Collect on vermiculite or absorbent

Waste Disposal Method: To be performed in compliance with all current local state and federal regulations.

We believe the data contained herein is factual, however, it is offered solely for your consideration, investigation, and verification. Do not take as a warranty.

MSDS for Acrylonitrile (page 1)

SECTION 6 **HEALTH HAZARD DATA**

Threshold Limit Value 2 ppm (skin)

TXDS: ihl-hmn TC_{LO}: 16 ppm/20M
orl-rat LD_{50}: 82 mg/kg

Effects of Overexposure May be fatal if inhaled, swallowed or absorbed through the skin. Inhalation causes tearing, headache, sneezing, nausea, weakness, dizziness, unconsciousness and possible death. Contact causes irritation of skin, eyes and mucous membranes.

First Aid Procedures Get immediate medical assistance for all cases of overexposure.
Skin: wash with soap/water.
Eyes: Flush thoroughly with water for at least 15 minutes.
Inhalation: remove to fresh air; give artificial respiration if necessary, when breathing restored administer amyl nitrite by inhalation for 3 minutes.
Ingestion: get medical attention at once; if conscious, induce vomiting.

SECTION 7 **SPECIAL PROTECTION INFORMATION**

Ventilation, Respiratory Protection, Protective Clothing, Eye Protection

Provide general and exhaust ventilation. Protect eyes and skin with safety goggles and gloves. Air supplied respirator should be worn if vapor concentration is above 2 ppm.

SECTION 8 **SPECIAL HANDLING AND STORING PRECAUTIONS**

Keep container closed in storage. Store in a cool, well-ventilated area away from alkaline or oxidizing materials. Do not store uninhibited material.
Do not breathe vapor. Do not get in eyes, on skin, or on clothing.

DOT Hazard Class - Flammable Liquid

SECTION 9 **HAZARDOUS INGREDIENTS**

(refer to section 3 through 8)

N/A

SECTION 10 **OTHER INFORMATION**

Tests on laboratory animals indicate material may be carcinogenic and cause mutagenic and adverse reproductive effects.

AUTHORIZED SIGNATURE J M Norman

DATE ISSUED: 11-81
DATE REVISED: 1-85

EM 1181 3/83

MSDS for Acrylonitrile (page 2)

STRAIGHT BILL OF LADING - SHORT FORM - ORIGINAL - Not Negotiable

RECEIVED, subject to the classification and lawfully filed tariffs in effect on the date of the issue of this Bill of Lading, the property described below, in apparent good order, except as noted (contents and condition of contents of packages unknown), marked, consigned, and destined as indicated below, which said carrier (the word carrier being understood throughout this contract as meaning any person or corporation in possession of the property under the contract) agrees to carry to its usual place of delivery at said destination, if on its route, otherwise to deliver to another carrier on the route to said destination. It is mutually agreed, as to each carrier of all or any of said property over all or any portion of said route to destination, and as to each party at any time interested in all or any of said property, that every service to be performed hereunder shall be subject to all the terms and conditions of the Uniform Domestic Straight Bill of Lading set forth (1) in Uniform Freight Classifications in effect on the date hereof, if this is a rail or a rail-water shipment, or (2) if the applicable motor carrier classification or tariff if this is a motor carrier's shipment.

Shipper hereby certifies that he is familiar with all of the terms and conditions of the said bill of lading, including those on the back thereof, set forth in the classification or tariff which governs the transportation of this shipment, and the said terms and conditions are hereby agreed to by the shipper and accepted for himself and his assigns.

From: AIR PRODUCTS AND CHEMICALS, INC.
INDUSTRIAL/SPECIALTY GASES

319 LINK LANE
FORT COLLINS, CO 80524

CORPORATE ID	SHIPPER NUMBER	DATE	BILL OF LADING NO.
5 6 5 *	S 5 6 5 0 0 5 0 6	06 2 6 85	06009 K
			THIS NUMBER MUST SHOW ON FREIGHT BILL

AIR PRODUCTS

NAME OF CARRIER	CONSIGNED TO
CONSOLIDATED FREIGHTWAYS	AIR PRODUCTS AND CHEMICALS, INC.

DESTINATION	STATE	ZIP CODE	DELIVERY ADDRESS
TAMAQUA	PA	1 8 2 5 2	R.D. 1

RAIL ROUTE	DELIVERING CARRIER	RAILCAR NO.

T Y P E   O R   U S E   B A L L   P O I N T   P E N   A N D   B E A R   D O W N

SPECIAL INSTRUCTIONS

— IMPORTANT —
In case of emergency, call toll free 800-523-9374 from anywhere in the continental United States, except Pennsylvania; 800-322-9092 for calls originating in Pennsylvania.

CHECK APPROPRIATE BOX

☐ COLLECT

☒ PREPAID

CHARGE NO. 565-9400-499

SEND PREPAID FREIGHT BILLS ONLY TO:

AIR PRODUCTS AND CHEMICALS, INC
P.O. BOX 2792, ALLENTOWN, PA 18105

NO. OF PKGS	DESCRIPTION OF HAZARDOUS MATERIALS ONLY	DOT E #	NMFC NO.	WEIGHT	RATE
	Acetylene, flammable gas, UN 1001, in cyls		85520		
	Air, compressed, nonflammable gas, UN 1002, in cyls		85540		
	Ammonia, anhydrous, nonflammable gas, UN 1005, in cyls		85560-2		
	RQ, Ammonia, anhydrous, nonflammable gas, UN 1005, in cyls		85560-2		
	Argon, compressed, nonflammable gas, UN 1006, in cyls		85880		
	Argon, refrigerated liquid, nonflammable gas, UN1951, in cyls		85880		
	Carbon dioxide, nonflammable gas, UN 1013, in cyls		85680-1		
	Carbon monoxide, flammable gas, UN 1016, in cyls		85730-2		
	Chlorine, nonflammable gas, UN 1017, in cyls		85740-1		
	RQ, Chlorine, nonflammable gas, UN 1017, in cyls		85740-1		
	Diborane mixtures, flammable gas, UN 1911, in cyls		85900		
	Ethane, compressed, flammable gas, UN 1035, in cyls		85980		
	Ethyl chloride, flammable liquid, UN1037, in cyls		44480-2		
	RQ, Ethyl chloride, flammable liquid, UN 1037, in cyls		44480-2		
	Ethylene, compressed, flammable gas, UN 1962, in cyls		85980		
	Fluorine, nonflammable gas, UN 1045, in cyls		85910		
	Helium, compressed, nonflammable gas, UN 1046, in cyls		85920-1		
	Helium, compressed, nonflammable gas, UN 1046, in alum cyls		85920-1		
	Helium, refrigerated liquid, nonflammable gas, UN 1963, in dewars		85915-A		
	Hydrogen, compressed, flammable gas, UN 1049, in cyls		86000		
	Hydrogen chloride, anhydrous, nonflammable gas, UN 1050, in cyls		85580-2		
	Hydrogen sulfide, flammable gas, UN 1053, in cyls		86020-2		
	RQ, Hydrogen sulfide, flammable gas, UN 1053, in cyls/tanks		86020-2		
	Liquefied petroleum gas, flammable gas, UN 1075, in cyls		86140		
	Methane, compressed, flammable gas, UN 1971, in cyls		85980		

NO. OF PKGS	DESCRIPTION OF HAZARDOUS MATERIALS ONLY	DOT E #	NMFC NO	WEIGHT	RATE
	RQ, Methyl chloride, flammable gas, UN 1063, in cyls/tanks		86040-2		
	Nitrogen, compressed, nonflammable gas, UN 1066, in cyls		86060		
	Nitrogen, refrigerated liquid, nonflammable gas, UN 1977, in cyls		86060		
	Oxygen, compressed, nonflammable gas, UN 1072, in cyls		86120		
	Oxygen, refrigerated liquid, nonflammable gas, UN1073, in cyls		86120		
	Sulfur dioxide, nonflammable gas, UN 1079, in cyls/tanks		86160-2		
	Sulfur hexafluoride, nonflammable gas, UN 1080, in cyls		85880		
	Compressed gas, NOS, flammable gas, UN 1954, in cyls		85880		
	Compressed gas, NOS, flammable gas, UN 1954, MIXTURE in cyls		85880		
	Compressed gas, NOS, nonflammable gas, UN 1956, in cyls		85880		
	Compressed gas, NOS, nonflammable gas, UN 1956, MIXTURE in cyls				
	Nitrous oxide, compressed, nonflammable gas, UN 1070, in cyls		86100-1		
	Poisonous liquid or gas, NOS, Poison A, NA 1955, MIXTURE in cyls		85900		
	Poisonous liquid or gas, flammable, NOS, Poison A, NA 1953, MIXTURE in cyls		85900		
	RESIDUE: last contained; helium, refrigerated liquid, nonflammable gas, UN1963, in dewars		41042-A		
2	TUNGSTEN HEXAFLUORIDE, CORROSIVE MATERIAL, UN 2196 IN STL. CYLS.		43940-2	240	
	DESCRIPTION OF ARTICLES - NON-HAZARDOUS				
2	PKGS. TOTALS LBS.			240	

Subject to section 7 of conditions of applicable Bill of Lading. If this shipment is to be delivered to the consignee without recourse on the consignor, the consignor shall sign the following statement.
The carrier shall not make delivery of this shipment without payment of freight and all other lawful charges.

FORM 3102-4 (REV. 10/88)

AIR PRODUCTS AND CHEMICALS, INC.
Signature of Consignor

PER ____________

This is to certify that the above-named materials are properly classified, described, packaged, marked and labeled, and are in proper condition for transportation according to the applicable regulations of the Department of Transportation.

CARRIER ACKNOWLEDGES RECEIPT OF PACKAGES AND REQUIRED PLACARDS FOR SHIPMENT(S) RECEIVED.

CARRIER ____________ NO. OF PKGS. ____________ DATE ____________

Example of a Bill of Lading

STRAIGHT BILL OF LADING - SHORT FORM - ORIGINAL - Not Negotiable

RECEIVED, subject to the classification and lawfully filed tariffs in effect on the date of the issue of this Bill of Lading, the property described below, in apparent good order, except as noted (contents and condition of contents of packages unknown), marked, consigned, and destined as indicated below, which said carrier (the word carrier being understood throughout this contract as meaning any person or corporation in possession of the property under the contract) agrees to carry to its usual place of delivery at said destination, if on its route, otherwise to deliver to another carrier on the route to said destination. It is mutually agreed, as to each carrier of all or any of said property over all or any portion of said route to destination, and as to each party at any time interested in all or any of said property, that every service to be performed hereunder shall be subject to all the terms and conditions of the Uniform Domestic Straight Bill of Lading set forth (1) in Uniform Freight Classifications in effect on the date hereof, if this is a rail or a rail-water shipment, or (2) if the applicable motor carrier classification or tariff if this is a motor carrier's shipment.

Shipper hereby certifies that he is familiar with all of the terms and conditions of the said bill of lading, including those on the back thereof, set forth in the classification or tariff which governs the transportation of this shipment, and the said terms and conditions are hereby agreed to by the shipper and accepted for himself and his assigns.

From: AIR PRODUCTS AND CHEMICALS, INC.
EMPTY CYLINDERS ONLY

BILL OF LADING NO. **4545H**

AIR PRODUCTS

CORPORATE ID	SHIPPER NUMBER	DATE	THIS NUMBER MUST SHOW ON FREIGHT BILL

NAME OF CARRIER		CONSIGNED TO
DESTINATION	STATE / ZIP CODE	DELIVERY ADDRESS
RAIL ROUTE	DELIVERING CARRIER	RAILCAR NO.

TYPE OR USE BALL POINT PEN AND BEAR DOWN

SPECIAL INSTRUCTIONS

— IMPORTANT —
In case of emergency, call toll free 800-523-9374 from anywhere in the continental United States, except Pennsylvania; 800-322-9092 for calls originating in Pennsylvania.

CHECK APPROPRIATE BOX

☐ COLLECT

☐ PREPAID

CHARGE NO. ______

SEND PREPAID FREIGHT BILLS ONLY TO:

AIR PRODUCTS AND CHEMICALS, INC
P.O. BOX 2792, ALLENTOWN, PA 18105

NO. OF PKGS	DESCRIPTION OF HAZARDOUS MATERIALS ONLY	DOT E #	NMFC NO.	WEIGHT	RATE
	RESIDUE: last contained: acetylene, flammable gas, UN 1001, in old stl cyls		41160-1-3		
	RESIDUE: last contained: air, compressed, nonflammable gas, UN 1002, in old cyls		41160-1-3		
	RESIDUE: last contained: ammonia, anhydrous, nonflammable gas, UN 1005, in old stl cyls		41160-1-3		
	RESIDUE: last contained: RQ, ammonia, anhydrous, nonflammable gas, UN 1005, in old stl cyls		41160-1-3		
	RESIDUE: last contained: argon, compressed, nonflammable gas, UN 1006, in old stl cyls		41160-1-3		
	RESIDUE: last contained: carbon dioxide, nonflammable gas, UN 1013, in old stl cyls		41160-1-3		
	RESIDUE: last contained: carbon monoxide, flammable gas, UN 1016, in old stl cyls		41160-1-3		
	RESIDUE: last contained: chlorine, nonflammable gas, UN 1017, in old stl cyls		41160-1-3		
	RESIDUE: last contained: RQ, chlorine, nonflammable gas, UN 1017, in old stl cyls		41160-1-3		
	RESIDUE: last contained: diborane mixtures, flammable gas, UN 1911, in old stl cyls		41160-1-3		
	RESIDUE: last contained: ethane, compressed, flammable gas, UN 1035, in old stl cyls		41160-1-3		
	RESIDUE: last contained: ethylene, compressed, flammable gas, UN 1962, in old stl cyls		41160-1-3		
	RESIDUE: last contained: fluorine, nonflammable gas, UN 1045, in old stl cyls		41160-1-3		
	RESIDUE: last contained: helium, compressed, nonflammable gas, UN 1046, in old stl cyls		41160-1-3		
	RESIDUE: last contained: helium, refrigerated liquid, nonflammable gas, UN 1963, in old stl cyls		41160-1-3		
	RESIDUE: last contained: hydrogen, compressed, flammable gas, UN 1049, in old stl cyls		41160-1-3		
	RESIDUE: last contained: hydrogen chloride, anhydrous, nonflammable gas, UN 1050, in old stl cyls		41160-1-3		
	RESIDUE: last contained: hydrogen sulfide, flammable gas, UN 1053, in old stl cyls		41160-1-3		
	RESIDUE: last contained: RQ, hydrogen sulfide, flammable gas, UN 1053, in old stl cyls/tanks		41160-1-3		
	RESIDUE: last contained: liquefied petroleum gas, flammable gas, UN 1075, in old stl cyls		41160-1-3		
	RESIDUE: last contained: methane, compressed, flammable gas, UN 1971, in old stl cyls		41160-1-3		
	RESIDUE: last contained: methyl chloride, flammable gas, UN 1063, in old stl cyls		41160-1-3		
	RESIDUE: last contained: nitrogen, compressed, nonflammable gas, UN 1066, in old stl cyls		41160-1-3		
	RESIDUE: last contained: nitrogen, refrigerated liquid, nonflammable gas, UN 1977, in old stl cyls		41160-1-3		
	RESIDUE: last contained: nitrous oxide, compressed, nonflammable gas, UN 1070, in old stl cyls		41160-1-3		

NO. OF PKGS	DESCRIPTION OF HAZARDOUS MATERIALS ONLY	DOT E #	NMFC NO.	WEIGHT	RATE
	RESIDUE: last contained: oxygen, compressed, nonflammable gas, UN 1072, in old stl cyls		41160-1-3		
	RESIDUE: last contained: sulfur dioxide, nonflammable gas, UN 1079 in old stl cyls		41160-1-3		
	RESIDUE: last contained: sulfur hexafluoride, nonflammable gas, UN 1080, in old stl cyls		41160-1-3		
	RESIDUE: last contained: compressed gas, NOS, flammable gas, UN 1954, in old stl cyls		41160-1-3		
	RESIDUE: last contained: compressed gas, NOS, flammable gas, UN 1954, MIXTURE(S) in old stl cyls		41160-1-3		
	RESIDUE: last contained: compressed gas, NOS, nonflammable gas, UN 1956, in old stl cyls		41160-1-3		
	RESIDUE: last contained: compressed gas, NOS, nonflammable gas, UN 1956, MIXTURE(S) in old stl cyls		41160-1-3		
	RESIDUE: last contained: poisonous liquid or gas, flammable, NOS, Poison A, NA 1953, arsine MIXTURE(S), in old stl cyls		41160-1-3		
	RESIDUE: last contained: poisonous liquid or gas, NOS, Poison A, NA 1955, nitric oxide MIXTURE(S), in old stl cyls		41160-1-3		
	RESIDUE: last contained: poisonous liquid or gas, NOS, Poison A, NA 1955, nitrogen dioxide MIXTURE(S), in old stl cyls		41160-1-3		
	RESIDUE: last contained: poisonous liquid or gas, flammable, NOS, Poison A, NA 1953, phosphine, MIXTURE(S), in old stl cyls		41160-1-3		
	RESIDUE: last contained: poisonous liquid or gas, NOS, Poison A, NA 1955, MIXTURE(S), in old stl cyls		41160-1-3		
	DESCRIPTION OF ARTICLES - NON-HAZARDOUS				
	Cylinders for shipping air, gas, or liquids under pressure, NOI, old, used, steel with no hazardous residue		41160-1-3		
PKGS. ←	TOTALS		→ LBS.		

Subject to section 7 of conditions of applicable Bill of Lading. If this shipment is to be delivered to the consignee without recourse on the consignor, the consignor shall sign the following statement. The carrier shall not make delivery of this shipment without payment of freight and all other lawful charges.

AIR PRODUCTS AND CHEMICALS, INC.
Signature of Consignor

FORM 3102-6 (REV. 8/87)

PER ______

This is to certify that the above-named materials are properly classified, described, packaged, marked and labeled, and are in proper condition for transportation according to the applicable regulations of the Department of Transportation.

CARRIER ACKNOWLEDGES RECEIPT OF PACKAGES AND REQUIRED PLACARDS FOR SHIPMENT(S) RECEIVED.

CARRIER | NO. OF PKGS. | DATE

Example of a Bill of Lading for Empty Containers

HAZARDOUS MATERIALS SHIPPING PAPER

AIR PRODUCTS

DATE	DRIVER/CUSTOMER	VEHICLE NUMBER	TRIP REPORT/SHIPPER NUMBER

NO. & TYPE CONTAINERS		RQ	DOT SHIPPING NAME	DOT HAZARD CLASS	UN/NA NUMBER	ADDITIONAL DESCRIPTIONS	DOT E NUMBER
	cyls		**Acetylene**	**Flammable Gas**	**UN 1001**		
	cyls		Air, Compressed	Nonflammable Gas	UN 1002		
	cyls		**Argon, Compressed**	**Nonflammable Gas**	**UN 1006**		
	cyls		Carbon Dioxide	Nonflammable Gas	UN 1013		
	cyls		**Carbon Dioxide, Refrigerated Liquid**	**Nonflammable Gas**	**UN 2187**		
	cyls		Carbon Dioxide-Oxygen Mixture	Nonflammable Gas	UN 1014		
	cyls		**Carbon Dioxide-Nitrous Oxide Mixture**	**Nonflammable Gas**	**UN 1015**		
	cyls		Fluorine	Nonflammable Gas	UN 1045		
	cyls		**Helium, Compressed**	**Nonflammable Gas**	**UN 1046**		
			Helium, Refrigerated Liquid	Nonflammable Gas	UN 1963		
	cyls		**Helium-Oxygen Mixture**	**Nonflammable Gas**	**NA 1980**		
	cyls		Hydrogen, Compressed	Flammable Gas	UN 1049		
	cyls		**Liquefied Petroleum Gas**	**Flammable Gas**	**UN 1075**		
	cyls		Methylacetylene-Propadiene, Stabilized	Flammable Gas	UN 1060		
	cyls		**Nitrogen, Compressed**	**Nonflammable Gas**	**UN 1066**		
	cyls		Nitrous Oxide, Compressed	Nonflammable Gas	UN 1070		
	cyls		**Nitrous Oxide, Refrigerated Liquid**	**Nonflammable Gas**	**UN 2201**		
	cyls		Oxygen, Compressed	Nonflammable Gas	UN 1072		
	cyls		**Oxygen, Compressed**	**Nonflammable Gas**	**UN 1072**		
	cyls		Poisonous Liquid or Gas, N.O.S.	Poison A	NA 1955	Arsine Mixture	
	cyls		**Poisonous Liquid or Gas, N.O.S.**	**Poison A**	**NA 1955**	**Nitric Oxide Mixture**	
	cyls		Poisonous Liquid or Gas, N.O.S.	Poison A	NA 1955	Nitrogen Dioxide Mixture	
	cyls		**Poisonous Liquid or Gas, Flammable, N.O.S.**	**Poison A**	**NA 1953**	**Phosphine Mixture**	
	cyls		Poisonous Liquid or Gas, N.O.S.	Poison A	NA 1955		
	cyls		**Poisonous Liquid or Gas, N.O.S.**	**Poison A**	**NA 1955**		
	cyls		Poisonous Liquid or Gas, Flammable, N.O.S.	Poison A	NA 1953	Arsine Mixture	
	cyls		**Compressed Gas, N.O.S.**	**Flammable Gas**	**UN 1954**		
	cyls		Compressed Gas, N.O.S.	Flammable Gas	UN 1954		
	cyls		**Compressed Gas, N.O.S.**	**Nonflammable Gas**	**UN 1956**		
	cyls		Compressed Gas, N.O.S.	Nonflammable Gas	UN 1956		
	cyls		**Argon, Refrigerated Liquid**	**Nonflammable Gas**	**UN 1951**		
	cyls		Nitrogen, Refrigerated Liquid	Nonflammable Gas	UN 1977		
	cyls		**Flammable Liquid, Corrosive, N.O.S. (Dichlorosilane)**	**Flammable Liquid**	**UN 2924**	**Poison–Inhalation Hazard**	**SA-870503**
			Oxygen, Refrigerated Liquid	Nonflammable Gas	UN 1073		
				Flammable Gas			
			Air, Refrigerated Liquid	Nonflammable Gas	UN 1003		
			Oxidizer, N.O.S.		**UN 1479**		
			Corrosive Liquid, N.O.S.	Corrosive Material	UN 1760		
			Flammable Liquid, N.O.S.		**UN 1993**		
				Oxidizer			
				Corrosive Material			
				Flammable Liquid			
		RQ	**Ammonia, Anhydrous**	**Nonflammable Gas**	**UN 1005**		
	cyls		Ammonia, Anhydrous	Nonflammable Gas	UN 1005		
		RQ	**Chlorine**	**Nonflammable Gas**	**UN 1017**		
	cyls		Chlorine	Nonflammable Gas	UN 1017		
		RQ	**Hydrogen Sulfide**	**Flammable Gas**	**UN 1053**		
	cyls		Hydrogen Sulfide	Flammable Gas	UN 1053		
		RQ	**Nitrogen Dioxide, Liquid**	**Poison A**	**UN 1067**		
	cyls		Nitrogen Dioxide, Liquid	Poison A	UN 1067		

NO. & TYPE CONTAINERS		RQ	DOT SHIPPING NAME	DOT HAZARD CLASS	UN/NA NUMBER	ADDITIONAL DESCRIPTIONS	WT./LBS.	DOT E NUMBER
	drms		**Calcium Carbide**	**Flammable Solid**	**UN 1402**	**Dangerous when wet**	**lbs.**	
	drms	RQ	Calcium Carbide	Flammable Solid	UN 1402	Dangerous when wet	lbs.	
	DF		**Helium, Refrigerated Liquid**	**Nonflammable Gas**	**UN 1963**		**lbs.**	
			Nitrogen, Refrigerated Liquid	Nonflammable Gas	UN 1977		lbs.	

Circle appropriate placards applied or supplied: Flammable Gas Nonflammable Gas Flammable Oxygen Oxidizer Flammable Solid Poison Gas Dangerous Corrosive Poison

"EMERGENCY RESPONSE NUMBERS": 1-800-523-9374 - outside PA; 1-800-322-9092 - in PA; 1-215-481-4911 - call collect outside continental U.S.A.

This is to certify that the above-named materials are properly classified, described, packaged, marked and labeled, and are in proper condition for transportation according to the applicable regulations of the Department of Transportation.

FORM 3419-1 (REV. 10/88)

CERTIFYING SIGNATURE

Example of a Hazardous Materials Shipping Paper

ICC 233 - 2 YR. RET.
ICC 237 - 3 YR. RET.
PLACE SPECIAL SERVICE PASTERS HERE

802 — UNION PACIFIC RAILROAD COMPANY — 802
FREIGHT AND TRANSIT WAYBILL

UNION PACIFIC

FORM 715
REV. 7-1

3

DANGEROUS D FOR SINGLE CONSIGNMENTS CARLOAD, LESS CARLOAD AND TOFC

CAR INITIAL	CAR NUMBER	KIND	WEIGHT IN TONS GROSS	TARE	NET	LENGTH / CAPACITY OF CAR ORDERED	FURNISHED	PREPAID FREIGHT BILL NUMBER	WAYBILL DATE	WAYBILL NO.
GATX	10874	T9			64				July 23, 80	74107

TRAILER / CONTAINER INITIAL NUMBER — PLAN — LENGTH

ORIGIN AND DATE, ORIGINAL CAR, TRANSFER FREIGHTBILL AND PREVIOUS WAYBILL REFERENCE AND ROUTING WHEN REBILLED

ORIGIN FIRM CODE

DESTN FIRM CODE

STOP THIS CAR AT STATION NO. — STATION — STATE
1ST
2ND
3RD

CONSIGNEE AND ADDRESS AT STOP — REASON FOR STOP
1ST
2ND
3RD

TO STATION NO. — STATION: SPRINGFIELD — STATE: OREGON

ROUTE: SHOW EACH JUNCTION AND CARRIER IN ROUTE ORDER TO DESTINATION OF WAYBILL. — CODE

UP EAST PORTLAND SP

SHIPPERS OR AGENTS ROUTING "S" OR "A": S

FROM NUMBER — STATION — STATE
ORIGIN 000090 Los Angeles California
BILLED AT

RECONSIGNED TO: AUTHORITY STATION NO. — STATION — STATE — RWC

SHIPPER AND COMPLETE ADDRESS — CODE

CANMORE MINES LTD.
5000 Ferguson Drive
LOS ANGELES, CALIFORNIA 90022

BILL OF LADING OR INVOICE NUMBER — ORDER NUMBER

CONSIGNEE AND COMPLETE ADDRESS — CODE

CALGAS LANE COUNTY, INC
SPRINGFIELD, OREGON 97101

SHOW TO WHOM FREIGHT BILL SHOULD BE BILLED NAME, ADDRESS, CITY, STATE, AND ZIP CODE

AGREEMENT WEIGHTS

FINAL DESTN.—ADDITIONAL ROUTING

SPRINGFIELD, OREGON

WEIGHED AT — AGREEMENT NUMBER

GROSS	TARE	ALLOWANCE	NET

INSTRUCTIONS: REGARDING PROTECTIVE SERVICE, MILLING, WEIGHING, ETC. SPECIFY TO WHOM THESE CHARGES, IF ANY SHOULD BE BILLED

CAR TRIP LEASED TO CONSIGNEE GROSS GAL. 25427
NET C GAL. 25198 US 30261 TEMP.
65 TARIFF 5.02 GRAVITY 503 INSPEC. CAPS
12/6/79 VALVES SAME

← WHEN SHIPPER IN THE UNITED STATES EXECUTES THE NO-RECOURSE CLAUSE OF SECTION 7 OF THE BILL OF LADING INSERT YES

PREPAID ← IF CHARGES ARE PREPAID, INSERT "PREPAID"

LEASED CAR — INCENTIVE OR LEASED CAR, OR UNIT TRAIN NUMBER

*If the shipment moves between two ports by a carrier by water, the law requires that the bill of lading shall state whether it is "carrier's or shipper's weight."

Note—Where the rate is dependent upon value, shippers are required to state specifically in writing the agreed or declared value of the property. The agreed or declared value of the property is hereby specifically stated by the shipper to be not exceeding PER

NO. PKGS.	DESCRIPTION OF ARTICLES (COMMODITY CODE NO. 49-057-81)	* WEIGHT	RATE	FREIGHT	ADVANCES	PREPAID
1 T/C	LIQUIFIED PETROLEUM GAS FLAMMABLE GAS UN 1075 PLACARDED: FLAMMABLE GAS	126,494 LBS	Weights and charges to follow. PREPAID.			

DESTINATION AGENT'S FREIGHT BILL NUMBER

FIRST JUNCTION	SECOND JUNCTION	THIRD JUNCTION	FOURTH JUNCTION	REPORTING AGENT WILL STAMP HEREIN STATION AND DATE REPORTED.

Example of a Rail Waybill

```
DETAIL CONSIST

01793    19          . ESTIMATED OUT MEMTENYD  AT 1200 19
   ETA SPRINGFIE AT 2200 19


4BML4111 793    19 01 ARRIVAL REQUEST
4DBL4111 793    19 01 WORK PERFORMED REQ    61271
4TFL4111 793    19 01 POWER WORK PERF REQ


 09 SETOUT 92239
 06  BN     8008 ENGI  OK11 1IB1308 SD402 05 0806        NE 69         06
                                                                    LEAD QUALIFIED
 06  BN     6923 ENGH  OK11 2IB0708 SD402 MT 0810        NW 69         06
                                                                    LEAD QUALIFIED
 09 SETOUT 92239
0 BN     12056 ECAB    027 92239 X    4BA 820006 P0095  B              04
 39  93385                                                   A
 BN  954243 EMHS    025 93385 A    4J4 832111  P8418   B              04
 BN  953838 EMHS    028 93385 A    4J4 853     P8419   B              04
 BN  966574 LMGT    065 93452 U    4JA LUMBER           B              04 BURLINNORTHE
     001 LDS  002 MTYS  00118 TONS  00127 FT
 09 SETOUT 92239
0 ADN    5796 LA5     102 92239 W    4BA PLY WD         C              04          GEORGIPACIFI
0 ADN    5653 LA5     105 97331 W    2Q6 PLY WD         C       090    04          GEORGIPACIFI
0 GATX 27433 LTS     132 92343 W    4AH NF GAS DAN     B       075    04          CARDOXDIVISI
0 UTLX 27471 LTS     131 92343 W    4AH NF GAS         B       075    04          CARDOXDIVISI
        /CNAEMERG NOTIFY CHEMTREC 800 424 9300            PP
0 UTLX 82088 LTS     123 92333 W    4AH NF GAS DAN     B       070    04          AIRCOINDGASE
        /CNAEMERG NOTIFY CHEMTREC 800 424 9300
0 UTLX900107 LTS     131 92333 W    4AH NF GAS DAN     C       070    04          AIRCOINDGASE
        /CNAEMERG NOTIFY CHEMTREC 800 424 9300
0 UTLX900106 LTS     131 92333 W    4AH NF GAS         C       070    04          AIRCOINDGASE
        /CNAEMERG NOTIFY CHEMTREC 800 424 9300
0 SOU 530691 LBE     105 92239 W    4BA PLY WD         C              04          GEORGIPACIFI
0 ADN    9457 LA5     108 92239 W    4BA BOARDS         C              04          GEORGIPACIFI
0 MB     4163 LB3     104 92415 W    4AK FIBBRD         C       085    04          JAMESRIVDIXN
0 UTLX 13581 LTS     130 97404 W    2Q6 C ACID DAN     B       090    04          FARMERCHEMIC
0 SOU   41112 LB3     077 92333 W    4AH PULP           C       070    04          SCOTT PAPER
0 CIRR 91126 LA5     087 92333 W    4AH PULP           C       070    04          SCOTT PAPER
0 BN   376740 EA5     035 40671 D    264 848111 P2105  B       556    04          XP-A5-FB5
0 BN   376676 EA5     036 40671 D    264 848111 P2105  B       556    04          XP-A5-FB5
0 ATSF621759 ER9C    047 97629 B    2D5 848222 US-GP  C       556    04          MIR280
0 SOU 529131 LB3     073 92343 W    4AH STONE          C       075    04          ROYAL SEEDS
     014 LDS  003 MTYS  01657 TONS  00972 FT

     015 LDS  006 MTYS  01802 TONS  01141 FT

     002 UNITS   06000 HORSEPOWER  00138 FT   01279 FT-TOTAL TRAIN LENGTH
END
```

Example of a Rail Consist

Form 5388 8-49

MARINE BRANCH

BARGE ~~LOADING~~ UNLOADING REPORT

Phillips Petroleum COMPANY

AT Freeport Texas

MAR 3 1 1980

DATE 3-19-76 COMMODITY Cyclohexane TICKET NO.

SHIPPED BY / ~~SHIPPED TO~~ / RECEIVED FROM ______ COMPANY, FOR ACCOUNT OF ______ COMPANY

DESTINATION / ORIGIN

TOWBOAT NAME ESTAY BARGES TCB-314 SHIPPED FROM / RECEIVED FROM

ARRIVED 7:20 A.M. 3-19-76 DOCKED 7:30 A.M. 3-19-76 READY TO LOAD ~~UNLOAD~~ 9:15 A.M. 3-19-76

STARTED LOADING / UNLOADING 9:15 A.M. 3-19-76 FINISHED LOADING ~~UNLOADING~~ 12:50 P.M. 3-19-76 BARGE RELEASED AT 1:10 P.M. 3-19-76

EXPLANATION OF ANY DELAY IN TIME:

SHORE LINE O.K. BY

CARGO O. K. BY ______ AT ______ M. ______ PIPE LEADING FROM TANK TO VESSEL: WHEN STARTED Full WHEN FINISHED Full

BARGE MEASUREMENTS

COMPT. NO.	Center ~~PORT~~ OPENING FT.	INS.	BS&W	TEMP.	CLOSING FT.	INS.	BS&W	TEMP.	STARBOARD OPENING FT.	INS.	BS&W	TEMP.	CLOSING FT.	INS.	BS&W	TEMP.
BARGE TCB-314																
1	Empty				13	1	0	66								
2					12	10¾	–	–								
3					12	11¼	–	–								
4																
FORE TRIM																
AFT TRIM																
BARGE																
1																
2																
3																
4																
FORE TRIM																
AFT TRIM																

COMPT. NO.	PORT OPENING FT.	INS.	BS&W	TEMP.	CLOSING FT.	INS.	BS&W	TEMP.	STARBOARD OPENING FT.	INS.	BS&W	TEMP.	CLOSING FT.	INS.	BS&W	TEMP.
BARGE																
1																
2																
3																
4																
FORE TRIM																
AFT TRIM																
BARGE																
1																
2																
3																
4																
FORE TRIM																
AFT TRIM																

SHORE TANK MEASUREMENTS

TANK NO.	OPENING GAUGE FT.	INCHES	BS&W IN.	TEMP.	GRAV.	CORR. FACTOR	GROSS BBLS.	NET BBLS.	CLOSING GAUGE FT.	INCHES	BS&W IN.	TEMP.	GRAV.	CORR. FACTOR	GROSS BBLS.	NET BBLS.	NET BARRELS DELIVERED / RECEIVED
106	28	0⅞	9	66					24	1⅜	9	67					

TOTAL GALLONS ______ TOTAL BARRELS ______ TOTAL TONS ______

THIS SHIPMENT COVERED BY TENDER NO. ______ ISSUED BY THE TEXAS RAILROAD COMMISSION ON ______ 19__

SIGNED: W. P. [illegible] SHIPS OFFICER

SIGNED: E. Huckaby

FOR ______ COMPANY

Example of a Water Barge Report

SHIPPER'S CERTIFICATION FOR RESTRICTED ARTICLES
(excluding radioactive materials)

Two completed and signed copies of this certification shall be handed to the carrier. (Use block letters)

WARNING: Failure to comply in all respects with the applicable regulations of the Department of Transportation, 49-CFR, CAB 82 and, for international shipments, the IATA Restricted Articles Regulations may be a breach of the applicable law, subject to legal penalties. This certification shall in no circumstance be signed by an IATA Cargo Agent or a consolidator for international shipments.

This shipment is within the limitations prescribed for: (*mark one*)

☒ passenger aircraft ☐ cargo-only aircraft

Number of Packages	Article Number (Int'l only See Section IV IATA RAR)	Proper Shipping Name of Articles as shown in Title 49 CFR, CAB 82 Tariff 6-D, and (for int'l shipments) the IATA Restricted Articles Regulations. Specify each article separately. Technical name must follow in parenthesis, the proper shipping name for N.O.S. items.	Class	IATA Packing Note No. Applied (int'l only)	Net Quantity per Package	Flash Point (closed cup) For Flammable Liquids	
						°C.	°F.
1	845	Gasoline Flammable Liquid	3	300	1 quart		-50

Special Handling Information:

I hereby certify that the contents of this consignment are fully and accurately described above by Proper Shipping Name and are classified, packed, marked, labelled and in proper condition for carriage by air according to applicable national governmental regulations, and for International Shipments, the current IATA Restricted Articles Regulations.

Name and full address of Shipper	Name and title of person signing Certification
ABC Refining Company	XYZ Laboratory
1494 Clean Creek Road	4630 Clean Air Road
Tulsa, Oklahoma	Houston, Texas
Date	**Signature of the Shipper (see WARNING above)**
3/15/80	Frank Smith

Air Waybill No.*	Airport of Departure*	Airport of Destination*
	Tulsa	Houston

Example of an Air Bill

Appendix C

HAZARDOUS MATERIALS INFORMATION RESOURCES

ELECTRONIC DATABASES

DATABASE	SOURCE
CAMEO (Computer-Aided Management of Emergency Operations)	U.S. Department of Commerce National Oceanographic and Atmospheric Agency (NOAA) Hazardous Materials Response Branch 7600 Sand Point Way, N.E., Bin C-17500 Seattle, WA 98115 (206) 526-6317
CHRIS (Chemical Hazards Response Information System) **OHM-TADS** (Oil and Hazardous Materials Technical Assistance Data System) **Over 30 different databases** with both scientific and regulatory chemical-related information.	Chemical Information Systems, Inc. 810 Glen Eagles Ct., Suite 300 Towson, MD 21286 (800) 247-8737 (410) 321-8440 (in Maryland)
EIS-C (Emergency Information Systems – Chemical Version)	Research Alternatives, Inc. 1401 Rockville Pike, Suite 500 Rockville, MD 20852 (301) 738-6900
HMIX (Hazardous Materials Information Exchange)	U.S. Department of Transportation Office of Hazardous Materials Initiatives and Training 400 Seventh Street, S.W. DHM-52, Room 5414A Washington, DC 20590 (202) 366-4900

DATABASE	SOURCE
HSDB (Hazardous Substances Databank) **MEDLINE** **RTECS** (Registry of Toxic Effects of Chemical Substances) **TOXLINE**	National Library of Medicine Medlars Management Section 38A, 4N421 8600 Rockville Pike Bethesda, MD 20894 (800) 638-8480

PRINTED REFERENCE SOURCES

(Selected Listing)

Cargo Tank Hazardous Materials Regulations
National Tank Truck Carriers
2200 Mill Road
Alexandria, VA 22314
(703) 838-1960

Chemical Data Guide for Bulk Shipment by Water – U.S. Coast Guard
Superintendent of Documents
U.S. Government Printing Office
Washington, DC 20402
(202) 512-1800

Chemical Hazards Response Information System (CHRIS manuals):
Condensed Guide to Chemical Hazards
Hazardous Chemical Data Guide
Superintendent of Documents
U.S. Government Printing Office
Washington, DC 20402
(202) 512-1800

Chemistry of Hazardous Materials, 2nd ed.
by Eugene Meyer
Prentice Hall
Prentice Hall Building
Englewood Cliffs, NJ 07632
(800) 223-2336

Code of Federal Regulations
(29 CFR 1910) General Industry Safety and Health Standards OSHA 2206
Superintendent of Documents
U.S. Government Printing Office
Washington, DC 20402
(202) 512-1800

Code of Federal Regulations
(49 CFR) Transportation Parts 100 to 199
Superintendent of Documents
U.S. Government Printing Office
Washington, DC 20402
(202) 512-1800

Condensed Chemical Dictionary, 12th ed.
Lab Safety Supply Company
P.O. Box 1368
Janesville, WI 53547
(608) 754-2345

Dangerous Properties of Industrial Materials (Sax), 8th ed.
Lab Safety Supply Company
P.O. Box 1368
Janesville, WI 53547
(608) 754-2345

Effects of Exposure to Toxic Gases and First Aid Medical Treatment
Matheson Gas Products
8800 Utica
Cucamonga, CA 91730
(909) 987-4611

Emergency Action Guides
Bureau of Explosives
Association of American Railroads
50 F Street, N.W.
Washington, DC 20001
(412) 772-2250

Emergency Care for Hazardous Materials Exposure, 2nd ed.
by Bronstein and Currance
C. V. Mosby Company
11830 Westline Industrial Drive
St. Louis, MO 63146
(314) 872-8370

Emergency Handling of Hazardous Materials in Surface Transportation
Bureau of Explosives
Association of American Railroads
50 F Street, N.W.
Washington, DC 20001
(412) 772-2250

Emergency Management of Hazardous Materials Incidents
National Fire Protection Association
1 Battery March Park
Quincy, MA 02269
(617) 770-3000

Emergency Medical Treatment for Poisoning
Pittsburgh Poison Center
3705 Fifth Avenue at DeSoto Street
Pittsburgh, PA 15213
(412) 692-5600

Emergency Response Guidebook for Selected Hazardous Materials
Superintendent of Documents
U.S. Government Printing Office
Washington, DC 20402
(202) 512-1800

Farm Chemicals Handbook
Meister Publishing Company
37733 Euclid Avenue
Willoughby, OH 44094
(216) 942-2000
Subject Area: Pesticides and agricultural chemicals

Fire Protections Guide to Hazardous Materials, 11th ed.
325 – *Fire Hazard Properties of Flammable Liquids, Gases and Volatile Solids* (1,300 flammable substances listed)
49 – *Hazardous Chemicals Data* (416 chemicals listed)
491M – *Manual of Hazardous Chemicals Reactions* (3,550 mixtures that may cause reaction)
704 – *Recommended System for the Identification of the Fire Hazards of Materials*
National Fire Protection Association
1 Battery March Park
Quincy, MA 02269
(617) 770-3000

The Firefighter's Handbook of Hazardous Materials
Maltese Enterprises, Inc.
P.O. Box 31009
Indianapolis, IN 46231
(800) 433-5664

Firefighter's Hazardous Materials Reference Book
by Daniel Davis and Grant Christianson
Van Nostrand Reinhold
115 Fifth Avenue
New York, NY 10003
(212) 254-3232

GATX Tank Car Manual, 6th ed.
General American Transportation Corporation
500 West Monroe
Chicago, IL 60661
(312) 621-6200
Subject Area: Railroad tank cars

Guidelines for the Selection of Chemical Protective Clothing, 2 vols., 3rd ed.
National Technical Information Service
5285 Port Royal Road
Springfield, VA 22161
(703) 487-4600
Subject Area: Chemical protective clothing

Hazardous Commodities Handbook
National Tank Truck Carriers
2200 Mill Road
Alexandria, VA 22314
(703) 838-1960

Hazardous Materials Emergency Planning Guide
U.S. Department of Transportation
Research and Special Programs
Hazardous Materials Training Office
400 Seventh Street, S.W.
DHM-51, Room 5414A
Washington, DC 20590
(202) 366-4000

Hazardous Materials Exposure: Emergency Response and Patient Care
by Borak, Callan, and Abbott
Prentice Hall
Prentice Hall Building
Englewood Cliffs, NJ 07632
(800) 223-2336

Hazardous Materials for the First Responder, 2nd ed.
IFSTA
Fire Service Publications
Oklahoma State University
930 North Willis
Stillwater, OK 74078
(405) 744-5723

Hazardous Materials Injuries: A Handbook for Pre-Hospital Care
Bradford Communications Corporation
10742 Tucker Street
Beltsville, MD 20705
(301) 345-0100
Subject Area: Emergency medical care

Hazardous Materials: Managing the Incident
IFSTA
Fire Service Publications
Oklahoma State University
930 North Willis
Stillwater, OK 74078
(405) 744-5723

Hazardous Materials Response Handbook, 2nd ed.
National Fire Protection Association
1 Battery March Park
Quincy, MA 02269
(617) 770-3000

How to Respond to Hazardous Chemical Spills
Noyes Publications
120 Mill Road
Park Ridge, NJ 07656
(201) 391-8484

Industrial Gases Data Book
Boc Gases
575 Mountain Avenue
Murray Hill, NJ 07974
(908) 464-8100

Manual for Spills of Hazardous Materials
(Environment Canada – Order #EN40-320-1948E)
Canadian Government Publishing Centre
Ottawa, Ontario K1A 0S9
CANADA
or:
International Specialized Book Services
5602 N.E. Hassalo Street
Portland, OR 97213
(503) 287-3093

National Fire Codes
National Fire Protection Association
1 Battery March Park
Quincy, MA 02269
(617) 770-3000

NFPA 471: Recommended Practice for Responding to Hazardous Material Incidents
National Fire Protection Association
1 Battery March Park
Quincy, MA 02269
(617) 770-3000

NFPA 472: Standard for Professional Competence of Responders to Hazardous Materials Incidents
National Fire Protection Association
1 Battery March Park
Quincy, MA 02269
(617) 770-3000

NFPA 473: Standard for Professional Competencies for EMS Personnel Responding to Hazardous Materials Incidents
National Fire Protection Association
1 Battery March Park
Quincy, MA 02269
(617) 770-3000

NIOSH Registry of Toxic Effects of Chemical Substances
Superintendent of Documents
U.S. Government Printing Office
Washington, DC 20402
(202) 512-1800

NIOSH/OSHA Pocket Guide to Chemical Hazards
Superintendent of Documents
U.S. Government Printing Office
Washington, DC 20402
(202) 512-1800

Threshold Limit Values and Biological Exposure Indices for 1994-1995
American Conference of Governmental Industrial Hygienists (ACGIH)
Kemper Woods Center
1330 Kemper Meadow Drive
Cincinnati, OH 45240
(513) 742-2020

Toxic and Hazardous Industrial Chemicals Safety Manual
The International Technical Information Institute
Toranomon-Tachikawa
Building 1-6-5
Nishi-Shimbashi
Tokyo 105 JAPAN, 1976

Transport of Radioactive Materials: Questions and Answers About Incident Response
Emergency Management Institute
16825 S. Seton Ave.
Emmitsburg, MD 21727

VERBAL INFORMATION SOURCES
(Selected Listing)

Important Numbers

CHEMTREC . . . Emergency number:	(800) 424-9300
Non-emergency number:	(800) 262-8200
CANUTEC	(613) 996-6666
National Response Center	(800) 424-8802
Department of Defense – Nuclear Accident Center	(703) 325-2102
Centers for Disease Control and Prevention	(404) 639-3311
Oil and Hazardous Material Technical Assistance Data System (OHM-TADS)	(800) 247-8737
Agency for Toxic Substances and Disease Registry (ATSDR)	(404) 639-6000
ACFX Rail Car Manufacturer	(716) 635-0222
GATX Rail Car Manufacturer	(312) 621-6200
UTLX Rail Car Manufacturer	(312) 431-3111

American Insurance Association (AIA)
(National Board of Fire Underwriters)
Engineering and Safety Service
1130 Connecticut Ave., N.W., Suite 1000
Washington, DC 20036
(202) 828-7100

American National Standards Institute (ANSI)
11 West 42nd St.
New York, NY 10036
(212) 642-4900

American Petroleum Institute (API)
Fire and Safety
1220 L Street, N.W.
Washington, DC 20005
(202) 682-8000

American Society of Mechanical Engineers (ASME)
United Engineering Center
345 East 47th Street
New York, NY 10017
(212) 705-7000

Ashland Chemical Company
3849 Fisher Road
Columbus, OH 43228
(614) 276-6143

Association of American Railroads (AAR)
3140 South Federal Street
Chicago, IL 60616
(312) 808-5800

Bureau of Explosives
Association of American Railroads
50 F Street, N.W.
Washington, DC 20001
(202) 639-2372

Chemical Manufacturers Association (CMA)
2501 M Street, N.W.
Washington, DC 20037
(202) 887-1100

Compressed Gas Association, Inc. (CGA)
1725 Jefferson Davis Boulevard, Suite 1004
Arlington, VA 22202
(703) 413-4341

Dow Chemical Company
Midland, MI 48640
(Emergency) (517) 636-4400

DuPont Company
1007 Market Street
Wilmington, DE 19898
(Emergency) (302) 774-7500

Factory Mutual Engineering Corporation Laboratories
1150 Boston-Providence Turnpike
Norwood, MA 02062
(617) 762-4300

The Fertilizer Institute (TFI)
501 Second Street, N.E.
Washington, DC 20002
(202) 675-8250

HELP (Hazardous Emergency Leak Procedures)
Union Carbide Corporation
South Charleston, WV 25303
(304) 744-3487

J. T. Baker Chemical Company
222 Red School Lane
Phillipsburg, NJ 08865
(908) 859-2151

Kerr-McGee Chemical Corp.
Kerr-McGee Center
Oklahoma City, OK 73125
(405) 270-1313

Mallinckrodt Group, Inc.
P.O. Box 5439
St. Louis, MO 63147
(314) 530-2000

National Fire Protection Association (NFPA)
1 Battery March Park
Quincy, MA 02269
(617) 770-3000

National Institute of Standards and Technology
Gaithersberg, MD 20899
(301) 975-2000

National Institute for Occupational Safety and Health (NIOSH)
Division of Technical Services
4676 Columbia Parkway
Cincinnati, OH 45226
(513) 533-8236

National Response Center
2100 2nd Street, S.W.
Washington, DC 20593
(202) 267-2185

National Safety Council
1121 Spring Lake Drive
Itasca, IL 60143
(708) 285-1121

National Tank Truck Carriers, Inc.
2200 Mill Road
Alexandria, VA 22314
(703) 838-1960

National Transportation Safety Board (NTSB)
490 L'Enfant Plaza East, S.W.
Washington, DC 20594
(202) 382-6600

Occupational Safety and Health Administration (OSHA)
U.S. Department of Labor
200 Constitution Ave., N.W.
Washington, DC 20210
(202) 219-8148

Superintendent of Documents
U.S. Government Printing Office
Washington, DC 20402
(202) 512-1800

Underwriters' Laboratories, Inc. (UL)
333 Pfingsten Road
Northbrook, IL 60062
(708) 272-8800

Union Carbide Corp.
P.O. Box 670
Bound Brook, NJ 08805
(908) 563-5000

U.S. Department of Energy
1000 Independence Avenue, S.W.
Washington, DC 20585
(202) 586-5000

U.S. Department of Transportation
Material Transportation Bureau
Office of Hazardous Materials Operations
Washington, DC 20590
(202) 366-4488

U.S. Nuclear Regulatory Commission
Washington, DC 20555
(301) 415-7000

Appendix D

COMMON COMPRESSED AND LIQUEFIED GASES

ACETYLENE

Chemical Formula .... C_2H_2
Synonyms .... Ethine, ethyne
DOT Classification .... Flammable gas
DOT Number 1001
Vapor Density .... 0.906 at 32°F (0°C) and 1 atm
Description .... Colorless, flammable gas; lighter than air; garlic odor.
Physiological Effects .... Non-toxic. Simple asphyxiant.

AIR

Synonyms .... Compressed air, atmospheric air, the atmosphere, grade D air, breathing air.
DOT Classification .... Non-flammable gas
DOT Number 1002
Vapor Density .... 1.00 at 70°F (21°C) and 1 atm
Description .... The natural atmosphere. Non-flammable, colorless, odorless gas.
Physiological Effects .... Non-toxic and non-flammable.

AMMONIA (ANHYDROUS)

Chemical Formula .... NH_3
DOT Classification .... Non-flammable gas (domestic U.S.); poison gas (international)
DOT Number 1005
Vapor Density .... 0.5970 at 32°F (0°C) and 1 atm
Description .... Pungent, colorless gas. Lighter than air.

Physiological Effects	Pungent odor affords a protective warning. Intense irritating effect on the mucous membranes of the eyes, nose, throat, and lungs. Physiological response to ammonia in air by volume is as follows: 20–50 ppm, readily detectable odor; 150–200 ppm, general discomfort and eye tearing; 400–700 ppm, severe irritation of the eyes, ears, nose, and throat; 1,700 ppm, coughing, bronchial spasms; 2,000–3,000 ppm, dangerous, less than 1/2 hour exposure may be fatal; 5,000–10,000 ppm, rapidly fatal; 10,000 ppm, immediately fatal.

ARGON

Chemical Symbol	Ar
DOT Classification	Non-flammable gas
DOT Number	1006
Vapor Density	1.38 at 70°F (21°C) and 1 atm
Description	Inert gas. Colorless, odorless, and tasteless. Non-toxic. Heavier than air.
Physiological Effects	Non-toxic. Simple asphyxiant.

BORON TRIFLUORIDE

Chemical Formula	BF_3
Synonym	Boron fluoride
DOT Classification	Poison gas
DOT Number	1008
Vapor Density	2.38 at 70°F (21°C) and 1 atm
Description	Colorless gas. Persistent, irritating, acid odor. Hydrolyzes in moist air to form dense white fumes. Heavier than air.
Physiological Effects	Irritates the nose, mucous membranes, and other parts of the respiratory system. Easily detected by sense of smell. Injurious if inhaled. Contact with skin can cause dehydrating burns similar to those inflicted by acid.

BUTADIENE (1,3-BUTADIENE)

Chemical Formula	C_4H_6
Synonyms	Vinylethylene, biethylene, erythrene, bivinyl, divinyl
DOT Classification	Inhibited form: Flammable gas
DOT Number	1010
Vapor Density	1.9153 at 60°F (16°C) and 1 atm
Description	Flammable, colorless gas with a mild aromatic odor. Heavier than air. ***Note:*** Information given here should not be assumed to apply to 1,2-butadiene nor to other butadienes.
Physiological Effects	Inhalation in excessive amounts leads to progressive anesthesia and eventual death.

CARBON DIOXIDE

Chemical Formula .... CO_2

DOT Classification .... Non-flammable gas

DOT Number 1013

Vapor Density .... 1.522 at 70°F (21°C) and 1 atm

Description .... Colorless, odorless gas. Heavier than air. Will not support combustion.

Physiological Effects .... Asphyxiant; non-toxic. Will not support life. Contact on skin can result in frostbite.

CARBON MONOXIDE

Chemical Formula .... CO

DOT Classification .... Poison gas. Flammable gas subsidiary hazard.

DOT Number 1016

Vapor Density .... 0.9676 at 70°F (21°C) and 1 atm

Description .... Colorless, odorless, flammable gas.

Physiological Effects .... Toxic chemical asphyxiant. Inhaled concentrations of 0.4% prove fatal in less than one hour. Higher concentrations can cause sudden collapse with little or no warning.

CHLORINE

Chemical Symbol Cl

DOT Classification .... Poison gas

DOT Number 1017

Vapor Density .... 2.482 at 32°F (0°C) and 1 atm

Description .... Greenish-yellow, non-flammable gas with a distinctive odor. Heavier than air. Will support combustion.

Physiological Effects .... Primarily a respiratory irritant. Readily detectable. Acute toxicity when sufficient concentrations are present. Liquid chlorine will cause skin and eye burns upon contact.

CYCLOPROPANE

Chemical Formula .... C_3H_6

Synonym .... Trimethylene

DOT Classification .... Flammable gas

DOT Number 1027

Vapor Density .... 1.48 at 70°F (21°C) and 1 atm

Description .... Colorless, flammable gas with a sweet, distinctive odor. Heavier than air.

Physiological Effects .... A general anesthetic. Concentrations (by volume) of 6–8% result in unconsciousness; 7–14%, moderate anesthesia; 14–23%, deep anesthesia. Concentrations of 23–40% are lethal through respiratory failure.

DIMETHYL ETHER

Chemical Formula .... C_2H_6O
Synonyms .... Methyl ether, methyl oxide, wood ether
DOT Classification .... Flammable gas
DOT Number 1033
Vapor Density .... 1.59 at 70°F (21°C) and 1 atm
Description .... Colorless, flammable gas. Faint sweetish odor. Heavier than air.
Physiological Effects .... Anesthetic; no permanent residual effects.

ETHANE

Chemical Formula .... C_2H_6
Synonyms .... Bimethyl, dimethyl, ethyl hydride, methyl-methane
DOT Classification .... Flammable gas
DOT Number 1035
Vapor Density .... 1.0469 at 60°F (16°C) and 1 atm
Description .... Colorless, odorless, flammable gas. Heavier than air.
Physiological Effects .... Non-toxic. Anesthetic effect. Simple asphyxiant.

ETHYLENE

Chemical Formula .... C_2H_4
Synonyms .... Ethene, blefiant gas, bicarburetted hydrogen, elayl, etherin
DOT Classification .... Flammable gas
DOT Number 1962
Vapor Density .... 0.978 at 32°F (0°C) and 1 atm
Description .... Colorless, flammable gas. Lighter than air. Faint sweet, musty odor.
Physiological Effects .... Non-toxic. Anesthetic. Simple asphyxiant.

FLUORINE

Chemical Formula .... F_2
DOT Classification .... Poison gas. Oxidizer subsidiary hazard.
DOT Number 1045
Vapor Density .... 1.31 at 70°F (21°C) and 1 atm
Description .... Pale yellow gas. Heavier than air. Most powerful oxidizing agent known. Reacts with practically all organic and inorganic substances.
Physiological Effects .... Highly toxic. Powerful caustic irritant. Burns on contact.

FLUOROCARBONS

Chemical Name	DOT Number
Bromotrifluoromethane	1009
Chlorodifluoroethane	2517
Chlorodifluoromethane	1018
Chloropentafluoroethane	1020
Chlorotrifluoromethane	1022
Dichlorodifluoromethane	1028
Difluoroethane	1030
Dichlorodifluoromethane and Difluoroethane mixture	2602
Chlorodifluoromethane and Chloropentafluoroethane mixture	1973
Trifluoromethane and Chlorotrifluoromethane mixture	1028
Difluoromethane and Chloropentafluoroethane mixture	1078
Dichlorodifluoromethane and Chlorofluoromethane mixture	1078
Chlorofluoromethane and Dichlorotetrafluoroethane mixture	1958

Synonym .... Freon

Description .... The fluorocarbons are relatively inert and, in general (with the exception of DOT numbers 2517 and 1030), are non-flammable in all concentrations in air under ordinary conditions. They are colorless and odorless in concentrations of less than 20% of volume in air. They are all heavier than air.

Physiological Effects .... The fluorocarbons have low levels of toxicity. Inhalation effects are similar to anesthetics. Contact with the boiling liquid can result in tissue freezing and frostbite.

HELIUM

Chemical Symbol .... He

DOT Classification .... Non-flammable gas

DOT Number 1046

Vapor Density .... 0.138 at 70°F (21°C) and 1 atm

Description .... Colorless, odorless, tasteless gas. Lighter than air.

Physiological Effects .... Non-toxic. Simple asphyxiant.

HYDROGEN

Chemical Formula .... H_2

DOT Classification .... Flammable gas

DOT Number 1049

Vapor Density .... 0.06950 at 32°F (0°C) and 1 atm

Description .... Colorless, odorless, tasteless, flammable gas. Lighter than air.

Physiological Effects .... Non-toxic. Simple asphyxiant.

HYDROGEN CHLORIDE (ANHYDROUS)

Chemical Formula .... HCl

Synonym .... Hydrochloric acid (anhydrous)

DOT Classification .... Poison gas. Corrosive subsidiary hazard.

DOT Number 1050

Vapor Density .... 1.268 at 32°F (0°C) and 1 atm

Description .... Colorless gas. Fumes strongly in moist air. Sharp, suffocating odor. Heavier than air.

Physiological Effects .... Toxic. Causes severe irritation to eyes, skin, and upper respiratory tract. Severe burning on contact. Prolonged inhalation may result in death.

HYDROGEN CYANIDE

Chemical Formula .... HCN

Synonym .... Formonitrile

DOT Classification .... Hydrogen cyanide, anhydrous, stabilized: Poison. Flammable liquid subsidiary hazard.

DOT Number 1051

Vapor Density .... 0.947 at 87.8°F (31°C) and 1 atm

Description .... Colorless gas, extremely dangerous. Flammable. Lighter than air.

Physiological Effects .... Highly toxic. Acute cyanide poisoning can rapidly render a person unconscious. Can enter the body through absorption by the skin, as well as by inhalation.

HYDROGEN SULFIDE

Chemical Formula .... H_2S

Synonym .... Sulfuretted hydrogen

DOT Classification .... Poison gas. Flammable gas subsidiary hazard.

DOT Number 1053

Vapor Density .... 1.2 at 59°F (15°C) and 1 atm

Description .... Colorless, flammable gas with an offensive odor smelling like rotten eggs. Heavier than air.

Physiological Effects .... A toxic and asphyxiant gas. Instantly fatal if inhaled in high concentrations. Irritates the eyes and upper respiratory tract. Can exhaust the sense of smell.

LIQUEFIED PETROLEUM GASES

Chemical Name	DOT Number
Butane	1011
Butylenes:	
1-Butene	1012
cis-2-Butene	1012
trans-2-Butene	1012
Isobutene	1012
Isobutane	1969
Petroleum Gas, Liquefied	1075
Propane	1978
Propylene	1077

DOT Classification .... Flammable gas

Description .... Flammable, colorless, non-corrosive gases. Heavier than air.

Physiological Effects .... Non-toxic. Prolonged inhalation of high concentrations has an anesthetic effect. Simple asphyxiant. Contact with the skin can cause burns.

METHANE

Chemical Formula .... CH_4

Synonyms .... Marsh gas, methyl nybride

DOT Classification .... Flammable gas

DOT Number 1971

Vapor Density .... 0.55491 at 60°F (16°C) and 1 atm

Description .... Colorless, odorless, tasteless, flammable gas. Lighter than air.

Physiological Effects .... Non-toxic. Simple asphyxiant.

METHYLAMINES (ANHYDROUS)

Chemical Name	Chemical Formula	DOT Number
Monomethylamine	CH_3NH_2	1061
Dimethylamine	$(CH_3)_2NH$	1032
Trimethylamine	$(CH_3)_3N$	1083

DOT Classification .... Flammable gas

Description .... Colorless, flammable gases. Heavier than air. Distinct and disagreeable fishy odor in concentrations up to 100 ppm. In higher concentrations these substances have an odor like ammonia.

Physiological Effects .... Toxic. Irritating to nose, throat, and eyes. Severe exposure of the eyes may lead to loss of sight.

METHYL CHLORIDE

Chemical Formula .... CH_3Cl

Synonym .... Chloromethane

DOT Classification .... Flammable gas

DOT Number 1063

Vapor Density .... 1.74 at 32°F (0°C) and 1 atm

Description .... Colorless, flammable gas. Heavier than air.

Physiological Effects .... Toxic. No pronounced smell. An anesthetic about one-fourth as potent as chloroform. Also acts as a narcotic. Inhalation of several hundred ppm can lead to death. Can have fatal after-effects.

METHYL MERCAPTAN

Chemical Formula .... CH_3SH

Synonym .... Methanethiol

DOT Classification .... Poison gas. Flammable gas subsidiary hazard.

DOT Number 1064

Vapor Density .... 1.66 at 49°F (9°C) and 1 atm

Description .... Colorless, flammable gas with extremely strong nauseating odor. Heavier than air.

Physiological Effects .... Toxic. The odor makes it necessary to evacuate premises in which leaks of any quantity have occurred. Acts on central nervous system after inhalation.

NITROGEN

Chemical Formula .... N_2

DOT Classification .... Non-flammable gas

DOT Number 1066

Vapor Density .... 0.967 at 70°F (21°C) and 1 atm

Description .... Colorless, odorless, tasteless, non-flammable gas. Will not support combustion or life. Lighter than air.

Physiological Effects .... Non-toxic. Simple asphyxiant.

NITROUS OXIDE

Chemical Formula .... N_2O

Synonyms .... Nitrogen monoxide, dinitrogen monoxide, laughing gas

DOT Classification .... Non-flammable gas

DOT Number 1070

Vapor Density .... 1.529 at 32°F (0°C) and 1 atm

Description .... Colorless, practically odorless, tasteless, non-flammable gas. Heavier than air. Mild oxidizing agent; will support combustion.

Physiological Effects .... Non-toxic, non-irritating. Used as a general anesthetic in medicine. Contact with liquid nitrous oxide will result in freezing or frostbite.

OXYGEN

Chemical Formula .... O_2

DOT Classification .... Non-flammable gas. Oxidizer subsidiary hazard.

DOT Number 1072

Vapor Density .... 1.1049 at 70°F (21°C) and 1 atm

Description .... Colorless, odorless, tasteless, non-flammable, elemental gas. Supports life and makes combustion possible. Heavier than air.

Physiological Effects .... Non-toxic. No harmful effects when inhaled.

PHOSGENE

Chemical Formula .... $COCl_2$

Synonyms .... Carbonyl chloride, carbon oxychloride

DOT Classification .... Poison gas. Corrosive subsidiary hazard.

DOT Number 1076

Vapor Density .... 3.4 at 68°F (20°C) and 1 atm

Description .... Non-flammable, colorless, extremely dangerous, poisonous gas. Heavier than air.

Physiological Effects .... Toxic. Attacks the respiratory system. 50 ppm proves rapidly fatal. Delayed action can be particularly injurious.

SULFUR DIOXIDE

Chemical Formula .... SO_2

Synonym .... Sulfurous acid anhydride

DOT Classification .... Poison gas

DOT Number 1079

Vapor Density .... 2.262 at 70°F (21°C) and 1 atm

Description .... Non-flammable, colorless gas with a pungent odor. Heavier than air.

Physiological Effects .... Extremely irritating to the respiratory tract. This in itself serves as a warning as to its presence. Exposure to high concentrations produces a suffocating effect.

SULFUR HEXAFLUORIDE

Chemical Formula .... SF_6

DOT Classification .... Non-flammable gas

DOT Number 1080

Vapor Density .... 1.0 at 68°F (20°C) and 1 atm

Description .... Colorless, odorless, non-flammable gas. Heavier than air.

Physiological Effects .... Non-toxic. Simple asphyxiant.

VINYL CHLORIDE

Chemical Formula .... C_2H_3Cl

Synonyms .... Chloroethylene, chloroethene

DOT Classification .... Flammable gas

DOT Number 1086

Vapor Density .... 2.15 at 59°F (15°C) and 1 atm

Description .... Colorless, flammable gas with a sweet ethereal odor. Heavier than air.

Physiological Effects .... Acts as a general anesthetic in concentrations of well over 500 ppm. Suspected of having a carcinogenic effect. Can irritate or damage the eyes on contact. Liquid can freeze the skin on prolonged contact.

VINYL METHYL ETHER

Chemical Formula .... C_3H_6O

Synonym .... Methyl vinyl ether

DOT Classification .... Inhibited form: Flammable gas

DOT Number 1087

Vapor Density .... 2.5 at 32°F (0°C) and 1 atm

Description .... Colorless, flammable gas with a sweet, pleasant odor. Heavier than air.

Physiological Effects .... Inhalation is thought to produce mild anesthesia. Contact with liquid can cause freezing of the skin or frostbite.

Appendix E

FLAMMABLE AND COMBUSTIBLE LIQUIDS

Following is a partial listing of Class I flammable liquids and Classes II and III combustible liquids:

CLASS I FLAMMABLE LIQUIDS

Acetaldehyde
Acetone
Acetonitrile
Acetyl chloride
Acetylene dichloride
Acrolein
Acrylonitrile
Allyl alcohol
Allylamine
Allyl chlorocarbonate
Allyl chloroformate
n-Amylamine
Amyl mercaptan
Benzene
Benzotrifluoride
Butyl acetate
Butyl alcohol
* tert-Butyl hydroperoxide
Butyllithium in hydrocarbon solvents
tert-Butyl peracetate
Butyraldehyde
Carbon disulfide
Chlorobenzene
Chlorohexane
Crotonaldehyde
Cyclohexane
Cyclohexylamine
Cyclopentane
* Dibutylperoxide
Diethylaluminum chloride
Diethylamine
Diethyldichlorosilane
Diisopropylamine
Diketene
1,1-Dimethylhydrazine
Dimethyl sulfide
Dioxane
Endrin
Ethyl acetate
Ethyl acrylate
Ethyl alcohol
Ethylamine
Ethylbenzene
Ethyl chloride
Ethylene dichloride
Ethyleneimine
Ethyl ether (ether, diethyl ether)
Ethyl nitrite

* Due to the severe hazards these materials pose, extra caution should be taken during mitigation efforts. Additional information on the above products can be obtained from the reference materials discussed in Chapter 1. Refer to page 215 for definitions of Class I, II, and III liquids.

Ethyltrichlorosilane
Gasoline
n-Hexane
Hydrogen cyanide
* Isoprene
Isopropyl ether
Methyl acetate
* Methyl acrylate
Methyl alcohol
Methyl ethyl ether
Methyl formate
Methyl isobutyl ketone
* Methyl methacrylate
Methyl oxide
Octane
Paraldehyde
Petroleum crude
Propargyl bromide
Propionaldehyde
n-Propylamine
* Propylene oxide
Propyl formate
* n-Propyl nitrate
Pyridine
* Styrene
Tetrahydrofuran
Toluene
Turpentine
Vinyl acetate
Vinyl ether
Vinylidene chloride
Xylene, m-xylene, o-xylene, p-xylene

CLASS II COMBUSTIBLE LIQUIDS

Acetic acid
Acetic anhydride
Acetyl oxide
Acetyl peroxide
Acrolein dimer
Acrylic acid
Cumene
n-Decane
Diamylamine
1,4-Dichlorobutane
Diesel fuel oil
Epichlorohydrin
Ethyl butyl acetate
Ethyl butyl alcohol
Ethyl butyl ketone
Fuel oil
Hydrazine
Kerosene
* Methyl ethyl ketone peroxide
Methyl parathion
Morpholine
* 1-Nitropropane, 2-Nitropropane
3-Pentanol
* Peracetic acid
Propionic acid
* Tetramethyl lead

CLASS III COMBUSTIBLE LIQUIDS

Acetone cyanohydrin
Acetophenone
Adiponitrile
Aniline
Benzoyl chloride
Benzyl chloride
Butyl benzoate
* tert-Butyl perbenzoate
* tert-Butyl peroxypivalate
Butyric acid
Cod liver oil
Corn oil
Cottonseed oil
Cresol
Cumene hydroperoxide
o-Dichlorobenzene
Diethylenetriamine
Diethyl phthalate
Diethyl sulfate
Diethylzinc
Dimethyl sulfate
Divinylbenzene
Ethanolamine
n-Ethylaniline
Ethyl benzoate
Ethylene cyanohydrin
Formic acid
Furfural
Isophorone
Lanolin
Lard oil
Lindane in solution
Linseed oil
Lubricating oil

* Methacrylic acid
Methyl salicylate
Mineral oil
Nitrobenzene
Oleic acid
Oleo oil
Parathion
Peanut oil
Phenylethanolamine
Quenching oil
Soybean oil
Sulfur monochloride
Tallow oil
Tetraethyl lead
Toluene-2,4-diisocyanate
o-Toluidine
Transformer oil
Triamylamine
Tri-n-butylamine
Vegetable oil

Appendix F

HAZARDOUS MATERIALS CLASSIFICATION

With the development of the Uniform Fire Code (UFC), Article 80, and the probability that this code may, in part, become a national standard, it is important that students be familiar with the haz-mat classifications used in this context. Appendix F gives an overview of oxidizers, peroxides, and unstable materials by class as listed in the UFC, Appendix VI-A. These hazard categories are based upon the Code of Federal Regulations, Title 29 (29 CFR). Where numerical classifications are included, they are in accordance with nationally recognized standards.

OXIDIZERS*

Oxidizers are available as:

- **gases** such as oxygen, ozone, fluorine, chlorine, and oxides of nitrogen
- **liquids** such as bromine, hydrogen peroxide, nitric acid, sulfuric acid, and perchloric acid
- **solids** such as chlorates, chromates, chromic acid, iodine, nitrates, nitrites, perchlorates, and peroxides

Class 4

A Class 4 oxidizer is an oxidizing material that can undergo an explosive reaction when catalyzed or exposed to heat, shock, or friction. Examples include:

- ammonium perchlorate
- ammonium permanganate
- guanidine nitrate
- hydrogen peroxide solutions, more than 91% by weight
- perchloric acid solutions, more than 72.5% by weight
- potassium superoxide

---

* Examples are based upon NFPA Standard No. 43-A.

Class 3

A Class 3 oxidizer is an oxidizing material that will cause a severe increase in the burning rate of combustible material with which it comes in contact. Examples include:

- ammonium dichromate
- calcium hypochlorite (except as provided under Class 2)
- hydrogen peroxide solutions, more than 52% to 91% by weight
- mono-(trichloro) tetra-(monopotassium dichloro)-penta-s-triazinetrione
- perchloric acid solutions, 60% to 72.5% by weight
- potassium dichloro-s-triazinetrione (potassium dichloroisocyanurate)
- sodium dichloro-s-triazine-2,4,6-trione (sodium dichloroisocyanurate)

Class 2

A Class 2 oxidizer is an oxidizing material that will moderately increase the burning rate or that may cause spontaneous ignition of combustible material with which it comes in contact. Examples include:

- calcium hypochlorite, 50% or less by weight or containing more than 35% calcium hypochlorite dihydrate by weight
- chromium trioxide (chromic acid)
- hydrogen peroxide solutions, more than 27.5% to 52% by weight
- nitric acid, more than 70%
- potassium perchlorate
- potassium permanganate
- sodium chlorite, more than 40% by weight
- sodium permanganate
- 1,3,5-trichloro-s-triazine-2,4,6-trione (trichloroisocyanuric acid)

Class 1

A Class 1 oxidizer is an oxidizing material that may increase the burning rate of combustible material with which it comes in contact. Examples include:

- aluminum nitrate
- ammonium persulfate
- barium chlorate
- barium nitrate
- barium perchlorate
- barium permanganate
- barium peroxide
- beryllium nitrate
- calcium chlorate
- calcium chlorite
- calcium citrate
- calcium nitrate
- calcium peroxide
- cobalt nitrate
- cupric nitrate
- ferric nitrate
- hydrogen peroxide solutions, more than 8% to 27.5% by weight
- lead nitrate
- lead peroxide
- lithium hypochlorite
- lithium peroxide
- magnesium nitrate
- magnesium perchlorate
- magnesium peroxide
- mercurous nitrate
- nickel nitrate
- nitric acid, 70% or less
- perchloric acid solutions, less than 60% by weight
- potassium chlorate
- potassium dichromate

- potassium nitrate
- potassium nitrite
- potassium persulfate
- silver nitrate
- sodium carbonate peroxide
- sodium chlorate
- sodium chlorite, 40% or less
- sodium dichloro-s-triazinetrione dihydrate
- sodium dichromate
- sodium nitrate
- sodium nitrite
- sodium perborate
- sodium perborate tetrahydrate
- sodium perchlorate monohydrate
- sodium persulfate
- strontium chlorate
- strontium nitrate
- strontium peroxide
- thorium nitrate
- uranium nitrate
- zinc chlorate
- zinc nitrate
- zinc peroxide
- zinc permanganate
- zirconium nitrate

ORGANIC PEROXIDES*

Organic peroxides are flammable compounds which contain the double oxygen or peroxy (–O–O–) group and are subject to explosive decomposition. Classification of organic peroxides is based on the relative hazard of the materials. Organic peroxides are available as:

- liquids
- pastes
- solids (usually finely divided powders)

Unclassified

Unclassified peroxides are those that are capable of detonation. These peroxides present an extremely high explosion hazard through rapid explosive decomposition and are regulated in accordance with the provisions of the Uniform Fire Code, Article 77, for Class A explosives.

Class I

Class I peroxides are capable of deflagration but not detonation. These peroxides present a high explosion hazard through rapid decomposition. Examples include:

- acetyl cyclohexane sulfonyl, 60% to 65% concentration by weight
- benzoyl peroxide, over 98% concentration
- t-butyl hydroperoxide, 90%
- t-butyl peroxyacetate, 75%
- t-butyl peroxyisopropylcarbonate, 92%
- diisopropyl peroxydicarbonate, 100%
- di-n-propyl peroxydicarbonate, 98%
- di-n-propyl peroxydicarbonate, 85%

---

* Examples are based upon NFPA Standard No. 43-B.

Class II

Class II peroxides burn very rapidly and present a severe reactivity hazard. Examples include:

- acetyl peroxide, 25%
- t-butyl hydroperoxide, 70%
- t-butyl peroxybenzoate, 98%
- t-butyl peroxy-2-ethylhexanoate, 97%
- t-butyl peroxyisobutyrate, 75%
- t-butyl peroxyisopropyl carbonate, 75%
- t-butyl peroxypivalate, 75%
- di-sec-butyl peroxydicarbonate, 98%
- di-sec-butyl peroxydicarbonate, 75%
- 1,1-di(t-butylperoxy)-3,5,5-trimethylcyclohexane, 95%
- di(2-ethylhexyl) peroxydicarbonate, 97%
- dibenzoyl peroxydicarbonate, 85%
- 2,5-dimethyl-2,5-di(benzoylperoxy) hexane, 92%
- peroxyacetic acid, 43%

Class III

Class III peroxides burn rapidly and present a moderate reactivity hazard. Examples include:

- acetyl cyclohexane sulfonyl peroxide, 29%
- benzoyl peroxide, 78%
- benzoyl peroxide paste, 55%
- benzoyl peroxide paste, 50%
- t-butyl peroxy-2-ethylhexanoate, 97%
- t-butyl peroxyneodecanoate, 75%
- cumene hydroperoxide, 86%
- decanoyl peroxide, 98.5%
- di-t-butyl peroxide, 99%
- di(4-butylcyclohexyl) peroxydicarbonate, 98%
- 1,1-di(t-butylperoxy)-3,5,5-trimethylcyclohexane, 75%
- 2,4-dichlorobenzoyl peroxide, 50%
- diisopropyl peroxydicarbonate, 30%
- 2,5-dimethyl-2,5-di(2-ethylhexanol peroxy)hexane, 90%
- 2,5-dimethyl-2,5-di(t-butylperoxy) hexane, 90%
- methyl ethyl ketone peroxide, 9% active oxygen

Class IV

Class IV peroxides burn in the same manner as ordinary combustibles and present a minimum reactivity hazard. Examples include:

- benzoyl peroxide, 70%
- benzoyl peroxide paste, 50%
- benzoyl peroxide slurry, 40%
- benzoyl peroxide powder, 35%
- t-butyl hydroperoxide, 70%
- t-butyl peroxy-2-ethylhexanoate, 50%
- dicumyl peroxide, 98%
- di(2-ethylhexyl) peroxydicarbonate, 40%
- lauroyl peroxide, 98%
- methyl ethyl ketone peroxide, 5.5% active oxygen
- methyl ethyl ketone peroxide, 9% active oxygen

Class V

Class V peroxides do not burn or present a decomposition hazard. Examples include:

- benzoyl peroxide, 35%
- 1,1-di(t-butylperoxy)-3,5,5-trimethyl-cyclohexane, 40%
- 2,5-di(t-butylperoxy)hexane, 47%
- 2,4-pentanedione peroxide, 4% active oxygen

UNSTABLE REACTIVE MATERIALS*

Class 4

Class 4 materials are readily capable of detonation, explosive decomposition, or explosive reaction at normal temperatures and pressures. This class includes materials that are sensitive to mechanical or localized thermal shock at normal temperatures and pressures. Examples include:

- acetyl peroxide
- dibutyl peroxide
- dinitrobenzene
- ethyl nitrate
- peroxyacetic acid
- picric acid, dry (trinitrobenzene)

Class 3

Class 3 materials are capable of detonation, explosive decomposition, or explosive reaction, but they require a strong initiating source which must be heated under confinement before initiation. This class includes materials that are sensitive to thermal or mechanical shock at elevated temperatures and pressures. Examples include:

- hydrogen peroxide, more than 52% by weight
- hydroxylamine
- nitromethane
- paranitroaniline
- perchloric acid
- tetrafluoroethylene monomer

Class 2

Class 2 materials are normally unstable and readily undergo violent chemical change but do not detonate. This class includes materials which can undergo chemical change with rapid release of energy at normal temperatures and pressures and which can undergo violent chemical change at elevated temperatures and pressures. Examples include:

- acrolein
- acrylic acid
- hydrazine
- methacrylic acid
- sodium perchlorate
- styrene
- vinyl acetate

Class 1

Class 1 materials are normally stable but can become unstable at elevated temperatures and pressures. Examples include:

- acetic acid
- hydrogen peroxide, 35% to 52%
- paraldehyde
- tetrahydrofuran

---

* Classification by degree of hazard shall be in accordance with UFC Standard No. 79-3. Also see NFPA Standard No. 49.

Appendix G

HAZ-MAT RESPONSE CHECKLISTS

RESPONDING TO EXPLOSIVE EMERGENCIES

HAZARD IDENTIFICATION

Before approaching the incident:

- ❒ Determine location of incident
- ❒ Access location preplans
- ❒ Identify markings and list below:

____

____

____

- ❒ Check shipping papers, MSDS, etc.
- ❒ Smoke/plume color: ________
- ❒ Sounds/odor
- ❒ Positively identify product(s):

____

____

- ❒ Call CHEMTREC / CANUTEC
- ❒ Location of leak or fire

____

____

- ❒ Cause of incident

____

____

- ❒ Extent and severity of injuries

____

____

____

- ❒ Position of container

____

____

- ❒ Integrity of container

____

____

- ❒ Fire exposure

____

____

- ❒ Environmental hazard

____

____

ACTION PLAN

NO FIRE:

- ❒ Determine type and size of container
- ❒ Find out quantity of product
- ❒ Determine container damage
- ❒ Is a fire possible?
- ❒ Prevent ignition sources
- ❒ Prevent product contamination
- ❒ Control additional product movement
- ❒ Set up control zones
- ❒ Determine evacuation distances
- ❒ Transfer and dispose of product if possible
- ❒ Cease radio usage if blasting caps are involved
- ❒ Anticipate resources

FIRE not involving product:

- ❒ Consider large scale evacuation
- ❒ Establish control zones
- ❒ Let product burn?
- ❒ Protect exposed cargo
- ❒ Extinguish fire
- ❒ Stabilize cargo
- ❒ Contain runoff
- ❒ Anticipate resources

FIRE within cargo area:

- ❒ Immediately withdraw personnel
- ❒ Evacuate a minimum distance of 2,500′
- ❒ Establish control zones
- ❒ Let product burn?
- ❒ Anticipate multiple explosions
- ❒ Anticipate resources

ZONING

- ❒ Check table of Initial Isolation and Protective Action Distances
- ❒ Check local protocol

NO FIRE:*

- ❒ 500′ evacuation corridor
- ❒ 200′ hot (restricted) zone
- ❒ Consider warm (limited access) zone

FIRE not involving product:*

- ❒ 1,000′ evacuation corridor
- ❒ 500′ hot zone
- ❒ Establish warm and cold (support) zones

FIRE within cargo area:*

- ❒ 2,500′ initial evacuation corridor
- ❒ 2,500′ hot zone
- ❒ Establish warm and cold zones

* All are recommended distances. Adjust as needed for the specific material, amount, and incident severity.

MANAGING THE INCIDENT

ATTACK STRATEGY:

- ❒ Upwind
- ❒ Initiate ICS
- ❒ Check SOPs
- ❒ Call in needed resources and specialized equipment
- ❒ Personal protective clothing
- ❒ Rescue
- ❒ EMS
- ❒ Decontamination
- ❒ Evacuate
- ❒ Crowd and traffic control
- ❒ Extinguishment
- ❒ Develop **secondary strategies** if fire is not rapidly extinguished
- ❒ Protect exposures
- ❒ Plan overhaul and clean-up

NO-ATTACK STRATEGY:

- ❒ Upwind
- ❒ Initiate ICS
- ❒ Check SOPs
- ❒ Evacuate civilians and response personnel
- ❒ Call in needed resources
- ❒ EMS
- ❒ Crowd and traffic control

Post-explosion scene:

- ❒ Anticipate a second explosion
- ❒ Call in needed resources
- ❒ Determine if the incident site is safe for entry
- ❒ Rescue
- ❒ EMS
- ❒ Decontamination
- ❒ Plan overhaul and clean-up

ASSISTANCE

Check agencies to contact:

- ❒ Dispatcher
- ❒ Law Enforcement
- ❒ Health Department
- ❒ CHEMTREC (800) 424-9300 or CANUTEC (613) 996-6666
- ❒ National Response Center
- ❒ EPA
- ❒ HMRT
- ❒ ATSDR (404) 639-6000
- ❒ Bomb specialists, list:

____

____

____

____

____

- ❒ Regional ordnance disposal team, list:

____

____

____

____

____

- ❒ Private contractors, list:

____

____

____

____

____

- ❒ Clean-up contractors, list:

____

____

____

____

____

TERMINATION

Check termination activities to complete:

- ❒ Decontamination
  Method (call ATSDR):
  ________ Dilution
  ________ Neutralization
  ________ Absorption
  ________ Isolation
  ________ Disposal
  Solution used: ________
- ❒ Rehabilitation
- ❒ Medical screening
- ❒ Exposure report
- ❒ Initial incident debriefing
- ❒ Post-incident analysis

RESPONDING TO COMPRESSED GAS EMERGENCIES

HAZARD IDENTIFICATION

Before approaching the incident:

- ❒ Determine location of incident
- ❒ Access location preplans
- ❒ Identify markings and list below:

____

____

- ❒ Check shipping papers, MSDS, etc.
- ❒ Sounds/odor/visual indications
- ❒ Positively identify product(s):

____

____

____

- ❒ Call CHEMTREC / CANUTEC
- ❒ List characteristics of product(s):
  - TLV ________
  - IDLH ________
  - LD_{50} or LC_{50} ________
  - Water soluble ________
  - Vapor density ________
  - Flammable range ________
  - Flash point ________
  - Oxidizer ________
  - Corrosive ________
  - Toxic by ________
- ❒ Location of leak or fire

____

- ❒ Extent and severity of injuries

____

- ❒ Position of container

____

- ❒ Integrity of container

____

- ❒ Fire exposure

____

- ❒ Environmental hazard

____

ACTION PLAN

NO LEAK or **FIRE:**

- ❒ Determine type and size of container
- ❒ Find out product level in container
- ❒ Determine container damage
- ❒ Check position of container
- ❒ Is a leak or fire possible?
- ❒ Move cylinder if it can be done safely
- ❒ Prevent ignition sources

LEAK but **NO FIRE:**

- ❒ Prevent ignition sources
- ❒ Determine number of leaking cylinders
- ❒ Use detection equipment
- ❒ Secure cylinders if possible
- ❒ Allow cylinder to bleed off?
- ❒ Move to outside location?
- ❒ Establish control zones
- ❒ Protect in place
- ❒ Evacuate
- ❒ Anticipate resources

LEAK and **FIRE:**

- ❒ Establish control zones
- ❒ Protect in place
- ❒ Evacuate
- ❒ Cylinder safety devices working properly?
- ❒ Determine if cylinders are secure
- ❒ Let product burn?
- ❒ Cool cylinder with water?
- ❒ Unmanned hose lines?
- ❒ Control gas flow after extinguishment
- ❒ Contain runoff
- ❒ Protect exposures
- ❒ Anticipate resources

ZONING

- ❒ Check table of Initial Isolation and Protective Action Distances
- ❒ Check local protocol

SMALL leak – **NO FIRE:***

- ❒ Use detection devices and adjust zones accordingly
- ❒ 250′ protective action distances
- ❒ 250′ hot zone

SMALL leak – **FIRE:***

- ❒ Use detection devices and adjust zones accordingly
- ❒ 500′ protective action distances
- ❒ 250′ hot zone
- ❒ Consider warm zone

LARGE leak – **NO FIRE:***

- ❒ Use detection devices and adjust zones accordingly
- ❒ 500′ protective action distances
- ❒ Downwind evacuation
- ❒ 250′ hot zone
- ❒ Establish warm and cold zones

LARGE leak – **FIRE:***

- ❒ Use detection devices and adjust zones accordingly
- ❒ 1,000′ initial protective action distances
- ❒ Downwind evacuation
- ❒ 500′ hot zone
- ❒ Establish warm and cold zones

* All are recommended distances. Adjust as needed for the specific material, amount, and incident severity.

MANAGING THE INCIDENT

ATTACK STRATEGY:

- ❒ Uphill/upwind
- ❒ Initiate ICS
- ❒ Check SOPs
- ❒ Never approach tank from ends
- ❒ Call in needed resources and specialized equipment
- ❒ Personal protective clothing
- ❒ Rescue
- ❒ EMS
- ❒ Decon prior to transporting victims
- ❒ Protect in place
- ❒ Evacuate
- ❒ Crowd and traffic control
- ❒ Extinguishment
- ❒ Protect exposures
- ❒ Plan overhaul and clean-up

NO-ATTACK STRATEGY:

- ❒ Uphill/upwind
- ❒ Initiate ICS
- ❒ Check SOPs
- ❒ Call in needed resources
- ❒ Personal protective clothing
- ❒ Rescue
- ❒ EMS
- ❒ Decon prior to transporting victims
- ❒ Evacuate
- ❒ Protect in place
- ❒ Protect exposures
- ❒ Crowd and traffic control
- ❒ Plan overhaul and clean-up

ASSISTANCE

Check agencies to contact:

- ❒ Dispatcher
- ❒ Law Enforcement
- ❒ Health Department
- ❒ CHEMTREC (800) 424-9300 or CANUTEC (613) 966-6666
- ❒ National Response Center
- ❒ EPA
- ❒ HMRT
- ❒ ATSDR (404) 639-6000
- ❒ Poison control center, if toxic:

____

____

- ❒ Technical specialists, list:

____

____

____

____

____

- ❒ Private contractors, list:

____

____

____

____

____

- ❒ Clean-up contractors, list:

____

____

____

____

____

TERMINATION

Check termination activities to complete:

- ❒ Decontamination
  - Method (call ATSDR):
  - ________ Dilution
  - ________ Neutralization
  - ________ Absorption
  - ________ Isolation
  - ________ Disposal
  - Solution used: ________
- ❒ Rehabilitation
- ❒ Medical screening
- ❒ Exposure report
- ❒ Initial incident debriefing
- ❒ Post-incident analysis

RESPONDING TO LIQUEFIED GAS EMERGENCIES

HAZARD IDENTIFICATION

Before approaching the incident:

- ❒ Determine location of incident
- ❒ Access location preplans
- ❒ Identify markings and list below:

____

____

- ❒ Check shipping papers, MSDS, etc.
- ❒ Sounds/odor/visual indications
- ❒ Positively identify product(s):

____

____

- ❒ Call CHEMTREC / CANUTEC
- ❒ List characteristics of product(s):
  - TLV ______
  - IDLH ______
  - LD_{50} or LC_{50} ______
  - Water soluble ______
  - Expansion ratio ______
  - Boiling point ______
  - Vapor density ______
  - Flammable range ______
  - Flash point ______
  - Oxidizer ______
  - Corrosive ______
  - Toxic by ______
- ❒ Location of leak or fire __________
- ❒ Extent and severity of injuries __________
- ❒ Position of container __________
- ❒ Integrity of container __________
- ❒ Cylinder impinged by fire? __________
- ❒ Environmental hazard __________

ACTION PLAN

NO LEAK or **FIRE:**

- ❒ Determine type and size of container
- ❒ Find out product level in container
- ❒ Determine container damage
- ❒ Check position of container
- ❒ Is a leak or fire possible?
- ❒ Move cylinder if it can be done safely
- ❒ Prevent ignition sources

LEAK but **NO FIRE:**

- ❒ Prevent ignition sources
- ❒ Use detection equipment
- ❒ Check wind direction
- ❒ Establish control zones
- ❒ Protect in place
- ❒ Evacuate
- ❒ Use fog stream to disperse vapors?
- ❒ Unmanned hose streams?
- ❒ Contain runoff?
- ❒ Determine volume of leak
- ❒ Shut valve?
- ❒ Anticipate resources

LEAK and **FIRE:**

- ❒ Establish control zones
- ❒ Evacuate
- ❒ Let product burn? Do not extinguish unless gas supply can be stopped
- ❒ Constantly monitor for BLEVE conditions:
  - ✓ How long has container been impinged?
  - ✓ Safety devices working?
  - ✓ Metal discoloring, bubbling, or bulging
- ❒ Sufficient water supply?
- ❒ Cool cylinder with water?
- ❒ Unmanned hose lines?
- ❒ Contain runoff
- ❒ Protect exposures
- ❒ Anticipate resources

ZONING

- ❒ Check table of Initial Isolation and Protective Action Distances
- ❒ Check local protocol

SMALL leak – **NO FIRE:***

- ❒ Use detection devices and adjust zones accordingly
- ❒ 250′ protective action distance
- ❒ 250′ hot zone

SMALL leak – **FIRE:***

- ❒ Use detection devices and adjust zones accordingly
- ❒ 500′ protective action distance
- ❒ 500′ hot zone
- ❒ Consider warm zone

LARGE leak – **NO FIRE:***

- ❒ Use detection devices and adjust zones accordingly
- ❒ 1,000′ protective action distance
- ❒ Downwind evacuation
- ❒ 500′ hot zone
- ❒ Establish warm and cold zones

LARGE leak – **FIRE:***

- ❒ Use detection devices and adjust zones accordingly
- ❒ 2,500′ initial protective action distance
- ❒ Downwind evacuation
- ❒ 2,500′ hot zone
- ❒ Establish warm and cold zones

* All are recommended distances. Adjust as needed for the specific material, amount, and incident severity.

MANAGING THE INCIDENT

ATTACK STRATEGY:

- ❒ Uphill/upwind
- ❒ Initiate ICS
- ❒ Check SOPs
- ❒ Never approach tank from ends
- ❒ Call in needed resources and specialized equipment
- ❒ Personal protective clothing
- ❒ Rescue
- ❒ EMS
- ❒ Decon prior to transporting victims
- ❒ Protect in place
- ❒ Evacuate
- ❒ Crowd and traffic control
- ❒ Extinguishment
- ❒ Protect exposures
- ❒ Diking/diversion/retention
- ❒ Plan overhaul and clean-up

NO-ATTACK STRATEGY:

- ❒ Uphill/upwind
- ❒ Initiate ICS
- ❒ Check SOPs
- ❒ Evacuate to at least 2,500′
- ❒ Call in needed resources
- ❒ Personal protective clothing
- ❒ Rescue
- ❒ EMS
- ❒ Decon prior to transporting victims
- ❒ Protect in place
- ❒ Protect exposures
- ❒ Crowd and traffic control
- ❒ Plan overhaul and clean-up

ASSISTANCE

Check agencies to contact:

- ❒ Dispatcher
- ❒ Law Enforcement
- ❒ Health Department
- ❒ CHEMTREC (800) 424-9300 or CANUTEC (613) 996-6666
- ❒ National Response Center
- ❒ EPA
- ❒ HMRT
- ❒ ATSDR (404) 639-6000
- ❒ Poison control center, if toxic:

____

____

- ❒ Technical specialists, list:

____

____

____

____

____

- ❒ Private contractors, list:

____

____

____

____

____

- ❒ Clean-up contractors, list:

____

____

____

____

____

TERMINATION

Check termination activities to complete:

- ❒ Decontamination
  - Method (call ATSDR):
    - ______ Dilution
    - ______ Neutralization
    - ______ Absorption
    - ______ Isolation
    - ______ Disposal
  - Solution used: ______
- ❒ Rehabilitation
- ❒ Medical screening
- ❒ Exposure report
- ❒ Initial incident debriefing
- ❒ Post-incident analysis

RESPONDING TO CRYOGENIC EMERGENCIES

HAZARD IDENTIFICATION

Before approaching the incident:

- ❒ Determine location of incident
- ❒ Access location preplans
- ❒ Identify markings and list below:

____

____

- ❒ Check shipping papers, MSDS, etc.
- ❒ Tank frosting?
- ❒ Sounds/odor/visual indications
- ❒ Positively identify product(s):

____

____

- ❒ Call CHEMTREC / CANUTEC
- ❒ List characteristics of product(s):
  - TLV ________
  - IDLH ________
  - LD_{50} or LC_{50} ________
  - Expansion ratio ________
  - Boiling point ________
  - Flammable range ________
  - Flash point ________
  - Oxidizer ________
  - Toxic by ________
- ❒ Location of leak or fire

____

- ❒ Extent and severity of injuries

____

- ❒ Position of container

____

- ❒ Integrity of container

____

- ❒ Cylinder impinged by fire?

____

- ❒ Environmental hazard

____

ACTION PLAN

NO LEAK or **FIRE:**

- ❒ Determine type and size of container
- ❒ Find out product level in container
- ❒ Determine container damage, especially note insulation damage
- ❒ Check position of container
- ❒ Is a leak or fire possible?
- ❒ Move cylinder if it can be done safely
- ❒ Prevent ignition sources
- ❒ Monitor container pressure

LEAK but **NO FIRE:**

- ❒ Prevent ignition sources
- ❒ Use detection equipment
- ❒ Check wind direction
- ❒ Establish control zones
- ❒ Protect in place
- ❒ Evacuate
- ❒ Do not apply water to product or container
- ❒ Use fog stream to disperse vapors?
- ❒ Unmanned hose streams?
- ❒ Isolate/contain product
- ❒ Contain runoff?
- ❒ Determine volume of leak
- ❒ Anticipate resources
- ❒ Monitor container pressure

LEAK and **FIRE:**

- ❒ Establish control zones
- ❒ Evacuate
- ❒ Let product burn?
- ❒ Constantly monitor for BLEVE conditions
- ❒ Sufficient water supply?
- ❒ Cool cylinder with water?
- ❒ Unmanned hose lines?
- ❒ Prevent product contamination
- ❒ Contain runoff
- ❒ Protect exposures
- ❒ Anticipate resources

ZONING

- ❒ Check table of Initial Isolation and Protective Action Distances
- ❒ Check local protocol

SMALL leak – **NO FIRE:***

- ❒ Use detection devices and adjust zones accordingly
- ❒ 250′ evacuation corridor
- ❒ 250′ hot zone

SMALL leak – **FIRE:***

- ❒ Use detection devices and adjust zones accordingly
- ❒ 500′ evacuation corridor
- ❒ 500′ hot zone
- ❒ Consider warm zone

LARGE leak – **NO FIRE:***

- ❒ Use detection devices and adjust zones accordingly
- ❒ 1,000′ evacuation corridor
- ❒ Downwind evacuation
- ❒ 500′ hot zone
- ❒ Establish warm and cold zones

LARGE leak – **FIRE:***

- ❒ Use detection devices and adjust zones accordingly
- ❒ 2,500′ initial evacuation corridor
- ❒ Downwind evacuation
- ❒ 2,500′ hot zone
- ❒ Establish warm and cold zones

* All are recommended distances. Adjust as needed for the specific material, amount, and incident severity.

MANAGING THE INCIDENT

ATTACK STRATEGY:

- ❒ Uphill/upwind
- ❒ Initiate ICS
- ❒ Check SOPs
- ❒ Never approach tank from ends
- ❒ Call in needed resources and specialized equipment
- ❒ Personal protective clothing
- ❒ Rescue
- ❒ EMS
- ❒ Decon prior to transporting victims
- ❒ Protect in place
- ❒ Evacuate
- ❒ Crowd and traffic control
- ❒ Extinguishment
- ❒ Protect exposures
- ❒ Diking/diversion/retention
- ❒ Plan overhaul and clean-up

NO-ATTACK STRATEGY:

- ❒ Uphill/upwind
- ❒ Initiate ICS
- ❒ Check SOPs
- ❒ Call in needed resources
- ❒ Personal protective clothing
- ❒ Rescue
- ❒ EMS
- ❒ Decon prior to transporting victims
- ❒ Protect in place
- ❒ Protect exposures
- ❒ Crowd and traffic control
- ❒ Plan overhaul and clean-up

ASSISTANCE

Check agencies to contact:

- ❒ Dispatcher
- ❒ Law Enforcement
- ❒ Health Department
- ❒ CHEMTREC (800) 424-9300 or CANUTEC (613) 996-6666
- ❒ National Response Center
- ❒ EPA
- ❒ HMRT
- ❒ ATSDR (404) 639-6000
- ❒ Poison control center, if toxic:

____

____

- ❒ Technical specialists, list:

____

____

____

____

____

- ❒ Private contractors, list:

____

____

____

____

____

- ❒ Clean-up contractors, list:

____

____

____

____

____

TERMINATION

Check termination activities to complete:

- ❒ Decontamination
  - Method (call ATSDR):
    - ________ Dilution
    - ________ Neutralization
    - ________ Absorption
    - ________ Isolation
    - ________ Disposal
  - Solution used: ________
- ❒ Rehabilitation
- ❒ Medical screening
- ❒ Exposure report
- ❒ Initial incident debriefing
- ❒ Post-incident analysis

RESPONDING TO FLAMMABLE AND COMBUSTIBLE LIQUID EMERGENCIES

HAZARD IDENTIFICATION

Before approaching the incident:

- ☐ Determine location of incident
- ☐ Access location preplans
- ☐ Identify markings and list below:

____

____

- ☐ Check shipping papers, MSDS, etc.
- ☐ Smoke/plume color: __________
- ☐ Sounds/odor
- ☐ Positively identify product(s):

____

____

- ☐ Call CHEMTREC / CANUTEC
- ☐ List characteristics of product(s):
  - Flammable range __________
  - Flash point __________
  - Water soluble __________
  - Specific gravity __________
  - Vapor density __________
  - TLV __________
  - IDLH __________
  - LD_{50} or LC_{50} __________
  - Toxic by __________
- ☐ Location of leak or fire

____

- ☐ Extent and severity of injuries

____

- ☐ Position of container

____

- ☐ Integrity of container

____

- ☐ Fire exposure

____

- ☐ Environmental hazard

____

ACTION PLAN

NO LEAK or **FIRE:**

- ☐ Determine type and size of container
- ☐ Find out product level in container
- ☐ Determine container damage
- ☐ Check position of container
- ☐ Is a leak or fire possible?
- ☐ Set up control zones
- ☐ Set up protective action distances
- ☐ Isolate/contain product
- ☐ Prevent ignition sources

LEAK but **NO FIRE:**

- ☐ Prevent ignition sources
- ☐ Establish control zones
- ☐ Protect in place
- ☐ Evacuate
- ☐ Apply foam or other agents as needed
- ☐ Control leak
- ☐ Isolate, contain, and/or confine product
- ☐ Transfer product to other container
- ☐ Anticipate resources
- ☐ Anticipate decon procedures

LEAK and **FIRE:**

- ☐ Establish control zones
- ☐ Protect in place
- ☐ Evacuate
- ☐ Control fuel flow if possible
- ☐ Let product burn?
- ☐ Determine best extinguishing agent
- ☐ Determine amount of extinguishing agent needed
- ☐ Unmanned hose lines?
- ☐ Protect exposures
- ☐ Contain runoff
- ☐ Anticipate resources
- ☐ Anticipate decon procedures

ZONING

- ☐ Check table of Initial Isolation and Protective Action Distances
- ☐ Check local protocol

SMALL spill – **NO FIRE:***

- ☐ Use detection devices and adjust zones accordingly
- ☐ 350′ protective action distance
- ☐ 200′ hot zone

SMALL spill – **FIRE:***

- ☐ Use detection devices and adjust zones accordingly
- ☐ 500′ protective action distance
- ☐ 200′ hot zone
- ☐ Consider warm zone

LARGE spill – **NO FIRE:***

- ☐ Use detection devices and adjust zones accordingly
- ☐ 1,000′ protective action distance
- ☐ 500′ hot zone
- ☐ Establish warm and cold zones

LARGE spill – **FIRE:***

- ☐ Use detection devices and adjust zones accordingly
- ☐ 1,000′ initial protective action distance
- ☐ 500′ hot zone
- ☐ Establish warm and cold zones

* All are recommended distances. Adjust as needed for the specific material, amount, and incident severity.

MANAGING THE INCIDENT

ATTACK STRATEGY:

- ☐ Uphill/upwind
- ☐ Initiate ICS
- ☐ Check SOPs
- ☐ Never approach tank from ends
- ☐ Call in needed resources and specialized equipment
- ☐ Personal protective clothing
- ☐ Rescue
- ☐ EMS
- ☐ Decon prior to transporting victims
- ☐ Evacuate
- ☐ Protect in place
- ☐ Crowd and traffic control
- ☐ Stop fuel flow
- ☐ Extinguishment
- ☐ Protect exposures
- ☐ Isolate and contain
- ☐ Diking/diversion/retention
- ☐ Plan overhaul and clean-up

NO-ATTACK STRATEGY:

- ☐ Uphill/upwind
- ☐ Initiate ICS
- ☐ Check SOPs
- ☐ Call in needed resources
- ☐ Personal protective clothing
- ☐ Rescue
- ☐ EMS
- ☐ Decon prior to transporting victims
- ☐ Evacuate
- ☐ Protect in place
- ☐ Protect exposures
- ☐ Crowd and traffic control
- ☐ Isolate and contain
- ☐ Diking/diversion/retention
- ☐ Plan overhaul and clean-up

ASSISTANCE

Check agencies to contact:

- ☐ Dispatcher
- ☐ Law Enforcement
- ☐ Health Department
- ☐ CHEMTREC (800) 424-9300 or CANUTEC (613) 996-6666
- ☐ National Response Center
- ☐ EPA
- ☐ HMRT
- ☐ ATSDR (404) 639-6000
- ☐ Foam suppliers, list:

____

____

____

____

____

- ☐ Technical specialists, list:

____

____

____

____

____

- ☐ Private contractors, list:

____

____

____

____

____

- ☐ Clean-up contractors, list:

____

____

____

____

____

TERMINATION

Check termination activities to complete:

- ☐ Decontamination
  - Method (call ATSDR):
  - __________ Dilution
  - __________ Neutralization
  - __________ Absorption
  - __________ Isolation
  - __________ Disposal
  - Solution used: __________
- ☐ Rehabilitation
- ☐ Medical screening
- ☐ Exposure report
- ☐ Initial incident debriefing
- ☐ Post-incident analysis

RESPONDING TO FLAMMABLE SOLID EMERGENCIES

HAZARD IDENTIFICATION

Before approaching the incident:

- ❒ Determine location of incident
- ❒ Access location preplans
- ❒ Identify markings and list below:

  ____
  ____

- ❒ Check shipping papers, MSDS, etc.
- ❒ Smoke color: ____________
- ❒ Positively identify product(s):

  ____
  ____

- ❒ Call CHEMTREC / CANUTEC
- ❒ List characteristics of product(s):
  - Water reactive ________
  - Spontaneously combustible ________
  - Pyrophoric ________
  - Melting point ________
  - Boiling point ________
  - Radioactive ________
  - Toxic by ________
- ❒ Location of leak or fire

  ____

- ❒ Extent and severity of injuries

  ____

- ❒ Position of container

  ____

- ❒ Integrity of container

  ____

- ❒ Fire exposure

  ____

- ❒ Environmental hazard

  ____

ACTION PLAN

NO LEAK or **FIRE:**

- ❒ Determine type and size of container
- ❒ Find out product level in container
- ❒ Determine container damage
- ❒ Is the container airtight?
- ❒ Is there water or an inert substance in container?
- ❒ Is water/inert substance leaking?
- ❒ Is a leak or fire possible?
- ❒ Set up control zones
- ❒ Set up protective action distances
- ❒ Isolate/contain product
- ❒ Prevent ignition sources

LEAK but **NO FIRE:**

- ❒ Establish control zones
- ❒ IEvacuate
- ❒ Protect in place
- ❒ Isolate, contain, or confine product
- ❒ Spontaneously combustible – keep covered
- ❒ Water reactive – keep dry/covered
- ❒ Pyrophoric – cover with sand/dirt
- ❒ Prevent ignition sources
- ❒ Anticipate resources
- ❒ Anticipate decon procedures

LEAK and **FIRE:**

- ❒ Establish control zones
- ❒ Evacuate
- ❒ Protect in place
- ❒ Let product burn?
- ❒ Determine best extinguishing agent
- ❒ Determine amount of extinguishing agent needed
- ❒ Small fires – cover with sand, dirt, or water spray
- ❒ Unmanned hose lines?
- ❒ Contain runoff
- ❒ Protect exposures
- ❒ Anticipate resources
- ❒ Anticipate decon procedures

ZONING

- ❒ Check table of Initial Isolation and Protective Action Distances
- ❒ Check local protocol

SMALL spill – **NO FIRE:***

- ❒ 200′ protective action distance
- ❒ 150′ hot zone

SMALL spill – **FIRE:***

- ❒ 500′ protective action distance
- ❒ 200′ hot zone
- ❒ Consider warm zone

LARGE spill – **NO FIRE:***

- ❒ 500′ protective action distance
- ❒ 300′ hot zone
- ❒ Consider warm and cold zones

LARGE spill – **FIRE:***

- ❒ 1,000′ initial protective action distance
- ❒ Downwind evacuation
- ❒ 500′ hot zone
- ❒ Establish warm and cold zones

* All are recommended distances. Adjust as needed for the specific material, amount, and incident severity.

MANAGING THE INCIDENT

ATTACK STRATEGY:

- ❒ Uphill/upwind
- ❒ Initiate ICS
- ❒ Check SOPs
- ❒ Call in needed resources and specialized equipment
- ❒ Personal protective clothing
- ❒ Rescue
- ❒ EMS
- ❒ Decon prior to transporting victims
- ❒ Evacuate
- ❒ Protect in place
- ❒ Crowd and traffic control
- ❒ Extinguishment
- ❒ Protect exposures
- ❒ Isolate and contain
- ❒ Diking/diversion/retention
- ❒ Plan overhaul and clean-up

NO-ATTACK STRATEGY:

- ❒ Uphill/upwind
- ❒ Initiate ICS
- ❒ Check SOPs
- ❒ Call in needed resources
- ❒ Personal protective clothing
- ❒ Rescue
- ❒ EMS
- ❒ Decon prior to transporting victims
- ❒ Evacuate
- ❒ Protect in place
- ❒ Protect exposures
- ❒ Crowd and traffic control
- ❒ Isolate and contain
- ❒ Diking/diversion/retention
- ❒ Plan overhaul and clean-up

ASSISTANCE

Check agencies to contact:

- ❒ Dispatcher
- ❒ Law Enforcement
- ❒ Health Department
- ❒ CHEMTREC (800) 424-9300 or CANUTEC (613) 996-6666
- ❒ National Response Center
- ❒ EPA
- ❒ HMRT
- ❒ ATSDR (404) 639-6000
- ❒ Technical specialists, list:

  ____
  ____
  ____
  ____
  ____

- ❒ Private contractors, list:

  ____
  ____
  ____
  ____
  ____

- ❒ Clean-up contractors, list:

  ____
  ____
  ____
  ____
  ____

TERMINATION

Check termination activities to complete:

- ❒ Decontamination
  - Method (call ATSDR):
    - ________ Dilution
    - ________ Neutralization
    - ________ Absorption
    - ________ Isolation
    - ________ Disposal
  - Solution used: ____________
- ❒ Rehabilitation
- ❒ Medical screening
- ❒ Exposure report
- ❒ Initial incident debriefing
- ❒ Post-incident analysis

RESPONDING TO OXIDIZER EMERGENCIES

HAZARD IDENTIFICATION

Before approaching the incident:

- ❐ Determine location of incident
- ❐ Access location preplans
- ❐ Identify markings and list below:

  ____

  ____

- ❐ Check shipping papers, MSDS, etc.
- ❐ Positively identify product(s):

  ____

  ____

- ❐ Call CHEMTREC / CANUTEC
- ❐ List characteristics of product(s):
  - Water reactive ________
  - Water soluble ________
  - Spontaneously combustible ________
  - Pyrophoric ________
  - SADT value ________
  - Melting point ________
  - Boiling point ________
  - Radioactive ________
  - Toxic by ________
- ❐ Location of leak or fire

  ____

- ❐ Cause of incident

  ____

- ❐ Extent and severity of injuries

  ____

- ❐ Position of container

  ____

- ❐ Integrity of container

  ____

- ❐ Fire exposure

  ____

- ❐ Environmental hazard

  ____

ACTION PLAN

NO LEAK or **FIRE:**

- ❐ Determine type and size of container
- ❐ Find out product level in container
- ❐ Determine container damage
- ❐ Is a leak or fire possible?
- ❐ Set up control zones
- ❐ Set up protective action distances
- ❐ Isolate/contain product
- ❐ Prevent ignition sources

LEAK but **NO FIRE:**

- ❐ Establish control zones
- ❐ Evacuate
- ❐ Protect in place
- ❐ Isolate, contain, and/or confine product
- ❐ Spontaneously combustible – keep covered
- ❐ Water reactive – keep dry
- ❐ Pyrophoric – cover with sand or dirt
- ❐ Prevent ignition sources
- ❐ Prevent environmental contamination
- ❐ Anticipate resources
- ❐ Anticipate decon procedures

LEAK and **FIRE:**

- ❐ Establish control zones
- ❐ Evacuate
- ❐ Protect in place
- ❐ Let product burn?
- ❐ Determine best extinguishing agent
- ❐ Determine amount of extinguishing agent needed
- ❐ Unmanned hose lines?
- ❐ Contain runoff
- ❐ Protect exposures
- ❐ Anticipate resources
- ❐ Anticipate decon procedures

ZONING

- ❐ Check table of Initial Isolation and Protective Action Distances
- ❐ Check local protocol

SMALL spill – **NO FIRE:***

- ❐ 250′ protective action distance
- ❐ 100′ hot zone

SMALL spill – **FIRE:***

- ❐ 500′ protective action distance
- ❐ 250′ hot zone
- ❐ Consider warm zone

LARGE spill – **NO FIRE:***

- ❐ 500′ protective action distance
- ❐ 300′ hot zone
- ❐ Consider warm and cold zones

LARGE spill – **FIRE:***

- ❐ 1,000′ initial protective action distance
- ❐ Downwind evacuation
- ❐ 500′ hot zone
- ❐ Establish warm and cold zones
- ❐ If deflagration is possible, evacuate civilians and response personnel to **at least** 2,500′

* All are recommended distances. Adjust as needed for the specific material, amount, and incident severity.

MANAGING THE INCIDENT

ATTACK STRATEGY:

- ❐ Uphill/upwind
- ❐ Initiate ICS
- ❐ Check SOPs
- ❐ Call in needed resources and specialized equipment
- ❐ Personal protective clothing
- ❐ Rescue
- ❐ EMS
- ❐ Decon prior to transporting victims
- ❐ Evacuate
- ❐ Protect in place
- ❐ Crowd and traffic control
- ❐ Extinguishment
- ❐ Check water supply
- ❐ Protect exposures
- ❐ Isolate and contain
- ❐ Diking/diversion/retention
- ❐ Plan overhaul and clean-up

NO-ATTACK STRATEGY:

- ❐ Uphill/upwind
- ❐ Initiate ICS
- ❐ Check SOPs
- ❐ Call in needed resources
- ❐ Personal protective clothing
- ❐ Rescue
- ❐ EMS
- ❐ Decon prior to transporting victims
- ❐ Evacuate
- ❐ Protect in place
- ❐ Protect exposures
- ❐ Crowd and traffic control
- ❐ Isolate and contain
- ❐ Diking/diversion/retention
- ❐ Plan overhaul and clean-up

ASSISTANCE

Check agencies to contact:

- ❐ Dispatcher
- ❐ Law Enforcement
- ❐ Health Department
- ❐ CHEMTREC (800) 424-9300 or CANUTEC (613) 996-6666
- ❐ National Response Center
- ❐ EPA
- ❐ HMRT
- ❐ ATSDR (404) 639-6000
- ❐ Technical specialists, list:

  ____

  ____

  ____

  ____

  ____

- ❐ Private contractors, list:

  ____

  ____

  ____

  ____

  ____

- ❐ Clean-up contractors, list:

  ____

  ____

  ____

  ____

  ____

TERMINATION

Check termination activities to complete:

- ❐ Decontamination

  Method (call ATSDR):
  - ________ Dilution
  - ________ Neutralization
  - ________ Absorption
  - ________ Isolation
  - ________ Disposal

  Solution used: ________

- ❐ Rehabilitation
- ❐ Medical screening
- ❐ Exposure report
- ❐ Initial incident debriefing
- ❐ Post-incident analysis

HAZARD IDENTIFICATION

Before approaching the incident:

- ❒ Determine location of incident
- ❒ Access location preplans
- ❒ Identify markings and list below:

______

______

- ❒ Check shipping papers, MSDS, etc.
- ❒ Positively identify product(s):

______

______

- ❒ Call CHEMTREC / CANUTEC
- ❒ List characteristics of product(s):
  - TLV ______
  - IDLH ______
  - LD_{50} or LC_{50} ______
  - Water soluble ______
  - Vapor density ______
  - Specific gravity ______
  - Flammable range ______
  - Flash point ______
- ❒ Toxic by inhalation
- ❒ Toxic by skin contact/absorption
- ❒ Toxic by ingestion
- ❒ Location of leak or fire

______

- ❒ Extent and severity of injuries

______

- ❒ Position of container

______

- ❒ Integrity of container

______

- ❒ Fire exposure

______

- ❒ Environmental hazard

______

ACTION PLAN

NO LEAK or **FIRE:**

- ❒ Determine type and size of container
- ❒ Find out product level in container
- ❒ Determine container damage
- ❒ Is a leak or fire possible?
- ❒ Set up control zones
- ❒ Set up protective action distances
- ❒ Isolate product
- ❒ Prevent ignition sources
- ❒ Cover container to prevent contamination

LEAK but **NO FIRE:**

- ❒ Establish control zones
- ❒ Evacuate
- ❒ Protect in place
- ❒ Isolate, contain, and/or confine product
- ❒ Cover product if possible
- ❒ Prevent ignition sources
- ❒ Anticipate resources
- ❒ Anticipate decon procedures

LEAK and **FIRE:**

- ❒ Establish control zones
- ❒ Evacuate
- ❒ Protect in place
- ❒ Let product burn?
- ❒ Determine best extinguishing agent
- ❒ Determine amount of extinguishing agent needed
- ❒ Unmanned hose lines?
- ❒ Contain runoff (use as little water as possible)
- ❒ Protect exposures
- ❒ Anticipate resources
- ❒ Anticipate decon procedures

ZONING

- ❒ Check table of Initial Isolation and Protective Action Distances
- ❒ Check local protocol

SMALL spill – **NO FIRE:***

- ❒ 250′ protective action distance
- ❒ 100′ hot zone

SMALL spill – **FIRE:***

- ❒ 500′ protective action distance
- ❒ 250′ hot zone
- ❒ Consider warm zone

LARGE spill – **NO FIRE:***

- ❒ 500′ protective action distance
- ❒ Downwind evacuation
- ❒ 250′ hot zone
- ❒ Establish warm and cold zones

LARGE spill – **FIRE:***

- ❒ 1,000′ initial protective action distance
- ❒ Downwind evacuation
- ❒ 500′ hot zone
- ❒ Establish warm and cold zones

* All are recommended distances. Adjust as needed for the specific material, amount, and incident severity.

MANAGING THE INCIDENT

ATTACK STRATEGY:

- ❒ Uphill/upwind
- ❒ Initiate ICS
- ❒ Check SOPs
- ❒ Call in needed resources and specialized equipment
- ❒ Personal protective clothing
- ❒ Rescue
- ❒ EMS
- ❒ Decon prior to transporting victims
- ❒ Evacuate
- ❒ Protect in place
- ❒ Crowd and traffic control
- ❒ Extinguishment
- ❒ Protect exposures
- ❒ Isolate and contain
- ❒ Diking/diversion/retention
- ❒ Plan overhaul and clean-up

NO-ATTACK STRATEGY:

- ❒ Uphill/upwind
- ❒ Initiate ICS
- ❒ Check SOPs
- ❒ Call in needed resources
- ❒ Personal protective clothing
- ❒ Rescue
- ❒ EMS
- ❒ Decon prior to transporting victims
- ❒ Evacuate
- ❒ Protect in place
- ❒ Protect exposures
- ❒ Crowd and traffic control
- ❒ Isolate and contain
- ❒ Diking/diversion/retention
- ❒ Plan overhaul and clean-up

ASSISTANCE

Check agencies to contact:

- ❒ Dispatcher
- ❒ Law Enforcement
- ❒ Health Department
- ❒ CHEMTREC (800) 424-9300 or CANUTEC (613) 966-6666
- ❒ Poison control center:

______

- ❒ ATSDR (404) 639-6000
- ❒ National Response Center
- ❒ EPA
- ❒ HMRT
- ❒ Technical specialists, list:

______

______

______

______

______

- ❒ Private contractors, list:

______

______

______

______

______

- ❒ Clean-up contractors, list:

______

______

______

______

______

TERMINATION

Check termination activities to complete:

- ❒ Decontamination
  - Method (call ATSDR):
  - ______ Dilution
  - ______ Neutralization
  - ______ Absorption
  - ______ Isolation
  - ______ Disposal
  - Solution used: ______
- ❒ Rehabilitation
- ❒ Medical screening
- ❒ Exposure report
- ❒ Initial incident debriefing
- ❒ Post-incident analysis

RESPONDING TO RADIOACTIVE MATERIAL EMERGENCIES

HAZARD IDENTIFICATION

Before approaching the incident:

- ☐ Determine location of incident
- ☐ Access location preplans
- ☐ Identify markings and list below:

____

____

- ☐ Check shipping papers, MSDS, etc.
- ☐ Positively identify product(s):

____

____

- ☐ Call CHEMTREC / CANUTEC
- ☐ List characteristics of product(s):
  - Water reactive ____________
  - Spontaneously combustible ____________
  - Pyrophoric ____________
  - Corrosive ____________
  - Explosive potential ____________
  - Toxic by ____________
- ☐ Amount of radioactivity (use appropriate monitoring equipment)

____

- ☐ Radioactive white label I
- ☐ Radioactive yellow label II
- ☐ Radioactive yellow label III
- ☐ Location of leak or fire

____

- ☐ Cause of incident

____

- ☐ Extent and severity of injuries

____

- ☐ Position of container

____

- ☐ Integrity of container

____

- ☐ Fire exposure

____

- ☐ Environmental hazard

____

ACTION PLAN

NO LEAK or **FIRE:**

- ☐ Use appropriate monitoring devices to determine hazards
- ☐ Determine type and size of container
- ☐ Determine container damage
- ☐ Is a leak or fire possible?
- ☐ Set up control zones
- ☐ Set up protective action distances
- ☐ Isolate product
- ☐ Contact radiological specialists
- ☐ Avoid container except when performing rescue
- ☐ Limit exposure times for personnel

LEAK but **NO FIRE:**

- ☐ Establish control zones
- ☐ Evacuate
- ☐ Protect in place
- ☐ Isolate product
- ☐ Contact radiological specialists
- ☐ Avoid the material
- ☐ Limit exposure times
- ☐ Anticipate resources
- ☐ Anticipate decon procedures

LEAK and **FIRE:**

- ☐ Establish control zones
- ☐ Evacuate
- ☐ Protect in place
- ☐ Isolate product
- ☐ Contact radiological specialists
- ☐ Avoid the material
- ☐ Limit exposure times
- ☐ Let product burn?
- ☐ Unmanned hose lines?
- ☐ Contain runoff
- ☐ Anticipate resources
- ☐ Anticipate extensive decon procedures

ZONING

- ☐ Check table of Initial Isolation and Protective Action Distances
- ☐ Check local protocol

LSA materials:*

- ☐ 500′ protective action distance
- ☐ 300′ hot zone (limit exposure to 2mR)
- ☐ 500′ warm zone (limit exposure to 1mR)

Hot source materials:*

- ☐ 1,500′ protective action distance
- ☐ 500′ hot zone (limit exposure to 1mR)
- ☐ 1,000′ warm zone
- ☐ Establish cold zone

Nuclear fuels, waste fuels, or weapons on **FIRE:***

- ☐ 3,000′ protective action distance
- ☐ Downwind evacuation
- ☐ Establish appropriate hot zone (limit exposure to 1mR)
- ☐ Establish appropriate warm and cold zones
- ☐ **DO NOT ENTER SITE**

* All are recommended distances. Adjust as needed for the specific material, amount, and incident severity.

MANAGING THE INCIDENT

ATTACK STRATEGY:

- ☐ Radiation exposure assessment
- ☐ Limit exposure times
- ☐ Keep to a safe distance
- ☐ Use shielding techniques
- ☐ Uphill/upwind
- ☐ Initiate ICS
- ☐ Check SOPs
- ☐ Call in needed resources and specialized equipment
- ☐ Personal protective clothing
- ☐ Rescue
- ☐ EMS
- ☐ Decon prior to transporting victims
- ☐ Evacuate
- ☐ Crowd and traffic control
- ☐ Extinguishment
- ☐ Protect exposures
- ☐ Isolate and contain
- ☐ Diking/diversion/retention
- ☐ Plan overhaul and clean-up

NO-ATTACK STRATEGY:

- ☐ Radiation exposure assessment
- ☐ Keep to a safe distance
- ☐ Uphill/upwind
- ☐ Initiate ICS
- ☐ Check SOPs
- ☐ Call in needed resources
- ☐ Personal protective clothing
- ☐ Rescue
- ☐ EMS
- ☐ Decon prior to transporting victims
- ☐ Evacuate
- ☐ Protect in place
- ☐ Protect exposures
- ☐ Crowd and traffic control
- ☐ Isolate and contain
- ☐ Diking/diversion/retention
- ☐ Plan overhaul and clean-up

ASSISTANCE

Check agencies to contact:

- ☐ Dispatcher
- ☐ Law Enforcement
- ☐ Health Department
- ☐ CHEMTREC (800) 424-9300 or CANUTEC (613) 996-6666
- ☐ Nuclear Regulatory Commission
- ☐ Department of Energy
- ☐ HMRT
- ☐ ATSDR (404) 639-6000
- ☐ National Response Center
- ☐ EPA
- ☐ Technical specialists, list:

____

____

____

____

____

- ☐ Private contractors, list:

____

____

____

____

____

- ☐ Clean-up contractors, list:

____

____

____

____

____

TERMINATION

Check termination activities to complete:

- ☐ Decontamination
  - Method (call ATSDR):
    - ________ Dilution
    - ________ Neutralization
    - ________ Absorption
    - ________ Isolation
    - ________ Disposal
  - Solution used: ____________
- ☐ Contamination monitoring
- ☐ Rehabilitation
- ☐ Medical screening
- ☐ Exposure report
- ☐ Initial incident debriefing
- ☐ Post-incident analysis

RESPONDING TO CORROSIVE EMERGENCIES

Hazard Identification

Before approaching the incident:

- ❐ Determine location of incident
- ❐ Access location preplans
- ❐ Identify markings and list below:

____

____

- ❐ Check shipping papers, MSDS, etc.
- ❐ Smoke/plume color: __________
- ❐ Positively identify product(s):

____

____

- ❐ Call CHEMTREC / CANUTEC
- ❐ List characteristics of product(s):
  - pH value ________
  - Strength ________
  - Concentration ________
  - Organic/inorganic ________
  - Oxidizer ________
  - Water soluble ________
  - Water reactivity ________
  - Flammable range ________
  - Flash point ________
  - Specific gravity ________
  - Vapor density ________
  - Toxic by ________
- ❐ Location of leak or fire

____

- ❐ Extent and severity of injuries

____

- ❐ Position of container

____

- ❐ Integrity of container

____

- ❐ Fire exposure

____

- ❐ Environmental hazard

____

Action Plan

NO LEAK or **FIRE:**

- ❐ Determine type and size of container
- ❐ Find out product level in container
- ❐ Determine container damage
- ❐ Is a leak or fire possible?
- ❐ Set up control zones
- ❐ Set up protective action distances
- ❐ Isolate/contain product
- ❐ Prevent ignition sources

LEAK but **NO FIRE:**

- ❐ Establish control zones
- ❐ Evacuate
- ❐ Protect in place
- ❐ Isolate, contain, and/or confine product
- ❐ Control leak
- ❐ Transfer product to other container
- ❐ Prevent ignition sources
- ❐ Anticipate resources
- ❐ Anticipate decon procedures

LEAK and **FIRE:**

- ❐ Establish control zones
- ❐ Evacuate
- ❐ Protect in place
- ❐ Let product burn?
- ❐ Evaluate hazards of inorganic vs. organic acids
- ❐ Determine best extinguishing agent
- ❐ Determine amount of extinguishing agent needed
- ❐ Unmanned hose lines?
- ❐ Protect exposures
- ❐ Contain runoff
- ❐ Anticipate resources
- ❐ Anticipate decon procedures

Zoning

- ❐ Check table of Initial Isolation and Protective Action Distances
- ❐ Check local protocol

SMALL spill – **NO FIRE:***

- ❐ 250′ protective action distance
- ❐ 150′ hot zone

SMALL spill – **FIRE:***

- ❐ 350′ protective action distance
- ❐ 200′ hot zone
- ❐ Consider warm zone

LARGE spill – **NO FIRE:***

- ❐ 1,000′ protective action distance
- ❐ 500′ hot zone
- ❐ Establish warm and cold zones

LARGE spill – **FIRE:***

- ❐ 1,000′ initial protective action distance
- ❐ Downwind evacuation
- ❐ 500′ hot zone
- ❐ Establish warm and cold zones

* All are recommended distances. Adjust as needed for the specific material, amount, and incident severity.

Managing the Incident

ATTACK STRATEGY:

- ❐ Uphill/upwind
- ❐ Initiate ICS
- ❐ Check SOPs
- ❐ Call in needed resources and specialized equipment
- ❐ Personal protective clothing
- ❐ Rescue
- ❐ EMS
- ❐ Decon prior to transporting victims
- ❐ Evacuate
- ❐ Protect in place
- ❐ Crowd and traffic control
- ❐ Extinguishment
- ❐ Protect exposures
- ❐ Isolate and contain
- ❐ Diking/diversion/retention
- ❐ Plan overhaul and clean-up

NO-ATTACK STRATEGY:

- ❐ Uphill/upwind
- ❐ Initiate ICS
- ❐ Check SOPs
- ❐ Call in needed resources
- ❐ Personal protective clothing
- ❐ Rescue
- ❐ EMS
- ❐ Decon prior to transporting victims
- ❐ Evacuate
- ❐ Protect in place
- ❐ Protect exposures
- ❐ Crowd and traffic control
- ❐ Isolate and contain
- ❐ Diking/diversion/retention
- ❐ Plan overhaul and clean-up

Assistance

Check agencies to contact:

- ❐ Dispatcher
- ❐ Law Enforcement
- ❐ Health Department
- ❐ CHEMTREC (800) 424-9300 or CANUTEC (613) 996-6666
- ❐ National Response Center
- ❐ EPA
- ❐ HMRT
- ❐ ATSDR (404) 639-6000
- ❐ Technical specialists, list:

____

____

____

____

____

- ❐ Private contractors, list:

____

____

____

____

____

- ❐ Clean-up contractors, list:

____

____

____

____

____

Termination

Check termination activities to complete:

- ❐ Decontamination
  - Method (call ATSDR):
  - ________ Dilution
  - ________ Neutralization
  - ________ Absorption
  - ________ Isolation
  - ________ Disposal
  - Solution used: __________
- ❐ Rehabilitation
- ❐ Medical screening
- ❐ Exposure report
- ❐ Initial incident debriefing
- ❐ Post-incident analysis

RESPONDING TO MISCELLANEOUS HAZARDOUS MATERIALS EMERGENCIES

HAZARD IDENTIFICATION

Before approaching the incident:

- ☐ Determine location of incident
- ☐ Access location preplans
- ☐ Identify markings and list below:

  ____

  ____

- ☐ Check shipping papers, MSDS, etc.
- ☐ Positively identify product(s):

  ____

  ____

- ☐ Call CHEMTREC / CANUTEC
- ☐ List characteristics of product(s):
  - TLV ________
  - IDLH ________
  - LC_{50} ________
  - Water soluble ________
  - Vapor density ________
  - Specific gravity ________
  - Flammable range ________
  - Toxic by ________

  Other: ________ ________

  ________ ________

  ________ ________

  ________ ________

- ☐ Location of leak or fire

  ____

- ☐ Extent and severity of injuries

  ____

- ☐ Position of container

  ____

- ☐ Integrity of container

  ____

- ☐ Fire exposure

  ____

- ☐ Environmental hazard

  ____

ACTION PLAN

NO LEAK or **FIRE:**

- ☐ Determine type and size of container
- ☐ Find out product level in container
- ☐ Determine container damage
- ☐ Is a leak or fire possible?
- ☐ Set up control zones
- ☐ Set up protective action distances
- ☐ Isolate/contain product
- ☐ Prevent ignition sources

LEAK but **NO FIRE:**

- ☐ Establish control zones
- ☐ Evacuate
- ☐ Protect in place
- ☐ Isolate, contain, and/or confine product
- ☐ Prevent ignition sources
- ☐ Anticipate resources
- ☐ Anticipate decon procedures

LEAK and **FIRE:**

- ☐ Establish control zones
- ☐ Evacuate
- ☐ Protect in place
- ☐ Let product burn?
- ☐ Determine best extinguishing agent
- ☐ Determine amount of extinguishing agent needed
- ☐ Unmanned hose lines?
- ☐ Contain runoff
- ☐ Protect exposures
- ☐ Anticipate resources
- ☐ Anticipate decon procedures

ZONING

- ☐ Check table of Initial Isolation and Protective Action Distances
- ☐ Check local protocol

SMALL spill – **NO FIRE:**

- ☐ ______′ protective action distance
- ☐ ______′ hot zone

SMALL spill – **FIRE:**

- ☐ ______′ protective action distance
- ☐ ______′ hot zone
- ☐ Consider warm zone

LARGE spill – **NO FIRE:**

- ☐ ______′ protective action distance
- ☐ Downwind evacuation
- ☐ ______′ hot zone
- ☐ Establish warm and cold zones

LARGE spill – **FIRE:**

- ☐ ______′ protective action distance
- ☐ Downwind evacuation
- ☐ ______′ hot zone
- ☐ Establish warm and cold zones

MANAGING THE INCIDENT

ATTACK STRATEGY:

- ☐ Uphill/upwind
- ☐ Initiate ICS
- ☐ Check SOPs
- ☐ Call in needed resources and specialized equipment
- ☐ Personal protective clothing
- ☐ Rescue
- ☐ EMS
- ☐ Decon prior to transporting victims
- ☐ Evacuate
- ☐ Protect in place
- ☐ Crowd and traffic control
- ☐ Extinguishment
- ☐ Protect exposures
- ☐ Isolate and contain
- ☐ Diking/diversion/retention
- ☐ Plan overhaul and clean-up

NO-ATTACK STRATEGY:

- ☐ Uphill/upwind
- ☐ Initiate ICS
- ☐ Check SOPs
- ☐ Call in needed resources
- ☐ Personal protective clothing
- ☐ Rescue
- ☐ EMS
- ☐ Decon prior to transporting victims
- ☐ Evacuate
- ☐ Protect in place
- ☐ Protect exposures
- ☐ Crowd and traffic control
- ☐ Isolate and contain
- ☐ Diking/diversion/retention
- ☐ Plan overhaul and clean-up

ASSISTANCE

Check agencies to contact:

- ☐ Dispatcher
- ☐ Law Enforcement
- ☐ Health Department
- ☐ CHEMTREC (800) 424-9300 or CANUTEC (613) 966-6666
- ☐ National Response Center
- ☐ EPA
- ☐ HMRT
- ☐ ATSDR (404) 639-6000
- ☐ Technical specialists, list:

  ____

  ____

  ____

  ____

  ____

- ☐ Private contractors, list:

  ____

  ____

  ____

  ____

  ____

- ☐ Clean-up contractors, list:

  ____

  ____

  ____

  ____

  ____

TERMINATION

Check termination activities to complete:

- ☐ Decontamination

  Method (call ATSDR):
  - ________ Dilution
  - ________ Neutralization
  - ________ Absorption
  - ________ Isolation
  - ________ Disposal

  Solution used: ____________

- ☐ Rehabilitation
- ☐ Medical screening
- ☐ Exposure report
- ☐ Initial incident debriefing
- ☐ Post-incident analysis

BIBLIOGRAPHY

Arthur, Robert A. *Chemistry for Today.* New York: Cambridge Book Company, 1957.

Bahme, Charles W. *Fire Officers Guide to Dangerous Chemicals.* Boston: National Fire Protection Association, 1978.

———. *Fire Officers Guide to Disaster Control.* Boston: National Fire Protection Association, 1978.

———. *Fire Officers Guide to Emergency Action.* Boston: National Fire Protection Association, 1976.

Borak, Jonathan, Michael Callan, and William Abbott. *Hazardous Materials Exposure: Emergency Response and Patient Care.* Englewood Cliffs, NJ: Prentice Hall, 1991.

Bronstein, Alvin C., and Phillip L. Currance. *Emergency Care for Hazardous Materials Exposure.* 2nd ed. St. Louis: Mosby Lifeline, 1994.

Bruegman, Randy R., and Glenn M. Levy. *Hazardous Materials: A Dispatcher's Seminar.* Greeley, CO: Aims Community College, 1984.

———. *Hazardous Materials Awareness.* Greeley, CO: Aims Community College, 1981.

———. *Hazardous Materials I.* Greeley, CO: Aims Community College, 1981.

———. *Hazardous Materials II.* Greeley, CO: Aims Community College, 1981.

Brunacini, Alan V. *Fire Command.* Boston: National Fire Protection Association, 1985.

Burlington Northern Railroad. *Emergency Handling of Hazardous Materials.* Denver, 1980.

———. *Pocket Guide to Tank Cars.* Denver, 1980.

The Chlorine Institute. *Chlorine Manual.* New York, 1979.

Code of Federal Regulations. Title 40 (Protection of the Environment), Parts 190–399. Washington, DC, 1993.

Code of Federal Regulations. Title 49 (Transportation), Parts 100–199. Washington, DC, 1993.

Colorado Department of Health. *Emergency Handling of Radioactive and Metallic Fires.* Denver, 1979.

Davis, Daniel J., Christianson, Grant T. *Firefighter's Hazardous Materials Reference Book.* New York: Van Nostrand Reinhold, 1991.

Edwards, Ron. *Fire Chemistry.* 2nd ed. Grand Rapids: S.A.F.E. Films, 1981.

Federal Emergency Management Association. *Fundamentals Course for Radiological Monitors.* Washington, DC, 1984.

Fisk, Franklin G., and Milo K. Blecha. *The Physical Sciences.* River Forest, IL: Laidlaw Brothers, 1971.

General American Transportation Corporation. *GATX Tank Car Manual.* Chicago, 1985.

Geomet Technologies. *Emergency Medical Treatment Needs: Chronic and Acute Exposure to Hazardous Material.* Rockville, MD, 1982.

Goodman, David B. *Hazardous Materials Compliance and Enforcement Course.* Washington, DC: U.S. Department of Transportation, 1984.

Grant, Harvey, and Robert Murray. *Emergency Care.* 2nd ed. Bowie, MD: Brady Company, 1978.

International Association of Fire Chiefs. *First Responder's Guide to Hazardous Materials.* Washington, DC, 1980.

International Conference of Building Officials. *Uniform Fire Code.* Whittier, CA, 1991.

International Fire Service Training Association. *Haz Mat Response Team.* Stillwater, OK: IFSTA and Fire Protection Publications, 1984.

———. *Hazardous Materials for First Responders.* 2nd ed. Stillwater, OK: IFSTA and Fire Protection Publications, 1994.

Isman, Warren, and Gene Carlson. *Hazardous Materials.* Encino, CA: Glencoe Press, 1980.

Mackison, Frank W., R. Scott Stricoff, and Lawrence J. Partridge, Jr., eds. *Occupational Health Guidelines for Hazardous Materials.* Washington, DC: A. D. Little and NIOSH/OSHA, 1981.

Meidl, James H. *Explosive and Toxic Hazardous Materials.* Encino, CA: Glencoe Press, 1972.

———. *Flammable Hazardous Materials.* 2nd ed. Encino, CA: Glencoe Press, 1970.

Meyer, Eugene. *Chemistry of Hazardous Materials.* 2nd ed. Englewood Cliffs, NJ: Prentice Hall, 1989.

Morgan, Donald P. *Recognition and Management of Pesticide Poisons.* 3rd ed. Washington, DC: U.S. Environmental Protection Agency, 1982.

National Fire Protection Association. *Fire Protection Guide on Hazardous Materials.* 11th ed. Quincy, MA, 1994.

———. *Flammable and Combustible Liquids Code.* Quincy, MA, 1993.

———. *Flashpoint of Trade Name Liquids.* Boston, 1978.

———. *Handling LNG Trucking Emergencies.* Boston, 1980.

———. *Hazardous Materials Response Handbook.* 2nd ed. Quincy, MA, 1993.

———. *Hazardous Materials Transportation Accidents.* Boston, 1978.

———. *Low Expansion Foam.* Quincy, MA, 1995.

———. *NFPA 471: Recommended Practice for Responding to Hazardous Materials Incidents.* Quincy, MA, 1992.

———. *NFPA 472: Standard for Professional Competence of Responders to Hazardous Materials Incidents.* Quincy, MA, 1992.

———. *NFPA 473: Standard for Professional Competencies for EMS Personnel Responding to Hazardous Materials Incidents.* Quincy, MA, 1992.

———. *Recognizing and Identifying Hazardous Materials.* Boston, 1984.

Noll, Gregory G., Michael S. Hildebrand, and James G. Yvorra. *Hazardous Materials: Managing the Incident,* 2nd ed. Stillwater, OK: IFSTA and Fire Protection Publications, 1995.

Occupational Safety and Health Administration. *Pocket Guide to Chemical Hazards.* Washington, DC: NIOSH/OSHA, 1978.

Pella, Milton O. *Physical Science for Progress.* Englewood Cliffs, NJ: Prentice Hall, 1970.

Philadelphia Electric Service Company. *The Fire Fighter and the Gas Company.* Philadelphia, n.d.

Sax, N. Irving, and Richard L. Lewis, Sr., eds. *Dangerous Properties of Industrial Materials.* 8th ed. New York: Van Nostrand Reinhold, 1992.

———. *Hawley's Condensed Chemical Dictionary.* 12th ed. New York: Van Nostrand Reinhold, 1992.

Schieler, Leroy, and Dennis Pauze. *Hazardous Materials.* Albany, NY: Delmar Publishers, 1976.

Student, Patrick J., ed. *Emergency Handling of Hazardous Materials in Surface Transportation.* Washington, DC: U.S. Bureau of Explosives, 1981.

Thomas, Clayton L., ed. *Taber's Cyclopedic Medical Dictionary.* 17th ed. Philadelphia: F. A. Davis Company, 1993.

Tokle, Gary, ed. *Hazardous Materials Response Handbook.* Quincy, MA: National Fire Protection Association, 1993.

Urbanski, Tadeusz. *Chemistry and Technology of Explosives.* 3 vols. New York: Pergamon Press, 1964.

U.S. Bureau of Explosives. *Hazardous Materials Regulation of the D.O.T.* Tariff No. B.O.E. Washington, DC, 1985.

U.S. Department of Health and Human Services. *NIOSH Pocket Guide to Chemical Hazards.* Washington, DC, 1994.

U.S. Department of Transportation. *All About Radioactive Materials.* Washington, DC, 1978.

———. *Emergency Action Guide for Selected Hazardous Materials.* Washington, DC, 1978.

———. *Emergency Response Guidebook.* Washington, DC, 1993.

———. *A Guide to the Federal Hazardous Materials Transportation Regulatory Program.* Washington, DC, 1983.

———. *Handling Pipeline Transportation Emergencies.* Washington, DC, 1980.

———. *Toward a Federal/State/Local Partnership in Hazardous Materials Transportation Safety.* Washington, DC, 1982.

———. *Transportation of Hazardous Materials.* Washington, DC, 1980.

U.S. Department of Transportation and U.S. Environmental Protection Agency. *Lessons Learned.* Washington, DC, 1985.

U.S. Environmental Protection Agency. *Oil and Hazardous Substance Pollution Contingency Plan.* Washington, DC, 1984.

———. *Pesticide Application Guide.* Washington, DC, 1980.

U.S. National Fire Academy. *Chemistry of Hazardous Materials.* Emmitsburg, MD, 1983.

———. *Hazardous Materials Incident Analysis.* Emmitsburg, MD, 1984.

———. *Hazardous Materials Substance Specialist.* Emmitsburg, MD, 1984.

———. *Hazardous Materials Tactical Considerations.* Emmitsburg, MD, 1980.

———. *The Pesticide Challenge.* Emmitsburg, MD, 1984.

———. *Pesticide Fire and Spill Control.* Emmitsburg, MD, 1980.

Washington Gas. *Gas Pipelines and Distribution Systems.* Washington, DC, n.d.

Wood, George. *Firefighters and Toxic Materials.* Chicago: Chicago Fire Academy, 1983.

INDEX

A

B

C

D

E

F

G

H

I

K

L

M

N

R

S

T

U

V

W

X

Z